ANNALS OF THE NEW YORK ACADEMY OF SCIENCES

Volume 893

EDITORIAL STAFF

Executive Editor
BARBARA M. GOLDMAN

Managing Editor
JUSTINE CULLINAN

Associate Editors
JOYCE HITCHCOCK
RICHARD STIEFEL

The New York Academy of Sciences
2 East 63rd Street
New York, New York 10021

THE NEW YORK ACADEMY OF SCIENCES
(Founded in 1817)

BOARD OF GOVERNORS, September 15, 1999 – September 15, 2000

BILL GREEN, *Chairman of the Board*
TORSTEN WIESEL, *Vice Chairman of the Board*
RODNEY W. NICHOLS, *President and CEO* [ex officio]

Honorary Life Governors
WILLIAM T. GOLDEN   JOSHUA LEDERBERG

JOHN T. MORGAN, *Treasurer*

Governors

D. ALLAN BROMLEY	LAWRENCE B. BUTTENWIESER	PRAVEEN CHAUDHARI
JOHN H. GIBBONS	RONALD L. GRAHAM	HENRY M. GREENBERG
ROBERT G. LAHITA	MARTIN L. LEIBOWITZ	JACQUELINE LEO
WILLIAM J. McDONOUGH	KATHLEEN P. MULLINIX	JOHN F. NIBLACK
SANDRA PANEM	RICHARD RAVITCH	RICHARD A. RIFKIND
	SARA LEE SCHUPF JAMES H. SIMONS	

ELEANOR BAUM, *Past Chairman of the Board*
HELENE L. KAPLAN, *Counsel* [ex officio]   PETER H. KOHN, *Secretary* [ex officio]

OXIDATIVE/ENERGY METABOLISM IN NEURODEGENERATIVE DISORDERS

ANNALS OF THE NEW YORK ACADEMY OF SCIENCES
Volume 893

OXIDATIVE/ENERGY METABOLISM IN NEURODEGENERATIVE DISORDERS

Edited by John P. Blass and Fletcher H. McDowell

The New York Academy of Sciences
New York, New York
1999

Copyright © 1999 by the New York Academy of Sciences. All rights reserved. Under the provisions of the United States Copyright Act of 1976, individual readers of the Annals *are permitted to make fair use of the material in them for teaching or research. Permission is granted to quote from the* Annals *provided that the customary acknowledgment is made of the source. Material in the* Annals *may be republished only by permission of the Academy. Address inquiries to the Executive Editor at the New York Academy of Sciences.*

Copying fees: *For each copy of an article made beyond the free copying permitted under Section 107 or 108 of the 1976 Copyright Act, a fee should be paid through the Copyright Clearance Center, Inc., 222 Rosewood Drive, Danvers, MA 01923. The fee for copying an article is $3.00 for nonacademic use; for use in the classroom, it is $0.07 per page.*

☉*The paper used in this publication meets the minimum requirements of the American National Standard for Information Sciences—Permanence of Paper for Printed Library Materials, ANSI Z39.48-1984.*

Library of Congress Cataloging-in-Publication Data

Oxidative/energy metabolism in neurodegenerative disorders / edited by John P. Blass and Fletcher H. McDowell
    p. ; cm. — (Annals of the New York Academy of Sciences, ISSN 0077-8923 ; v. 893)
    Includes bibliographical references and index.
    ISBN 1-57331-209-6 (cloth : alk. paper) — ISBN 1-57331-210-X (paper : alk. paper)
    1. Nervous system—Degeneration—Pathophysiology—Congresses. 2. Nervous system—Metabolism—Congresses. 3. Oxidation, Physiological—Congresses. 4. Energy metabolism—Congresses. 5. Active oxygen—Pathophysiology—Congresses. I. Title: Oxidative energy metabolism in neurodegenerative disorders. II. Blass, John P. III. McDowell, Fletcher H., 1923– IV. New York Academy of Sciences. V. Series.
    [DNLM: 1. Neurodegenerative Diseases—Congresses. 2. Energy Metabolism— Congresses. 3. Oxidative Stress—Congresses. WL 359 O98 1999]
    Q11 .N5 vol. 893
    [RC394.D35]]
    500 s—dc21
    [616.8'047]
                                                                                                          99-054405

GYAT / PCP
Printed in the United States of America
ISBN 1-57331-209-6 (cloth)
ISBN 1-57331-210-X (paper)
ISSN 0077-8923

ANNALS OF THE NEW YORK ACADEMY OF SCIENCES
Volume 893

OXIDATIVE/ENERGY METABOLISM IN NEURODEGENERATIVE DISORDERS[a]

Editors and Conference Chairs
JOHN P. BLASS AND FLETCHER H. MCDOWELL

CONTENTS

Preface. *By* JOHN P. BLASS	xi
Part I. Fundamental Aspects of Energy Metabolism and Free Radicals	
Glutamate Excitotoxity and Neuronal Energy Metabolism. *By* DAVID G. NICHOLLS, SAMANTHA L. BUDD, ROGER F. CASTILHO, AND MANUS W. WARD	1
Fundamental Aspects of Reactive Oxygen Species, or What's the Matter with Oxygen? *By* IRWIN FRIDOVICH	13
Part II. Mitochondrial Calcium Metabolism and the Permeability Transition	
Ca^{2+}-Mediated Mitochondrial Dysfunction and the Protective Effects of Bcl-2. *By* ANNE N. MURPHY	19
Mitochondrial Membrane Potential and the Permeability Transition in Excitotoxicity. *By* IAN J. REYNOLDS	33
Part III. Abnormalities in mtDNA and gDNAs	
Oxidative Phosphorylation Disease Diagnosis. *By* JOHN M. SHOFFNER	42
The α-Ketoglutarate Dehydrogenase Complex. *By* KWAN-FU REX SHEU AND JOHN P. BLASS	61

[a]This volume is the result of a conference entitled **Oxidative/Energy Metabolism in Neurodegenerative Disorders**, which was sponsored by the New York Academy of Sciences and the Winifred Masterson Burke Medical Research Institute, Inc. and held on March 19–22, 1999 in New York City.

Part IV. Experimental Models of Human Diseases

Oxidative Stress and a Key Metabolic Enzyme in Alzheimer Brains, Cultured Cells, and an Animal Model of Chronic Oxidative Deficits. *By* GARY E. GIBSON, LARRY C.H. PARK, HUI ZHANG, SANDRO SORBI, AND NOEL Y. CALINGASAN 79

The Use of Transgenic and Mutant Mice to Study Oxygen Free Radical Metabolism. *By* TING-TING HUANG, ELAINE J. CARLSON, INES RAINERI, ANNE MARIE GILLESPIE, HEATHER KOZY, AND CHARLES J. EPSTEIN 95

Part V. Physiological Aspects of Energy Metabolism

The Blood-Brain Barrier and Cerebrovascular Pathology in Alzheimer's Disease. *By* RAJ N. KALARIA 113

Neurological Changes Induced by Stress in Streptozotocin Diabetic Rats. *By* LAWRENCE P. REAGAN, ANA MARIA MAGARIÑOS, AND BRUCE S. MCEWEN 126

Part VI. Alzheimer's Disease

Functional Brain Imaging in the Resting State and during Activation in Alzheimer's Disease: Implications for Disease Mechanisms Involving Oxidative Phosphorylation. *By* STANLEY I. RAPOPORT 138

Cellular and Molecular Mechanisms Underlying Perturbed Energy Metabolism and Neuronal Degeneration in Alzheimer's and Parkinson's Diseases. *By* MARK P. MATTSON, WARD A. PEDERSEN, WENZHEN DUAN, CARSTEN CULMSEE, AND SIMONETTA CAMANDOLA 154

Use of Cytoplasmic Hybrid Cell Lines for Elucidating the Role of Mitochondrial Dysfunction in Alzheimer's Disease and Parkinson's Disease. *By* SOUMITRA S. GHOSH, RUSSELL H. SWERDLOW, SCOTT W. MILLER, BRINA SHEEMAN, W. DAVIS PARKER, JR., AND ROBERT E. DAVIS 176

Part VII. Huntington's Disease

Polyglutamine Domain Proteins with Expanded Repeats Bind Neurofilament, Altering the Neurofilament Network. *By* YOSHITAKA NAGAI, OSAMU ONODERA, WARREN J. STRITTMATTER, AND JAMES R. BURKE 192

Bioenergetics in Huntington's Disease. *By* THOMAS GRÜNEWALD AND M. FLINT BEAL 203

Part VIII. Cerebrovascular Disease

An Integrated Strategy for Evaluation of Metabolic and Oxidative Defects in Neurodegenerative Illness Using Magnetic Resonance Techniques. *By* B.G. JENKINS, Y.I. CHEN, E. KUESTERMANN, N.M. MAKRIS, T.V. NGUYEN, E. KRAFT, A.L. BROWNELL, H.D. ROSAS, D.N. KENNEDY, B.R. ROSEN, W.J. KOROSHETZ, AND M.F. BEAL 214

Apoptosis and Necrosis in Cerebrovascular Disease. *By* B. JOY SNIDER, FRANK J. GOTTRON, AND DENNIS W. CHOI 243

Part IX. Poster Papers

Effects of Ebselen, a Glutathione Peroxidase Mimic, in Several Models of Mitochondrial Dysfunction. *By* ALAIN BOIREAU, PIERRE DUBEDAT, FRANÇOISE BORDIER, MADELEINE COIMBRA, MIREILLE MEUNIER, ASSUNTA IMPERATO, AND SALIHA MOUSSAOUI 254

EDTA-Induced Monovalent Fluxes through the Ca^{2+} Uniporter in Brain Mitochondria. *By* NICKOLAY BRUSTOVETSKY AND JANET M. DUBINSKY   258

Signaling Events in NMDA Receptor–Induced Apoptosis in Cerebrocortical Cultures. *By* SAMANTHA L. BUDD AND STUART A. LIPTON   261

In Vitro and *in Vivo* Protein Oxidation Induced by Alzheimer's Disease Amyloid β-Peptide (1-42). *By* D. ALLAN BUTTERFIELD, SERVET M. YATIN, AND CHRISTOPHER D. LINK   265

Depolarization of *in Situ* Mitochondria by Hydrogen Peroxide in Nerve Terminals. *By* CHRISTOS CHINOPOULOS AND VERA ADAM-VIZI   269

Monoamine Oxidase Inhibits Mitochondrial Respiration. *By* GERALD COHEN AND NATASA KESLER   273

Enhanced Acetate and Glucose Utilization during Graded Photic Stimulation: Neuronal-Glial Interactions *in Vivo*. *By* GERALD A. DIENEL, KENIAN LIU, DAVID POPP, AND NANCY F. CRUZ   279

Calcium Overload Triggers Rod Photoreceptor Apoptotic Cell Death in Chemical-Induced and Inherited Retinal Degenerations. *By* DONALD A. FOX, ANN T. POBLENZ, AND LIHUA HE   282

Molecular Mechanisms of Free Radical Production and Protective Efficacies of Antioxidants in *in Vitro* Ischemia-Reperfusion. *By* MARINA V. FRANTSEVA, PETER L. CARLEN, AND JOSE L. PEREZ VELAZQUEZ   286

A Disturbance in the Neuronal Insulin Receptor Signal Transduction in Sporadic Alzheimer's Disease. *By* L. FRÖLICH, D. BLUM-DEGEN, P. RIEDERER, AND S. HOYER   290

Modulation of Presenilin-1 Processing by Nitric Oxide during Apoptosis Induced by Serum Withdrawal and Glucose Deprivation. *By* LAURA GASPARINI, ROBERTA GHIDONI, ANTONELLA C. ALBERICI, LUISA BENUSSI, DANIELE MORATTO, MARCO TRABUCCHI, JOHN H. GROWDON, ROGER M. NITSCH, AND GIULIANO BINETTI   294

Metabolic and Glutamatergic Disturbances in the Huntington's Disease Transgenic Mouse *By* DONALD S. HIGGINS, KARI R. HOYT, CORINNE BAIC, JESSICA VENSEL, AND MATTHEW SULKA   298

Inhibition of the Neuronal Insulin Receptor Causes Alzheimer-like Disturbances in Oxidative/Energy Brain Metabolism and in Behavior in Adult Rats. *By* SIEGFRIED HOYER AND HEINRICH LANNERT   301

Temporary Axonal Conduction Block and Axonal Loss in Inflammatory Neurological Disease: A Potential Role for Nitric Oxide? *By* R. KAPOOR, M. DAVIES, AND K.J. SMITH   304

Maturational Changes in Rabbit Brain Phosphocreatine and Creatine Kinase. *By* T. KEKELIDZE, I. KHAIT, A.TOGLIATTI, J. BENZYCRY, R. MULKERN, AND D. HOLTZMAN   309

Neurotoxicity and Oxidative Damage of Beta Amyloid 1-42 versus Beta Amyloid 1-40 in the Mouse Cerebral Cortex. *By* AUTUMN M. KLEIN, NEIL W. KOWALL, AND ROBERT J. FERRANTE   314

Ganglioside GD3, the Mitochondrial Permeability Transition, and Apoptosis. *By* BRUCE S. KRISTAL AND ABRAHAM M. BROWN   321

Interactions between Melatonin, Reactive Oxygen Species, and Nitric
  Oxide. *By* D.K. LAHIRI AND C. GHOSH 325

Effect of Oxidative Stress on DNA Damage and β-Amyloid Precursor
  Proteins in Lymphoblastoid Cell Lines from a Nigerian Population. *By*
  D.K. LAHIRI, Y. XU, J. KLAUNIG, O. BAIYEWU, A. OGUNNIYI,
  K. HALL, H. HENDRIE, AND A. SAHOTA 331

Evidence for Energy Failure following Irreversible Traumatic Brain Injury.
  By STEFAN M. LEE, MONICA D. WONG, AMIR SAMII, AND
  DAVID A. HOVDA 337

Regulation of Mitochondrial Gene Expression in Differentiated PC12 Cells.
  By LI-ING LIU, S.I. RAPOPORT, AND K. CHANDRASEKARAN 341

Does Dopamine Contribute to Striatal Damage Caused by Impaired Mitochondrial Function? *By* WILLIAM F. MARAGOS, REBEKAH J. JAKEL,
  M. DATHAN CHESNUT, JAMES W. GEDDES, AND LINDA P. DWOSKIN 345

Discordance between Traditional Pathologic and Energy Metabolic Changes
  in Very Early Alzheimer's Disease: Pathophysiological Implications.
  By SATOSHI MINOSHIMA, DONNA J. CROSS, NORMAN L. FOSTER,
  THOMAS R. HENRY, AND DAVID E. KUHL 350

Low Frequencies of Mitochondrial DNA Mutations Cause Cardiac Disease
  in the Mouse. *By* J.L. MOTT, D. ZHANG, P.L. FARRAR, S.W. CHANG,
  AND H.P. ZASSENHAUS 353

Effects of the Solvent 1,2,4-Trimethylcyclohexane on Respiratory Burst in
  Human Neutrophil Granulocytes: A Chemiluminescence and Electron
  Paramagnetic Resonance Spectrometry Study. *By* O. MYHRE,
  T. A. VESTAD, E. SAGSTUEN, H. AARNES, AND F. FONNUM 358

Neuronal RNA Oxidation in Alzheimer's Disease and Down's Syndrome.
  By AKIHIKO NUNOMURA, GEORGE PERRY, KEISUKE HIRAI,
  GJUMRAKCH ALIEV, ATSUSHI TAKEDA, SHIGERU CHIBA, AND
  MARK A. SMITH 362

Ca^{2+}-Dependent Permeability Transition and Complex I Activity in Lym-
  phoblast Mitochondria from Normal Individuals and Patients with
  Huntington's or Alzheimer's Disease. *By* ALEXANDER PANOV,
  TRACY OBERTONE, JULIE BENNETT-DESMELIK, AND
  J. TIMOTHY GREENAMYRE 365

Mitochondrial Porin, a Novel Target to Prevent Ischemia-Induced Neurodegeneration? *By* J.L. PEREZ VELAZQUEZ, M.V. FRANTSEVA,
  D. HUZAR, C. GUEZURIAN, AND P.L. CARLEN 369

Bcl-2 and p53: Role in Dopamine-Induced Apoptosis and Differentiation. *By*
  S. PORAT AND R. SIMANTOV 372

Mitochondria Mediate Nitric Oxide–Induced Cell Death. *By* ANNA
  BAL-PRICE, VILMA BORUTAITE, AND GUY C. BROWN 376

Induction of Apoptosis and Necrosis in Human Neuroblastoma Cells by
  Cholesterol Oxides. *By* MARIE LUISE RAO, DIETER LÜTJOHAN,
  MICHAEL LUDWIG, AND HEIKE KÖLSCH 379

Measurement of Oxidative DNA Damage in the Human p53 and PGK1 Gene
  at Nucleotide Resolution. *By* HENRY RODRIGUEZ AND
  STEVEN A. AKMAN 382

The Glucose Paradox in Cerebral Ischemia: New Insights. *By* AVITAL SCHURR, RALPHIEL S. PAYNE, MICHAEL T. TSENG, JAMES J. MILLER, AND BENJAMIN M. RIGOR 386

Interactions of Chloromethyltetramethylrosamine (Mitotracker Orange™) with Isolated Mitochondria and Intact Cells. *By* LUCA SCORRANO, VALERIA PETRONILLI, RAFFAELE COLONNA, FABIO DI LISA, AND PAOLO BERNARDI 391

Attenuation of Neuronal Death by NMDA and Oxygen-Glucose Deprivation in Cortical Neurons Maintained in High Glucose. *By* SO YOUNG SEO, EUN YOUNG KIM, HARRIET KIM, ILO JOU, AND BYOUNG JOO GWAG 396

Astrocyte Nitric Oxide Causes Neuronal Mitochondrial Damage, but Antioxidant Release Limits Neuronal Cell Death. *By* R. STONE, V.C. STEWART, R.D. HURST, J.B. CLARK, AND S.J.R. HEALES 400

Mechanisms of Selective Neuronal Cell Death due to Thiamine Deficiency. *By* KATHRYN G. TODD AND ROGER F. BUTTERWORTH 404

Inhibition of α-Ketoglutarate Dehydrogenase due to H_2O_2-Induced Oxidative Stress in Nerve Terminals. *By* LASZLO TRETTER AND VERA ADAM-VIZI 412

ATPases of Synaptic Plasma Membranes and Vesicles from Rat Cerebral Cortex during Aging and Hypoxia. *By* R.F. VILLA, A. D'ANGELO, AND A. GORINI 417

Mitochondrial Compartmentation at the Cellular Level: Astrocytes and Neurons. *By* HELLE S. WAAGEPETERSEN, URSULA SONNEWALD, HONG QU, AND ARNE SCHOUSBOE 421

Detection of Respiratory Chain Defects in Cultivated Skin Fibroblasts and Skeletal Muscle of Patients with Parkinson's Disease. *By* FALK ROBERT WIEDEMANN, KIRSTIN WINKLER, HARTMUT LINS, CLAUS-W. WALLESCH, AND WOLFRAM S. KUNZ 426

Prevention of Neurodegeneration by a Neuroprotective Radical Scavenger. *By* HIROSHI YASUDA 430

Conference Summary: Mitochondria, Neurodegenerative Diseases, and Selective Neuronal Vulnerability. *By* JOHN P. BLASS 434

Index of Contributors 441

Financial assistance was received from:

Joint Sponsor
- THE WINIFRED MASTERSON BURKE MEDICAL RESEARCH INSTITUTE, INC.

Major Funder
- NATIONAL INSTITUTE ON AGING–NATIONAL INSTITUTES OF HEALTH

Supporter
- MITOKOR

Contributors
- AMERICAN FEDERATION FOR AGING RESEARCH
- PARKE-DAVIS PHARMACEUTICAL RESEARCH, WARNER-LAMBERT COMPANY

The New York Academy of Sciences believes it has a responsibility to provide an open forum for discussion of scientific questions. The positions taken by the participants in the reported conferences are their own and not necessarily those of the Academy. The Academy has no intent to influence legislation by providing such forums.

Preface

JOHN P. BLASS
Burke Medical Research Institute, Cornell University, 785 Mamaroneck Avenue, White Plains, New York 10605, USA

Scientific ideas can become fashionable, fall out of fashion, and then become relevant again. Ideas often fall out of fashion because the new information that can be gleaned from them becomes limited by the techniques and knowledge available at a given time. Old ideas often become popular again when new techniques and knowledge to investigate them become available. This sequence appears to have occurred in the study of oxygen/energy metabolism in neurodegenerative diseases.

From the 1930s to the 1950s, the study of glucose oxidation in the brain was a "hot topic," particularly in neurological and psychiatric disorders. There were several reasons. The delineation and control of the major pathways of oxidative and energy metabolism were then at the forefront of biomedical research. These pathways included glycolysis, the Krebs tricarboxylic acid cycle, electron transport, and the "pentose phosphate pathway." The high oxygen consumption of the brain under physiological conditions was recognized, as was the second-to-second dependence of the brain on glucose oxidation for its function. The coupling of the rate of oxygen/energy metabolism to the intensity of neuronal function was recognized. The impairments in judgment and other aspects of brain function with even relatively mild decreases in oxygen supply came under intensive study, in part because of the military implications for the more powerful aircraft being developed for the Second World War. Hypoglycemic (insulin) shock therapy was widely used to treat psychoses, and the purportedly beneficial effects called attention to the importance of glucose metabolism in brain diseases. Glucose tolerance tests were found to give unusual—perhaps abnormal—results in patients with a variety of psychoses. The vitamins thiamin (B_1) and niacin (B_3) were found to play critical roles in both oxygen/energy metabolism and in brain diseases. These discoveries had direct medical significance. Thiamine was found to effectively prevent and treat the "Wernicke-Korsakoff" syndrome, a form of nutritional dementia seen most often in alcoholics. Niacin was found to prevent and treat pellagra, which has a wide variety of nervous system manifestations including psychotic behavior. Enrichment of flour with niacin largely eradicated pellagra in the USA, a major public health triumph.

By the late 50s and through to the 80s interest in the role of oxidative/energy metabolism in the brain and specifically in brain diseases declined. The problems that could be answered with the techniques then available had largely been solved. The invasive methods then used to study cerebral oxygen/energy metabolism in humans *in vivo* did not identify abnormalities in "schizophrenia," in affective disorders, or in a number of other conditions. Abnormalities in brain metabolism were documented in various types of comas ("metabolic encephalopathies") and in dementias, notably in Alzheimer's dsease (then called "senile dementia"). Recognition that oxygen extraction fraction and cerebrovascular autoregulation were normal even in Alzheimer patients with reduced brain metabolism led to the widely accepted proposal that reductions in oxygen/energy metabolism in comas and dementias were secondary to reduced brain activity in these conditions. It was increasingly recognized that many factors can influence glucose tolerance tests, and the earlier results with this technique in psychotic patients came to be believed to be nonspecific. Genetic disorders that interfered significantly with oxygen/energy metabolism came to be widely considered to be incompatible with extrauterine life. Research on the effects of impaired oxidative/energy metabolism continued to be done, however, for cerebrovascular diseases.

By the early 1970s, evidence began to accumulate that inborn abnormalities of enzymes of oxygen/energy metabolism can occur in children and even in people who reach adult life. These abnormalities were typically partial deficiencies—e.g., "leaky mutations." Abnormalities were found in the pyruvate dehydrogenase complex, in the Krebs tricarboxylic acid cycle, and in the electron transport chain. Although patients with such "inborn errors" were relatively rare, they established a "proof of principle" that partial deficiencies in oxygen/energy metabolism *are* compatible with extrauterine life and are typically associated with disease of the nervous system. The more severe the biochemical deficiency, the earlier the onset and the more severe the manifestations of the neurological disorders tend to be (see chapter by Sheu & Blass). More recently, inherent abnormalities in the biochemical apparatus for oxygen/energy metabolism have been described in common diseases of the brain including Alzheimer's disease (see chapter by Gibson *et al.*) and Parkinson's disease (see chapter by Ghosh *et al.*), as well as in such well-known neurological diseases as Huntington's disease (see chapter by Grünewald & Beal), motor neuron disease (see chapter by Huang *et al.*), and Friedreich's ataxia (see chapter by Huang *et al.*). Studies, not covered in this volume, have used modern techniques of molecular genetics to associate psychoses with relatively specific abnormalities of oxygen/energy metabolism.[1–3]

Examination of human brain metabolism with modern techniques such as PET scanning and fMRI has documented that regional impairments in brain oxygen/energy metabolism occur in many diseases of the nervous system (see chapter by Jenkins *et al.*). For many of these disorders, the decreases in regional brain metabolism are assumed to reflect—to "map"—the regions of brain involved. However, the impairments in brain metabolism in Alzheimer patients *precede* the onset of detectable abnormalities in neuropsychological function or any evidence of brain atrophy by MRI scan (see chapter by Jenkins *et al.*). These studies were made possible by advances in the molecular genetics of Alzheimer's disease that allowed the identification of individuals at very high risk to develop that disorder—specifically homozygotes for the ε-4 allele of the *APOE* gene. Documentation by modern imaging techniques of reduced brain metabolism without other evidence of brain disease disproves the assumption that the decreases in brain metabolism *in vivo* in Alzheimer's disease are secondary to "slowed thinking" or brain atrophy.

Studies over the last 20 years have also demonstrated mechanisms by which relatively mild impairments of brain oxygen/energy metabolism can impair brain function without killing brain cells (see chapter by Gibson *et al.*). For instance, relatively mild impairments of brain oxygen/energy metabolism can cause profound alterations in neurotransmitter function, reducing release of acetylcholine and increasing release of potentially excitotoxic neurotransmitters such as glutamate. These changes in neurotransmitter metabolism are clearly relevant to delirium and to the delirium-dementia spectrum of cognitive impairment.

One of the most important reasons for the reawakening of interest in oxygen/energy metabolism is the recognition that oxidative stress is important in the pathophysiology of many diseases, including many diseases of the nervous system (see chapters by Murphy; Grünewald & Beal; Ghosh *et al.*; and Kristal & Brown). The term *oxidative stress* refers to damage to tissues by free radicals, particularly by reactive oxygen species (ROS). A major source of ROS is mitochondria, which are also the main site of oxidation reactions in energy metabolism. Oxidative stress occurs secondarily to many kinds of tissue reactions, including inflammation and cell death. Extensive experimental evidence proves that oxidative stress can damage cells and can lead to their death. Extensive studies of chemical neuropathology indicate that oxidative stress is a component of many diseases, including such common disorders as Alzheimer's disease and stroke. There is a wide consensus that oxidative stress, often arising from abnormalities in mitochondrial metabolism, is important to the pathophysiology of many diseases of the nervous system. Treatments of these disorders with "free radical scavengers" are being extensively investigated. How frequently oxidative stress acts as a primary trigger for cell damage in human illnesses remains to be determined.

The symposium represented by this volume was convened because of the accelerating accumulation of new data on oxygen/energy metabolism in brain disease. The symposium allowed workers in this field to come together to discuss their results and ideas. This volume is meant to provide a convenient "state-of-the-art" summary for workers in this field or coming into this field. The symposium and this volume concentrate on new data obtained by current techniques. More general reviews including older data are available elsewhere.

It is a pleasure to thank the contributors for the elegant science they presented at the meeting and in the chapters they wrote for this volume. I also wish to thank my colleagues at the Burke Medical Research Institute and the New York Academy of Sciences for their critical help in organizing the symposium and in preparing this volume.

REFERENCES

1. LINDHOLM, E., L. CAVELIER, W.M. HOWELL, I. ERIKSSON, P. JALONEN, R. ADOLFSON, D.H. BLACKWOOD, W.J. MUIR, A.J. BROOKES, U. GYLLENSTEN & E.E. JAZIN. 1997. Mitochondrial sequence variants in patients with schizophrenia. Eur. J. Hum. Genet. **5:** 406–412.
2. WHATELY, S.A., D. CURTI & R.M. MARCHBANKS. 1996. Mitochondrial involvement in schizophrenia and other functional psychoses. Neurochem. Res. **21:** 995–1004.
3. PRINCE, J.A., K. BLENNOW, C.G. GOTTFRIES, I. KARLSSON & L. ORELAND. 1999. Mitochondrial function is differentially altered in the basal ganglia of chronic schizophrenics. Neuropsychopharmacology **21:** 372–379.

Glutamate Excitotoxicity and Neuronal Energy Metabolism

DAVID G. NICHOLLS, SAMANTHA L. BUDD[a] ROGER F. CASTILHO,[b] AND MANUS W. WARD

Department of Pharmacology and Neuroscience, University of Dundee, Dundee DD1 9SY, Scotland, UK

ABSTRACT: The bioenergetic properties of the *in situ* mitochondria play a central role in controlling the susceptibility of neurons to acute or chronic neurodegenerative stress. The mitochondrial membrane potential, $\Delta\psi_m$, is the parameter that controls three interrelated mitochondrial functions of great relevance to neuronal survival: namely, ATP synthesis, Ca^{2+} accumulation, and superoxide generation. The *in vitro* model we study is the rat cerebellar granule cell in primary culture and its susceptibility to NMDA receptor–mediated necrosis, which is preceded by a delayed failure of cytoplasmic Ca^{2+} homeostasis ("delayed Ca^{2+} deregulation," DCD). DCD is not caused by a failure of mitochondrial ATP synthesis since it also occurs in cells maintained purely by glycolysis. The *in situ* mitochondria maintain a $\Delta\psi_m$ sufficient for ATP synthesis throughout the exposure of the cells to glutamate until DCD occurs. Even at that stage it appears that mitochondrial depolarization may be an effect of DCD rather than a primary cause. This somewhat unorthodox view resolves a number of apparent paradoxes, such as observations of enhanced superoxide generation by *in situ* mitochondria during excitotoxic exposure, since isolated mitochondria generate superoxide only under conditions of high $\Delta\psi_m$. Mitochondrial depolarization by selective inhibitors that do not deplete cellular ATP is acutely neuroprotective.

Many aspects of the delayed neuronal death that occurs *in vivo* following focal ischemia or transient global ischemia can be modeled *in vitro* with dissociated primary neuronal cultures. For several years a focus of research has been the investigation of the molecular mechanisms that underlie the necrotic or apoptotic death of cultured neurons exposed to glutamate. This mimics the pathological activation of glutamate receptors by the amino acid that is released as a consequence of the bioenergetic crisis induced by deprivation of oxygen and/or glucose during an ischemic episode.

Exposure of cultured neurons to glutamate can result in a time-dependent, stochastic failure of cytoplasmic Ca^{2+} homeostasis, termed by Tymianski "delayed Ca^{2+} deregulation" or DCD,[1] which precedes and reliably predicts the subsequent,

---

[a]Present address: Center for Neuroscience and Aging, The Burnham Institute, 10901 North Torrey Pines Road, La Jolla, CA 92037, USA.

[b]Present address: Departamento de Patologia Clinica, Faculdade de Ciencias Medicas, Universidade Estadual de Campinas, Campinas, SP 13083-970, Brazil.

predominantly necrotic, death of the neuron. This glutamate-mediated cell death is largely, but not entirely, mediated by Ca^{2+} and Na^+ permeant NMDA-selective glutamate receptors. There is discussion as to whether the high sensitivity of neurons to NMDA receptor activation is due to the focusing of Ca^{2+} onto some undefined excitotoxic locus within the cell[2] or selective accumulation by the mitochondria,[3] or whether it is simply a reflection of the amount of Ca^{2+} [4,5] and Na^+ [6] entering the cell. In either case the pathological entry of Ca^{2+} initiates the processes leading to necrotic cell death; and this chapter will focus on the bioenergetic consequences of this ion flux, with particular reference to the mitochondrion. There has been a tremendous upsurge of interest in mitochondrial function, with the realization that, far from being perfect, infallible machines, the organelles are fragile structures operating near their physicochemical limits. In addition to cerebral ischemia, their dysfunction appears to underlie a host of degenerative disease states in the brain, as well as key aspects of programmed cell death (reviewed in Ref. 7).

Our laboratory has used the cultured rat cerebellar granule cell preparation to investigate the response of intraneuronal mitochondria to excitotoxic glutamate.[8–10] After 6–8 days *in vitro* cultures respond to maximal NMDA receptor activation (100 μM glutamate/10 μM glycine in the absence of added Mg^{2+}) with an initial peak elevation in cytoplasmic free Ca^{2+} ($[Ca^{2+}]_c$) that recovers to a plateau and remains stable until individual cells undergo stochastic DCD (FIG. 1). Several studies have shown that the glutamate-induced elevation in $[Ca^{2+}]_c$ in a variety of cultured neurons can exceed the mitochondrial Ca^{2+} "set-point" at which the organelles behave as buffers of cytoplasmic Ca^{2+},[11] rendering the organelles potentially capable of accumulating massive amounts of the cation.[3,8,12–19]

From first principles, mitochondrial Ca^{2+} accumulation could be beneficial to the cell, accumulating the excess Ca^{2+} and aiding the maintenance of cytoplasmic Ca^{2+} homeostasis. Alternatively, mitochondrial Ca^{2+} loading might endanger the cell, by leading to a failure of oxidative phosphorylation, bioenergetic collapse—for example, as a consequence of the "mitochondrial permeability transition"[20]—disruption of the redox balance of the cell, and generation of excessive amounts of reactive oxygen species (ROS).[21–23] The approach taken in our laboratory has been to adapt techniques developed for isolated mitochondria, in an attempt to dissect out the role of these diverse functions in the survival or death of the glutamate-exposed neuron.

Mitochondrial and cellular bioenergetics are closely integrated in the neuron (FIG. 1). As with other cells, ATP synthesis occurs by both substrate-level and oxidative phosphorylation, although the glycolytic capacity of the granule cells is sufficient to allow glycolysis alone to support the cell in the presence of inhibitors of the mitochondrial ATP synthase such as oligomycin.[8,24] This opens up the possibility of using classical cell-permeant mitochondrial inhibitors and ionophores to in-

---

FIGURE 1. Schematic mitochondrial/cytoplasmic bioenergetic interactions in neurons during NMDA receptor activation. **A**, In the absence of mitochondrial inhibitors the proton circuit drives both ATP synthesis and Ca^{2+} accumulation. The individual $[Ca^{2+}]_c$ traces are from single granule cell somata. Note that DCD is apparent in some cells within 20 min of glutamate addition. The *histogram* shows the proportion of cells that have deregulated ($[Ca^{2+}]_c > 1.5$ μM) after 60 min of glutamate exposure. **B**, In the presence of oligomycin, the cells are supported by glycolysis. Under these conditions individual cells undergo DCD to the same extent as in the absence of inhibitor *(histograms)*.

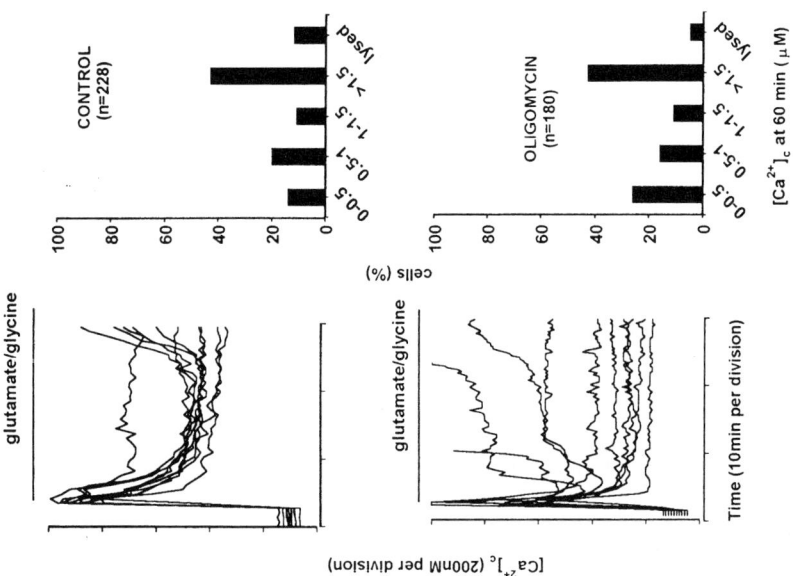

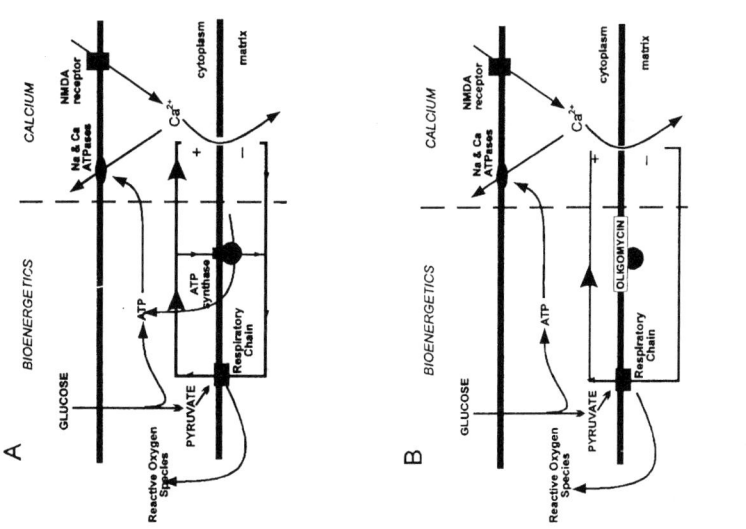

vestigate and dissect out individual mitochondrial functions to establish their contribution to DCD.

The simplest hypothesis is that Ca^{2+} loading leads to mitochondrial depolarization. The resulting reversal of the ATP synthase and ATP depletion following collapse of the membrane potential ($\Delta\psi_m$) would leave insufficient ATP for the operation of the plasma membrane ion pumps responsible for maintaining ionic homeostasis. This can readily be modeled by exposing granule cells to glutamate/glycine in the presence of a protonophore such as FCCP (FIG. 2A). This leads to immediate ATP depletion[24] and failure of Ca^{2+} homeostasis. In contrast, cells incubated in the presence of oligomycin (allowing neither mitochondrial synthesis nor hydrolysis of ATP) respond with a peak and plateau $[Ca^{2+}]_c$ indistinguishable from that in the absence of the inhibitor (FIG. 1B). Most importantly, the timing and extent of DCD is not significantly affected by the inhibitor.[8] Two conclusions can be drawn from this: first, a failure of oxidative phosphorylation cannot be the cause of the deregulation, since DCD occurs in cells whose mitochondria are not generating ATP; second, DCD is not brought forward by restricting the total ATP generating capacity of the cell with oligomycin. It is unlikely that DCD is due to an overwhelming increase in ATP demand by the processes initiated by cellular Ca^{2+} loading.

The remaining mitochondrial functions (Ca^{2+} accumulation, redox balance, ROS generation) are controlled by $\Delta\psi_m$, which is unaffected, or rather slightly increased, by oligomycin. In order to establish whether any of these parameters affects DCD, the mitochondrion can be virtually shut down by the combination of a respiratory chain inhibitor and oligomycin (FIG. 2B) Initial assumptions that the abolition of the main intracellular sink for Ca^{2+} would lead to a massive rise in $[Ca^{2+}]_c$ following NMDA receptor activation were confounded. For reasons that are still being investigated,[8,9] the initial and plateau cytoplasmic $[Ca^{2+}]_c$ responses to glutamate are decreased when the mitochondria are thus inhibited. Most importantly, cell survival prior to DCD is greatly enhanced.[8] No glutamate-dependent component of the eventual death of these neurons could be detected in the presence of the two inhibitors.[8]

Since mitochondrial deenergization causes a decrease in $[Ca^{2+}]_c$ (compare FIGS. 1A and 2B), it is necessary to establish whether the enhanced cell survival can be accounted for by a decreased cytoplasmic free Ca^{2+}. The simplest way to manipulate $[Ca^{2+}]_c$ is by altering external Ca^{2+} concentrations.[9] In the presence of a doubled external Ca^{2+}, granule cells with polarized mitochondria undergo very rapid DCD (FIG. 3A). Rotenone/oligomycin is still protective (FIG. 3B) even though $[Ca^{2+}]_c$ is as high during the plateau as in cells exposed to normal Ca^{2+} in the absence of inhibitors, showing that the plateau $[Ca^{2+}]_c$ does not per se control DCD.

The existence of a temporary, stable $[Ca^{2+}]_c$ plateau in granule cells exposed to glutamate demonstrates that total cytoplasmic Ca^{2+} must be in a steady state, with the rate of influx via the NMDA receptors balanced by the rate at which the cation

---

FIGURE 2. Schematic mitochondrial/cytoplasmic bioenergetic interactions in neurons during NMDA receptor activation. **A**, In the presence of a protonophore the ATP synthase reverses, rapidly depleting ATP. The $[Ca^{2+}]_c$ traces show that all cells deregulate immediately upon addition of glutamate. **B**, The combination of rotenone and oligomycin depolarizes the mitochondrion, preventing it from accumulating Ca^{2+}. Addition of glutamate results in a *diminished* $[Ca^{2+}]_c$ elevation, and no DCD is seen.

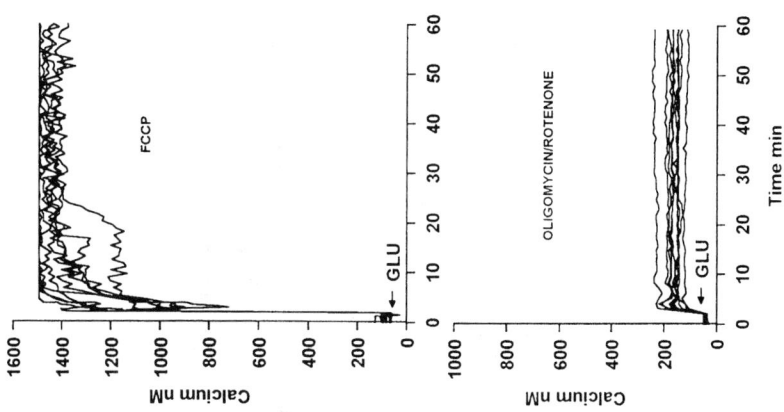

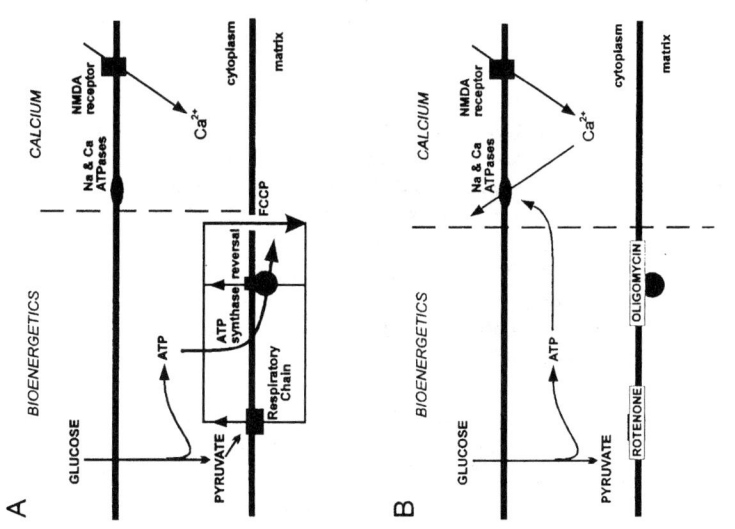

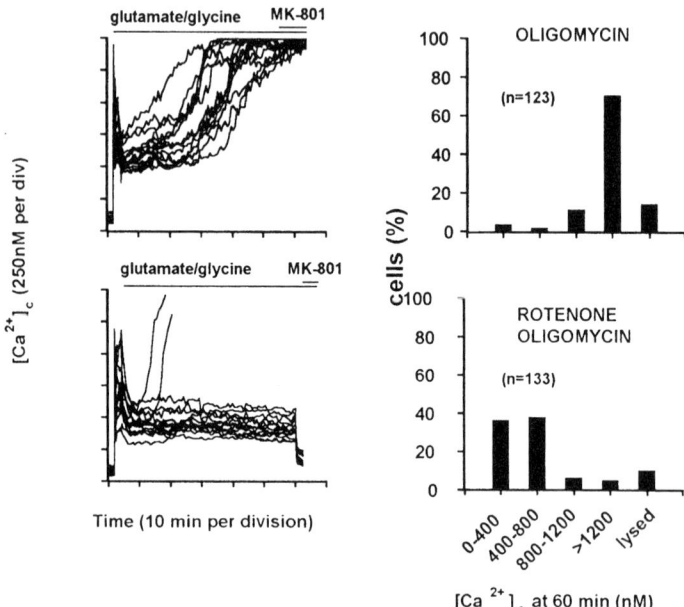

FIGURE 3. DCD is not due to an elevated cytoplasmic Ca^{2+}. Protection afforded by mitochondrial depolarization with rotenone/oligomycin is still seen when glutamate is added to granule cells incubated in doubled extracellular Ca^{2+} (2.6 mM), even though the plateau $[Ca^{2+}]_c$ is raised.

is either extruded or accumulated into the mitochondrion. The plateau is retained, and even lowered and prolonged, in the presence of rotenone/oligomycin, which indicates that the plasma membrane efflux pathways are sufficiently active alone in order to balance the rate of Ca^{2+} entry into the cell. The appearance, after a time interval of 2-60 min, of DCD suggests that a time-dependent deterioration in a component of this balance must be occurring, mediated by some mitochondrial function. Since the plasma membrane $3Na^+/Ca^{2+}$ exchanger would not be thermodynamically competent to extrude Ca^{2+} in the face of the greatly decreased Na^+ electrochemical gradient existing across the plasma membrane during NMDA receptor activation,[25,26] the simplest explanations for DCD would be that it is the consequence of a time-dependent increase in NMDA receptor activity, or a corresponding decrease in the capacity of the plasma membrane Ca^{2+}-ATPase to extrude Ca^{2+}. Detailed electrophysiological investigation of these alternatives has yet to be performed. The failure to detect an increased NMDA receptor-mediated rate of entry of the Ca^{2+} surrogate Mn^{2+} during DCD (27) argues for the Ca^{2+}-ATPase as a locus for the time-dependent failure of cytoplasmic Ca^{2+} homeostasis.

That the mitochondrion is the source of this deterioration is supported by the protection afforded by mitochondrial deenergization (FIG. 2B).[8] FIGURE 4 shows further that the proportion of oligomycin-treated cells that have undergone DCD during 60 min of continuous glutamate exposure is dependent on the time for which the mitochondria are functional. The addition of rotenone to depolarize the cells after

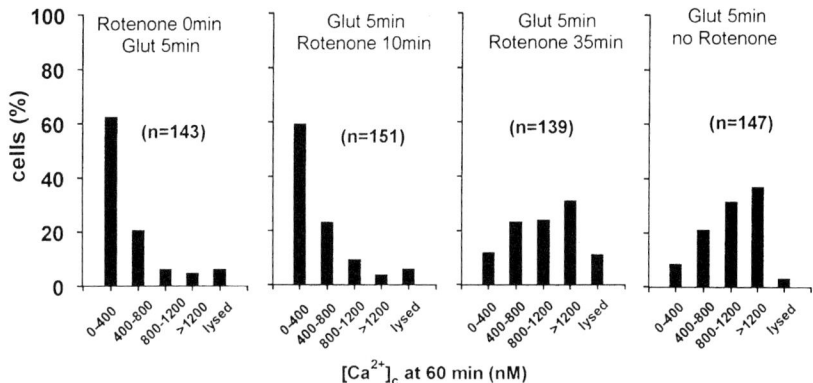

FIGURE 4. DCD is a function of the time for which mitochondria remain polarized (and Ca^{2+} loaded) in the presence of glutamate. Granule cells were exposed to glutamate/glycine for 60 min in the presence of oligomycin. Mitochondria were depolarized by the further addition of rotenone at defined times. The proportion of cells displaying DCD after 60 min was a function of the period for which the mitochondria were polarized during that period.

30 min of glutamate exposure leads to extensive cytoplasmic Ca^{2+} deregulation, perhaps as a consequence of the resulting flood of Ca^{2+} out of the mitochondrion into the cytoplasm.

The implication of the previous paragraph is that mitochondria within glutamate-exposed cells remain polarized, at least throughout the plateau period. Since this is contrary to the consensus that glutamate depolarizes $\Delta\psi_m$,[16,18,28–34] some evidence is required. We have utilized the cell-permeant cationic fluorescent probe tetramethyl rhodamine methyl ester ($TMRM^+$) to monitor changes in potential upon glutamate addition. The probe, introduced by Loew and colleagues,[35] responds to changes in both plasma membrane potential ($\Delta\psi_p$) and $\Delta\psi_m$. Since the activation of glutamate receptors will lead to plasma membrane depolarization, it is essential to control for this. At single-cell resolution, using concentrations of $TMRM^+$ sufficient to induce fluorescent quenching within the matrix, mitochondrial hyperpolarization—for example, induced by oligomycin addition to induce the classic "state 3–state 4 transition" seen with isolated mitochondria—results in a small but easily detectable decrease in whole-cell fluorescence due to the enhanced uptake into the matrix (FIG. 5A). Conversely, addition of protonophore leads to a rapid dequenching of the signal as the probe is released into the cytoplasm followed by a slower decline as the probe effluxes across the plasma membrane.

The accurate quantification of $\Delta\psi_m$ for *in situ* mitochondria is virtually impossible. Rather than a numerical value for the potential, a more important and accessible operational definition is whether $\Delta\psi_m$ is sufficient for ATP synthesis under the conditions pertaining in that actual cell. The oligomycin-induced state 3–4 transition allows us to use a null-point analysis of the mitochondrial bioenergetic condition. Hyperpolarization implies that the mitochondria were net generators of ATP prior to addition of the inhibitor, and "damaged" mitochondria, maintaining a potential by virtue of hydrolyzing cytoplasmic ATP, will depolarize on addition of oligomycin.

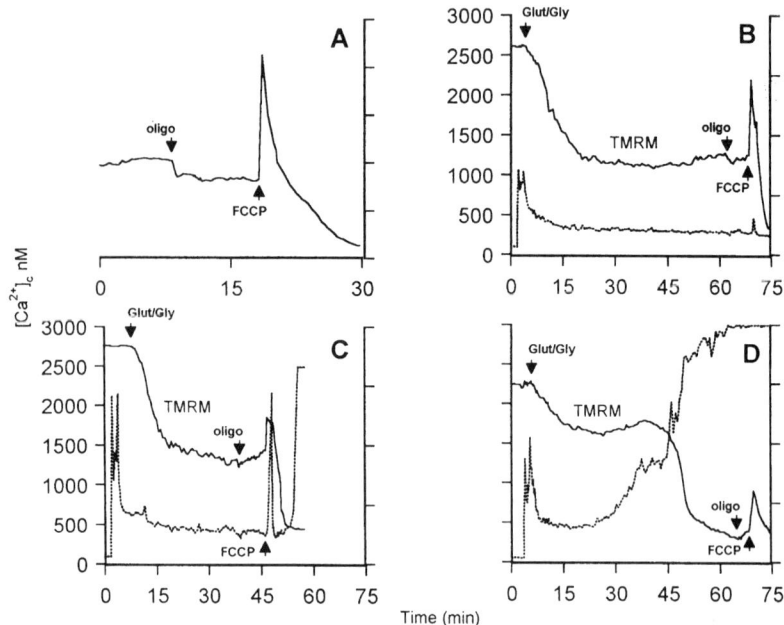

FIGURE 5. Membrane potentials *(solid traces)* and $[Ca^{2+}]_c$ *(dotted traces)* during glutamate exposure. The cationic TMRM$^+$ was loaded at 50 nM, sufficient to produce quenching of fluorescence within the mitochondrial matrix. **A,** Oligomycin-induced hyperpolarization of control mitochondria is detected as a fluorescent quenching. Depolarization with FCCP causes rapid dequenching as the probe is released into the cytoplasm, followed by loss across the plasma membrane. **B,** An individual cell exposed to glutamate for 60 min retains mitochondrial polarization, seen by the responses to oligomycin and FCCP. The initial slow fluorescent decrease is due to plasma membrane. Note that little Ca^{2+} is released into the cytoplasm on addition of FCCP depolarization. **C,** This cell shows a slow fluorescence increase with oligomycin, indicating that the mitochondria have a lowered $\Delta\psi_m$ that was maintained by ATP hydrolysis prior to oligomycin addition. Note that the mitochondria release more Ca^{2+} on addition of FCCP. **D,** This cell undergoes early DCD, accompanied by a biphasic dequenching and loss of TMRM$^+$ fluorescence. However some $\Delta\psi_m$ remains even after 60 min from the responses to oligomycin and FCCP.

FIGURE 5B–D shows a gallery of responses where $[Ca^{2+}]_c$ and TMRM$^+$ fluorescence were measured in parallel. In each case control experiments confirm that the first slow phase of fluorescence decline, complete within 20 min, is almost entirely accounted for by the plasma membrane depolarization. FIGURE 5B shows a cell that survives 60 min of glutamate/glycine exposure. The mitochondria were generating ATP prior to oligomycin addition as shown by the decrease in whole-cell fluorescence with the inhibitor. The spike and decline in TMRM$^+$ fluorescence with FCCP are also characteristic of polarized mitochondria (compare FIG. 5A). Note that the mitochondria in this cell have accumulated a modest amount of Ca^{2+} as evidenced by the small spike of $[Ca^{2+}]_c$ when the matrix Ca^{2+} pool is released by FCCP. FIGURE 5C shows a cell that releases far more matrix Ca^{2+} with FCCP and then proceeds to undergo DCD. Note that the addition of oligomycin causes an *increase* in TMRM+ fluorescence, suggesting that the mitochondria in this cell had become net hydrolyzers of cytoplasmic ATP at this stage. FIGURE 5D shows a cell that had early DCD. Mitochondrial depolarization, visualized as a slow dequenching of fluorescence followed by a precipitate decline, appears to occur only after $[Ca^{2+}]_c$ has started to rise. Even at 60 min, some 20 min after DCD, some residual mitochondrial potential is apparent from the effects of oligomycin and FCCP.

The conclusion is that mitochondria, far from depolarizing at an early stage of glutamate exposure, remain sufficiently polarized to generate ATP until DCD is actually under way. While definitive proof is still lacking, the indication is that mitochondrial depolarization is an effect of DCD rather than a cause. In this context, no significant retardation of DCD is observed in these cells in the presence of established inhibitors of the mitochondrial permeability transition, including bongkrekic acid, cyclosporin A and *N*-methylvaline-cyclosporin.[9,10] Confirmation that mitochondria can generate ATP during glutamate exposure is given in FIGURE 6, where cells are maintained by the nonglycolytic substrates lactate or pyruvate.[9] That DCD occurs at the same time and to the same extent as in cells maintained by glucose (in the presence or absence of oligomycin) indicates that the source of ATP is unimportant for DCD and argues against a specific metabolic failure.

What, then, is the mechanism by which the mitochondria within glutamate-exposed granule cells induce DCD at the plasma membrane? The strongest protection against DCD that we have observed in cells with polarized mitochondria has been with the cell-permeant antioxidant Mn(TBAP) (FIG. 7). The antioxidant also greatly improved the ability of the cells to withstand mitochondria dumping of Ca^{2+} into the cytoplasm after 30 min of glutamate exposure.[10] Interestingly the antioxidant allowed the cells to maintain a more stable plateau $[Ca^{2+}]_c$ in the presence of glutamate (FIG. 7), consistent with an enhanced capacity of the plasma membrane to extrude Ca^{2+}.

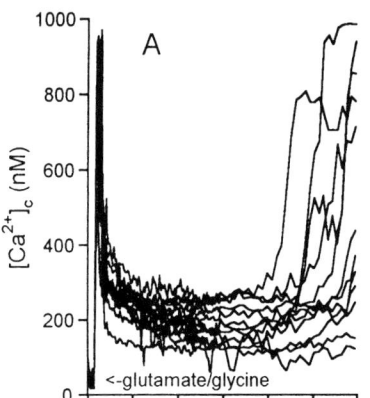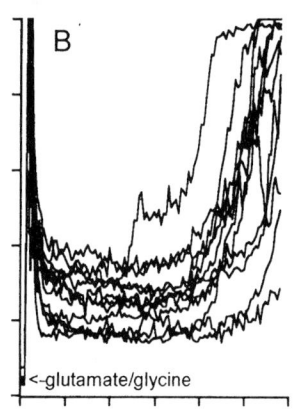

FIGURE 6. Cells maintain a Ca^{2+} plateau maintained by oxidative phosphorylation and undergo DCD as in the presence of glucose. Cells were maintained in the absence of glucose by 10 mM lactate (**A**) or 10 mM pyruvate (**B**).

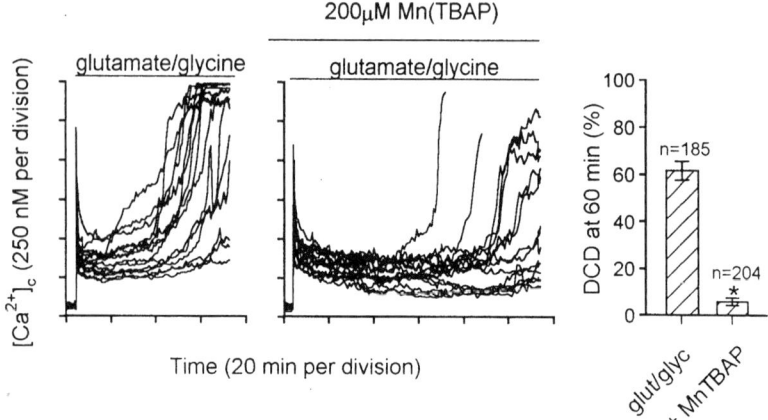

FIGURE 7. The cell-permeant antioxidant Mn(TBAP) prolongs the time for which cells can be exposed to glutamate before the onset of DCD. The *histogram* shows the proportion of cells that have undergone DCD by 60 min in the presence and absence of Mn(TBAP).

While the molecular mechanism by which mitochondrial Ca^{2+} loading enhances the generation of ROS is unclear, studies with isolated mitochondria demonstrate that the maintenance of a high $\Delta\psi_m$ is a prerequisite for radical production.[36–38] We have suggested that earlier reports of depolarization-*enhanced* ROS generation might have been due to the fluorescent products of the ROS assays being in themselves probes of $\Delta\psi_m$;[39] our demonstration of retained mitochondrial polarization during glutamate exposure, and the protective effects of mitochondrial depolarization, suggest that ROS generation by the Ca^{2+}-loaded, polarized mitochondria may be responsible for oxidative damage at the plasma membrane culminating in delayed Ca^{2+} deregulation.

ACKNOWLEDGMENTS

This work was supported by the Wellcome Trust and the Medical Research Council.

REFERENCES

1. TYMIANSKI, M., M.P. CHARLTON, P.L. CARLEN & C.H. TATOR. 1993. Source specificity of early calcium neurotoxicity in cultured embryonic spinal neurons. J. Neurosci. **13:** 2085–2104.
2. SATTLER, R., M.P. CHARLTON, M. HAFNER & M. TYMIANSKI. 1998. Distinct influx pathways, not calcium load, determine neuronal vulnerability to calcium neurotoxicity. J. Neurochem. **71:** 2349–364.
3. PENG, T.I. & J.T. GREENAMYRE. 1998. Privileged access to mitochondria of calcium influx through *N*-methyl-D-aspartate receptors. Mol. Pharmacol. **53:** 974–980.

4. HARTLEY, D.M., M.C. KURTH, L. BJERKNESS, J.H. WEISS & D.W. CHOI. 1993. Glutamate receptor–induced $^{45}Ca^{2+}$ accumuation in cortical cell culture correlated with subsequent neuronal degeneration. J. Neurosci. **13:** 1993–2000.
5. EIMERL, S. & M. SCHRAMM. 1994. The quantity of calcium that appears to induce neuronal death. J. Neurochem. **62:** 1223–1226.
6. KIEDROWSKI, L., J.T. WROBLEWSKI & E. COSTA. 1994. Intracellular sodium concentration in cultured cerebellar granule cells challenged with glutamate. Mol. Pharmacol. **45:** 1050–1054.
7. BEAL, M.F., N. HOWELL & I. BODIS-WOLLNER. 1997. Mitochondria and Free Radicals in Neurodegenerative Disease. Wiley-Lis. New York.
8. Budd, S.L. & D.G. Nicholls. 1996. Mitochondrial calcium regulation and acute glutamate excitotoxicity in cultured cerebellar granule cells. J. Neurochem. **67:** 2282–2291.
9. CASTILHO, R.F., O. HANSSON, M.W. WARD, S.L. BUDD & D.G. NICHOLLS. 1998. Mitochondrial control of acute glutamate excitotoxicity in cultured cerebellar granule cells. J. Neurosci. **18:** 10277–10286.
10. Nicholls, D.G., S.L. BUDD, M.W. WARD & R.F. CASTILHO. 1999. Excitotoxicity and mitochondria. In Mitochondria in the Life and Death of the Cell. G.C. Brown, D.G. Nicholls & C. Cooper, Eds. Portland Press. London. In press.
11. NICHOLLS, D.G. 1978. The regulation of extra-mitochondrial free Ca by rat liver mitochondria. Biochem. J. **176:** 463–474.
12. WERTH, J.L. & S.A. THAYER. 1994. Mitochondria buffer physiological calcium loads in cultured rat dorsal root ganglion neurons. J. Neurosci. **14:** 346–356.
13. WHITE, R.J. & I.J. REYNOLDS. 1995. Mitochondria and Na^+/Ca^{2+} exchange buffer glutamate-induced calcium loads in cultured cortical neurons. J. Neurosci. **15:** 1318–1328.
14. KIEDROWSKI, L. & E. COSTA. 1995. Glutamate-induced destabilization of intracellular calcium concentration homeostasis in cultured cerebellar granule cells: role of mitochondria in calcium buffering. Mol. Pharmacol. **47:** 140–147.
15. WANG, G.J. & S.A. THAYER. 1996. Sequestration of glutamate-induced Ca^{2+} loads by mitochondria in cultured rat hippocampal neurons. J. Neurophysiol. **76:** 1611–1621.
16. SCHINDER, A.F., E.C. OLSON, N.C. SPITZER & M. MONTAL. 1996. Mitochondrial dysfunction is a primary event in glutamate excitotoxicity. J. Neurosci. **16:** 6125–6133.
17. KHODOROV, B.I., V.G. PINELIS, T. STOROZHEVYKH, O.V. VERGUN & N.P. VINSKAYA. 1996. Dominant role of mitochondria in protection against a delayed neuronal Ca overload induced by endogenous excitatory amino acids following a glutamate pulse. FEBS Lett. **393:** 135–138.
18. ISAEV, N.K., D.B. ZOROV, E.V. STELMASHOOK, R.E. UZBEKOV, M.B. KOZHEMYAKIN & I.V. VICTOROV. 1996. Neurotoxic glutamate treatment of cultured cerebellar granule cells induces Ca^{2+-}dependent collapse of mitochondrial membrane potential and ultrastructural alterations of mitochondria. FEBS Lett. **392:** 143–147.
19. WHITE, R.J. & I.J. REYNOLDS. 1997. Mitochondria accumulate Ca^{2+} following intense glutamate stimulation of cultured rat forebrain neurones. J. Physiol. (Lond.) **498:** 31–47.
20. BERNARDI, P., E. BASSO, R. COLONNA, P. COSTANTINI, F. DI LISA, O. ERIKSSON, E. FONTAINE, M. FORTE, F. ICHAS, S. MASSARI, A. NICOLLI, V. PETRONILLI & L. SCORRANO. 1998. Perspectives on the mitochondrial permeability transition. Biochim. Biophys. Acta Bio-Energetics **1365:** 200–206.
21. REYNOLDS, I.J. & T.G. HASTINGS. 1995. Glutamate induces the production of reactive oxygen species in cultured forebrain neurons following NMDA receptor activation. J. Neurosci. **15:** 3318–3327.
22. DUGAN, L.L., S.L. SENSI, L.M.T. CANZONIERO, S.D. HANDRAN, S.M. ROTHMAN, T.S. LIN, M.P. GOLDBERG & D.W. CHOI. 1995. Mitochondrial production of reactive oxygen species in cortical neurons following exposure to NMDA. J. Neurosci. **15:** 6377–6388.

23. BINDOKAS, V.P., J. JORDAN, C.C. LEE & R.J. MILLER. 1996. Superoxide production in rat hippocampal neurons: selective imaging with hydroethidine. J. Neurosci. **16:** 1324–1336.
24. BUDD, S.L. & D.G. NICHOLLS. 1996. A re-evaluation of the role of mitochondria in neuronal calcium homeostasis. J. Neurochem. **66:** 403–411.
25. KIEDROWSKI, L., G. BROOKER, E. COSTA & J.T. WROBLEWSKI. 1994. Glutamate impairs neuronal calcium extrusion while reducing sodium gradient. Neuron **12:** 295–300.
26. PINELIS, V.G, M. SEGAL, V. GREENBERGER & B.I. KHODOROV. 1994. Changes in cytosolic sodium caused by a toxic glutamate treatment of cultured hippocampal neurons. Biochem. Mol. Biol. Int. **32:** 475–482.
27. KHODOROV, B.I., D.A. FAYUK, S.G. KOSHELEV, O.V. VERGUN, V.G. PINELIS, N.P. VINSKAYA, T.P. STOROZHEVYKH, E.N. ARSENYEVA, L.G. KHASPEKOV, A.P. LYZHIN, N. ISAEV, I.V. VICTOROV & J.M. DUBINSKY. 1996. Effect of a prolonged glutamate challenge on plasmalemmal calcium permeability in mammalian central neurones. Mn^{2+} as a tool to study calcium influx pathways. Int. J. Neurosci. **88:** 215–241.
28. ANKARCRONA, M., J.M. DYPBUKT, E. BONFOCO, B. ZHIVOTOVSKY, S. ORRENIUS, S.A. LIPTON & P. NICOTERA. 1995. Glutamate-induced neuronal death: a succession of necrosis or apoptosis depending on mitochondrial function. Neuron **15:** 961–973.
29. WHITE, R.J. & I.J. REYNOLDS. 1996. Mitochondrial depolarization in glutamate-stimulated neurons: an early signal specific to excitotoxin exposure. J. Neurosci. **16:** 5688–5697.
30. KHODOROV, B.I., V. PINELIS, O. VERGUN, T. STOROZHEVYKH & N. VINSKAYA. 1996. Mitochondrial deenergization underlies neuronal calcium overload following a prolonged glutamate challenge. FEBS Lett. **397:** 230–234.
31. ANKARCRONA, M., J.M. DYPBUKT, S. ORRENIUS & P. NICOTERA. 1996. Calcineurin and mitochondrial function in glutamate-induced neuronal cell death. FEBS Lett. **394:** 321–324.
32. PREHN, J.H.M. 1998. Mitochondrial transmembrane potential and free radical production in excitotxic neurodegeneration. Naunyn-Schmiedebergs Arch. Pharmacol. **357:** 316–322.
33. KIEDROWSKI, L. 1998. The difference between mechanisms of kainate and glutamate excitotoxicity in vitro: osmotic lesion versus mitochondrial depolarization. Restor. Neurol. Neurosci. **12:** 71–79.
34. KEELAN, J., O. VERGUN, B.I. KHODOROV & M.R. DUCHEN. 1998. Calcium-dependent mitochondrial depolarization specific to toxic glutamate depolarization in cultured rat hippocampal neurones. J. Physiol. (Lond) **506:** 75P.
35. EHRENBERG, B., V. MONTANA, M.D. WEI, J.P. WUSKELL & L.M. LOEW. 1988. Membrane potential can be determined in individual cells from the nernstian distribution of cationic dyes. Biophys. J. **53:** 785–794.
36. BOVERIS, A., N. OSHINO & B. CHANCE. 1972. The cellular production of hydrogen peroxide. Biochem. J. **128:** 617–630.
37. NEGRE-SALVAYRE, A., C. HIRTZ, G. CARRERA, R. CAZENAVE, M. TROLY, R. SALVAYRE, L. PÉNICAUD & L. CASTEILLA. 1997. Role for uncoupling protein-2 as a regulator of mitochondrial hydrogen peroxide generation. FASEB J. **11:** 809–815.
38. ZHANG, L., L.D. YU & C.A. YU. 1998. Generation of superoxide anion by succinate-cytochrome c reductase from bovine heart mitochondria. J. Biol. Chem. **273:** 33972–3976.
39. BUDD, S.L., R.F CASTILHO & D.G. NICHOLLS. 1997. Mitochondrial membrane potential and hydroethidine-monitored superoxide generation in cerebellar granule cells. FEBS Lett. **415:** 21–24.

Fundamental Aspects of Reactive Oxygen Species, or What's the Matter with Oxygen?

IRWIN FRIDOVICH[a]

Department of Biochemistry, Duke University Medical Center, Durham, North Carolina 27710, USA

ABSTRACT: A byproduct of normal aerobic metabolism is the generation of dangerously reactive intermediates of the reduction of O_2. These include O_2^-, H_2O_2, and $HO\cdot$ and arise because of the predisposition of O_2 for univalent reductions. These reactive oxygen species (ROS), and others that they can engender, threaten all cellular macromolecules, and defenses are needed. Among those known to date are: superoxide dismutases to convert O_2^- into $O_2 + H_2O_2$; catalases to dismute H_2O_2 into $O_2 + H_2O$; and peroxidases to reduce H_2O_2 into $2H_2O$) and to reduce ROOH into ROH and H_2O. These defenses are aided by enzymes that repair or recycle oxidatively damaged nucleic acids and proteins. A role for such oxidative damage in aging and neurodegenerative diseases is well supported.

WHY IS MOLECULAR OXYGEN DANGEROUS?

The dangers imposed by O_2, upon all life forms, arise from its electronic structure, which predisposes it to reduction by a univalent pathway. The paramagnetism of O_2 in the ground state indicates that it contains two unpaired electrons with parallel spin states. In contrast all the electrons in stable organic molecules are arranged in pairs with antiparallel spin states. Moving a pair of electrons from any stable organic compound into O_2 will require the inversion of an electronic spin state, since two electrons cannot occupy the same orbital unless their spins are antiparallel. This may be illustrated as follows:

a.

Inversion of electronic spin states, by interaction with nuclear spins, does occur, but on a time scale several orders of magnitude longer than the lifetime of collisional complexes. It is thus unlikely to occur while the O_2 and the reductant are in contact. But it can occur during the intervals between collisions; so the electrons are preferentially transferred to O_2 one at a time. Four electrons (and four protons) are needed to reduce O_2 to two molecules of water. Hence the univalent pathway necessitates

---

[a]Phone: 919-684-5122; fax: 919-684-8885.

intermediates, as follows:

b. $$O_2 \xrightarrow{e^-} O_2^- \xrightarrow[2H^+]{e^-} H_2O_2 \xrightarrow{e^-} OH^- + HO^\bullet \xrightarrow[2H^+]{e^-} 2H_2O$$

PROPERTIES OF THE INTERMEDIATES

The superoxide anion radical can act both as a reductant and as an oxidant. In the reducing environment of the cell interior it is primarily an oxidant. Among its targets are small molecules such as catecholamines, thiols, tetrahydropterins, ascorbate, and leukoflavins. O_2^- is the conjugate base of the hydroperoxyl radical the pK_a of which is 4.8 and which is a much stronger oxidant than O_2^-. Hence it is possible that the oxidations attributed to O_2^- may actually be caused by $HO_2^\bullet$, even though at physiological pH it represents somewhat less than 1% of the O_2^-.

Dehydratases containing [4Fe-4S] clusters are oxidized by O_2^- with rate constants in the range $10^6 \to 10^7$ M^{-1} s$^-$. Aconitases, fumarases A and B, dihydroxy acid dehydratase, and 6-phosphogluconate dehydratase are in this category; and their oxidation by O_2^- renders the [4Fe-4S] clusters unstable such that they lose iron. These enzymes are thus rapidly inactivated by O_2^- in a way that releases iron in a form that mediates production of hydroxyl and alkoxyl radicals as follows:

c. $$Fe(II) + HOOH \text{ (or ROOH)} \to Fe(III) + HO^\bullet \text{ (or RO}^\bullet\text{)} + OH^-$$

The HO$^\bullet$ and RO$^\bullet$ thus produced are very potent oxidants and can oxidize nucleic acids, proteins, and phospholipids. The Fe(III) formed in reaction c will be reduced by the thiols and enediols that are so abundant in the cell, and can thus repeatedly participate in reaction c.

THE NO CONNECTION

NO, produced by the action of NO synthases on L-arginine, serves as a smooth muscle relaxant, a neurotransmitter, and a component of the offensive armamentarium of macrophages. NO is thus produced in locales where it may encounter O_2^-. NO and O_2^- react at ~10^{10} M^{-1} s^{-1}, yielding peroxynitrite (ONOO$^-$), which, like all peracids, is a potent oxidant. Here, then, is another means by which O_2^- can damage cells.

BIOMARKERS OF OXIDATIVE DAMAGE

Oxidation of nucleic acids by hydroxyl or alkoxyl radicals can produce a plethora of products, including hydroxylated bases, and can also cause chain scission. Such hydroxylated bases can be measured in DNA and are also excreted in urine following their excision by endonucleases. O_2^-, or species derived therefrom, are known to be mutagenic. Protein oxidation produces ortho and metatyrosine, dityrosine, methionine sulfoxide, and poorly defined protein carbonyls. All of these have been detected. Oxidation of membrane lipids also produces a mixture of products including ethane, pentane, malondialdehyde, and various other carbonyl-bearing products, all of which have been detected. Peroxynitrite can cause nitration of tyrosine residues, and nitrotyrosine has been detected in blood plasma, brain, and atheromatous plaque. It is clear that the dangers imposed by O_2 and its progeny are very real and that substantial damage is sustained. How do we survive it?

DEFENSES, AVOIDANCE

Many biological oxidases utilize mechanisms that accomplish divalent and even tetravalent reduction of O_2, without releasing intermediates. Cytochrome c oxidase is such an enzyme; by virtue of two heme and two Cu(II) prosthetic groups it can accumulate four reducing equivalents, which are then transferred to bound O_2, via a bound peroxo intermediate, to produce two molecules of water. So the first line of defense is avoidance, which is very effective since in most cells under normal conditions only 0.1–1.0% of the electrons transferred to O_2 produce O_2^-. The rate of O_2 consumption is such that even a 0.1% yield of O_2^- would impose an intolerable oxidative burden were there not additional defenses.

SUPEROXIDE DISMUTASES

The superoxide dismutases (SODs) catalyze the conversion of O_2^- into H_2O_2 + O_2 and do so with unmatched efficiency, being limited only by the rate of diffusion. Since the creation of an O_2-rich atmosphere, by photosynthetic organisms, must have imposed a common selection pressure on a varied anaerobic biota, it is not surprising that several families of SODs currently exist. In eukaryotes there is a MnSOD in the mitochondria, a CuZnSOD in the cytosol and in the other organelles, and a different CuZnSOD in the extracellular spaces. Plants have a MnSOD, a CuZnSOD, and also a FeSOD. Prokaryotes have thus far yielded a FeSOD, a MnSOD, and, in the periplasm of gram-negative bacteria, a CuZnSOD. Recently a NiSOD has been found in some fungi (*Streptomyces*). In all the active site metal of the SOD is reduced by one O_2^- and then reoxidized by the next. In this way it transfers an electron from one O_2^- to the next while avoiding the electrostatic repulsion that would hinder direct electron transfer between one O_2^- and another.

CATALASES AND PEROXIDASES

The H_2O_2 produced from O_2^-, and directly by specific oxidases, must be eliminated. This is accomplished by catalases and peroxidases. The catalases carry out a dismutation reaction in which one H_2O_2 oxidizes another such that the one is converted to O_2 and the other to two molecules of water. Most catalases are heme-containing enzymes; but *Lactobacilli* and related bacteria, which cannot synthesize heme, can produce a very effective catalase the prosthetic group of which is a binuclear cluster of Mn atoms. Peroxidases use a variety of reductants to reduce H_2O_2 to $2 H_2O$. Thus there are peroxidases that utilize glutathione, ascorbate, ferrocytochrome c, reduced thioredoxin, NADH, and even iodide and chloride. Plant peroxidases are notoriously broad in their utilization of reductants. Most peroxidases are heme-based enzymes; but the prosthetic group of the GSH peroxidases, so important in higher animals, is a selenocysteine residue. The GSH peroxidases deserve special mention for another reason and that is their ability to catalyze the reduction of alkylhydroperoxides, such as those generated by the oxidation of polyunsaturated lipids, to alcohols.

LOW MOLECULAR WEIGHT ANTIOXIDANTS

When a radical oxidizes any stable organic compound the first product is another radical derived from that organic compound. This can lead to chain reactions in which one initiation event can lead to the oxidation of many molecules. This can be illustrated for the case of polyunsaturated lipids (LH_2) by the following reactions:

c. $\quad LH_2 + HO^{\bullet} \rightarrow LH^{\bullet} + OH^- + H^+ \quad \}$ initiation

d. $\quad LH^{\bullet} + O_2 \rightarrow LHOO^{\bullet}$

e. $\quad LHOO^{\bullet} + LH_2 \rightarrow LHOOH + LH^{\bullet}$ $\quad \Big\}$ propagation

In a condensed phase, such as the hydrophobic interior of a biological membrane, such a chain reaction can greatly amplify the consequences of the initiating event. Vit E (tocopherol, vitamin E) acts to terminate such chain reactions. Vit E is a hindered phenol that is anchored to the membrane by a long hydrophobic phytyl tail. When the phenol group of Vit E encounters the lipid peroxyl radical ($LHOO^{\bullet}$), it reduces it to the hydroperoxide and is itself oxidized to the phenoxyl radical. The Vit E phenoxyl radical is not very reactive, because of steric hindrance, and lasts long enough to be reduced by ascorbate in the adjacent aqueous phase. This accomplishes the effective export of the radical from the membrane to the aqueous phase. The ascorbyl radical can there be reduced by GSH. Tocopherol and ascorbate are thus antioxidants, the one hydrophobic and the other hydrophilic. Other antioxidants include urate, flavonoids, dihydroubiquinone, and dihydrolipoate.

REPAIR

Despite the multiple defenses already listed, some oxidative damage will be done. The accumulation of this damage is minimized by turnover and/or repair. There are methionine sulfoxide reductases to reverse the oxidation of methionine residues and proteases to hydrolyze oxidized proteins. There are endonucleases to remove oxidized bases from nucleic acids and so begin the process of repair, and there are phospholipases that serve to hydrolyze oxidized phospholipids. Yet oxidative damage to such vital cellular components has been shown to accumulate as a function of age and to do so more rapidly in short-lived than in long-lived species.

NEURONAL DAMAGE

Since we are here specially interested in neurodegenerative diseases, this essay should end with a consideration of oxidative damage to neuronal tissues. O_2^-, NO, and $ONOO^-$ are produced in brain, and their production is increased during reperfusion following 30 min of global ischemia.[1] The brains of transgenic mice, expressing severalfold higher levels of CuZnSOD, were more resistant than normal mice to blunt trauma.[2] Cold injury to the frontal cortex of rat brain increased production of ascorbyl radical and release of lactate dehydrogenase, and this was attenuated by pretreatment with SOD and catalase.[3] MPTP was much more damaging to the brains of MnSOD-deficient mice than it was to mice overexpressing this enzyme.[4] Applying glutamate to brain by microdialysis, as a means of modeling excitotoxicity, increased $HO^\bullet$ production.[5] The role of oxygen-derived free radicals in brain reperfusion injury has been reviewed.[6]

Neurons that produce NO were very sensitive to excitotoxicity imposed by N-methyl-D-aspartate (NMDA), and MnSOD was protective.[7] Sympathetic neurons in culture were seen to die when nerve growth factor was withdrawn from the culture medium, and this was prevented by CuZnSOD.[8] The survival of transplanted neurons was increased fourfold when they overexpressed CuZnSOD.[9] Kainate toxicity to cortical neurons was attenuated by SOD plus catalase, and by a chemical scavenger of O_2^-.[10] The relationship between excitotoxicity and oxidative stress in neural tissue has been reviewed.[11] The amyloid beta peptide from senile plaques in Alzheimer's disease has been seen to diminish endothelium-dependent vessel wall relaxation caused by acetylcholine or bradykinin, while it was without effect on the endothelium-independent relaxation caused by nitroprusside. This endothelial dysfunction was prevented by SOD.[12] Finally it should be mentioned that approximately 25% of cases of familial amyotrophic lateral sclerosis have been associated with point mutations in the CuZnSOD. This appears to be due to some toxic gain of function by the mutated forms of CuZnSOD rather than to deficiency of SOD activity.[13] The nature of this gain of function remains unknown. This cursory examination of some relevant literature makes it clear that neurons are not exempt from the deleterious effects of the intermediates of oxygen reduction.

The basic chemistry and biology of the oxygen-derived radicals has been reviewed recently.[14–16]

REFERENCES

1. FORMAN, L.J., P. LIU, R.G. NAGELE & P.Y. WONG. 1998. Augmentation of nitric oxide, superoxide, and peroxynitrite production during cerebral ischemia and reperfusion in the rat. Neurochem. Res. **23:** 141–148.
2. MIKAWA, S., H. KINOUCHI, H. KAMII, G.T. GOBBEL, S.F. CHEN, E. CARLSON, C.J. EPSTEIN & P.H. CHAN. 1996. Attenuation of acute and chronic damage following

traumatic brain injury in copper, zinc superoxide dismutase transgenic mice. J. Neurosurg. **85:** 885–891.
3. KIHARA, T., S. SAKATA & M. IKEDA. 1995. Direct detection of ascorbyl radical in experimental brain injury: microdialysis and an electron spin resonance study. J. Neurochem. **65:** 282–286.
4. CORTOPASSI, G. & E. WANG. 1995. Modelling the effects of age-related mtDNA mutation accumulation; complex I deficiency, superoxide, and cell death. Biochim. Biophys. Acta **1271:** 171–176.
5. YANG, C.S., P.J. TSAI, N.N. LIN, L. LIU & J.S. KUO. 1995. Elevated extracellular glutamate levels increased the formation of hydroxyl radical in the striatum of anesthetized rat. Free Radic. Biol. Med. **19:** 453–459.
6. MOORE, L.E. & R.J. TRAYSTMAN. 1994. Role of oxygen free radicals and lipid peroxidation in cerebral reperfusion injury. Adv. Pharmacol. **31:** 565–576.
7. GONZATEZ-ZULUETA, M., L.M. ENSZ, G. MUKHINA, R.M. LEBOVITZ, R.M. ZWACKA, J. F. ENGELHARDT, L.W. OBERLEY, V.L. DAWSON & T.M. DAWSON. 1998. Manganese superoxide dismutase protects nNOS neurons from NMDA and nitric oxide-mediated neurotoxicity. J. Neurosci. **18:** 2040–2055.
8. JORDAN, J., G.D. GHADGE, J.H. PREHN, P.T. TOTH, R.P. ROOS & R.J. MILLER. 1995. Expression of human copper/zinc superoxide dismutase inhibits death of rat sympathetic neurons caused by withdrawal of nerve growth factor. Mol. Pharmacol. **47:** 1095–1100.
9. NAKAO, N., E.M. FRODL, H. WIDNER, E. CARLSON, F.A. EGGERDING, C.J. EPSTEIN & P. BRUNDIN. 1995. Overexpressing Cu/Zn superoxide dismutase enhances the survival of transplanted neurons in a rat model of Parkinson's disease. Nature Med. **1:** 226–231.
10. CHENG, Y. & A.Y. SUN. 1994. Oxidative mechanisms involved in kainate-induced cytotoxicity in cortical neurons. Neurochem. Res. **19:** 1557–1564.
11. BONDY, S.C. & C.P. LEBEL. 1993. The relationship between excitotoxicity and oxidative stress in the central nervous system. Free Radic. Biol. Med. **14:** 633–642.
12. PRICE, J.M., E.T. SUTTON, A. HELLERMAN & T. THOMAS. 1997. Beta amyloid induces cerebrovascular dysfunction in the rat brain. Neurol. Res. **19:** 534–538.
13. GURNEY, M.E., F.B. CUTTING, P. ZHAI, P.K. ANDRUS & E.D. HALL. 1996. Pathogenic mechanisms in familial amyotrophic lateral sclerosis. Pathol. Biol. (Paris) **44:** 51–56.
14. FRIDOVICH, I. 1998. An overview of oxy radicals in medical biology. Advan. Mol. Cell. Biol. **25:** 1–14.
15. FRIDOVICH, I. 1997. Superoxide anion radical (O_2^-), superoxide dismutases, and related matters. J. Biol. Chem. **272:** 18515–18517.
16. HASSAN, H.M. 1997. Cytoxicity of oxy radicals and the evolution of superoxide dismutases. Lung Biol. Health Dis. **105:** 27–47.

Ca^{2+}-Mediated Mitochondrial Dysfunction and the Protective Effects of Bcl-2

ANNE N. MURPHY[a]

MitoKor, 11494 Sorrento Valley Road, San Diego, California 92121, USA

ABSTRACT: Mitochondrial Ca^{2+} sequestration likely contributes to cell death in excitotoxicity and ischemia reperfusion injury, and may also be involved in chronic forms of neurodegeneration in which a compromise in bioenergetic function alters cellular Ca^{2+} homeostasis. Bcl-2 overexpression is known to protect against Ca^{2+}-mediated death; the mechanism of protection remains unresolved. Our data of the ability of Bcl-2 to potentiate mitochondrial Ca^{2+} uptake capacity and resistance to Ca^{2+}-induced damage is discussed in light of current information on apoptotic signaling pathways.

Remarkable progress has been made in understanding various aspects of human disease. It is therefore somewhat surprising that we do not have a clear understanding of how cells die—an event that is fundamental to pathophysiology. For instance, it has been appreciated for many years that an excessive load of intracellular Ca^{2+} can induce fairly rapid necrotic cell death. Such increases in cytoplasmic Ca^{2+} have been theorized to underlie cell death in response to a number of pathophysiological events including ischemia/reperfusion injury, exposure of many cell types to pro-oxidants and certain toxins, and excessive stimulation of NMDA receptors in glutamate-induced neurotoxicity.[1,2] The mediators of such death have historically been identified as Ca^{2+}-stimulated proteases, phospholipases, and endonucleases that lead to generalized cell disruption. Although Ca^{2+}-stimulated cytosolic events may vary with the type of inducer of an increase in Ca^{2+}, recent work has indicated that excitotoxic neuronal death may be wholly (or at least partly) dependent upon changes in mitochondrial behavior as a result of excessive mitochondrial Ca^{2+} uptake.[3,4] Such a shift in the Ca^{2+} toxicity paradigm may seem to have been slow to evolve, but advances in imaging technology have only recently made such studies of *in situ* mitochondrial function possible. The following paragraphs will address mitochondrial Ca^{2+} sequestration under physiological and pathophysiological conditions, and the role of Bcl-2 in prevention of excessive Ca^{2+} sequestration-mediated damage to mitochondria.

MITOCHONDRIAL Ca^{2+} SEQUESTRATION

Like with many biological functions, Ca^{2+} transport into mitochondria meets a physiological need; but when it is excessive, this transport process poses a liability

[a]Phone: 858-509-5636; fax: 858-793-7805.
e-mail: murphya@mitokor.com

for the cell. Ca^{2+} transport into mitochondria at physiological levels serves a normal signaling function, and mitochondrial Ca^{2+} transporters likely exist for this function.[6,7] Physiologic Ca^{2+} increases in the cytoplasm that are transduced into increases in matrix-free Ca^{2+} stimulate mitochondrial ATP production, producing more chemical energy at a time when the cell is in need (for review, see Refs. 8 and 9). As well, mitochondria appear to actively participate in propagation of a Ca^{2+} signal across a cell in a manner that appears to depend upon the cell type and conditions (for review, see Refs. 9–11). The degree to which mitochondria accumulate Ca^{2+} in response to a hormone or neurotransmitter may serve to coordinate the temporal conductance of a Ca^{2+} signal and/or modulate the speed with which cytoplasmic Ca^{2+} returns to baseline.

This Ca^{2+} signaling process is dependent upon sequestration of the ion across the inner mitochondrial membrane by the Ca^{2+} uniporter and possibly by RAM, an even more rapid mode of Ca^{2+} sequestration.[12,13] It is unclear at this time whether these two transport activities represent separate molecular species, or whether RAM is mediated by the uniporter operating in an altered manner. Uniporter-mediated transport, which demonstrates a low affinity for Ca^{2+} in comparison to endoplasmic reticular sequestration, is driven by the electrical portion ($\Delta\psi$) of the electrochemical gradient of protons at the inner mitochondrial membrane. The membrane potential is established through electron transport under normal circumstances, or it can be generated via reversal of the proton translocating ATP synthetase when electron transport is inhibited.[14] The kinetics of mitochondrial Ca^{2+} uptake and efflux mechanisms allow for net mitochondrial Ca^{2+} sequestration in response to a chronic or transient rise in cytoplasmic Ca^{2+} or cytoplasmic Ca^{2+} oscillations.[6–13]

Mitochondrial Ca^{2+} efflux mechanisms are also energy dependent, and include Na^+-dependent (in electrically excitable tissues) and Na^+-independent transport. Efflux mechanisms demonstrate less rapid maximal transport kinetics than the uptake mechanisms; thus net sequestration occurs when extramitochondrial Ca^{2+} is sufficiently high to stimulate uptake.[6,7] It has been suggested that mitochondria lie in close apposition to sites of endoplasmic reticular Ca^{2+} release,[10,15] thus facilitating mitochondrial involvment in inositol trisphosphate-mediated Ca^{2+} signals. Spatial proximity to the plasma membrane, the site of Ca^{2+} entry through voltage-dependent Ca^{2+} channels during potassium-induced neuronal depolarization, is a primary determinant of the extent of Ca^{2+} sequestration.[16] These results imply that highly heterogeneous microdomains of Ca^{2+} exist that potentiate the extent of mitochondrial Ca^{2+} loading in response to physiological stimuli.

In contrast to normal levels of Ca^{2+} sequestration, excessive uptake can result in mitochondrial injury, and either necrotic or apoptotic death. The proposed events that follow excessive Ca^{2+} uptake include stimulation of mitochondrial free radical production, induction of the membrane permeability transition (the opening of a nonselective proteinaceous pore at the inner mitochondrial membrane),[17] and Ca^{2+}-induced inhibition to components of oxidative phosphorylation.[18,19] Massive Ca^{2+} influx into cells occurs as an initiating event in ischemia/reperfusion and excitotoxic damage, trauma, and exposure to some toxins. Compelling new studies clearly implicate mitochondrial Ca^{2+} sequestration as a necessary event in glutamate toxicity in primary cultures of neurons.[3,4] Ca^{2+} sequestration during glutamate exposure was inhibited by depolarization of the mitochondrial membrane, effectively removing the driving force for Ca^{2+} sequestration. Budd and Nicholls[3] induced mitochondrial depolarization by treatment of cerebellar granule cells with rotenone (an inhibitor of complex I of the mitochondrial electron transport chain) and oligomycin (an inhibitor of the mitochondrial ATP synthetase that prevents this enzyme from functioning in reverse to establish a mitochondrial membrane potential). These cells are capable of generating sufficient ATP for survival by anaerobic glycolysis with no inherent toxicity of mitochondrial depolarization. Stout and colleagues[4] prevented mitochondrial Ca^{2+} uptake by transient treatment of cultured forebrain neurons with respiratory uncoupler (FCCP) during glutamate exposure, and similarly prevented glutamate toxicity. This study revealed that inhibition of mitochondrial Ca^{2+} sequestration results in significantly higher cytosolic Ca^{2+} loads in response to glutamate, from which it can be deduced that cytosolic Ca^{2+} per se is not the mediator of cell death. The mitochondrial event responsible for triggering excitotoxic neuronal death in response to Ca^{2+} sequestration remains elusive, but may involve Ca^{2+}-induced mitochondrial free radical generation that inhibits Ca^{2+} extrusion from the cell.[20]

Are there other instances in which mitochondrial Ca^{2+} loading is causally linked to cell death? Certainly mitochondrial Ca^{2+} loading is evident following ischemia reperfusion injury,[19,21–24] which is associated with both necrotic and apoptotic death.[25] There have been reports that ruthenium red, an inhibitor of mitochondrial Ca^{2+} uptake, can protect against ischemia/reperfusion injury *in vivo* and in isolated perfused hearts,[26–29] although it is appreciated that this agent can inhibit other Ca^{2+} channels. Indirect evidence that mitochondrial Ca^{2+} sequestration is important in ischemia/reperfusion injury is provided by multiple studies documenting the protective effect of cyclosporin A *in vivo* (for examples see Refs. 30–33). Cyclosporin A is an inhibitor of the permeability transition, a downstream consequence of excessive Ca^{2+} sequestration; however this agent is also not specific, as its use also results in inhibition of calcineurin (see below, Refs. 18,34). It also seems likely that mitochondrial Ca^{2+} uptake is involved in apoptotic death in response to Ca^{2+} ionophore or the endoplasmic reticular Ca^{2+} ATPase inhibitor thapsigargin. Other apoptotic paradigms, those that seemingly involve an alteration in protein kinase signaling, may involve a shift of intracellular Ca^{2+} stores that is correlated with death (with or without a dramatic measureable increase in cytosolic Ca^{2+}).[35–38] These studies imply that an alteration in the phosphorylation state of a signaling molecule ultimately results in mitochondrial Ca^{2+} loading; but a primary effect of alteration of these kinase pathways on the phosphorylation state or subcellular localization of apoptotic regulators such as BAD, caspase-9, and BAX is more likely to be causally linked to cell death in these instances.[39–42] The significance of mitochondrial Ca^{2+} sequestration therefore is less clear in these apoptosis models.

Whether mitochondrial Ca^{2+} loading is involved in the death of neurons in chronic forms of neurodegeneration is unknown. Indirect evidence suggests that Ca^{2+}-mediated triggering may be involved in neuronal death in Alzheimer's disease. Specifically, overexpression of calbindin can block amyloid β-peptide–induced apoptosis of PC12 cells that also overexpress mutant presenilin-1.[43] In addition, these PS-1 mutant cells are more sensitive to metabolic inhibition that results in increased cytoplasmic Ca^{2+}.[44] Hippocampal neurons cultured from PS-1 knock-in mice demonstrate enhanced sensitivity to excitotoxic stress.[45] In each of these models, evidence of mitochondrial dysfunction was provided. It is possible that bioenergetic failure owing to potential mild defects in mitochondrial electron transport activity associated with Alzheimer's, Parkinson's, and Huntington's diseases[19,46] predisposes neurons to potential Ca^{2+} overload during normal neurotransmitter signaling. If the maximal capacity of mitochondria to generate ATP is compromised, then the bioenergetic challenge to maintain normal ATP levels following neurotransmitter-stimulated depolarization may be greater. The compromise in ATP levels could decrease the rate of Ca^{2+} extrusion from the cells and the rate of neurotransmitter reuptake; therefore, mitochondria may be exposed to higher levels of cytosolic Ca^{2+} for a more prolonged period of time following stimulation. The repeated cycling of this event as a result of neural transmission may result in a chronic increase in mitochondrial Ca^{2+} loads. Alternatively, Brini *et al.*[47] have suggested that a mitochondrial respiratory defect will be toxic because it limits mitochondrial Ca^{2+} uptake in response to an agonist. This, in turn, dampens the resulting response of the mitochondria to upregulate ATP production. This study, however, did not address the effect of repeated agonist stimulation on mitochondrial Ca^{2+} levels. It is also difficult to extrapolate these results to events in neurons given that the process was studied in transformed cells, which have a robust capacity to maintain cellular ATP levels by glycolysis. Nonetheless, the study raises the intriguing possibility that chronic neuronal dysfunction (as opposed to neuronal death) associated with cognitive or motor dysfunction may arise from a defect in normal mitochondrial Ca^{2+} signaling.

BCL-2 PROTECTION AGAINST CA^{2+}-INDUCED MITOCHONDRIAL INJURY

Overexpression of the antiapoptotic protein Bcl-2 inhibits neuronal death *in vivo* in response to ischemia/reperfusion injury or treatment with a mitochondrial complex I inhibitor,[48–50] implying an enhanced resistance to Ca^{2+} toxicity. Yet the mechanism(s) of action of this protein remains unclear. In fact, there is evidence for multiple, potentially independent actions of Bcl-2.[51] The most abundant data suggests that Bcl-2 is capable of inhibiting the release of cytochrome *c* from the intermembrane space of mitochondria in response to apoptotic stimuli (Refs. 52–54 and many others), but whether it does so by a uniform mechanism or via multiple modes of action is currently unknown. Cytochrome *c* released from the intermembrane space can facilitate formation of an apoptosome complex comprising APAF-1, caspase 9, and dATP[55] that ultimately activates caspase 3.[55,56] The extent to which this pathway of cytochrome *c* release/caspase 3 activation is conserved in all forms of apoptotic death is unclear. There is evidence for cell type- and apoptotic signal-specific exceptions to dependence of the death pathway on mitochondrial cytochrome *c* release.[57,58]

In search of a mechanism by which Bcl-2 can block Ca^{2+}-mediated cell death, we asked what effect Bcl-2 overexpression has on the ability of mitochondria to accumulate large loads of Ca^{2+} and the ability of mitochondria to resist Ca^{2+}-induced respiratory inhibition. These experiments were based on the hypothesis that the mitochondrial capacity to accumulate Ca^{2+} and to continue normal respiratory function may be a critical determinant in neural cell viability following ischemia/reperfusion or excitotoxic challenges. Using stable *bcl-2* transfectants of GT1-7 hypothalamic tumor cells, we either isolated mitochondria or used digitonin-permeabilized cells to measure the maximal quantity of Ca^{2+} the mitochondria could accumulate and retain.[59] In the course of these studies, it was evident that Bcl-2 has no effect on respiratory characteristics of these cells under control conditions in which no exogenous Ca^{2+} was added to the incubation medium.[59–61] We found that Bcl-2 enhanced the maximal Ca^{2+} uptake capacity of mitochondria in digitonin-permeabilized cells by 1.7-fold with glutamate and malate as the respiratory substrates, and 1.5-fold with succinate as the substrate.[59] Using isolated mitochondria, the Ca^{2+} uptake capacity was 3.9-fold greater in Bcl-2 overexpressors with glutamate and malate as the substrates. It is unlikely that these differences were attributable to differences in mitochondrial volume, because the cytochrome oxidase activity and mitochondrial DNA content were equivalent in the control and *bcl-2* overexpressing cells.[59] Most important, this enhanced ability of the Bcl-2 mitochondria to accumulate large loads of Ca^{2+} correlated with increased resistance to Ca^{2+}-induced respiratory inhibition, which is evident at a load of Ca^{2+} that is near the maximal uptake capacity of the control mitochondria but below the capacity of the Bcl-2 mitochondria.[59] This protection of electron transport by Bcl-2 could be overcome by concentrations of Ca^{2+} that exceeded the Ca^{2+} uptake capacity of the Bcl-2 mitochondria. Subsequent experiments have indicated that the likely mechanism by which Bcl-2 expression potentiates Ca^{2+} uptake capacity is by allowing mitochondria to reestablish a membrane potential following Ca^{2+}-induced depolarization (V. Moota, C. Richards, R. Balaban & A. Murphy, unpublished observations). We, as well as others, have noted a similar ability of Bcl-2 to protect mitochondrial respiratory function in response to other toxic stimuli, including cyanide/aglycemia, pro-oxidants, and Fas ligand.[59–63] We have also found that Bcl-2 can inhibit Ca^{2+}-induced release of cytochrome *c* using digitonin-permeabilized cells.[64]

These data argue that given an equivalent load of sequestered Ca^{2+}, Bcl-2–containing mitochondria will be better able to establish a membrane potential and execute oxidative phosphorylation, and will release less cytochrome *c* than mitochondria with no Bcl-2. These data *do not* necessarily argue that given an equivalent cytoplasmic Ca^{2+} concentration, Bcl-2 mitochondria would accumulate more Ca^{2+}. Our data would predict that in response to an increase in cytoplasmic Ca^{2+}, Bcl-2–containing mitochondria would accumulate more Ca^{2+} than non-Bcl-2–containing mitochondria only if the total Ca^{2+} load was beyond the uptake capacity of the control mitochondria. It is important to point out that our experiments have not directly addressed the effect of Bcl-2 on the *affinity* of mitochondria (or specifically the Ca^{2+} uniporter) for Ca^{2+}. In fact, the fluorescence tracings[59] indicate that following cell permeabilization and an initial pulse of Ca^{2+}, Bcl-2 and control mitochondria buffer the extramitochondrial Ca^{2+} concentration to similar levels, although the sensitivity of the dye is not ideal for quantitation at these low levels of Ca^{2+}. The ability of Bcl-2 to protect from Ca^{2+}-induced mitochondrial damage also correlates with Bcl-2 protection against ionomycin-induced delayed death of GT1-7 cells,[65] but only when ionomycin concentrations are relatively low.

Other studies have confirmed that Bcl-2 potentiates the ability of cells and/or mitochondria to deal with large loads of Ca^{2+}.[66,67] Ichimiya *et al.*[66] have demonstrated that overexpression of Bcl-2 in a normal rat kidney cell line protects them from death in response to hydrogen peroxide that is dependent upon an increase in cytoplasmic

Ca^{2+}. Bcl-2 delayed the rise in cytoplasmic Ca^{2+} in response to hydrogen peroxide, although this effect was only moderate. Bcl-2 was also shown to speed the recovery of cytoplasmic Ca^{2+} down to baseline levels following ionomycin exposure. These results would be in keeping with our theory that Bcl-2 allows mitochondria to continue oxidative phosphorylation in the face of Ca^{2+} stress, providing the ATP required to extrude Ca^{2+} from the cell. Interestingly, they also found that endogenous stores of Ca^{2+} in the mitochondria of unstimulated Bcl-2–overexpressing cells were higher than in controls, as evidenced by a more pronounced increase in cytoplasmic Ca^{2+} upon treatment with uncoupler. Evtodienko *et al.*[67] have also described an enhanced ability of hepatoma mitochondria to accumulate Ca^{2+} when compared to normal rat liver mitochondria, which is associated with increased levels of Bcl-2.

In summary of these data, and in light of the studies described above that provide evidence that mitochondrial Ca^{2+} uptake is critical to inducing excitotoxic death,[3,4] our data would argue that Bcl-2 protects mitochondria from the death-triggering event(s) that result from excessive mitochondrial Ca^{2+} sequestration.

MECHANISTIC QUESTIONS

Bcl-2 may have multiple mechanisms of action, some of which extend beyond regulation of cytochrome *c* release and mitochondrial function to regulation of cell cycle progression potentially via calcineurin.[51,68] Although heterodimerization is not required for antiapoptotic activity, the ability of pro- and antiapoptotic members of the Bcl-2 family to bind to one another to regulate death is still a recurring theme in mechanistic descriptions. For instance, growth factor–induced phosphorylation of BAD via the Akt pathway or Raf-1 blocks binding to Bcl-x_L, thereby inhibiting its antiapoptotic effects.[39–41,69,70]

Recent additional hints have been provided regarding the potential mechanism of Bcl-2/Bcl-x_L-mediated protection against death induced by sustained increases in intracellular Ca^{2+}.[70–72] When colon, prostate, or neural cell lines are treated with Ca^{2+} ionophore or thapsigargin, BAD becomes dephosphorylated and associates with mitochondria. Coexpression of BAD with constitutively active calcineurin (a Ca^{2+}-dependent serine-theonine phosphatase) enhances cell death, and dominant negative calcineurin inhibits the loss in viability.[70] If mutant BAD that is incapable of binding Bcl-x_L is overexpressed, the magnitude of cell death is diminished, again suggesting that the mechanism of induction of death by BAD is via inhibition of Bcl-x_L activity.

Calcineurin may also play a role in the death of neurons.[71,73] Wang and colleagues[70] found that when primary cultures of hippocampal neurons were treated with a high concentration (1 mM) of glutamate, dephosphorylated BAD redistributed to mitochondria, where it bound to Bcl-x_L. These events correlated with cell death, and dominant negative calcineurin or pharmacologic inhibitors of calcineurin (FK506 or cyclosporin A) attenuated death. Although BAD translocation was evident, calcineurin activity did not appear to be involved in staurosporine-induced death; yet others have found a protective effect of expression of calbindin in hippocampal neurons exposed to staurosporine, potentially implicating a separate Ca^{2+}-mediated dephosphorylation of BAD in neurons or the lack of relevance of BAD translocation in this pathway. The authors commented that at even higher doses of glutamate (10 mM), hippocampal cell death was not blocked by dominant negative calcineurin, potentially suggesting that necrotic death in response to glutamate is not blocked by calcineurin inhibition. The studies that implicate mitochondrial Ca^{2+} sequestration as the key determinant in the excitotoxic death of neurons[3,4] employ models in which the death is primarily necrotic. Ankarcrona and coworkers[74] have suggested that those cerebellar granule cells treated with a high concentration of glutamate in which mitochondria depolarize and never recover membrane potential go on to die by necrosis. In contrast, those neurons that lose mitochondrial membrane potential during glutamate exposure, but recover $\Delta\psi$ within 30 minutes after removal of excitotoxin die by apoptosis. Together, these studies suggest that the apoptotic component of excitotoxic death may be mediated by calcineurin-activated translocation of BAD to mitochondria. Necrotic excitotoxic death is purely a function of excessive mitochondrial Ca^{2+} uptake from which mitochondria may not recover. Still unanswered, however, is the mechanism by which BAD binding to Bcl-x_L stimulates a mitochondrial event that leads to death.

The mechanism by which Bcl-2 potentiates mitochondrial Ca^{2+} uptake capacity and confers resistance to Ca^{2+}-induced respiratory damage also remains uncertain. Although Bcl-2 has a similar effect on mitochondrial Ca^{2+} uptake in others cells,[59] we do not know whether Bcl-x_L induces similar changes in mitochondrial Ca^{2+} handling. The most obvious answer to the mechanistic question is that Bcl-2 blocks the membrane permeability transition;[75] and, in turn, Bcl-2 prevents cytochrome c release that would otherwise be associated with pore-induced matrix swelling and rupture of the outer mitochondrial membrane. Although Bcl-2 may inhibit the transition, this may be an oversimplification of the mechanism on multiple levels. If this is the only activity involved in the mechanism of action of Bcl-2 in response to Ca^{2+} sequestration, then a permeability transition inhibitor should mimic the activity of Bcl-2. In contrast, when we treated digitonin-permeabilized cells with cyclosporin A, the total capacity for Ca^{2+} uptake increased in both Bcl-2 overexpressors and their controls such that a difference in maximal uptake capacity persisted between the two types of mitochondria.[59] This either means that Bcl-2 is a more effective inhibitor of pore formation than cyclosporin A, or that Bcl-2 has other protective effects related to Ca^{2+} sequestration that indirectly provide resistance to pore formation.

Further evidence that Bcl-2 can function to do more than simply inhibit the transition involves consideration of the mechanism(s) of cytochrome c release. It has been argued that Bcl-2 functions to inhibit cytochrome c release by inhibiting transition-induced mitochondrial swelling. Mitochondrial swelling does result from large-amplitude pore opening when mitochondria are suspended in the buffers used for *in vitro* studies; it is unclear that pore formation *in vivo* induces mitochondrial swelling because of the presence in the cytoplasm of macromolecules that are too large to permeate the pore.[18,34,76] In other words, the osmotic force that potentiates swelling under conditions *in vitro* may not exist *in vivo*. Recent data has been presented that cytochrome c release may be induced by Ca^{2+} in brain mitochondria in the absence of the permeability transition when the mitochondria are in buffer containing ATP and Mg^{2+}, which are effective inhibitors of the transition. We, as well, have observed cytochrome c release from GT1-7 mitochondria in response to Ca^{2+} that is not associated with large-amplitude swelling, cannot be inhibited by cyclosporin A,[64] but is inhibited by Bcl-2. In addition, we have observed Bcl-2 inhibition of cytochrome c release in response to osmotic stress in the absence of Ca^{2+},[64] a condition that is unlikely to involve the permeability transition. Taken together, the current results suggest a mechanism of protection from cytochrome c release by Bcl-2 that is independent of conventional permeability pore formation.

As mentioned earlier, we as well as others have documented an ability of Bcl-2 to maintain electron transport in response to various forms of stress. This may be either the cause or the effect of inhibition of the permeability pore. If Bcl-2 acts to maintain electron transport in response to oxidant or Ca^{2+} stress, then the propensity to form the transition pore would be decreased due potentially to maintenance of a higher membrane potential. If, on the other hand, the permeability pore is directly inhibited by Bcl-2, then electron transport may be maintained by virtue of retention of cytochrome c and possibly retention of factors necessary to maintain normal rates of electron flow such as pyridine nucleotides or glutathione. It does seem logical that prevention of cytochrome c release and the resultant continuation of normal electron flow to cytochrome oxidase is the mechanism by which Bcl-2 seemingly acts as an antioxidant.[77,78] By preventing the loss of cytochrome c, Bcl-2 may prevent a shift in the redox status of ubiquinone to a more reduced state that would otherwise potentiate mitochondrial free radical production.[18,79] The observations of a chronic drop in $\Delta\psi$ in certain cells undergoing apoptosis[80] may, in fact, result from cytochrome c efflux that is sufficient to compromise electron transport to the extent that the membrane potential is decreased. It is unclear at this point whether the transition is responsible for cytochrome Ca^{2+} release under all the conditions tested, as the use of fluorescent indicators of mitochondrial membrane potential are fraught with error.[18,34]

Potential related mechanisms at the mitochondria involve interaction of pro- and antiapoptotic Bcl-2 family members with mitochondrial membrane protein components. With regard to the proapoptotic proteins, Bax and Bak have been proposed to interact with the components of the permeability pore, including the adenine nucleotide translocator of the inner membrane[81] and the voltage-dependent anion channel (VDAC) of the outer membrane.[82,83] There is discrepant data on whether Bax functions to open a conventional CsA-sensitive permeability pore, possibly due to the highly varied assay conditions and concentrations of Bax employed in each of the studies. In media comprising physiological potassium chloride concentrations, Mg^{2+}-dependent Bax induced cytochrome c release that was not blocked by conventional inhibitors of the permeability pore.[84] In mannitol and/or sucrose-containing media, Bax-induced cytochrome c release is sensitive to conventional inhibitors.[82,85] Pastorino and colleaugues[86] have found that overexpresion of BAX in Jurkat T cells induces a death pathway that is associated with cytochrome c release and that is blocked by the combination of permeability pore inhibitors, CsA plus aristolochic acid (a phospholipase A_2 inhibitor). Calcineurin inhibitors (including FK506) could not substitute for CsA, arguing that the effect was via permeability pore inhibition.

With regard to proapoptotic binding to components of the permeability pore, Shimizu and coworkers[83] have claimed that Bcl-x_L directly interacts with VDAC to decrease its permeability. The authors propose that opening of VDAC is the mechanism by which cytochrome c leaves the intermembrane space in response to Bax or Bak. It is difficult, however, to envisage how decreased permeability of VDAC could be advantageous to mitochondria given that this is the mechanism by which oxidizable substrates and ADP are delivered to mitochondria. This would predict different maximal rates of respiration, which are not observed.[59]

There also have been reports of channel formation by the antiapoptotic proteins Bcl-x_L and Bcl-2.[87–90] It is difficult to theorize how formation of a channel might be to the functional advantage of mitochondria. Formation of a passive channel would dissipate $\Delta\psi$ and create a liability for the mitochondria with respect to both normal metabolic function and susceptibility to permeability pore formation. On the other hand, if ion conductance is linked to an energy source, mitochondria may benefit. For instance, the energy of the pH gradient across in the inner mitochondrial membrane is used to drive K^+ extrusion by the K^+/H^+ antiporter, resulting in maintenance of mitochondrial volume. It seems more rational to propose that Bax or other BH3-only family members either form a pore[91] or facilitate pore formation via binding to other mitochondrial proteins (discussed above).

CONCLUSION

Significant strides have been made in elucidation of the biochemical pathways of apoptotic death involving regulators that lie both upstream and downstream of mitochondria. At the same time, important new studies have uncovered a critical role for mitochondrial Ca^{2+} sequestration in the excitoxic death of neurons. The extent to which mitochondrial Ca^{2+} loading is important in apoptotic pathways remains unresolved. New revelations that Ca^{2+}-stimulated dephosphorylation of BAD results in inhibition of Bcl-x_L protection are potentially important.[70] Combining this with data that Bcl-2 provides mitochondrial protection against Ca^{2+} sequestration–induced injury,[59] and making the assumption that Bcl-2 and Bcl-x_L have similar functional properties, it can be theorized that Ca^{2+}-stimulated calcineurin ultimately inactivates Bcl-2/Bcl-x_L; this inactivation, in turn, may predispose mitochondria to Ca^{2+} sequestration–induced loss of apoptogenic factors such as cytochrome c. The validity of such a proposal will await further investigation of the direct mechanism of Bcl-2/Bcl-x_L inhibition of Ca^{2+}-induced depolarization, permeability transition, and/or cytochrome c release.

ACKNOWLEDGMENTS

Portions of this work were supported by grant NS 34154 to A.N.M.

REFERENCES

1. ORRENIUS, S., M.J. BURKITT, G.E.N. KASS, J.M. DYPBUKT & P. NICOTERA. 1992. Calcium ions and oxidative cell injury. Ann. Neurol. **32:** S33–S42.
2. CHOI, D.W. 1987. Calcium: still center stage in hypoxic-ischemic neuronal death. J. Neurosci. **7:** 369–369.
3. BUDD, S.L. & D.G. NICHOLLS. 1996. Mitochondria, calcium regulation, and acute glutamate excitotoxicity in cultured cerebellar granule cells. J. Neurochem. **67:** 2282–2291.

4. STOUT, A.K., H.M. RAPHAEL, B.I. KANTEREWICA, E. KLANN & I.J. REYNOLDS. 1998. Glutamate-induced neuron death requires mitochondrial calcium uptake. Nature Neurosci. **1:** 366–373.
5. NICHOLLS, D.G. & S. L. BUDD. 1998. Mitochondria and neuronal glutamate excitotoxicity. Biochim. Biophys. Acta **1366:** 97–112.
6. GUNTER, T.E. & D.R. PFEIFFER. 1990. Mechanisms by which mitochondria transport calcium. Am. J. Physiol. **258:** C755–C786.
7. GUNTER, T.E., K.K. GUNTER, S.-S. SHEU & C.E. GAVIN. 1994. Mitochondrial calcium transport: physiological and pathological relevance. Am. J. Physiol. **267:** C313–C339.
8. HANSFORD, R.G. 1994. Physiological role of mitochondrial Ca^{2+} transport. J. Bioenerg. Biomembr. **26:** 495–508.
9. ROBB-GASPERS, L.D., G.A. RUTTER, P. BURNETT, G. HAJINÓCZKY, R.M. DENTON & A.P. THOMAS. 1998. Coupling between cytosolic and mitochondrial calcium oscillations: role in the regulation of hepatic metabolism. Biochim. Biophys. Acta **1366:** 17–32.
10. SIMPSON, P.B. & J.T. RUSSELL. 1998. Role of mitochondrial Ca^{2+} regulation in neuronal and glial cell signalling. Brain Res. Rev. **26:** 72–81.
11. DUCHEN, M.R. 1999. Contributions of mitochondria to animal physiology: from homeostatic sensor to calcium signalling and cell death. J. Physiol. **516:** 1–17.
12. SPARAGNA, G.C., K.K. GUNTER, S.S. SHEU & T.E. GUNTER. 1995. Mitochondrial calcium uptake from physiological-type pulses of calcium. A description of the rapid uptake mode. J. Biol. Chem. **270:** 27510–27515.
13. GUNTER, T.E., L. BUNTINAS, G.C. SPARAGNA & K.K. GUNTER. 1998. The Ca^{2+} transport mechanisms of mitochondria and Ca^{2+} uptake from physiological-type Ca^{2+} transients. Biochim. Biophys. Acta **1366:** 5–15.
14. BUCHET, K. & C. GODINOT. 1998. Functional F1-ATPase essential in maintaining growth and membrane potential of human mitochondrial DNA-depleted ρ° cells. J. Biol. Chem. **273:** 22983–22989.
15. RIZZUTO, R., P. PINTON, W. CARRINGTON, F.S. FAY, K.E. FOGARTY, L.M. LIFSHITZ, R.A. TUFT & T. POZZAN. 1998. Close contacts with the endoplasmic reticulum as determinants of mitochondrial Ca^{2+} responses. Science **280:** 1763–1766.
16. PIVOVAROVA, N.B., J.HONGPAISAN, S.B. ANDREWS & D.D. FRIEL. 1999. Depolarization-induced mitochondrial Ca accumulation in sympathetic neurons: spatial and temporal characteristics. J. Neurosci. **19:** 6372–6384.
17. ZORATTI, M. & I. SZABÒ. 1995. The mitochondrial permeability transition. Biochim. Biophys. Acta **1241:** 139–176.
18. MURPHY, A.N., G. FISKUM & M.F. BEAL. 1999. Mitochondria in neurodegeneration: bioenergetic function in cell life and death. J. Cereb. Blood Flow Metab. **19:** 231–245.
19. FISKUM, G., A.N. MURPHY & M.F. BEAL. 1999. Mitochondria in neurodegeneration: acute ischemia and chronic neurodegenerative diseases. J. Cereb. Blood Flow Metab. **19:** 351–370.
20. CASTILHO, R.F., M.W. WARD & D.G. NICHOLLS. 1999. Oxidative stress, mitochondrial function, and acute glutamate excitotoxicity in cultured cerebellar granule cells. J. Neurochem. **72:** 1394–1401.
21. SCIAMANNA, M.A., J. ZINKEL, A.Y. FABI & C.P. LEE. 1992. Ischemic injury to rat forebrain mitochondria and cellular calcium homeostasis. Biochem. Biophys. Acta **1134:** 223–232.
22. ZAIDAN, E. & N.R. SIMS. 1994. The calcium content of mitochondria from brain subregions following short-term forebrain ischemia and recirculation in the rat. J. Neurochem. **63:** 1812–1819.
23. HOSSMANN, K.-A., B.G. OPHOFF, R. SCHMIDT-KASTNER & U. OSCHLIES. 1985. Mitochondrial calcium sequestration in cortical and hippocampal neurons after prolonged ischemia of the cat brain. Acta Neuropathol. **68:** 230–238.
24. DUX, E., G. MIES, K.A. HOSSMAN & L. SIKLÓS. 1987. Calcium in the mitochondria following brief ischemia of gerbil brain Neurosci. Lett. **78:** 295–300.

25. MACMANUS, J.P. & M.D. LINNIK. 1997. Gene expression induced by cerebral ischemia: an apoptotic perspective. J. Cereb. Blood Flow Metab. **17:** 815–832.
26. SMITH, H.J. 1980. Depressed contractile function in reperfused canine myocardium: metabolism and response to pharmacological agents. Cardiovasc. Res. **14:** 458–468.
27. STONE. D., V. DARLEY-USMAR, D.R. SMITH & V. O'LEARY. 1989. Hypoxia-reoxygenation induced increase in cellular Ca^{2+} in myocytes and perfused hearts: the role of mitochondria. J. Mol. Cell. Cardiol. **21**(10): 963–973.
28. FIGUEREDO, V.M, K.P. DRESDNER, JR., A.C. WOLNEY & A.M. KELLER. 1991. Postischaemic reperfusion injury in the isolated rat heart: effect of ruthenium red. Cardiovasc. Res. **25:** 337–342.
29. MIYAMAE, M., S.A. CAMACHO, M.W. WEINER & V.M. FIGUEREDO. 1996. Attenuation of postischemic reperfusion injury is related to prevention of $[Ca^{2+}]_m$ overload in rat hearts. Am. J. Physiol. **271:** H2145–2153.
30. HALESTRAP, A.P., C.P. CONNERN, E.J. GRIFFITHS & P.M. KERR. 1997. Cyclosporin A binding to mitochondrial cyclophilin inhibits the permeability transition pore and protects hearts from ischaemia/reperfusion injury. Mol. Cell. Biochem. **174:** 167–172.
31. UCHINO, H., E. ELMER, K. UCHINO, P.A. LI, Q.P. HE, M.L. SMITH & B.K. SIESJO. 1998. Amelioration by cyclosporin A of brain damage in transient forebrain ischemia in the rat. Brain Res. **812:** 216–226
32. BORUTAITE, V., R. MORKUNIENE & G.C. BROWN. 1999. Release of cytochrome c from heart mitochondria is induced by high Ca2+ and peroxynitrite and is responsible for Ca^{2+}-induced inhibition of substrate oxidation. Biochim. Biophys. Acta **1453:** 41–48.
33. MATSUMOTO, S., H. FRIBERG, M. FERRAND-DRAKE & T. WIELOCH. 1999. Blockade of the mitochondrial permeability transition pore diminishes infarct size in the rat after transient middle cerebral artery occlusion. J. Cereb. Blood Flow Metab. **19:** 736–741.
34. BERNARDI, P., L. SCORRANO, R. COLONNA, V. PETRONILLI, & F. DI LISA 1999. Mitochondria and cell death: mechanistic aspects and methodological issues. Eur. J. Biochem. **264:** 687–701.
35. BAFFY, G., T. MIYASHITA, J.R. WILLIAMSON & J.C. REED. 1993. Apoptosis induced by withdrawal of interleukin-3 (IL-3) from an IL-3–dependent hematopoietic cell line is associated with repartitioning of intracellular calcium and is blocked by enforced Bcl-2 oncoprotein production. J. Biol. Chem. **268:** 6511–6519.
36. WADIA, J.S., R.M.E. CHLMERS-REDMAN, W.J.H. JU, G.W. CARLILE, J.L. PHILLIPS, A.D. FRASER & W.G. TATTON. 1998. Mitochondrial membrane potential and nuclear changes in apoptosis caused by serum and nerve growth factor withdrawal: time course and modification by (-)-deprenyl. J. Neurosci. **18:** 932–947.
37. PREHN, J.H.M., J. JORDÁN, G.D. GHADGE, E. PREIS, M.F. GALINDO, R.P. ROOS, J. KRIEGLSTEIN & R.J. MILLER. 1997. Ca^{2+} and reactive oxygen species in staurosporine-induced neuronal apoptosis. J. Neurochem. **68:** 1679–1685.
38. KRUMAN, I., Q. GUO & M.P. MATTSON. 1998. Calcium and reactive oxygen species mediate staurosporine-induced mitochondrial dysfunction and apoptosis in PC12 cells. J. Neurosci. Res. **51:** 293–308.
39. ZHA, J., H. HARADA, E. YANG, J. JOCKEL & S.J. KORSMEYER. 1996. Serine phosphorylation of death agonist BAD in response to survival factor results in binding to 14-3-3 not BCL-X. Cell **87:** 619–628
40. DEL PESO, L., M. GONZALEZ-GARCIA, C. PAGE, R. HERRERA & G. NUNEZ. 1997. Interleukin-3–induced phosphorylation of BAD through the protein kinase Akt. Science **278:** 687–689.
41. DATTA, S.R., H. DUDEK, X. TAO, S. MASTERS, H. FU, Y. GOTOH & M.E. GREENBERG. 1997. Akt phosphorylation of BAD couples survival signals to the cell-intrinsic death machinery. Cell **91:** 231–241.

42. PUTCHA, G.V., M. DESHMUKH & E.M. JOHNSON, JR. 1999. BAX translocation is a critical event in neuronal apoptosis: regulation by neuroprotectants, BCL-2, and caspases. J. Neurosci. **19:** 7476–7485.
43. GUO, Q., S. CHRISTAKOS, N. ROBINSON & M.P. MATTSON. 1998. Calbindin D28k blocks the proapoptotic actions of mutant presenilin 1: reduced oxidative stress and preserved mitochondrial function. Proc. Natl. Acad. Sci. USA **95:** 3227–3232.
44. GUO, Q., W. FU, B.L. SOPHER, M.W. MILLER, C.B. WARE, G.M. MARTIN & M.P. MATTSON. 1999. Increased vulnerability of hippocampal neurons to excitotoxic necrosis in presenilin-1 mutant knock-in mice. Nat. Med. **5:** 101–106.
45. KELLER, J.N., Q. GUO, F.W. HOLTSBERG, A.J. BRUCE-KELLER & M.P. MATTSON. 1998. Increased sensitivity to mitochondrial toxin-induced apoptosis in neural cells expressing mutant presenilin-1 is linked to perturbed calcium homeostasis and enhanced oxyradical production. J. Neurosci. **18:** 4439–4450.
46. HENNEBERRY, R.C. 1997. Excitotoxicity as a consequence of impairment of energy metabolism: the energy-linked excitotoxic hypothesis. *In* Mitochondria and Free Radicals in Neurodegenerative Diseases. M. F. Beal, N. Howell & I. Bodis-Wollner, Eds.: 111–143. Wiley. New York
47. BRINI, M., P. PINTON, M.P. KING, M. DAVIDSON, E.A. SCHON & R. RIZZUTO. 1999. A calcium signaling defect in the pathogenesis of a mitochondrial DNA inherited oxidative phosphorylation deficiency. Nat. Med. **5:** 951–954.
48. MARTINOU, J.-C., M. DUBOIS-DAUPHIN, J.K. STAPLE, I. RODRIGUEZ, H. FRANKOWSKI, M. MISSOTTEN, P. ALBERTINI D. TALABOT, S. CATSICAS, C. PIETRA & J. HUARTE. 1994. Overexpression of BCL-2 in transgenic mice protects neurons from naturally occurring cell death and experimental ischemia. Neuron **13:** 1017–1030.
49. LINNIK, M.D., P. ZAHOS, M.D. GESCHWIND & H.J. FEDEROFF. 1995. Expression of bcl-2 from a defective herpes simplex virus–1 vector limits neuronal death in focal cerebral ischemia. Stroke **26:** 1670–1675.
50. YANG, L., R.T. MATTHEWS, J.B. SCHULZ, T. KLOCKGETHER, A.W. LIAO, J.-C. MARTINOU, J.B. PENNEY, JR., B.T. HYMAN & M.F. BEAL 1998. 1-Methyl-4-phenyl-1,2,3,6-tetrahydropyride neurotoxicity is attenuated in mice overexpressing Bcl-2. J. Neurosci. **18:** 8145–8152.
51. REED, J.C. 1997. Double identity for proteins of the Bcl-2 family. Nature **387:** 773–776.
52. KLUCK, R.M., E. BOSSY-WETZEL, D.R. GREEN & D.D. NEWMEYER. 1997. The release of cytochrome c from mitochondria: a primary site for Bcl-2 regulation of apoptosis. Science **275:** 1132–1136.
53. YANG, J., X. LIU, K. BHALLA, C.N. KIM, A.M. IBRADO, J. CAI, T.-I. PENG, D.P. JONES & X. WANG. 1997. Prevention of apoptosis by Bcl-2: release of cytochrome c from mitochondria blocked. Science **275:** 1129–1132.
54. ELLERBY, H.M., S.J. MARTIN, L.M. ELLERBY, S.S. MAIEM, S. RABIZADEH, G.S. SALVESEN, C.A. CASIANO, N.R. CASHMAN, D.R. GREEN & D.E. BREDESEN. 1997. Establishment of a cell-free system of neuronal apoptosis: comparison of premitochondrial, mitochondrial, and postmitochondrial phases. J. Neurosci. **17:** 6165–6178.
55. LI, P., D. NIJHAWAN, I. BUDIHARDJO, S.M. SRINIVASULA, M. AHMAD, E.S. ALNEMRI. & X. WANG. 1997. Cytochrome c and dATP-dependent formation of apaf-1/caspase-9 complex inititates an apoptotic protease cascade. Cell **91:** 479–489.
56. ROSEN, A. & L. CASCIOLA-ROSEN. 1997. Macromolecular substrates for the ICE-like proteases during apoptosis. J. Cell. Biochem. **64:** 50–54.
57. LI, F., A. SRINIVASAN, Y. WANG, R.C. ARMSTRONG, K.J. TOMASELLI & L.C. FRITZ. 1997. Cell-specific induction of apoptosis by microinjection of cytochrome c. Bcl-x_L has activity independent of cytochrome c release. J. Biol. Chem. **272:** 30299–30305.
58. TANG, D.G., L. LI, Z. ZHU & B. JOSHI. 1998. Apoptosis in the absence of cytochrome c accumulation in the cytosol. Biochem. Biophys. Res. Commun. **242:** 380–384.

59. MURPHY, A.N., D.E. BREDESEN, G. CORTOPASSI, E. WANG & G. FISKUM. 1996. Bcl-2 potentiates the maximal calcium uptake capacity of neural cell mitochondria. Proc. Natl. Acad. Sci. USA **93:** 9893–9898.
60. MYERS, K.M., G. FISKUM, Y. LIU, S.J. SIMMENS, D.E. BREDESEN & A.N. MURPHY. 1995. Bcl-2 protects neural cells from cyanide/aglycemia induced lipid oxidation, mitochondrial injury, and loss of viability. J. Neurochem. **65:** 2432–2440.
61. MURPHY, A.N., K.M. MYERS & G. FISKUM. 1996. Bcl-2 and N-acetylcysteine inhibition of mitochondrial respiratory impairment following exposure of neural cells to chemical hypoxia/aglycemia. *In* Pharmacology of Cerebral Ischemia: 163–172. Wissenschaftliche Verlugsgesellschaft. Stuttgart.
62. SHIMIZU, S., Y. EGUCHI, W. KAMIIKE, Y. FUNAHASHI, A. MIGNON, V. LACRONIQUE, H. MATSUDA & Y. TSUJIMOTO. 1998. Bcl-2 prevents apoptotic mitochondrial dysfunction by regulating proton flux. Proc. Natl. Acad. Sci USA **95:** 1455–1459.
63. ADACHI, S., A.R. CROSS, B.M. BABIOR & R.A. GOTTLIEB. 1997. Bcl-2 and the outer mitochondrial membrane in the inactivation of cytochrome *c* during Fas-mediated apoptosis. J. Biol. Chem. **272:** 21878–21882.
64. MURPHY, A.N., G. WANG & C.M. RICHARDS. 1998. Further characterization of mitochondrial cytochrome c release and inhibition by Bcl-2. Nov. 7–12, Los Angeles, CA. Soc. Neurosci. Abstr. **24**(2): 1945.
65. MURPHY, A.N. & G. FISKUM. 1999. Bcl-2 and Ca^{2+}-mediated mitochondrial dysfunction in neural cell death. *In* Mitochondria and Cell Death. G.C. Brown, D. Nicholls & C. Cooper, Eds.: 33–41. Portland Press. London.
66. ICHIMIYA, M., S.H. CHANG, H. LIU, I.K. BEREZESKY, B.F. TRUMP & P. AMSTAD. 1998. Effect of Bcl-2 on oxidant-induced cell death and intracellular Ca^{2+} mobilization. Am. J. Physiol. **275:** C832–C839.
67. EVTODIENKO, Y., V.V. TEPLOVA, T.S. AZARACHVILY, A. KUDIN, O. PRUSAKOVA, I. VIRTANEN & N.-E. SARIS. 1999. The Ca^{2+} threshold for the mitochondrial permeability transition and the content of proteins related to Bcl-2 in rat liver and Zajdela hepatoma mitochondria. Mol. Cell. Biochem. **194:** 251–256.
68. ADAMS, J.M. & S. CORY. 1998. The Bcl-2 protein family: arbiters of cell survival. Science **281:** 1322–1326.
69. WANG, H.-G., U.R. RAPP & J.C. REED. 1996. Bcl-2 targets the protein kinase Raf-1 to mitochondria. Cell **87:** 629–638.
70. WANG, H.-G., N. PATHAN, I.M. ETHELL, S. KRAJEWSKI, Y. YAMAGUCHI, F. SHIBASAKI, F. MCKEON, T. BOBO, T.F. FRANKE & J.C. REED. 1999. Ca^{2+}-induced apoptosis through calcineurin dephosphorylation of Bad. Science **284:** 339–343.
71. SHIBASAKI, F. & F. MCKEON. 1995. Calcineurin functions in Ca^{2+}-activated cell death in mammalian cells. J. Cell Biol. **131:** 735–743.
72. SHIBASAKI, F., E. KONDO, T. AKAGI & F. MCKEON. 1997. Suppression of signalling through transcription factor NF-AT by interactions between calcineurin and Bcl-2. Nature **386:** 728–731
73. ANKARCRONA, M., J.M. DYPBUKT, S. ORRENIUS & P. NICOTERA. 1996. Calcineurin and mitochondrial function in glutamate-induced neuronal cell death. FEBS. Lett. **394:** 321–324.
74. ANKARCRONA, M., J.M. DYPBUKT, E. BONFOCO, B. ZHIVOTOVSKY, S. ORRENIUS, S.A. LIPTON & P. NICOTERA. 1995. Glutamate-induced neuronal death: a succession of necrosis or apoptosis depending on mitochondrial function. Neuron **15:** 961–973.
75. MARZO, I., C. BRENNER, N. ZAMZAMI, S.A. SUSIN, G. BEUTNER, D. BRDICZKA, R.REMY, Z.-H. XIE, J.C. REED & G. KROEMER. 1998. The permeability transition pore complex: a target for apoptosis regulation by caspases and Bcl-2–related proteins. J. Exp. Med. **187:** 1261–1271.

76. BERNARDI, P. 1999. Mitochondrial transport of cations: channels, exchangers, and permeability transition. Physiol. Rev. **79:** 1127–1155.
77. KANE, D.J., T.A. SARAFIAN, R. ANTON, H. HAHN, E.B. GRALLA, J.S. VALENTINE, T. ÖRD & D.E. BREDESEN. 1993. Bcl-2 inhibition of neural death: decreased generation of reactive oxygen species. Science **262:** 1274–1277.
78. HOCKENBERY, D., Z.N. OLTVAI, X. YIN, C.L. MILLIMAN et al. 1993. Bcl-2 functions in an antioxidant pathway to prevent apoptosis. Cell **75:** 241–251.
79. CAI, J. & D.P. JONES. 1998. Superoxide in apoptosis: mitochondrial generation triggered by cytochrome c loss. J. Biol. Chem. **273:** 11401–11404
80. KROEMER, G., N. ZAMZAMI & S.A. SUSIN. 1997. Mitochondrial control of apoptosis. Immunol. Today **18:** 44–51.
81. MARZO, I., C. BRENNER, N. ZAMZAMI, J.M. JURGENSMEIER, S.A. SUSIN, H.L. VIEIRA, M.C. PREVOST, Z. XIE, S. MATSUYAMA, J.C. REED & G. KROEMER. 1998. Bax and adenine nucleotide translocator cooperate in the mitochondrial control of apoptosis. Science **281:** 2027–2031.
82. NARITA, M., S. SHIMIZU, T. ITO, T. CHITTENDEN, R.J. LUTZ, H. MATSUDA & Y. TSUJIMOTO. 1998. Bax interacts with the permeability transition pore to induce permeability transition and cytochrome c release in isolated mitochondria. Proc. Natl. Acad. Sci. USA **95:** 14681–14686.
83. SHIMIZU, S., M. NARITA & Y. TSUJIMOTO. 1999. Bcl-2 family proteins regulate the release of apoptogenic cytochrome c by the mitochondrial channel VDAC. Nature **399:** 483–487.
84. ESKES, R., B. ANTONSSON, A. OSEN-SAND, S. MONTESSUIT, C. RICHTER, R. SADOUL, G. MAZZEI, A. NICHOLS & J.C. MARTINOU. 1998. Bax-induced cytochrome C release from mitochondria is independent of the permeability transition pore but highly dependent on Mg^{2+} ions. J. Cell Biol. **143:** 217–224
85. JÜRGENSMEIER, J.M., Z. XIE, Q. DEVERZUX, L. ELLERBY, D. BREDESEN & J.C. REED. 1998. Bax directly induces release of cytochrome c from isolated mitochondria. Proc. Natl. Acad. Sci. **95:** 4997–5002.
86. PASTORINO, J.G., S.-T. CHEN, M. TAFANI, J.W. SNYDER & J.L. FARBER. 1998. The overexpression of Bax produces cell death upon induction of the mitochondrial permeability transition. J. Biol. Chem. **273:** 7770–7775.
87. MUCHMORE, S.W., M. SATTLER, H. LIANG, R.P. MEADOWS, J.E. HARLAN, H.S. YOON, D. NETTESHEIM, B.S. CHANG, C.B. THOMPSON, S.-L. WONG, S.-C. NG & S.W. FESIK. 1996. X-ray and NMR structure of human Bcl-X_L, and inhibitor of programmed cell death. Nature **381:** 335–341.
88. MINN, A.J., P. VÈLEZ, S.L. SCHENDEL, H. LIANG, S.W. MUCHMORE, S.W. FESIK, M. FILL & C.B. THOMPSON. 1997. Bcl-x_L forms an ion channel in synthetic lipid membranes. Nature **385:** 353–357.
89. SCHENDEL, S.L., Z. XIE, M.O. MONTAL, S. MATSUYAMA, M. MONTAL & J.C. REED. 1997. Channel formation by antiapoptotic protein Bcl-2. Proc. Natl. Acad. Sci. USA **94:** 5113–5118.
90. SCHLESINGER, P.H., A. GROSS, X.-M. YIN, K. YAMAMOTO, M. SAITO, G. WAKSMAN & S.J. KORSMEYER. 1997. Comparison of the ion channel characteristics of proapoptotic BAX and antiapoptotic BCL-2. Proc. Natl. Acad. Sci. USA **94:** 11357–11362.
91. BASEÑEZ, G.A., O. NECHUSHTAN, A. DROZHININ, E. CHANTURIYA, S. CHOE, K.A. TUT, Y.-T. WOOD, J. HSU, J. ZIMMERBERG & R.J. YOULE. 1999. Bax, but not Bcl-x_L, decreases the lifetime of planar phospholipid bilayer membranes at subnamomolar concentrations. Proc. Natl. Acad. Sci. USA **96:** 5492–5497.

Mitochondrial Membrane Potential and the Permeability Transition in Excitotoxicity

IAN J. REYNOLDS[a]

Department of Pharmacology, University of Pittsburgh, E1354 Biomedical Science Tower, Pittsburgh, Pennsylvania 15261, USA

> ABSTRACT: Acute neuronal injury caused by activation of glutamate receptors in neurons, or excitotoxicity, can be triggered by the activation of N-methyl-D-aspartate receptors and the entry of large amounts of Ca^{2+}. Recent studies have suggested that mitochondria have a critical role in the excitotoxicity injury mechanism. Mitochondria accumulate large amounts of Ca^{2+} following glutamate stimulation, and also generate reactive oxygen species. Moreover, the prevention of mitochondrial Ca^{2+} accumulation protects neurons from injury. The target for the actions of Ca^{2+} in the mitochondrial matrix has not yet been established. The permeability transition pore has the characteristics of a mechanism that is well suited to mediate neuronal injury. However, evidence for activation of the permeability transition pore in intact neurons is rather indirect, and these data suffer from some ambiguities that make it difficult to conclude that permeability transition is a critical contributor to mitochondrially mediated neuronal injury.

INTRODUCTION

There is considerable evidence for a role of the neurotransmitter glutamate in acute neurodegenerative disease. Early observations argued that excessive stimulation of neurons by glutamate was injurious.[1] Subsequent *in vivo* studies have shown that extracellular glutamate concentrations are elevated during ischemia[2] and that glutamate receptor antagonists are neuroprotective.[3] It is also possible to recapitulate glutamate-mediated injury in cell culture,[4,5] and these *in vitro* studies demonstrated that glutamate alone is sufficient to kill neurons, even without the additional burden of ischemia. Glutamate can activate any one of an extended family of receptors that either are ion channels (ionotropic receptors) or are coupled through G-proteins to second messenger pathways (metabotropic receptors). Although the latter may have important modulatory functions, the direct toxic effects of glutamate are mediated by ionotropic receptors. The characteristics of toxicity vary somewhat, depending on the receptor activated. The most rapidly induced toxicity occurs when Ca^{2+}-permeable receptors are activated.[6,7] These receptors are typically N-methyl-D-aspartate (NMDA) receptors, although Ca^{2+}-permeable AMPA receptors also effectively mediate neuronal injury.[8] This rapid form of injury is dependent on Ca^{2+} entry for expression, and there may be additional important influences exerted by the concomitant Na^+ load[9] and K^+ loss.[10]

---

[a] Phone: 412-648-2134; fax: 412-648-1945.
e-mail: iannmda@pop.pitt.edu

Perhaps surprisingly, the cellular target(s) for the neurotoxic effects of Ca^{2+} have not been conclusively established, even though the cytotoxic effects of Ca^{2+} are widely recognized and occur in many cell types in addition to neurons.[11] There are a number of possible Ca^{2+}-dependent mechanisms that have been proposed to contribute to injury. These include the Ca^{2+}-mediated activation of nitric oxide synthase,[12] and also the activation of Ca^{2+}-dependent proteases such as calpain.[13] However, none of the proposed mechanisms fully account for the toxicity of Ca^{2+}. It is also difficult to establish the amount of Ca^{2+} that is toxic. There are very large cytosolic Ca^{2+} changes associated with toxic NMDA receptor activation,[14,15] but the location of the Ca^{2+} entry may also be a critical consideration.[16,17]

Recent studies from this laboratory and others have raised the possibility of a role of mitochondria in the acute, Ca^{2+}-mediated toxicity of glutamate in neurons. The appreciation of the mitochondrial role allows the question of the target of Ca^{2+} to be redirected from the cytoplasmic compartment to the mitochondrial matrix. The identification of the critical target for Ca^{2+} in the matrix remains a key question, so the suggestion that this target is, in fact, the permeability transition pore (PTP) is intriguing. In this article, the evidence for a key role for mitochondria in excitotoxicity will be reviewed. In addition, the "PTP hypothesis" for excitotoxicity will be presented and evaluated.

MITOCHONDRIA IN EXCITOTOXICITY

Ischemic brain injury is obviously a tremendously complex process, and it is extremely difficulty to gain mechanistic insights while studying the complete process in the intact brain. However, it is possible to recapitulate some features of this process in a greatly simplified form by using primary cultures of neurons.[4,5] The process of NMDA receptor–mediated excitotoxic injury requires a relatively short exposure to an NMDA receptor agonist. The resulting Ca^{2+} load is typically handled by neurons quite effectively, so that cytosolic Ca^{2+} concentrations return to normal levels after the cessation of the stimulus. Loss of viability then occurs 6–24 h later. We reasoned that following the disposition of the Ca^{2+} might provide clues to help identify the target of Ca^{2+} in toxicity. In a series of studies we established that the primary mechanisms for clearing glutamate-induced cytoplasmic Ca^{2+} loads were the plasma membrane Na^+/Ca^{2+} exchange (NCEp) and mitochondrial Ca^{2+} accumulation.[18,19] Moreover, the mitochondrial route became progressively more important as the intensity of the NMDA receptor activation approached that which would ultimately prove toxic to neurons.[19] The mitochondrial involvement in Ca^{2+} homeostasis is not at all surprising. The early studies of Nicholls and colleagues[20] demonstrated the importance of mitochondria in buffering Ca^{2+} loads imposed on synaptosomes, and the ability of mitochondria to transport large amounts of Ca^{2+} has been established in several different tissues.[21,22] It is also clear that mitochondria have an important role in shaping cytosolic Ca^{2+} transients, because the Ca^{2+} that is accumulated following a stimulus is retained relatively briefly, and the Ca^{2+} that is subsequently released from mitochondria dominates the recovery of cytosolic Ca^{2+} concentrations to baseline. The magnitude of this component varies between neuron types, but the mitochondrial component of the recovery phase is particularly evident in dorsal root ganglion neurons following Ca^{2+} channel activation.[23] It can also be demonstrated in neurons by blocking the major mitochondrial Ca^{2+} efflux pathway, the mitochondrial Na^+/Ca^{2+} exchange (NCEm).[19,21,24]

Further evidence for a role of mitochondria comes from observations of the generation of reactive oxygen species during glutamate exposure. Oxidative stress is a feature of many forms of neurodegenerative disease, both acute and chronic[25–27] and has been associated with glutamate-mediated events in ischemic and epileptic injury processes. There are several potential mechanisms for elevated generation of reactive oxygen species (ROS) in association with hypoxia, ischemia, and reperfusion.[28] However, NMDA receptor activation can increase ROS generation without alterations in the supply of oxygen. Lafon-Cazal and collagues[29] showed that cerebellar granule cells stimulated with NMDA receptor agonists increase ROS generation. Dykens established that brain mitochondria presented with elevated Ca^{2+} and Na^+ increase ROS generation.[30] The development of techniques using oxidant-sensitive fluorescent dyes in cultured neurons facilitated these studies. Studies using these dyes, including dichlorofluorescin, dihydroethidium, and dihydrorhodamine,[31–33] found that the ROS generation was Ca^{2+} dependent and required mitochondrial Ca^{2+} accumulation. In addition, the generation of ROS was blocked by rotenone, which is also consistent with a mitochondrial source for the oxidants. Although there are a number of potential limitations in the use of these dyes, including the pH sensitivity and the impact of mitochondrial membrane potential on the signal generated by the dyes,[31,32,34] there was, nevertheless, a clear consensus on the mitochondrial source of the ROS between the different techniques.

These results demonstrated that mitochondria accumulate Ca^{2+} and subsequently generate ROS specifically in association with NMDA receptor activation that would be toxic. Stimuli that are sufficient to elevate cytosolic Ca^{2+} and result in a presumably lower level of mitochondrial Ca^{2+} accumulation do not result in ROS generation, do not depolarize mitochondria (see below), and are not acutely toxic.[35] This suggests that there may be a critical level of mitochondrial Ca^{2+} accumulation necessary to trigger injury. Several studies have attempted to measure matrix Ca^{2+} concentrations in neurons directly, using the Ca^{2+} indicator rhod-2,[36,37] and have measured Ca^{2+} transients that appear to be associated with mitochondrial Ca^{2+} accumulation. Quantitation of these results remains a challenge, given that the Ca^{2+} accumulation by mitochondria may be substantial, and given that the Ca^{2+} in the matrix may not remain in solution. Nevertheless, these studies do provide direct evidence for some level of mitochondrial Ca^{2+} accumulation during an excitotoxic stimulus, and are consistent with ultrastructural studies that have also seen Ca^{2+} accumulation in association with neuronal injury.[38,39] A key prediction of the suggestion that mitochondrial Ca^{2+} accumulation is a critical event in excitotoxicity is that prevention of Ca^{2+} uptake should be protective. Unfortunately, blocking the calcium uniporter, the major mitochondrial Ca^{2+} uptake pathway, is difficult to accomplish pharmacologically. There are rather few well-established inhibitors of the Ca^{2+} uniporter. Perhaps the most well known is ruthenium red. However, this agent penetrates the cell membrane poorly, and also has absorbance properties that interfere with commonly used Ca^{2+} indicators. A more potent component of ruthenium red has been isolated. Ru 360 has subnanomolar potency in blocking Ca^{2+} uptake by isolated mitochondria. However, it is some 10^4-fold less potent in intact myocytes,[40] suggesting that cell penetration is still a major problem. A less direct approach involves depolarizing mitochondria, which eliminates the driving force for Ca^{2+} accumulation. Using rotenone and oligomycin to accomplish this, Budd and Nicholls[41] showed that prevention of mitochondrial Ca^{2+} accumulation temporarily protected cerebellar granule neurons from NMDA-stimulated injury. In subsequent studies, we took advantage of the reversibility of the uncoupler FCCP to obtain a more robust neuroprotection. Interestingly, in these studies the prevention of mitochondrial Ca^{2+} accumulation resulted in exceptionally large increases in cytosolic Ca^{2+} concentrations that were not associated with injury.[42] In addition, some cytoplasmic, Ca^{2+}-mediated events like nitric oxide synthesis continued unabated, even though the neurons were protected from injury. This further supports the suggestion that mitochondrial Ca^{2+}–sensitive processes, rather than cytoplasmic mechanisms, are the key targets for Ca^{2+} in excitotoxicity.

These *in vitro* studies that place mitochondria at the center of the pathophysiological mechanism of excitotoxic injury are also broadly consistent with many previous findings *in vivo*. For example, many studies have shown that mitochondria accumulate Ca^{2+} in response to injury.[38,43,44] Alterations in mitochondrial morphology are also considered to be a hallmark of necrotic neuronal injury (see, for example, Ref. 45). It has also been shown that respiratory mechanisms are impaired by ischemic injury.[46,47] Indeed, there is likely to be a complex relationship between ischemia and the concomitant respiratory failure, glutamate release and the subsequent excitotoxic injury, because mitochondria that are impaired by oxygen and glucose deprivation presumably will not transport Ca^{2+} as effectively. Whether Ca^{2+} becomes more toxic in ischemic mitochondria or whether other mechanisms come into play has not yet been established. It is also likely that there is more than one mechanism by which neurons die in response to glutamate, even in the relatively simple culture systems described here. Indeed, it has been suggested that mitochondria play a key role in defining the type of neuronal injury that will be expressed.[48]

THE PERMEABILITY TRANSITION IN NEURONAL INJURY

Establishing the centrality of the role of mitochondria in neuronal injury is a valuable step in the search for the target for the toxic effects of excessive intracellular Ca^{2+}. However, the identity of this target is still not clear. An intriguing possibility has been raised by these and other studies, which is the suggestion that the permeability transition pore (PTP) is a key component of the pathophysiological mechanism that links mitochondrial Ca^{2+} accumulation to neuronal death. The PTP is a well-characterized phenomenon in isolated mitochondria.[49,50] It is likely to be comprised of a number of proteins, perhaps the most important of which is a large, nonselective ion channel in the inner mitochondrial membrane. It has been suggested that the pore reflects a pathophysiological state of operation of the adenine nucleotide translocase, which is located in the inner membrane, and shares some of the same pharmacological properties.[51] The PTP is fundamentally a voltage-sensitive ion channel, but one that would not normally open because the depolarization requirements are too great.[50] However, modulators like Ca^{2+} and ROS alter the voltage dependence to greatly increase the probability of pore activation. PTP activation is likely to be catastrophic, because it essentially short circuits the gradients that drive ATP synthesis. Indeed, collapsing the proton gradient is likely to result in the reversal of the F_1F_O ATPase, so that ATP is consumed under conditions of PTP activation. The failure of ATP synthesis and the increased consumption of ATP derived from glycolysis should have a profound impact on cell function.

Specifically within the context of excitotoxicity, the putative sequence of events leading to neuronal death could be stated as follows. The activation of glutamate receptors, and NMDA receptors in particular, results in a substantial influx of Ca^{2+} into the cytoplasm. This Ca^{2+} is then transported into the mitochondria by the Ca^{2+} uniporter or related mechanism. The Ca^{2+} can contribute to at least three events that promote PTP activation: direct activation of the PTP, promotion of PTP activation by increased ROS generation, and depolarization as the result of normal Ca^{2+} cycling. The combination of these three effects results in PTP activation, loss of ATP, loss of neuronal ion homeostasis, and finally the loss of neuronal viability.

The findings discussed in the previous section provide the evidence for mitochondrial Ca^{2+} accumulation and ROS generation, so the key issue to be resolved is whether PTP is activated during the induction of injury, and whether PTP contributes to death. Most studies that have examined PTP in neuronal culture have relied upon membrane potential measurements to infer the activation of PTP. Thus, several groups have found an NMDA receptor–mediated, Ca^{2+}-dependent mitochondrial depolarization using JC-1 or one of the rhodamine family of membrane potential indicators.[52–55] These depolarizations are at least partially sensitive to cyclosporin A, which is perhaps the most potent PTP inhibitor in intact cells. In addition, these studies have reported that cyclosporin A is neuroprotective, which further supports the PTP hypothesis.

A closer look at the issue does not, however, result in more compelling support for the hypothesis. Perhaps the greatest difficulty lies in the direct measurement of PTP in intact cells in general, and in intact neurons in particular. Most studies of PTP have relied on membrane potential measurements, but membrane potential changes are not associated exclusively with transition. As already noted, there is a net positive charge movement into the matrix associated with one Ca^{2+} ion entering the matrix through the uniporter and then leaving via Na^+/Ca^{2+} exchange.[21] Thus, Ca^{2+} cycling should be associated with depolarization. In addition, the synthesis of ATP uses the proton gradient. The increase in demand for ATP in association with the cytoplasmic ion load caused by glutamate stimulation would also be expected to depolarize mitochondria, but this should not be considered pathophysiological. It is also possible that the PTP opens rather transiently.[56] While producing depolarization, it is not clear that this phenomenon would be injurious to neurons. Thus, although mitochondrial depolarization should be associated with PTP activation, it is not clear that depolarization alone is sufficient to define the involvement of PTP. Other methods have been described for measuring PTP that are independent of membrane potential, but still rely on the alteration in inner membrane permeability that should be associated with this phenomenon.[57,58] However, these methods have not yet been applied to neurons.

Conclusions based on pharmacological approaches are also difficult to interpret. Several studies have reported that cyclosporin A protects neurons against glutamate-triggered injury.[53,54] However, it has previously been suggested that cyclosporin A is protective as the result of inhibition of calcineurin,[12] which can be accomplished at lower concentrations of the immunosuppressive. There are a number of drugs that inhibit PTP in isolated mitochondria,[49] but many of these agents lack the specificity to be useful for defining phenomena such as PTP in intact neurons. For example, we investigated the impact of trifluoperazine and dibucaine on glutamate-induced mitochondrial depolarization. Although these drugs delayed depolarization, there may have been additional effects on mitochondrial Na^+/Ca^{2+} exchange.[59] In addition, a series of agents that putatively inhibit PTP do not appear to exert any neuroprotective effects against excitotoxic stimuli. These include dibucaine and trifluoperazine,[59] the antiestrogen tamoxifen, which blocks transition in liver mitochondria (K.R. Hoyt et al., submitted for publication),[60] and a series of ubiquinone analogues that effectively suppress transition in permeabilized skeletal muscle.[61,62]

The studies discussed here argue strongly in favor of a central role for mitochondria in excitotoxic neuronal injury. There is good evidence to suggest that the Ca^{2+} load that is imposed by intense glutamate receptor activation results in substantial mitochondrial Ca^{2+} accumulation. It is also evident that enhanced ROS generation is associated temporally and perhaps spatially with the Ca^{2+} accumulation by mitochondria. The generation of ROS alone might be sufficient to lethally injure neurons by a variety of mechanisms, and this is a suggestion that has not yet been specifically evaluated. However, the observation of cyclosporin A-sensitive injury has promoted the concept that PTP might be a critical component of the injury cascade. This is an exciting possibility, because it would offer a new target for neuroprotective agents that would presumably have a great deal of specificity for the pathophysiological state. However, it remains difficult to study transition in intact neurons, and the agents used to define PTP pharmacologically lack sufficient potency and specificity to provide a compelling argument for or against the hypothesis. Thus, while the involvement of mitochondria in injury is clearly important, the role of PTP in this process remains to be defined.

REFERENCES

1. OLNEY, J.W. 1969. Brain lesions, obesity and other disturbances in mice treated with monosodium glutamate. Science **164:** 719–721.
2. BENVENISTE, H., J. DREJER, A. SCHOUSBOE & N.H. DIEMER. 1984. Elevation of the extracellular concentrations of glutamate and aspartate in rat hippocampus during transient cerebral ischemia monitored by intracerebral microdialysis. J. Neurochem. **43:** 1369–1374.
3. SIMON, R.P., J.H. SWAN, T. GRIFFITHS & B.S. MELDRUM. 1984. Blockade of N-methyl-D-aspartate receptors may protect against ischemic damage in the brain. Science **226:** 850–852.
4. ROTHMAN, S.M. 1984. Synaptic release of excitatory amino acid neurotransmitter mediates anoxic neuronal death. J. Neurosci. **4:** 1884–1891.
5. CHOI, D.W., M. MAULUCCI-GEDDE & A.R. KRIEGSTEIN. 1987. Glutamate neurotoxicity in cortical cell culture. J. Neurosci. **7:** 357–368.
6. CHOI, D.W. 1987. Ionic dependence of glutamate neurotoxicity. J. Neurosci. **7:** 369–379.
7. KOH, J.Y., M.P. GOLDBERG, D.M. HARTLEY & D.W. CHOI. 1990. Non-NMDA receptor mediated neurotoxicity in cortical culture. J. Neurosci. **10:** 693–705.
8. LU, Y.M., H.Z. YIN, J. CHIANG & J.H. WEISS. 1996. Ca^{2+}-permeable AMPA/kainate and NMDA channels: high rate of Ca^{2+} influx underlies potent induction of injury. J. Neurosci. **16:** 5457–5465.

9. KIEDROWSKI, L., J.T. WROBLEWSKI & E. COSTA. 1994. Intracellular sodium concentration in cultured cerebellar granule cells challenged with glutamate. Mol. Pharmacol. **45:** 1050–1054.
10. YU, S.P., C.P. YEH, S. SENSI, B.J. GWAG, L.M. CANZONIERO, Z.S. FARHANGRAZI, H.S. YING, M. TIAN, L.L. DUGAN & D.W. CHOI. 1997. Mediation of neuronal apoptosis by enhancement of outward potassium current. Science **278:** 114–117.
11. ORRENIUS, S., D.J MCCONKEY, G. BELLOMO & P. NICOTERA. 1989. Role of Ca^{2+} in toxic cell killing. Trends Pharmacol. Sci. **10:** 281–285.
12. DAWSON, T.M., J.P STEINER, V.L DAWSON, J.L. DINERMAN, G.R. UHL & S.H. SNYDER. 1993. Immunosuppressant FK506 enhances phosphorylation of nitric oxide synthase and protects against glutamate neurotoxicity. Proc. Natl. Acad. Sci. USA **90:** 9808–9812.
13. SIMAN, R., J.C. NOSZEK & C. KEGERISE. 1989. Calpain I activation is specifically related to excitatory amino acid induction of hippocampal damage. J. Neurosci. **9:** 1579–1590.
14. HYRC, K., S.D. HANDRAN, S.M. ROTHMAN & M.P. GOLDBERG. 1997. Ionized intracellular calcium concentration predicts excitotoxic neuronal death: observations with low affinity fluorescent calcium indicators. J. Neurosci. **17:** 6669–6677.
15. STOUT, A.K. & I.J. REYNOLDS. 1999. High-affinity calcium indicators underestimate increases in intracellular calcium concentrations associated with excitotoxic glutamate stimulations. Neuroscience **89:** 91–100.
16. TYMIANSKI, M., M.P. CHARLTON, P.L. CARLEN & C.H. TATOR. 1993. Source specificity of early calcium neurotoxicity in cultured embryonic spinal neurons. J. Neurosci. **13:** 2085–2104.
17. SATTLER, R., M.P. CHARLTON, M. HAFNER & M. TYMIANSKI. 1998. Distinct influx pathways, not calcium load, determine neuronal vulnerability to calcium neurotoxicity. J. Neurochem. **71:** 2349–2364.
18. WHITE, R.J. & I.J. REYNOLDS. 1995. Mitochondria and Na^+/Ca^{2+} exchange buffer glutamate-induced calcium loads in cultured cortical neurons. J. Neurosci. **15:** 1318–1328.
19. WHITE, R.J. & I.J. REYNOLDS. 1997. Mitochondria accumulate Ca^{2+} following intense glutamate stimulation of cultured rat forebrain neurones. J. Physiol. (Lond.) **498:** 31–47.
20. NICHOLLS, D.G. & K.E.O. AKERMAN. 1982. Mitochondrial calcium transport. Biochim. Biophys. Acta **683:** 57–88.
21. GUNTER, T.E., K.K. GUNTER, S.-S. SHEU & C.E. GAVIN. 1994. Mitochondrial calcium transport: physiological and pathological relevance. Am. J. Physiol. Cell Physiol. **267:** C313–C339.
22. GUNTER, T.E & D.R. PFEIFFER. 1990. Mechanisms by which mitochondria transport calcium. Am. J. Physiol. Cell Physiol. **258:** C755–C786.
23. THAYER, S.A. & R.J. MILLER. 1990. Regulation of the intracellular free calcium concentration in single rat dorsal root ganglion neurones in vitro. J. Physiol. (Lond.) **425:** 85–115.
24. BARON, K.T. & S.A. THAYER. 1997. CGP 37157 modulates mitochondrial Ca^{2+} homeostasis in cultured rat dorsal root ganglion neurons. Eur. J. Pharmacol. **340:** 295–300.
25. GÖTZ, M.E., G. KÜNIG, P. RIEDERER & M.B.H. YOUDIM. 1994. Oxidative stress: free radical production in neural degeneration. Pharmacol. Ther. **63:** 37–122.
26. BRAUGHLER, J.M. & E.D. HALL. 1989. Central nervous system trauma and stroke: I. Biochemical considerations for oxygen radical formation and lipid peroxidation. Free Radic. Biol. Med. **6:** 301–389.
27. HALLIWELL, B. 1992. Reactive oxygen species in the central nervous system. J. Neurochem. **59:** 1609–1623.
28. HALL, E.D. & J.M. BRAUGHLER. 1989. Central nervous system trauma and stroke: II.

Physiological and pharmacological evidence for involvement of oxygen radicals and lipid peroxidation. Free Radic. Biol. Med. **6:** 303–313.
29. LAFON-CAZAL, M., S. PIETRI, M. CULCASI & J. BOCKAERT. 1993. NMDA-dependent superoxide production and neurotoxicity. Nature **364:** 535–537.
30. DYKENS, J.A. 1994. Isolated cerebral and cerebellar mitochondria produce free radicals when exposed to elevated Ca^{2+} and Na^+: implications for neurodegeneration. J. Neurochem. **63:** 584–591.
31. REYNOLDS, I.J. & T.G. HASTINGS. 1995. Glutamate induces the production of reactive oxygen species in cultured forebrain neurons following NMDA receptor activation. J. Neurosci. **15:** 3318–3327.
32. DUGAN, L.L., S.L. SENSI, L.M.T. CANZONIERO, S.D. HANDRAN, S.M. ROTHMAN, T.-S. LIN, M.P. GOLDBERG & D.W CHOI. 1995. Mitochondrial production of reactive oxygen species in cortical neurons following exposure to N-methyl-D-aspartate. J. Neurosci. **15:** 6377–6388.
33. BINDOKAS, V.P., J. JORDAN, C.C. LEE & R.J. MILLER. 1996. Superoxide production in rat hippocampal neurons: selective imaging with hydroethidine. J. Neurosci. **16:** 1324–1336.
34. BUDD, S.L., R.F. CASTILHO & D.G. NICHOLLS. 1997. Mitochondrial membrane potential and hydroethidine-monitored superoxide generation in cultured cerebellar granule cells. FEBS Lett. **415:** 21–24.
35. HOYT, K.R., A.K. STOUT, J.M. CARDMAN & I.J. REYNOLDS. 1998. An evaluation of intracellular sodium and mitochondria in the buffering of kainate-induced intracellular free calcium changes in rat forebrain neurons. J. Physiol. (Lond.) **509:** 103–116.
36. PENG, T.I., M.J. JOU, S.-S. SHEU & J.T. GREENAMYRE. 1998. Visualization of NMDA receptor–induced mitochondrial calcium accumulation in striatal neurons. Exp. Neurol. **149:** 1–12.
37. PENG, T.I. & J.T. GREENAMYRE. 1998. Privileged access to mitochondria of calcium influx through N-methyl-D-aspartate receptors. Mol. Pharmacol. **53:** 974–980.
38. TAYLOR, C.P., M.L. WEBER, C.L. GAUGHAN, E.J. LEHNING & R.M. LOPACHIN. 1999. Oxygen/glucose deprivation in hippocampal slices: altered intraneuronal elemental composition predicts structural and functional damage. J. Neurosci. **19:** 619–629.
39. STYS, P.K., E. LEHNING, A.J. SAUBERMANN & R.M. LOPACHIN, JR. 1997. Intracellular concentrations of major ions in rat myelinated axons and glia: calculations based on electron probe x-ray microanalyses. J. Neurochem. **68:** 1920–1928.
40. MATLIB, M.A., Z. ZHOU, S. KNIGHT, S. AHMED, K.M. CHOI, J. KRAUSE-BAUER, R. PHILLIPS, R. ALTSCHULD, Y. KATSUBE, N. SPERELAKIS & D.M. BERS. 1998. Oxygen-bridged dinuclear ruthenium amine complex specifically inhibits Ca^{2+} uptake into mitochondria in vitro and in situ in single cardiac myocytes. J. Biol. Chem. **273:** 10223–10231.
41. BUDD, S.L. & D.G. NICHOLLS. 1996. A reevaluation of the role of mitochondria in neuronal Ca^{2+} homeostasis. J. Neurochem. **66:** 403–411.
42. STOUT, A.K., H.M. RAPHAEL, B.I. KANTEREWICZ, E. KLANN & I.J. REYNOLDS. 1998. Glutamate-induced neuron death requires mitochondrial calcium uptake. Nature Neurosci. **1:** 366–373.
43. SCIAMANNA, M.A., J. ZINKEL, A.Y. FABI & C.P. LEE. 1992. Ischemic injury to rat forebrain mitochondria and cellular calcium homeostasis. Biochim. Biophys. Acta Mol. Cell Res. **1134:** 223–232.
44. KRISTIÁN, T., Y. OUYANG & B.K. SIESJÖ. 1996. Calcium-induced neuronal cell death in vivo and in vitro: are the pathophysiologic mechanisms different? Adv. Neurol. **71:** 107–118.
45. FRIBERG, H., M. FERRAND-DRAKE, F. BENGTSSON, A.P. HALESTRAP & T. WIELOCH. 1998. Cyclosporin A, but not FK 506, protects mitochondria and neurons against

hypoglycemic damage and implicates the mitochondrial permeability transition in cell death. J. Neurosci. **18:** 5151–5159.
46. SCIAMANNA, M.A., J. ZINKEL, A.Y. FABI & C.P. LEE. 1992. Ischemic injury to rat forebrain mitochondria and cellular calcium homeostasis. Biochim. Biophys. Acta Mol. Cell Res. **1134:** 223–232.
47. ROSENTHAL, R.E., F. HAMUD, G. FISKUM, P.J. VARGHESE & S. SHARPE. 1987. Cerebral ischemia and reperfusion: prevention of brain mitochondrial injury by lidoflazine. J. Cereb. Blood Flow Metab. **7:** 752–758.
48. ANKARCRONA, M., J.M. DYPBUKT, E. BONFOCO, B. ZHIVOTOVSKY, S. ORRENIUS, S.A. LIPTON & P. NICOTERA. 1995. Glutamate-induced neuronal death: a succession of necrosis or apoptosis depending on mitochondrial function. Neuron **15:** 961–973.
49. ZORATTI, M. & I. SZABO. 1995. The mitochondrial permeability transition. Biochim. Biophys. Acta **1241:** 139–176.
50. PETRONILLI, V., A. NICOLLI, P. CONSTANTI, R. COLONNA & P. BERNARDI. 1994. Regulation of the permeability transition pore, a voltage dependent mitochondrial channel inhibited by cyclosporin A. Biochim. Biophys. Acta **1187:** 255–259.
51. HALESTRAP, A.P. & A.M. DAVIDSON. 1990. Inhibition of Ca^{2+} induced high amplitude swelling of liver and heart mitochondria by cyclosporin is probably caused by the inhibitor binding to mitochondrial matrix peptidyl-prolyl cis-trans isomerase and preventing it interacting with the adenine nucleotide translocase. Biochem. J. **268:** 153–160.
52. WHITE, R.J. & I.J. REYNOLDS. 1996. Mitochondrial depolarization in glutamate-stimulated neurons: an early signal specific to excitotoxin exposure. J. Neurosci. **16:** 5688–5697.
53. SCHINDER, A.F., E.C. OLSON, N.C. SPITZER & M. MONTAL. 1996. Mitochondrial dysfunction is a primary event in glutamate neurotoxicity. J. Neurosci. **16:** 6125–6133.
54. NIEMINEN, A.-L., T.G. PETRIE, J.J. LEMASTERS & W.R. SELMAN. 1996. Cyclosporin A delays mitochondrial depolarization induced by N-methyl-D-aspartate in cortical neurons: evidence of the mitochodrial permeability transition. Neuroscience **75:** 993–997.
55. ANKARCRONA, M., J.M. DYPBUKT, S. ORRENIUS & P. NICOTERA. 1996. Calcineurin and mitochondrial function in glutamate-induced neuronal cell death. FEBS Lett. **394:** 321–324.
56. ICHAS, F. & J.P. MAZAT. 1998. From calcium signaling to cell death; two conformations for the mitochondrial permeability transition pore. Switching from low- to high-conductance state. Biochim. Biophys. Acta **1366:** 33–50.
57. NIEMINEN, A.-L., A.K. SAYLOR, S.A. TESFAI, B. HERMAN & J.J. LEMASTERS. 1995. Contribution of the mitochondrial permeability transition to lethal injury after exposure of hepatocytes to t-butylhydroperoxide. Biochem. J. **307:** 99–106.
58. ERIKSSON, O., E. FONTAINE & P. BERNARDI. 1998. Chemical modification of arginines by 2,3-butanedione and phenylglyoxal causes closure of the mitochondrial permeability transition pore. J. Biol. Chem. **273:** 12669–12674.
59. HOYT, K.R., T.A. SHARMA & I.J. REYNOLDS. 1997. Trifluoperazine and dibucaine inhibit glutamate-induced mitochondrial depolarization in cultured rat forebrain neurones. Br. J. Pharmacol. **122:** 803–808.
60. CUSTODIO, J.B., A.J. MORENO & K.B. WALLACE. 1998. Tamoxifen inhibits induction of the mitochondrial permeability transition by Ca^{2+} and inorganic phosphate. Toxicol. Appl. Pharmacol. **152:** 10–17.
61. FONTAINE, E., F. ICHAS & P. BERNARDI. 1998. A ubiquinone binding site regulates the mitochondrial permeability transition pore. J. Biol. Chem. **273:** 25734–25740.
62. SCANLON, J.M. & I.J. REYNOLDS. 1999. Soc. Neurosci. (Abstr.) **25:** 2131.

Oxidative Phosphorylation Disease Diagnosis

JOHN M. SHOFFNER[a]

Molecular Medicine Laboratory, Children's Healthcare of Atlanta, 5455 Meridian Mark Road, NE, Suite 530, Atlanta, Georgia 30342, USA

> ABSTRACT: Although the mtDNA encodes only 13 polypeptide subunits of the OXPHOS enzymes, approximately 1,000 proteins are estimated to be necessary for proper OXPHOS function. Over the past 10 years a wide variety of adult and pediatric OXPHOS diseases were found to be caused by or associated with mitochondrial DNA (mtDNA) mutations and nuclear DNA mutations. These advances enhanced the ability to definitively diagnose patients, develop management plans, and provide genetic counseling. Recently described nuclear DNA and mtDNA mutations are enhancing our understanding of this complex group of diseases. The impact of these advances on our understanding of OXPHOS disease pathogenesis will be reviewed.

Contemporary concepts of oxidative phosphorylation (OXPHOS) diseases originated with the description of a multisystem disorder called Kearns-Sayre syndrome, which was characterized by chronic progressive external ophthalmoplegia, retinitis pigmentosa, and mitochondrial myopathy; and of a rare hypermetabolic disorder called Luft's disease, which was characterized by increased body temperature, structural abnormalities in mitochondria, and abnormal OXPHOS function. Early descriptions of OXPHOS diseases played an important role in formulating the criteria used by clinicians to diagnose OXPHOS diseases for over three decades. Patient diagnosis depended on the recognition of characteristic phenotypes, the identification of histologic and ultrastructural abnormalities in mitochondria, and the identification of abnormalities in OXPHOS enzyme activities. Each of these approaches is complex, making the diagnosis of nonclassical cases difficult. Although the mtDNA encodes only 13 polypeptide subunits of the OXPHOS enzymes, approximately 1,000 proteins are estimated to be necessary for proper OXPHOS function. This degree of complexity makes OXPHOS disease diagnosis a challenging and time-consuming process. The term *mitochondrial medicine* emerged to encompass the complex synthesis of clinical, biochemical, pathological, and genetic information required for patient diagnosis. This article provides a clinical overview of OXPHOS diseases and presents an evaluation algorithm.

OXIDATIVE PHOSPHORYLATION BIOCHEMISTRY AND GENETICS

OXPHOS consists of five protein-lipid enzyme complexes that are located in the mitochondrial inner membrane (FIG. 1). These enzymes contain flavins, coenzyme Q_{10} (ubiquinone), iron-sulfur clusters, hemes, and protein-bound copper. Simplified

---

[a]Phone: 404-250-2650 or 888-448-1495; fax: 404-250-2660.

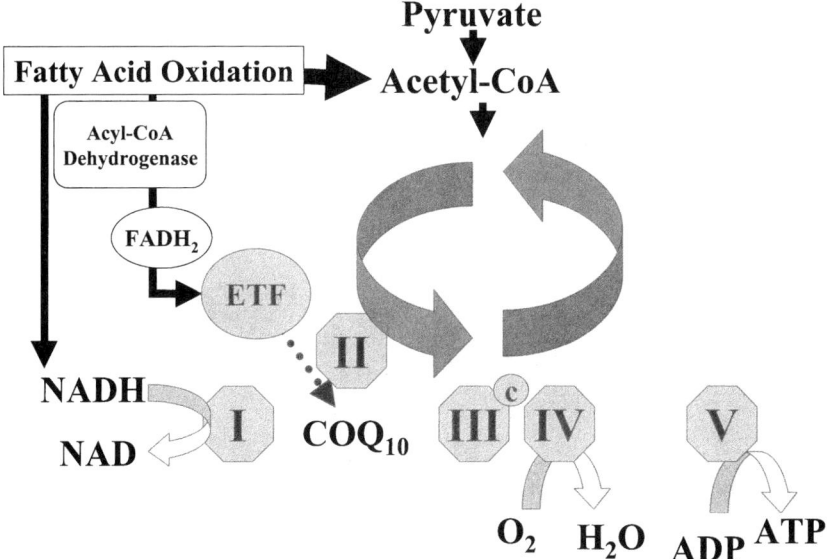

FIGURE 1. Oxidative phosphorylation, tricarboxylic acid cycle, and fatty acid oxidation.

designations are complex I, complex II, complex III, complex IV (cytochrome *c* oxidase), and complex V (ATP synthase). Complexes I and II collect electrons from the catabolism of fats, proteins, and carbohydrates and transfer them sequentially to coenzyme Q_{10}, complex III, and complex IV. Complexes I, III, and IV utilize the energy in electron transfer to pump protons across the inner mitochondrial membrane, producing a proton gradient that is used by complex V to condense ADP and inorganic phosphate into ATP. The adenine nucleotide translocase (ANT) delivers ATP to the cytoplasm in exchange for ADP.

The clinical genetics of OXPHOS diseases is characterized by maternal and Mendelian inheritance patterns. The human mtDNA is contained within mitochondria that are located in the cytoplasm of the cell. The mtDNA is a 16,569-nucleotide pair, double-stranded, circular molecule that codes for two ribosomal RNAs (rRNA), 22 transfer RNAs (tRNA), and 13 polypeptides that together with polypeptides coded by the nuclear DNA form complexes I, III, IV, and V (FIG. 2). Only complex II is encoded entirely by the nuclear DNA. The cytoplasmic location of the mtDNA is associated with a unique inheritance pattern called *maternal inheritance,* which refers to the exclusive transmission of mtDNAs from a mother to her children. When a pathogenic mtDNA mutation is present, the consequences of maternal transmission are influenced by whether the mtDNA is *homoplasmic* or *heteroplasmic*. The mtDNAs within a cell or tissue are referred to as homoplasmic when all the mtDNAs share the same sequence and are referred to as heteroplasmic when mtDNAs with different sequences coexist. Normal and mutant mtDNA sequences differ only at the nucleotide or nucleotides that have been mutated. Pathogenic mtDNA mutations can be either homoplasmic or heteroplasmic, but polymorphisms are almost always homoplasmic. When heteroplasmy exists, the normal and mutant mtDNAs segregate

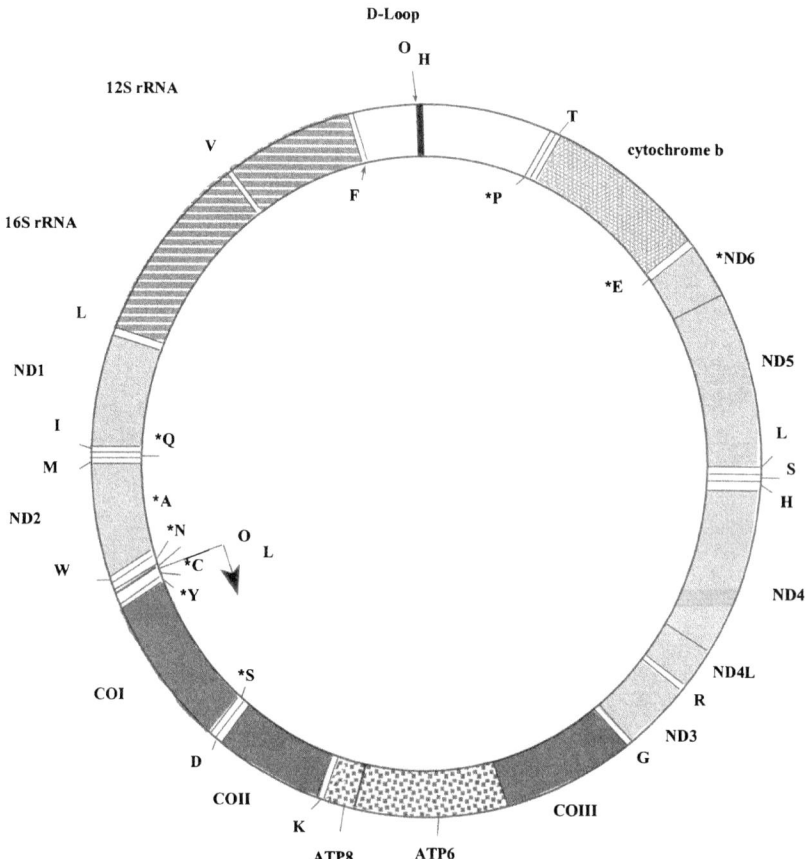

FIGURE 2. Human mtDNA. Locations of polypeptide genes encoding complex I subunits (ND1, ND2, ND3, ND4, ND4L, ND5, ND6); the complex III subunit (cytochrome b); the complex IV subunits (COI, COII, COIII); and the complex V subunits (ATP6, ATP8); the intervening transfer RNAs; and the ribosomal RNA genes (12S rRNA and 16S rRNA). Abbreviations: O_L, origin of light strand replication; O_H, origin of heavy strand replication. Transfer RNAs: A, alanine; R. arginine; N, asparagine; D, aspartate; C, cysteine; E, glutamate; Q, glutamine; G, glycine; H, histidine; I, isoleucine; L, leucine; K, lysine; M, methionine; F, phenylalanine; p, proline; S, serine; T, threonine; W, tryptophan; Y, tyrosine; V, valine.

randomly during cytokinesis to the daughter cells. Once the mutant mtDNAs reach a critical level, cellular phenotype changes rapidly from normal to abnormal. The relationship between genotype and phenotype is more complex for pathogenic mtDNA mutations that are homoplasmic. Disease expression appears to be influenced by poorly understood genetic and environmental interactions.

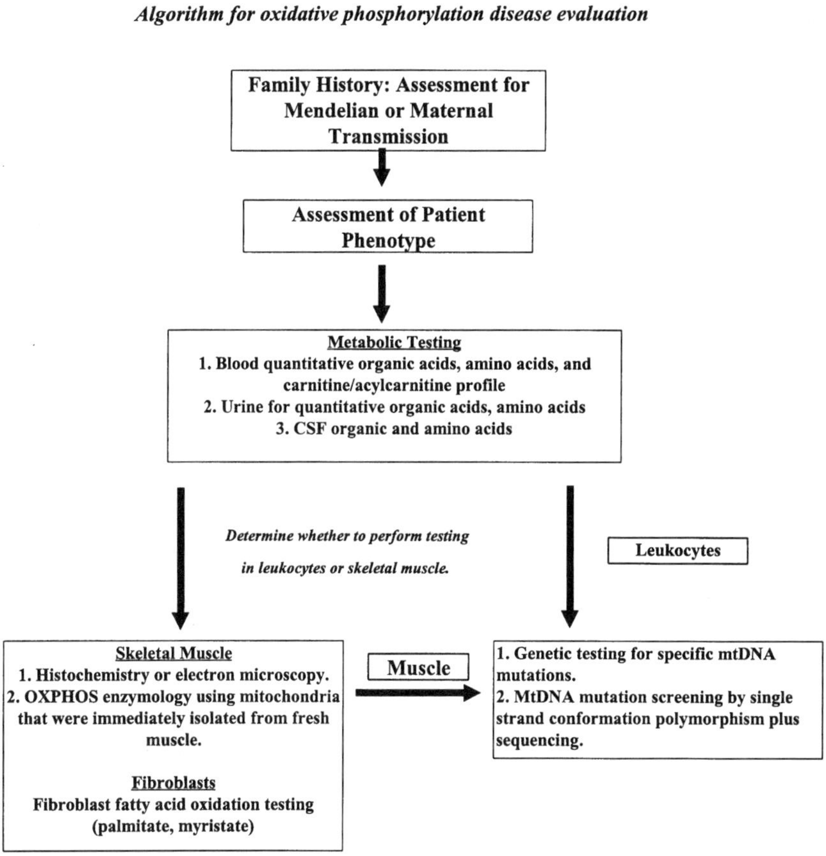

Algorithm for oxidative phosphorylation disease evaluation

FIGURE 3. Diagnostic algorithm for assessment of oxidative phosphorylation diseases.

AN ALGORITHM FOR PATIENT DIAGNOSIS

OXPHOS defects can result from mutations in any of the mitochondrial genes or in any of the nuclear OXPHOS genes. The basic elements of the algorithm are phenotype recognition, metabolic testing, assessment for skeletal muscle pathology, OXPHOS biochemistry, and genetic testing of the mtDNA (FIG. 3). As a consequence of rapid advances in genetic testing, the entire mtDNA can be screened using single-strand conformation polymorphism (SSCP) analysis and direct sequencing of SSCP variants. This approach to patient evaluation permits assignment of the patient's phenotype to the nuclear DNA or to the mtDNA, which is important for genetic counseling of families.

TABLE 1. OXPHOS disease phenotypes[a]

Mitochondrial DNA Mutations

Mitochondrial DNA deletions and duplications
  Kearns-Sayre Syndrome
  Chronic Progressive External Ophthalmoplegia Syndromes
  Pearson's Syndrome
  Diabetes mellitus and deafness
  Mitochondrial myopathy
  Fahr's syndrome variants (complex phenotypes with prominent cerebral calcifications)

Missense mutations
  Leber's hereditary optic neuropathy
  Leber's hereditary optic neuropathy plus dystonia
  Leigh's disease
  Pigmentary retinopathy, ataxia, and neuropathy syndromes

Transfer RNA mutations
  Myoclonic epilepsy and ragged-red fiber disease
  Mitochondrial encephalomyopathy, lactic acidosis, and stroke-like episodes
  Diabetes mellitus (usually with deafness)
  Hypertrophic cardiomyopathy plus mitochondrial myopathy
  Mitochondrial myopathy

Ribosomal RNA mutation
  Maternally inherited deafness with aminoglycoside sensitivity
  Cardiomyopathy

Nuclear DNA Mutations

Abnormal mtDNA copy number regulation
  mtDNA depletion diseases

Nuclear DNA mutations causing multiple mtDNA rearrangements
  Autosomal dominant chronic progressive external ophthalmoplegia
  Autosomal recessive chronic progressive external ophthalmoplegia
  Mitochondrial neurogastrointestinal encephalomyopathy
  Wolfram syndrome
  MERRF variant

Complex I (AQDQ 18-kDa subunit)
  Leigh disease

Complex I (51-kDa subunit, NDUFV1)
  Leigh disease
  Alexander disease

Complex I (NDUFS8 subunit)
  Leigh disease

Complex II (flavoprotein subunit)
  Leigh disease

Complex IV assembly (SURF1 gene)
  Leigh disease

Possible abnormal nuclear regulation of mitochondrial protein assembly (?chaperonin, ?proteolytic functions)
  Hereditary spastic paraplegia (paraplegin)

Abnormal mitochondrial iron homeostasis
  Friedreich ataxia

Abnormal mitochondrial copper homeostasis
  Wilson disease

[a]OXPHOS diseases present with a broad array of phenotypes. Mutations can occur spontaneously as in the mtDNA rearrangements, exhibit maternal inheritance, or exhibit Mendelian inheritance patterns.

Phenotype Recognition

Due to the large number of phenotypes that are described, physician awareness of these disorders is limited. TABLE 1 outlines major classes of OXPHOS diseases.

Metabolic Testing in OXPHOS Diseases

Abnormalities in oxidative phosphorylation can produce identifiable defects in related metabolic pathways such as glycolysis, pyruvate metabolism, the tricarboxylic acid cycle, protein catabolism, and fatty acid oxidation. Although the quantitation of organic acids and amino acids in blood, urine, and cerebrospinal fluid can provide useful diagnostic information, normal values for metabolic tests do not exclude the diagnosis. Metabolic acidosis as well as elevations of lactate, pyruvate, lactate/pyruvate ratio (>20), alanine, tricarboxylic acid cycle intermediates, dicarboxylic acids, and/or a generalized amino aciduria can be important diagnostic clues to the presence of an OXPHOS disease. Other metabolites that may also be increased are tiglylglycine, ethylmalonic acid, 3-methylglutaconic acid, 2-ethylhydracrylic acid, 2-methylsuccinate, butyrylglycine, isovaleryl glycine, and ammonia. Excretion of carnitine esters may be associated with reduced blood and tissue carnitine levels. A 24-hour urine collection is useful since it can provide an integrated evaluation of organic and amino acids as well as insight into the function of the highly OXPHOS-dependent proximal renal tubules. Although this is easily accomplished in adults, a 24-hour urine collection is difficult in pediatric patients, and spot urine collection is used. Analysis of organic and amino acids in venous blood can be complicated by technical factors such as duration of tourniquet application, activity such as recent seizures or vigorous crying and struggling that occurs in some children during venipuncture, and delays in sample processing. In order to enhance the accuracy of quantitative organic and amino acid analysis as well as our ability to reliably compare serial determinations, the blood is collected as a morning sample after an overnight fast. The sample is immediately deproteinized with 1:1 dilution of 7% perchloric acid to prevent artifactual changes in organic acid and amino acid levels associated with delayed processing.

Skeletal Muscle Pathology in OXPHOS Diseases

Most patients who are suspected of having an OXPHOS disease will require a muscle biopsy. Pathological analysis of the muscle biopsy by histochemistry and electron microscopy can be helpful in supporting the diagnosis of an OXPHOS disease. *In situ* hyybridization of muscle or other tissues with a mtDNA-specific probe is useful in screening patients for disorders caused by reduced levels of mtDNA, called *mtDNA depletion syndromes*. OXPHOS diseases that have abnormal mitochondrial protein synthesis caused by mtDNA rearrangements and mitochondrial tRNA mutations can be distinguished pathologically by the proliferation of subsarcolemmal mitochondria and varying degrees of degeneration of the muscle fibers as detected by the modified Gomori trichrome stain, by the more sensitive succinate dehydrogenase stain, and by an abnormal cytochrome *c* oxidase reaction. Ultrastructural analysis of the muscle may reveal structurally abnormal mitochondria with paracrystalline inclusions, which are intermembranous condensations of mitochondrial creatine kinase. Thus, detection of muscle mitochondrial abnormalities can provide useful clues as to the class of mtDNA mutation a patient harbors. Unfortunately, OXPHOS diseases caused by mtDNA missense mutations and by nuclear DNA mutations rarely show diagnostic histologic or ultrastructural features. The muscle pathology may be normal; may show neurogenic changes, myofiber hypertrophy, or hypotrophy; or may show nonspecific changes such as variable degrees of myofiber variability, atrophy, or hypotrophy of Type I fibers, accumulations of lipid, and mild increases in glycogen. Patients with OXPHOS defects (mtDNA or nuclear DNA) do not usually display dystrophic changes in muscle such as increased connective tissue or significant myonecrosis. This observation can be important in distinguishing patients with OXPHOS diseases from other classes of patients with neuromuscular diseases.

OXPHOS Biochemistry

The presence of an OXPHOS disease can be confirmed by biochemical and genetic testing. When specific mutations are not implicated by the clinical examination, OXPHOS enzyme analysis in skeletal muscle mitochondria can be used to classify a disorder as an OXPHOS disease. In order to perform accurate assessments of this delicate enzyme system, we recommend the immediate isolation of mitochondria from fresh muscle biopsies. This approach avoids artifacts in OXPHOS enzyme analysis that can be associated with freezing the biopsy prior to mitochondrial isolation. Although it is now possible to achieve a precise diagnosis of certain OXPHOS diseases by DNA analysis alone, OXPHOS enzymology is necessary in many cases. To determine the specific activities of OXPHOS enzymes, the complex I, complex III, and complex IV assays are used to assess electron flow across single OXPHOS complexes and the complex I+III and complex II+III assays assess the movement of electrons between complexes (FIG. 1). The specificity of these assays is demonstrated by using specific respiratory inhibitors, and the proper functioning of the reagents used in these assays is insured by performing each enzyme assay in mitochondria isolated from mouse or rat skeletal muscle in parallel with the patient assays.

Complex I defects are extremely difficult to identify. They are commonly observed in patient samples. Distinguishing between pathogenic defects and those produced by technical factors can be difficult. In order to assess complex I function, three assays are used that employ different electron acceptors: (1) n-decylCoQ as the electron acceptor, (2) CoQ1 as the electron acceptor, and (3) the traditional complex I+III assay. The first two assays are the most specific for mitochondrial complex I activity, but CoQ reduction probably occurs at different sites due to the more hydrophilic nature of CoQ1 and the more lipophilic nature of n-decylCoQ. The complex I+III assays measure the rate of electron flow between complexes I and III. However, approximately 50% of the observed activity is nonmitochondrial and must be accounted for in the interpretation. Due to the complexities associated with oxidative phosphorylation assessment, corroborative data are sought by testing for abnormalities in skin fibroblast β-oxidation. β-Oxidation of substrates like palmitate (C16:0) and myristate (C14:0) is often reduced, thus providing support for the diagnosis of an OXPHOS defect. During the first step of β-oxidation, a double bond is added to the fatty acid, electrons are transferred to the electron transfer flavoprotein via $FADH_2$, and electrons are transferred to complex I via NADH. OXPHOS defects, particularly those involving complex I, reduce the oxidation of palmitate and myristate to levels that are approximately 40%–60% of the control mean. This contrasts with diseases like carnitine palmitoyl transferase deficiency and medium-chain acyldehydrogenase deficieny, which reduce the oxidation of these fatty acids to <10% and <20% of the control means, respectively. Assessment of long-chain fatty acid oxidation by the trifunctional protein is normal in patients who harbor OXPHOS defects.

Genetic Testing of OXPHOS Diseases

At the time of muscle biopsy, a small portion of the biopsy is frozen in liquid nitrogen for DNA isolation. In our experience the integrated clinical-genetic, metabolic, and biochemical-genetic protocol increases the probability of reaching the correct biochemical and genetic diagnosis that is necessary for accurate genetic counseling and effective patient management. FIGURE 3 is designed to assist physicians in recognizing genotype and phenotype associations. It is important to remember that over 50 mtDNA mutations are known. Most mtDNA mutations are private or semiprivate mutations (i.e., occurring in relatively few families). Therefore, in order to exclude a mtDNA mutation as a cause for symptoms in a proband, a comprehensive analysis of the mtDNA by SSCP and sequencing is important. This approach permits assignment of the patient's disease manifestations to the nuclear DNA or to the mtDNA.

MITOCHONDRIAL DNA MUTATIONS: FREQUENTLY ENCOUNTERED OXPHOS DISEASES

Three classes of pathogenic mtDNA mutations exist: (1) mtDNA rearrangements, in which mtDNA genes are deleted or duplicated; (2) mtDNA point mutations in tRNA or ribosomal RNA genes, resulting in defects in mitochondrial protein synthesis; and (3) missense mutations that change an amino acid, thus altering a critical function of an OXPHOS polypeptide. A comprehensive review of OXPHOS diseases and mtDNA mutations is presented in Reference 1. The OXPHOS diseases known to be caused by nuclear DNA mutations are inherited in an autosomal-dominant or autosomal-recessive pattern (TABLE 1). A brief synopsis of important OXPHOS diseases is given below.

Kearns-Sayre and Chronic Progressive External Ophthalmoplegia (CPEO) Syndromes

Ptosis, ophthalmoplegia, and a ragged-red fiber myopathy represent a clinical triad that is highly predictive for the presence of a mtDNA mutation. Patients with these manifestations can be classified into one of three groups according to their age at onset and the severity of their clinical symptoms. The most severe variant is the Kearns-Sayre syndrome, which is characterized by infantile, childhood, or adolescent onset of disease manifestations and significant multisystem involvement that can include cardiac abnormalities (cardiomyopathies and cardiac conduction defects), diabetes mellitus, cerebellar ataxia, deafness, and evidence of multifocal neurodegeneration. Some patients will present in infancy with an atypical variant called Pearson's syndrome. These individuals manifest anemia, leukopenia, and thrombocytopenia, resulting in frequent transfusions. Exocrine pancreas dysfunction is an important manifestation of this disease. Patients with Pearson's syndrome may have severe systemic manifestations or may be oligosymptomatic. However, if patients survive infancy and early childhood, Kearns-Sayre syndrome develops. *CPEO plus* refers to a disorder of intermediate severity that has an adolescent or adult onset and variable involvement of tissues other than the eyelids and eye muscles. The mildest variant is *isolated CPEO*, in which clinical signs and symptoms develop during adulthood and are limited to the eyelids and eye muscles. In each of these classification groups, patients worsen with age. Individuals who are initially classified as isolated CPEO can progress to CPEO plus, and patients with Kearns-Sayre syndrome often develop more severe multisystem involvement.

The most common cause for Kearns-Sayre and CPEO syndromes is mtDNA rearrangements that consist of mtDNA deletion mutations and the mtDNA duplication mutations.[2–4] The mtDNA deletion mutation has the simplest structure and consists of a mtDNA molecule that is missing contiguous tRNA and OXPHOS polypeptide genes, thus yielding a mtDNA molecule that is smaller than the normal 16.6-kb mtDNA. The structurally more complex mtDNA duplication mutation produces a mtDNA molecule that is larger than the normal mtDNA and contains two tandemly arranged mtDNA molecules consisting of a full-length 16.6-kb mtDNA coupled to a mtDNA deletion mutation.[3,4] Leukocytes and platelets containing mtDNA rearrangements tend to be lost from the circulation. Assessment for mtDNA deletions in blood samples is probably the most common mistake made by physicians when requesting mtDNA genetic testing. In most cases, the analysis is uninformative. Skeletal muscle is optimal for detection of mtDNA rearrangements due to its ability to retain mtDNA mutations and its easy accessibility.

Approximately 80% of patients with Kearns-Sayre syndrome, 70% with CPEO plus, and 40% with CPEO harbor mtDNA rearrangements.[5,6] In most patients with mtDNA rearrangements, the mutation was not inherited, but appears to be a spontaneous event that occurred after fertilization of the oocyte. Due to replicative segregation of mutant and normal mtDNAs, the identification of maternal inheritance of a mtDNA rearrangement by clinical criteria can be difficult and often requires analysis of skeletal muscle mtDNA from maternal lineage relatives of the proband. Of the two classes of mtDNA rearrangements, the mtDNA duplication mutation has the greatest probability of being maternally transmitted. Point mutations in mitochondrial tRNA genes are also an important cause for Kearns-Sayre and CPEO syndromes. Characterization of the mtDNA mutation in a patient with either Kearns-Sayre or CPEO syndrome is important for genetic counseling of the patient and family members.

Myoclonic Epilepsy and Ragged-red Fiber Disease (MERRF)

MERRF can begin at any age, ranging from late childhood to adulthood. The clinical features that are most predictive for a diagnosis of MERRF are epilepsy (myoclonic epilepsy, generalized seizures, or focal seizures), cerebellar ataxia, and a ragged-red fiber myopathy. Other manifestations, including dementia, corticospinal tract degeneration, peripheral neuropathy, optic atrophy, and deafness, are identified in conjunction with multisystem involvement that includes myopathy, proximal renal tubule dysfunction, cardiomyopathy, and lactic acidemia plus hyperalaninemia.

Myoclonic jerks occur at rest and increase in frequency and amplitude with movement. The myoclonus in MERRF patients is best categorized as cortical reflex myoclonus and can be associated with epileptiform discharges and photic sensitivity with large-amplitude occipital wave forms on EEG as well as giant cortical somatosensory evoked repsonses.[7–9] As many as 80% to 90% of the MERRF cases are caused by an A-to-G mutation that alters a conserved nucleotide in the TψC loop of the tRNA^{Lysine} at position 8344 of the mtDNA (A8344G).[10] A small percentage of MERRF patients harbor a T-to-C mutation at position 8356 of the mtDNA.[11, 12]

Mitochondrial Encephalomyopathy, Lactic Acidosis, and Stroke-like Episodes (MELAS)

Disease manifestations in patients with MELAS can appear at essentially any age. Patients are generally below 45 years of age and are characterized as "stroke in the young." They present with a large- or small-vessel stroke that can be associated with a migraine headache and/or seizures. Delineating this presentation from the long list of other causes of stroke in the young can be difficult and is assisted by recognizing myopathy, ataxia, cardiomyopathy, diabetes mellitus, retinitis pigmentosa, proximal renal tubule defects, or lactic acidemia and hyperalaninemia. Since these systemic manifestations are not present in all patients, biochemical and genetic studies are essential in establishing the diagnosis. Cerebellar ataxia is often observed in patients with MELAS and may precede the development of stroke by many years. However, careful patient evaluation usually reveals manifestations in other organs, thus distinguishing these patients from other classes of cerebellar ataxia.

An A-to-G mutation in the tRNA$^{Leucine(UUR)}$ gene (A3243G) accounts for approximately 80% of the MELAS cases. Other mtDNA mutations associated with MELAS are listed in TABLE 2. A mutation at position 8356 of the tRNA^{Lysine} gene was associated with features of both MERRF and MELAS.[12] An important feature of the A3243G mutation[13] as well as of some mtDNA rearrangements[14] is that they significantly increase the risk of developing diabetes mellitus. As many as 1% of randomly selected patients with adult-onset diabetes mellitus may harbor the A3243G mutation.[15] OXPHOS diseases are important considerations in the differential diagnosis of patients with diabetes mellitus and stroke. The A3243G mutation is an important cause for Kearns-Sayre and CPEO syndromes and should be considered in the differential diagnosis of these disorders. The identification of this mutation is important in these patients since it is maternally inherited and is associated with a greater risk of stroke than the mtDNA rearrangements.

Leigh Disease and Cerebellar Ataxia plus Pigmentary Retinopathy Syndromes

Leigh disease or subacute necrotizing encephalopathy is suspected when cranial nerve abnormalities, respiratory dysfunction, and ataxia are observed in conjunction with bilateral hyperintense signals on T2-weighted MRI images in the basal ganglia, cerebellum, or brainstem. The age of onset for disease manifestations is usually infancy or early childhood. Two mtDNA mutations, a T-to-G[16,17] (T8993G) or a T-to-C[18] mutation in the ATPase 6 gene at position 8993 (T8993C), are important causes for Leigh disease. The T8993G mutation is the most frequently encountered of the two mutations and changes an evolutionarily conserved leucine to an arginine of the

ATPase 6 polypeptide, thus replacing a neutral amino acid with a basic amino acid within the proton channel of complex V and impairing ATP synthesis.[19,20] The T8993G mutation is the most frequently encountered of these two mutations and was originally identified in patients with retinitis pigmentosa plus cerebellar ataxia syndromes.[21]

The T8993G mutation acts in a recessive manner. Patients generally have no manifestations when the levels of the T8993G mutation in tissues is less than approximately 60% to 70% of the total mtDNA. Patients that harbor between approximately 70% and 90% mutant mtDNAs in their tissues have highly variable disease manifestations. In mildly affected individuals, a pigmentary retinopathy can be the only clinical manifestation. In more severely affected individuals, cerebellar ataxia and retinitis pigmentosa are commonly observed together. Brain imaging of these patients can show isolated cerebellar atrophy or more extensive cerebellar and brainstem involvement with olivopontocerebellar atrophy. Additional manifestations that can be observed are hypertrophic cardiomyopathy, sensory and motor neuropathies, muscle weakness, and elevated lactate or alanine levels in blood or urine. Approximately 7% to 20% of patients with Leigh's disease harbor the T8993G mutation[16,19,20,22] (and unpublished results). See TABLE 2 for other mutations associated with Leigh disease.

Leber's Hereditary Optic Neuropathy (LHON)

LHON was the first disease that was found to be caused by a mtDNA point mutation.[23] LHON presents with acute or subacute, painless loss of central visual acuity that usually occurs between 12 and 30 years of age.[24–26] The typical ophthalmoscopic features of acute LHON include circumpapillary telangiectatic microangiopathy and swelling of the nerve fiber layer around the optic disc.[27,28] However, since the advent of genetic testing, it has become clear that these characteristic retinal changes are not present in all patients.

Three mtDNA mutations account for approximately 80% to 90% of the cases of LHON. Although the clinical presentations of patients is similar for individuals who harbor these mutations, the probability that the patient will experience some degree of clinical recovery shows important differences. The most common cause of LHON is an A-to-G mutation at position 11,778 (A11778G) that changes a highly conserved arginine to a histidine at the 340th amino acid of the ND4 gene.[23] This mutation accounts for approximately 50% of the cases in Europe and over 90% of the cases in Japan.[29] Once blindness occurs, recovery of vision is uncommon and is observed in only about 5% to 8% of individuals harboring the A11778G mutation.[26,30–32] Approximately 50% to 80% of males within Caucasian pedigrees that harbor this mutation become blind, whereas only 8% to 32% of females are affected.[26] Due to the high penetrance of the A11778G mutation in males, it has also been hypothesized that an X-chromosome locus might be important in disease expression. X-chromosome contribution to LHON is not likely based on molecular studies.[33–35] The second most common cause for LHON is a G-to-A transition mutation in the ND1 gene at position 3,460 (G3460A) that changes an alanine to a threonine at amino acid 52 of the ND1 polypeptide and accounts for approximately 15% of cases.[36,37] Both the G3460A and A11778G mutations preferentially affect males and are found in pedigrees showing maternal transmission of the disease as well as in singleton cases.[23,26,36–38] Recovery of vision has been observed in approximately 22 percent of patients with this mutation.[39] A third mutation associated with LHON is a T-to-C mutation at position 14,484 of the ND6 gene that changes a methionine to a valine (T14484C). This mutation is observed in about 15% of patients with LHON and is associated with the highest probability of vision recovery.[40] Since approximately 40% of patients with the T14484C mutation experience vision recovery to 20/60 or more,[31,40] this mutation may be less pathogenic than the A11778G and G3460A mutations. A heterogeneous array of mtDNA mutations with complex interactions have been proposed to account for most of the remaining cases of LHON. See Reference 1 for review of these mutations and their characteristics.

NUCLEAR DNA MUTATIONS AND OXPHOS DISEASE

mtDNA Depletion Diseases

mtDNA depletion diseases are an important group of disorders affecting infants and neonates in which a quantitative reduction in mtDNA copy number exists within various tissues.[41–45] Patients have variable combinations of mitochondrial myopathy with cytochrome c oxidase negative fibers, hypotonia, hepatopathy, progressive external ophthalmoplegia, and severe lactic acidosis. The diagnosis is made using quantitative Southern blot analysis, which demonstrates that the copy number of the mtDNA is greatly reduced in affected tissues. Interestingly, the unaffected tissues of some patients may show normal levels of mtDNA. The disorder is transmitted in an autosomal recessive fashion.

Kearns-Sayre and Chronic Progressive External Ophthalmoplegia (CPEO) Syndromes

Kearns-Sayre and chronic progressive external ophthalmoplegia syndromes can be transmitted in an autosomal-dominant[46–50] or autosomal-recessive diseases. mtDNA analysis of affected individuals in these families revealed that each harbors an array of deleted mtDNAs.[46] Clinical manifestations include ophthalmoplegia, proximal muscle weakness, sensorineural hearing loss and abnormal vestibular responses, tremor, ataxia, and sensorimotor neuropathy.[51] Although multiple mtDNA deletions accumulated in various tissues of some patients, clinical manifestations within the same pedigree is often highly variable, ranging from individuals with severe manifestations to individuals who are asymptomatic. In one family with this disorder, the male proband exhibited the manifestations of Kearns-Sayre syndrome and Leigh disease.[52] Elevations in blood lactate, a ragged-red fiber myopathy, and OXPHOS defects primarily affecting complexes I and IV occur. The biochemical abnormalities are typical of mutations that cause defects in mitochondrial protein synthesis. The mtDNA deletions are best detected in skeletal muscle biopsies.[46] These mutations are generally absent in populations of rapidly dividing cells such as cultured fibroblasts, peripheral blood cells, cultured myoblasts,[53] myotubes, or *in vitro* innervated muscle cells.[51] Autosomal-dominant forms of the disease map to chromosome 10q23.3-24.3[54] and chromosome 3p14.1-21.2.[55]

Myoneurogastrointestinal Disorder and Encephalopathy (MNGIE)

MNGIE is an autosomal recessive disorder characterized by a progressive external ophthalmoplegia, dementia with a progressive leukodystrophy, mitochondrial myopathy, peripheral neuropathy, and prominent involvement of the gastrointestinal tract.[56–62] The gastrointestinal manifestations are heralded by significant diarrhea, malabsorption, and weight loss, with normal pancreatic function. Radiologic investigations may show marked thickening of the small intestines, which reflects the pathological findings of extensive mural thickening and fibrosis of the submucosa and subserosa. Lactate may be elevated along with other tricarboxylic acid cyle intermediates. This disorder is linked to chromosome 22q13.32-qter.[63]

Wolfram Syndrome

Wolfram syndrome is characterized by diabetes insipidus, insulin-dependent diabetes mellitus, optic atrophy, and deafness. In a small percentage of cases, multiple mtDNA deletions are observed. Multiple mtDNA deletions are observed in tissues of these individuals. This autosomal recessive disorder is linked to chromsome 4p16.[64, 65]

Leigh Disease

Although Leigh disease can be caused by defects in a variety of metabolic pathways, OXPHOS defects are the most commonly identified abnormality in this group of patients. To date, all nuclear OXPHOS gene mutations discovered were transmitted in an autosomal recessive fashion. Complex I, complex IV, and complex V defects are important causes of Leigh's disease.[66–79]

Four mutations in nuclear encoded OXPHOS subunits were identified in Leigh disease patients. One is in a nuclear OXPHOS gene mutation in the flavoprotein subunit of complex II.[80] The other three mutation groups involve complex I subunits. Another is a mutation in the 18-kDa (AQDQ) complex I subunit, which maps to chromosome 5.[81] A third represents mutations in the NDUSF9 (TYKY) subunit of complex I.[82] A fourth represents mutations in the 51-kDa subunit of complex I (NDUFV1).[83] Each of these mutations is transmitted in an autosomal recessive fashion. The complex I mutation in the 18-kDa subunit showed normal organic and amino acids in skeletal muscle light microscopy and electron microscopy. The complex I defect was present in both skeletal muscle and in fibroblasts. This patient provides genetic confirmation for the common observation that complex I defects generally do not produce detectable metabolic abnormalities. Additional phenotypic heterogeneity was observed with the 51-kDa subunit mutations. One individual was diagnosed as having Alexander disease, which is characterized by megalencephaly with progressive spasticity and dementia.[83]

Although complex IV defects are frequently observed, mutations affecting mtDNA-encoded or nuclear-encoded subunits of complex IV are rare. Mutations in an evolutionarily highly conserved gene, the SURF1 gene (chromosome 9q34), were recognized as a cause for systemic cytochrome *c* oxidase (complex IV) deficiency.[84,85] These individuals had Leigh disease with early onset hypotonia, ataxia, brainstem abnormalities, regression, and the characteristic bilateral basal ganglia lesions found in Leigh disease. The SURF1 gene appears to be essential for complex

IV assembly. Mutations in the SURF1 gene are heterogeneous, consisting of large deletions, nonsense mutations, and donor–splice site mutants. Compound heterozygotes are common. The use of functional complementation to discover this gene defect promises to be a powerful tool in uncovering novel mechanisms for OXPHOS disease pathogenesis.

A group of Leigh disease patients referred to as the Saguenay Lac-Saint-Jean type shows complex IV deficiency.[86–89] Although phenotypically similar to the patients harboring mutations in the SURF1 gene, this recessively transmitted disorder maps to chromosome 2. Whereas the complex IV defect in the group with SURF1 mutations is systemic, the Saguenay Lac-Saint-Jean group has 50% activity in muscle, fibroblasts, and amniocytes; less than 10% activity in brain and liver; and normal activity in kidney and heart.

Hereditary Spastic Paraplegia with Ragged-red Fiber Myopathy

An autosomal recessive form of spastic paraparesis was identified at chromosome 16q24.3.[90] Patients experience progressive weakness, spasticity, and mild decreases in vibratory sensation as their major manifestations. Dysphagia, scoliosis, and optic nerve atrophy have also occurred. This unique form of hereditary spastic paraplegia is caused by mutations in the gene called *paraplegin*, which is localized to the mitochondria.[91] Paraplegin has a high degree of homology with a subclass of ATPases called the *AAA family*. This group of ATPases are metalloproteases with proteolytic and chaperonin functions. Patients have ragged-red fibers and cytochrome c oxidase–deficient fibers in their skeletal muscle. As noted above, these observations suggest that paraplegin may in some fashion be important to mitochondrial protein synthesis.

Friedreich Ataxia

Friedreich ataxia was recently discovered to be a mitochondrial disease. Clinical manifestations are systemic and include hypoactive or absent deep tendon reflexes, ataxia, corticospinal tract dysfunction, impaired vibratory and prioprioceptive function, hypertrophic cardiomyopathy, and diabetes mellitus. This autosomal recessive disorder was mapped to chromosome 9q13. This disease is caused by a GAA trinucleotide repeat expansion in the first intron of the frataxin gene.[92] Frataxin is a mitochondrial protein[93] that is involved in iron homeostasis. Frataxin gene mutations result in impaired activity of the iron-sulfur–containing enzymes within the mitochondria: complex I, complex II, complex III, and aconitase.[94]

Wilson Disease

Basal ganglia degeneration and cirrhosis (i.e., hepatolenticular degeneration) due to abnormal copper accumulation are the major clinical manifestations of Wilson disease. This disease is autosomal recessive and is mapped to chromosome 13q14-q21. Abnormal export of copper from the cell is caused by mutations in the ATP7B gene.[95,96] This genes encodes a protein that is localized to the mitochondria, thus implicating an important role in mitochondrial copper metabolism for this copper-dependent ATPase.[97] This disturbance of mitochondrial copper metabolism is likely to account for the defects in the copper-containing OXPHOS enzyme, cytochrome c oxidase (complex IV), observed in patients with Wilson disease.

SUMMARY

Physicians in all specialties are becoming increasingly aware of OXPHOS diseases. Although the prevalence of OXPHOS diseases in the general population is unknown, the number of requests for pediatric and adult evaluations are increasing rapidly. A basic awareness of OXPHOS disease phenotypes as well as of the essential elements of patient evaluation are important for appropriate patient management and referrals. Centers that specialize in OXPHOS disease evaluations can be instrumental in working with referring physicians to develop a cost-effective diagnostic plan that is individualized to suit the patient's needs. Comprehensive mtDNA analysis by SSCP and sequencing is important in defining whether a family harbors a mtDNA mutation or is likely to harbor a nuclear DNA mutation. After a complete evaluation, genetic counseling based on Mendelian principles or mtDNA principles of inheritance can be applied. Although approaches that assess patients for mtDNA mutations are evolving rapidly, significant ambiguity in patient diagnosis often remains even after detailed testing is complete. Advances in our understanding of mutations in nuclear OXPHOS genes will provide a powerful addition to our ability to diagnose, manage, and counsel patients with these disorders.

ACKNOWLEDGMENTS

This work was conducted under the auspices of Scottish Rite Children's Medical Center. It was supported by NIH Grant NS33999 and a grant from the United Mitochondrial Disease Association awarded to J.M.S.

REFERENCES

1. SHOFFNER, J.M. & D.C. WALLACE. 1995. Oxidative phosphorylation diseases. *In* The Metabolic and Molecular Bases of Inherited Disease: 1535–1610.
2. HOLT, I.J. *et al.* 1988. Deletions of muscle mitochondrial DNA in patients with mitochondrial myopathies. Nature **331:** 717–719.
3. POULTON, J. *et al.* 1989. Tandem direct duplications of mitochondrial DNA in mitochondrial myopathy: analysis of nucleotide sequence and tissue distribution. Nucleic Acids Res. **17:** 10223–10229.
4. POULTON, J. *et al.* 1993. Families of mtDNA re-arrangements can be detected in patients with mtDNA deletions: duplications may be a transient intermediate form. Hum. Mol. Genet. **2:** 23–30.
5. HOLT, I.J. *et al.* 1989. Mitochondrial myopathies: clinical and biochemical features of 30 patients with major deletions of muscle mitochondrial DNA. Ann. Neurol. **26:** 699–708.
6. MORAES, C.T. *et al.* 1989. Mitochondrial DNA deletions in progressive external ophthalmoplegia and Kearns-Sayre syndrome. N. Engl. J. Med. **320:** 1293–1299.
7. ROSING, H.S. *et al.* 1985. Maternally inherited mitochondrial myopathy and myoclonic epilepsy. Ann. Neurol. **17:** 228–237.

8. WALLACE, D.C. et al. 1988. Familial mitochondrial encephalomyopathy (MERRF): genetic, pathophysiological, and biochemical characterization of a mitochondrial DNA disease. Cell **55:** 601–610.
9. THOMPSON, P.D. et al. 1994. Cortical reflex myoclonus in patients with the mitochondrial DNA transfer RNA-Lys(8344) (MERRF) mutation. J. Neurol. **241:** 335–340.
10. SHOFFNER, J.M. et al. 1990. Myoclonic epilepsy and ragged-red fiber disease (MERRF) is associated with a mitochondrial DNA tRNA(Lys) mutation. Cell **61:** 931–937.
11. SILVESTRI, G. et al. 1992. A new mutation in the tRNA-Lys gene associated with myoclonic epilepsy and ragged-red fibers (MERRF). Am. J. Hum. Genet. **51:** 1213–1217.
12. ZEVIANI, M. et al. 1993. A MERRF/MELAS overlap syndrome associated with a new point mutation in the mitochondrial DNA tRNA(Lys) gene [published erratum appears in Eur. J. Hum. Genet. 1993;1(2):124]. Eur. J. Hum. Genet. **1:** 80–87.
13. VAN DEN OUWELAND, J.M. et al. 1994. Maternally inherited diabetes and deafness is a distinct subtype of diabetes and associates with a single point mutation in the mitochondrial tRNA(Leu(UUR)) gene. Diabetes **43:** 746–751.
14. BALLINGER, S.W. et al. 1994. Mitochondrial diabetes revisited. Nature Genet. **7:** 458–459.
15. OTABE, S. et al. 1994. The high prevalence of the diabetic patients with a mutation in the mitochondrial gene in Japan. J. Clin. Endocrinol. Metab. **79:** 768–771.
16. SHOFFNER, J.M. et al. 1992. Subacute necrotizing encephalopathy: oxidative phosphorylation defects. Neurology **42:** 2168–2174.
17. TATUCH, Y. et al. 1992. Heteroplasmic mitochondrial DNA mutation (T to G) at 8993 can cause Leigh disease when the percentage of abnormal mtDNA is high. Am. J. Hum. Genet. **50:** 852–858.
18. SANTORELLI, F.M. et al. 1994. A T to C mutation at nt 8993 of mitochondrial DNA in a child with Leigh syndrome. Neurology **44:** 972–974.
19. TATUCH, Y. et al. 1994. The 8993 mtDNA mutation: heteroplasmy and clinical presentation in three families. Eur. J. Hum. Genet. **2:** 35–43.
20. TROUNCE, I. et al. 1994. Cytoplasmic transfer of the mtDNA nt 8993 T-->G (ATP6) point mutation associated with Leigh syndrome into mtDNA-less cells demonstrates cosegregation with a decrease in state III respiration and ADP/O ratio. Proc. Natl. Acad. Sci. USA **91:** 8334–8338.
21. HOLT, I.J. et al. 1990. A new mitochondrial disease associated with mitochondrial DNA heteroplasmy. Am. J. Hum. Genet. **46:** 428–433.
22. SANTORELLI, F.M. et al. 1993. The mutation at nt 8993 of mitochondrial DNA is a common cause of Leigh's syndrome. Ann. Neurol. **34:** 827–834.
23. WALLACE, D.C. et al. 1988. Mitochondrial DNA mutation associated with Leber's hereditary optic neuropathy. Science **242:** 1427–1430.
24. NEWMAN, N.J. & D.C. WALLACE. 1990. Mitochondria and Leber's hereditary optic neuropathy. Am. J. Ophthalmol. **109:** 726–730.
25. NEWMAN, N.J. 1991. Leber's hereditary optic neuropathy. Ophthalmol. Clinics North Am. **4:** 431–447.
26. NEWMAN, N.J. et al. 1991. The clinical characteristics of pedigrees of Leber's hereditary optic neuropathy with the 11,778 mutation. Am. J. Ophthalmol. **111:** 750–762.
27. SMITH, J.L. et al. 1973. Ocular fundus in acute Leber optic neuropathy. Arch. Ophthalmol. **90:** 349–354.
28. SMITH, D. et al. 1990. Clinical spectrum of Leber's congenital amaurosis in the second to fourth decades of life. Ophthalmology **97:** 1156–1161.
29. NAKAMURA, M. et al. 1992. High frequency of mitochondrial ND4 gene mutation in Japanese pedigrees with Leber hereditary optic neuropathy. Jpn. J. Ophthalmol. **36:** 56–61.

30. HOLT, I.J. et al. 1989. Genetic heterogeneity and mitochondrial DNA heteroplasmy in Leber's hereditary optic neuropathy. J. Med. Genet. **26:** 739–743.
31. OOSTRA, R.J. et al. 1994. Leber's hereditary optic neuropathy: correlations between mitochondrial genotype and visual outcome. J. Med. Genet. **31:** 280–286.
32. STONE, E.M. et al. 1992. Visual recovery in patients with Leber's hereditary optic neuropathy and the 11778 mutation. J. Clin. Neuroophthalmol. **12:** 10–14.
33. SWEENEY, M.G. et al. 1992. Evidence against an X-linked locus close to DXS7 determining visual loss susceptibility in British and Italian families with Leber hereditary opatic neuropathy. Am. J. Hum. Genet. **51:** 741–748.
34. CHEN, J.D. et al. 1989. Preliminary exclusion of an X-linked gene in Leber optic atrophy by linkage analysis. Hum. Genet. **82:** 203–207.
35. CHEN, J.-D. & M. DENTON. 1991. X-chromosome gene in Leber hereditary optic neuropathy. Am. J. Hum. Genet. **48:** 692–693.
36. HOWELL, N. et al. 1991. Leber hereditary optic neuropathy: identification of the same mitochondrial ND1 mutation in six pedigrees. Am. J. Hum. Genet. **49:** 939–950.
37. HUOPONEN, K. et al. 1991. A new mtDNA mutation associated with Leber hereditary optic neuropathy. Am. J. Hum. Genet. **48:** 1147–1153.
38. LOTT, M.T. et al. 1990. Variable genotype of Leber's hereditary optic neuropathy patients. Am. J. Ophthalmol. **109:** 625–631.
39. JOHNS, D.R. et al. 1992. Leber's hereditary optic neuropathy. Clinical manifestations of the 3460 mutation. Arch. Ophthalmol. **110:** 1577–1581.
40. JOHNS, D.R. et al. 1993. Leber's hereditary optic neuropathy. Clinical manifestations of the 14484 mutation. Arch. Ophthalmol. **111:** 495–498.
41. TELERMAN-TOPPET, N. et al. 1992. Fatal cytochrome c oxidase-deficient myopathy of infancy associated with mtDNA depletion. Differential involvement of skeletal muscle and cultured fibroblasts. J. Inherit. Metab. Dis. **15:** 323–326.
42. TRITSCHLER, H.J. et al. 1992. Mitochondrial myopathy of childhood associated with depletion of mitochondrial DNA. Neurology **42:** 209–217.
43. MORAES, C.T. et al. 1991. mtDNA depletion with variable tissue expression: a novel genetic abnormality in mitochondrial diseases. Am. J. Hum. Genet. **48:** 492–501.
44. BOUSTANY, R.N. et al. 1983. Mitochondrial cytochrome deficiency presenting as a myopathy with hypotonia, external ophthalmoplegia, and lactic acidosis in an infant and as fatal hepatopathy in a second cousin. Ann. Neurol. **14:** 462–470.
45. FIGARELLA-BRANGER, D. et al. 1992. Defects of the mitochondrial respiratory chain complexes in three pediatric cases with hypotonia and cardiac involvement. J. Neurol. Sci. **108:** 105–113.
46. ZEVIANI, M. et al. 1989. An autosomal dominant disorder with multiple deletions of mitochondrial DNA starting at the D-loop region. Nature **339:** 309–311.
47. BASTIAENSEN, L.A. et al. 1979. Ophthalmoplegia-plus. Doc. Ophthalmol. **46:** 365–380.
48. BERENBERG, R.A. et al. 1977. Lumping or splitting? "Ophthalmoplegia-plus" or "Kearns-Sayre syndrome." Ann. Neurol. **1:** 37–54.
49. BARRON, S.A. et al. 1979. A familial mitochondrial myopathy with central defect in neural transmission. Arch. Neurol. **36:** 553–556.
50. MCAULEY, F.D. 1956. Progressive external ophthalmoplegia. Br. J. Ophthalmol. **40:** 686–690.
51. ZEVIANI, M. 1992. Nucleus-driven mutations of human mitochondrial DNA. J. Inherit. Metab. Di.s **15:** 456–471.
52. CORMIER, V. et al. 1991. Autosomal dominant deletions of the mitochondrial genome in a case of progressive encephalomyopathy. Am. J. Hum. Genet. **48:** 643–648.
53. SERVIDEI, S. et al. 1991. Dominantly inherited mitochondrial myopathy with multiple deletions of mitochondrial DNA: clinical, morphologic, and biochemical studies. Neurology **41:** 1053–1059.

54. SUOMALAINEN, A. et al. 1995. An autosomal locus predisposing to deletions of mitochondrial DNA. Nature Genet. **9:** 146–151.
55. KAUKONEN, J.A. et al. 1996. An autosomal locus predisposing to multiple deletions of mtDNA on chromosome 3p. Am. J. Hum. Genet. **58:** 763–769.
56. BARDOSI, A. et al. 1987. Myo-, neuro-, gastrointestinal encephalomyopathy (MNGIE syndrome) due to partial deficiency of cytochrome c oxidase. A new mitochondrial multisystem disorder. Acta Neuropathol. **74:** 248–258.
57. IONASESCU, V. et al. 1984. Late-onset oculogastrointestinal muscular dystrophy. Am. J. Med. Genet. **18:** 781–788.
58. IONASESCU, V. et al. 1983. Inherited ophthalmoplegia with intestinal pseudo-obstruction. J. Neurol. Sci. **59:** 215–228.
59. IONASESCU, V. 1983. Oculogastrointestinal musclular dystrophy. Am. .J Med. Genet. **15:** 103–112.
60. SIMON, L.T. et al. 1990. Polyneuropathy, ophthalmoplegia, leukoencephalopathy, and intestinal pseudo-obstruction: POLIP syndrome. Ann. Neurol. **28:** 349–360.
61. BLAKE, D. et al. 1990. MNGIE syndrome: report of 2 new patients. Neurology **40**(Suppl. 1): 294.
62. ROWLAND, L.P. 1992. Progressive external ophthalmoplegia and ocular myopathies. Handbook Clin. Neurol. **18**(62): 287–329.
63. HIRANO, M. et al. 1998. Mitochondrial neurogastrointestinal encephalomyopathy syndrome maps to chromosome 22q13.32-qter. Am. J. Hum. Genet. **63:** 526–533.
64. BARRIENTOS, A. et al. 1996. Autosomal recessive Wolfram syndrome associated with an 8.5-kb mtDNA single deletion. Am. J. Hum. Genet. **58:** 963–970.
65. BARRIENTOS, A. et al. 1996. A nuclear defect in the 4p16 region predisposes to multiple mitochondrial DNA deletions in families with Wolfram syndrome. J. Clin. Invest. **97:** 1570–1576.
66. ROBINSON, B.H. et al. 1987. Clinical presentation of mitochondrial respiratory chain defects in NADH-coenzyme Q reductase and cytochrome oxidase: clues to pathogenesis of Leigh disease. J. Pediatr. **110:** 216–222.
67. HOPPEL, C.L. et al. 1987. Deficiency of the reduced nicotinamide adenine dinucleotide dehydrogenase component of complex I of mitochondrial electron transport: fatal infantile lactic acidosis and hypermetabolism with skeletal-cardiac myopathy and encephalopathy. J. Clin. Invest. **80:** 71–77.
68. VAN ERVEN, P.M.M. et al. 1987. Intravenous pyruvate loading test in Leigh syndrome. J. Neurol. Sci. **77:** 217–227.
69. GLERUM, M. et al. 1987. Abnormal kinetic behavior of cytochrome oxidase in a case of Leigh disease. Am. J. Hum. Genet. **41:** 584–593.
70. HOGANSON, G.E. et al. 1984. Deficiency of muscle cytochrome c oxidase in Leigh's disease. Pediatr. Res. **18:** 222A.
71. MIYABAYASHI, S. et al. 1983. Two siblings with cytochrome c oxidase deficiency. J. Inherited Metab. Dis. **6:** 121–122.
72. MIYABAYASHI, S. et al. 1984. Cytochrome c oxidase deficiency in two siblings with Leigh encephalomyelopathy. Brain Dev. **6:** 362–372.
73. MIYABAYASHI, S. et al. 1985. Biochemical study in 28 children with lactic acidosis, in relation to Leigh's encephalopathy. Eur. J. Pediatr. **143:** 278–283.
74. MIYABAYASHI, S. et al. 1987. Immunochemical study in three patients with cytochrome c oxidase deficiency presenting as Leigh's encephalomyopathy. J. Inherited Metab. Dis. **10:** 289–292.
75. WILLEMS, J.L. et al. 1977. Leigh's encephalomyelopathy in a patient with cytochrome c oxidase deficiency in muscle tissue. Pediatrics **60:** 850–857.
76. DIMAURO, S. et al. 1987. Cytochrome c oxidase deficiency in Leigh syndrome. Ann. Neurol. **22:** 498–506.

77. BERKOVIC, S.F. *et al.* 1987. Cytochrome c oxidase deficiency: a remarkable spectrum of clinical and neuropathogical findings in a single family. Neurology **37**(Suppl. 1): 223.
78. ARTS, W.F. *et al.* 1987. Cytochrome c oxidase deficiency in subacute necrotizing encephalomyelopathy. J. Neurol. Sci. **77**: 103–115.
79. MIRANDA, D.F. *et al.* 1989. Cytochrome c oxidase (COX) deficiency in Leigh's syndrome: genetic evidence for a nuclear DNA-encoded mutation. Neurology **39**: 697–702.
80. BOURGERON, T. *et al.* 1995. Mutation of a nuclear succinate dehydrogenase gene results in mitochondrial respiratory chain deficiency. Nature Genet. **11**: 144–149.
81. VAN DEN HEUVEL, L. *et al.* 1998. Demonstration of a new pathogenic mutation in human complex I deficiency: a 5-bp duplication in the nuclear gene encoding the 18-kD (AQDQ) subunit. Am. J. Hum. Genet. **62**: 262–268.
82. LOEFFEN, J. *et al.* 1998. The first nuclear-encoded compelx I mutation in a patient with Leigh syndrome. Am. J. Human. Genet. **63**: 1598–1608.
83. SCHUELKE, M. *et al.* 1999. Mutant NDUFV1 subunit of mitochondrial complex I causes leukodystrophy and myoclonic epilepsy. Nature Genet. **21**: 260–261.
84. TIRANTI, V. *et al.* 1998. Mutations of SURF-1 in Leigh disease associated with cytochrome c oxidase deficiency. Am. J. Hum. Genet. **63**: 1609–1621.
85. ZHU, Z. *et al.* 1998. SURF1, encoding a factor involved in the biogenesis of cytochrome c oxidase. Nature Genet. **20**: 337–343.
86. MERANTE, F. *et al.* 1993. A biochemically distinct form of cytochrome oxidase (COX) deficiency in the Saguenay-Lac-Saint-Jean region of Quebec. Am. J. Hum. Genet. **53**: 481–487.
87. HEYER, E. 1995. Mitochondrial and nuclear genetic contribution of female founders to a contemporary population in northeast Quebec. Am. J. Hum. Genet. **56**: 1450–1455.
88. LEE, N. *et al.* 1998. Saguenay Lac Saint Jean cytochrome oxidase deficiency: sequence analysis of nuclear encoded COX subunits, chromosomal localization and a sequence anomaly in subunit VIc. Biochim. Biophys. Acta **27**: 1–4.
89. MORIN, C. *et al.* 1993. Clinical, metabolic, and genetic aspects of cytochrome C oxidase deficiency in Saguenay-Lac-Saint-Jean. Am. J. Hum. Genet. **53**: 488–496.
90. DE MICHELE, G. *et al.* 1998. A new locus for autosomal recessive hereditary spastic paraplegia maps to chromosome 16q24.3. Am. J. Hum. Genet. **63**: 135–139.
91. CASARI, G. *et al.* 1998. Spastic paraplegia and OXPHOS impairment casued by mutations in paraplegin, a nuclear encoded mitochondrial metalloprotease. Cell **93**: 973–983.
92. CAMPUZANO, V. *et al.* 1996. Friedreich's ataxia: autosomal recessive disease caused by an intronic GAA triplet repeat expansion. Science **271**: 1423–1427.
93. KOUTNIKOVA, H. *et al.* 1997. Studies of human, mouse and yeast homologues indicate a mitochondrial function for frataxin. Nature Genet. **16**: 345–351.
94. ROTIG, A. *et al.* 1997. Aconitase and mitochondrial iron-sulphur protein deficiency in Friedreich ataxia. Nature Genet. **17**: 215–217.
95. TANZI, R.E. *et al.* 1993. The Wilson disease gene is a copper transporting ATPase with homology to the Menkes disease gene. Nature Genet. **5**: 344–350.
96. BULL, P.C. *et al.* 1993. The Wilson disease gene is a putative copper transporting P-type ATPase similar to the Menkes gene. Nature Genet. **5**: 327–337.
97. LUTSENKO, S. & M.J. COOPER. 1998. Localization of the Wilson's disease protein to mitochondria. Proc. Natl. Acad. Sci. USA **95**: 6004–6009.

The α-Ketoglutarate Dehydrogenase Complex

KWAN-FU REX SHEU[a] AND JOHN P. BLASS[b]

Dementia Research Service, Burke Medical Research Institute, Weill Medical College of Cornell University, 785 Mamaroneck Avenue, White Plains, New York 10605, USA

> ABSTRACT: The α-ketoglutarate dehydrogenase complex (KGDHC) is an important mitochondrial constituent, and deficiency of KGDHC is associated with a number of neurological disorders. KGDHC is composed of three proteins, each encoded on a different and well-characterized gene. The sequences of the human proteins are known. The organization of the proteins into a large, ordered multienzyme complex (a "metabolon") has been well studied in prokaryotic and eukaryotic species. KGDHC catalyzes a critical step in the Krebs tricarboxylic acid cycle, which is also a step in the metabolism of the potentially excitotoxic neurotransmitter glutamate. A number of metabolites modify the activity of KGDHC, including inactivation by 4-hydroxynonenal and other reactive oxygen species (ROS). In human brain, the activity of KGDHC is lower than that of any other enzyme of energy metabolism, including phosphofructokinase, aconitase, and the electron transport complexes. Deficiencies of KGDHC are likely to impair brain energy metabolism and therefore brain function, and lead to manifestations of brain disease. In general, the clinical manifestations of KGDHC deficiency relate to the severity of the deficiency. Several such disorders have been recognized: infantile lactic acidosis, psychomotor retardation in childhood, intermittent neuropsychiatric disease with ataxia and other motor manifestations, Friedreich's and other spinocerebellar ataxias, Parkinson's disease, and Alzheimer's disease (AD). A KGDHC gene has been associated with the first two and last two of these disorders. KGDHC is not uniformly distributed in human brain, and the neurons that appear selectively vulnerable in human temporal cortex in AD are enriched in KGDHC. We hypothesize that variations in KGDHC that are not deleterious during reproductive life become deleterious with aging, perhaps by predisposing this mitochondrial metabolon to oxidative damage.

INTRODUCTION

The α-ketoglutarate dehydrogenase complex (KGDHC) is a well-characterized mitochondrial component that has been found to be abnormal in a number of neurodegenerative disorders including Alzheimer's disease (AD). KGDHC is a component of the Krebs tricarboxylic acid cycle. The step it catalyzes is also important in the removal of glutamate, a potentially excitotoxic neurotransmitter. The discussion below reviews the normal genetics and biochemistry of KGDHC and then the evidence that KGDHC is involved in several diseases including AD.

---

[a]Deceased.
[b]Phone: 914-597-2359 or 914-597-2356; fax: 914-597-2757.
e-mail: jpblass@mail.med.cornell.edu

TABLE 1. Genes for human KGDHC[a]

Component	Name of Gene	Location	Size (approx.)	Exons	Introns
E1k	OGDH	7p13-14	85 kb	22	21
E2k	DLST	14q24.3	23 kb	15	14
E2k	pseudogene	1p31	2.3kb	1	0
E3	DLD	7q31-32	20 kb	14	13

[a]See text for references.

GENETICS OF KGDHC

Studies, notably by Lester Reed and his colleagues[1–4] and others[5] demonstrated that KGDHC consists of three proteins arranged in a complex array. In humans and other mammals, each of these proteins is coded on a separate gene (TABLE 1). The genes for these enzyme proteins have been characterized in prokaryotic and eukaryotic organisms, but the discussion below concentrates on the human genes. This is done partly for reasons of space and partly because the human genes and proteins are of most interest for human neurodegenerative diseases. The thiamine-pyrophosphate–containing protein that interacts directly with the α-ketoglutarate substrate is 2-oxoglutarate dehydrogenase (E1k). It is encoded on the *OGDH* gene, which is located on chromosome 7p13-p14. The core protein, dihydrolipoyl succinyl transferase (E2k) is encoded on the *DLST* gene, which in humans is located on chromosome 14q24.3. The dihydrolipoamide dehydrogenase component (E3) is encoded on the *DLD* gene, which is on chromosome 7q31.1-7q32.

The Human OGDH Gene

The human *OGDH* gene has been characterized by Koike and coworkers.[6] The gene is large—approximately 85 kb. It is located on chromosome 7p13-p14. There are 22 exons. A consensus thiamin diphosphate (TDP) binding region is present. The 5'-flanking region of the *OGDH* gene contains an inverted GC box but no TATA or CAAT boxes.[7] The −53 to −44 and −33 to −24 regions are associated with positive regulation and the −93 to −84 sequences to negative regulation. Promotor activity is upregulated by the substrate, α-ketoglutarate, and by a combination of α-ketoglutarate and its transamination product, the neurotransmitter amino acid glutamate.[6,8] There are two 10-bp *cis* acting elements (−53 to −44 and −33 to −24) and two *trans*-acting elements (−536 to −496 and −93 to −84). An as-yet-unidentified nuclear factor binds to nt −63 to −24. These findings are consistent with the *OGDH* having characteristics of both a housekeeping and an inducible enzyme. Although thiamine diphosphate is a cofactor for the enzyme encoded on *OGDH*, thiamine deficiency did not lower the levels of the mRNA for this component in several tissue culture systems.[9]

The Human DLST Gene

The human DLST gene and a pseudogene have been described by Nakano and coworkers.[10] The gene is approximately 23 kb long. There are 15 exons varying in length from 34 to 1552 bp, and 14 introns varying in length from 127 to 3961 bp.

The sequence of the human gene showed minor variations from that of the previously published cDNAs.[11,12] This variation is in accordance with direct demonstration of polymorphic variations in the human *DLST* gene, in several populations.[13–18] Intron 1 is G+C rich. The sequence encoded by exons 8 and 9 encodes the region from part of the inner core-catalytic domain to part of the lipoyl domain. Human *DLST*, like that from other mammals, does not contain the consensus sequence found in other organisms for binding the other two gene products in KGDHC (namely the E1k and E3 proteins; see below).

A single transcription initiation site was found for *DLST*.[10] The promotor-regulatory region contained a TCAAT sequence (−374 to −370) but no TATA box. Possible binding sites were found for Sp1 (−25 to −20), cAMP receptor AP-2 (−426 to −420), and for glucocorticoid-responsive elements (−526 to −520). Southern blotting revealed a single functional *DLST* gene.

A pseudogene of *DLST* has also been found, on chromosome 1p31.[10,19] The pseudogene has 93%[10]–99%[19] homology with the actual gene. The pseudogene is transcribed.[10,19] Whether the pseduogene is partially translated has not been studied in detail, nor has its potential role in physiology or pathology.[10,19,20] Specifically, there are no studies to determine if the product of the "pseudogene" might play a role in determining the binding of E3 to the E2k core of KGDHC, similar to the role played by "protein X" in the PDHC complex.[21] The E2k pseudogene can be a source of artifacts in studies, leading for instance to the finding of "apparent" mutations due to simultaneous PCR amplification of both the gene and the pseudogene. More recent studies utilize PCR primers that include enough intronic sequence to preclude effectively the coamplification of the pseudogene.[14,15]

The Human DLD *Gene*

The human DLD gene has been characterized by Feigenbaum and Robinson.[22] It is approximately 20 kb long and contains 14 exons, which vary from 69 to 780 bp in length. The 13 introns vary greatly in size, from 93 bp to 7.0 kb. In the gene itself, the FAD functional domain (aa residues 8–38) is encoded on exons 3–4, including the pyrophosphate binding loop and redox-active cystein groups (aa 45 and 50). The NAD-binding domain (aa residues 180–210) is encoded by exons 8–10. The consensus sequence (aa 125–201) spans exons 8 and 9.[23] The proposed active proton acceptor/donor, histidine residue 452, is on exon 13. Exons 13 and 14 encode a region important for enzyme activity and for the binding of dihydrolipoamide.

Reverse transcription revealed three or more transcription initiation sites.[22] All of these are upstream of the published cDNA sequences.[24,25] It is not certain how many of the three potential transcription products are functionally important, and whether or not they are all edited to the same form before translation.

In the promotor region of *DLD*,[22,26] a CAAT box-like sequence was found (−129 to −124), as was a possible CREB site (−106 to −101). No TATA box was found.[26] The existence of an Sp1 site is debated.[22,26] NRF-1, an enhancing element upstream of subunit IV of cytochrome *c* oxidase and other nuclear-encoded mitochondrial enzymes, was identified at −89.[22] The sequence from −1262 through −1252 matches that for a negative regulatory element (NRE) for the human insulin gene.[22] Deletion of an element from −1223 to −769 increased expression of a reporter gene ligated to the *DLD* promotor-regulatory region.[26] These findings are consistent with the *DLD*

TABLE 2. Proteins of Human KGDHC

Component	Name	EC Number	Amino Acids	Leader (AA)	Cofactors
E1k	2-oxoglutarate decarboxylase (2-oxoacid decarboxylase)	EC 1.2.4.2	962	40	thiamine diphosphate
E2k	dihydrolipoamide succinyltransferase	EC 2.3.1.61	453	?	thioctic acid, coenzyme A
E3	dihydrolipoamide dehydrogenase	EC 1.8.1.4	464	35	FAD, NAD(H)

promotor having housekeeping and facultative characteristics. The effects of physiological or pathological states on the transcription of DLD have not been described in detail in the mammalian brain.

STRUCTURE OF KGDHC

The biochemistry of KGDHC is complex. KGDHC can correctly be thought of as an enzyme, as a complex composed of three different enzymes, and as a submitochondrial particle (i.e., a "metabolon"). As expected, KGDHC varies in primary sequence and structure between bacteria and mammals and to a lesser extent among mammals. The following discussion concentrates on human KGDHC, since that complex is particularly relevant for human diseases. Where human data is unavailable, information on the complex from other large mammals is used.

The three enzyme proteins that make up the complex are conveniently referred to as E1k, E2k, and E3 (TABLE 2). The cDNA for each of the human enzymes has been isolated and sequenced, so that the amino acid sequence (primary structure) for each protein is known. E1k is 2-oxoglutarate decarboxylase, and is also referred to as 2-oxoacid decarboxylase (EC 1.2.4.2). The human protein contains 962 amino acids linked to a 40-amino acid leader sequence.[6–8] E2k is dihydrolipoamide succinyltransferase, also referred to as dihydrolipoamide acyltransferase (EC 2.3.1.61). The human protein is 453 amino acids long,[10] but the leader sequence is not well defined. E3 is dihydrolipoamide dehydrogenase (EC 1.8.1.4.). The human protein is 464 amino acids long, linked to a 35-amino acid leader sequence. The primary structure (amino acid sequence) has >95% homology with the reported amino acid sequence of pig heart E3.[24,25] E1k and E2k are specific for KGDHC. E3, however, is a component not only of KGDHC but also of the pyruvate dehydrogenase (PDHC) and branched-chain dehydrogenase (BCDHC) complexes and of the glycine cleavage system.[27] After salt induced dissociation of E3 from the E1k/E2k complex, the E3 from the same species as the E1k/E2k core is much more effective in reconstituting KGDHC activity than that from other species; for instance, bovine E3 to bovine E1k/E2k is much more effective than pig E3 or yeast E3 to bovine E1k/E2k.[21] This observation is surprising, because of the high sequence homology among mammalian E3s.[21]

The structure of human KGDHC has not been studied in detail, but is believed by analogy to resemble that of other mammals. The core of mammalian KGDHC is composed of 24 E2k (transacylase) units arranged as a cube with octahedral (432) symmetry.[28] The structure and mechanism of action of the bacterial E2k core has been worked out in detail.[29–31] E1k and E3 units adhere, noncovalently, to the E2k core. In mammalian KGDHC, some studies suggest that the E1k protein appears to bind more tightly to the E2k core than does the E3 protein.[28] Other studies suggest that single copies of E1 and E3 form homodimers by a high-affinity reaction, with the homodimers binding to the E2k core through the N-terminal region of E1.[28,32] Human E2k lacks the consensus binding domains for E1 and E3 that have been identified in other mammalian E2ks,[10,11] and the mechanism of association of the three proteins in human KGDHC has not yet been specified.

ENZYMOLOGY OF KGDHC AND ITS COMPONENTS

The overall reaction catalyzed by KGDHC is:

$$^-O_2CCH_2CH_2COCO_2^- + CoASH + NAD^+ \alpha\ ^-O_2CCH_2CH_2CO-ScoA + CO_2 + NADH + H^+$$

where $^-O_2CCH_2CH_2COCO_2^-$ is α-ketoglutarate, CoASH is reduced coenzyme A, and $^-O_2CCH_2CH_2COSCoA$ is succinyl-coenzymeA (FIG. 1). The overall reaction is irreversible because of removal of the CO_2 generated in the first, E1k-catalyzed step.

E1k utilizes thiamine diphosphate (TDP) as a cofactor. The TDP is tightly but not covalently bound to this protein; it can be removed by treatment with first acid and then alkaline ammonium sulfate. E1k catalyzes the oxidative decarboxylation of α-ketoglutarate, and the subsequent binding of the resulting succinic acid fragment to a sulfur residue of a lipoic acid on E2k, with concomitant regeneration of the TDP on E1k. This reaction is physiologically irreversible, due to the diffusion away of the CO_2 product.

The core enzyme, E2k, catalyzes the transfer of the four carbon fragments from the TDP on E1k to coenzyme A, producing succinyl-CoA. It also transfers reducing equivalents to the flavoprotein moiety attached to the E3 protein. E2k contains one or more dihydrolipoic acid residues covalently bound in amide linkages to the amino groups of one or more lysines. This linkage provides a flexible arm about 140 µm long, which can rotate among the E1k, E2k, and E3 subunits, as a "swinging arm." A "multiple random coupling mechanism" has been proposed, based on computer modeling of KGDHC from *E. coli*.[29]

E3 is a flavoprotein. It catalyzes the two-electron transfer of reducing equivalents from E2k to produce NADH and H^+. Site-directed mutagenesis of human E3 has shown that lysine-54 (K54) is necessary for the the protein-FAD interaction and for the catalytic efficiency of the enzyme, and that glutamate-192 (E192) is involved in maintaining the appropriate orientation of K54 during catalysis.[33] E3 can also act as a diaphorase, carrying out single-electron transfers from a variety of electron donors to a variety of electron acceptors.[28] The R-enantiomer is >25-fold more active as a substrate than the S-enantiomer.[34] The amount of E3 found in mitochondria typically exceeds the amount required for the activity of the intramitochondrial complexes of which it is known to be a part (namely, KGDHC, PDHC, BCDHC). This finding has led to speculation that E3 may also have other roles in the transfer of electrons among substrates within mitochondria[28] including ascorbic acid[35] and ubiquinone.[36] E3 proteins isolated from pig heart and cow intestine bind to G4-DNA of tet-

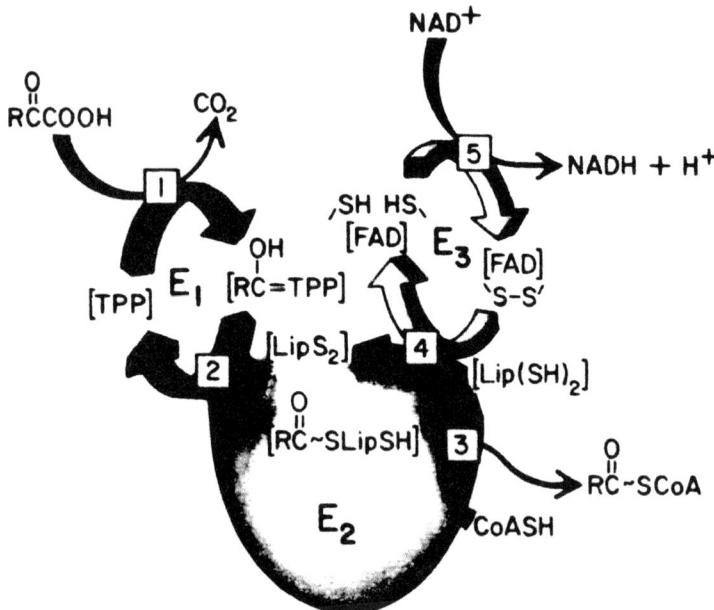

FIGURE 1. The KGDHC reaction. The drawing illustrates the reaction catalyzed by KGDHC. For information about E1k, E2k, and E3, see TABLE 2. (From Reed.[1] Reprinted by permission.)

rahymina thermophila.[37] Binding of E3 to human or other mammalian DNA has not been described. E3 from yeast has been crystallized.[38]

The reactions catalyzed by E2k and E3 are reversible under physiological conditions and are influenced by the principles of mass action. Major factors controlling the activity of KGDHC are the CoASH/succinyl-CoA and the NAD+/NADH ratios. In addition, the concentration of divalent cations modulates the activity of KGDHC.[39–41] Ca^{2+} (A_{50} = 1 µM) and Mg^{2+} (A_{50} = 25 µM) together synergisitcally reduced the Km for α-ketoglutarate by over 10-fold, from 4 ± 1 mM to 0.3 mM.[39] These concentrations of the divalent cations can probably occur in mitochondria under pathological conditions.[39,41] KGDHC activity is stimulated by decreases in pH within the physiological pH range, and is also enhanced by the peptide spermine.[10]

A number of inhibitors reduce the activity of KGDHC. Several reactive oxygen species (ROS) can inactivate the complex, including 4-hydroxy-2-nonenal,[42,43] superoxide anion,[44] generators of NO·,[44,45] and hydrogen peroxide.[46] Hydroxynonenal[47] and perhaps hydrogen peroxide[46] may reduce the activity of KGDHC relatively selectively compared to other mitochondrial components of energy metabolism. Inactivation of KGDHC by ROS including hydroxynonenal has been proposed to play a role in the increased sensitivity of aging tissues to ischemia-reperfusion injury.[47,48] Treating E3 with ROS in the presence of myeloperoxidase reduces its activity as as a lipoamide dehydrogenase but increases its activity as a diaphorase.[49] Methotrexate[50] and some environmental toxins[48] can reduce the activity of KGDHC.

TABLE 3. Activities of enzymes of energy metabolism in human brain[a]

Glycolysis[91]	
Hexokinase (soluble)	39
Phosphohexose isomerase	581
Phosphofructokinase	**9.9**
Aldolase	35
Triose phosphate isomerase	7,472
Glyceraldehyde-3-phosphate dehydrogenase[92]	17
Phosphoglyceromutase	414
Pyruvate kinase	703
Lactate dehydrogenase	662
Pyruvate Dehydrogenase[53]	11
Krebs Tricarboxylic Acid Cycle	
Citrate synthetase[93]	40 (120 in a biopsy sample[94])
Aconitase[95]	9
Isocitrate dehdrogenase (NAD-linked)[94]	10
KGDHC[53]	**2.3**
Succinic Dehydrogenase (Complex II)[93]	98
Fumarase[96]	180
Malate Dehydrogenase (NAD-linked)[97]	2,491
Electron Transport[93]	
Complex I	**18.4**
Complex II/III	97.9
Complex IV	285

[a] Values are nmol/min/mg whole brain protein. Samples analyzed were usually from cerebral cortex. See references for details.

TABLE 4. Relatively slow reactions of energy metabolism in human brain

Pathway	Enzyme(s)	Activity[a]
Glycolysis	Phosphofructokinase	9.9
	Glyceraldehyde-3-phosphate dehydrogenase	17
Pyruvate Oxidation	Pyruvate Dehydrogenase Complex	11
Krebs Tricarboxylic Acid Cycle	Aconitase	8.8
	Isocitrate Dehydrogenase	10
	α-Ketoglutarate Deyhydrogenase Complex	**2.3**
Electron Transport	Complex I	18

[a] Activity is in nmol/min/mg protein. Values for other enzymes of energy metabolism in human brain are at least an order of magnitude greater than for the α-ketoglutarate dehydrogenase complex (KGDHC). See TABLE 3 and the references therein for details.

NEUROBIOLOGY OF KGDHC

As far as is known, the proteins comprising KGDHC and the structure of the complex itself do not differ significantly among brain and other tissues. There is no evidence for brain-specific forms of KGDHC, at either the gene or protein levels. Therefore, conclusions drawn about the structure and regulation of KGDHC in other tissues are likely to apply to brain as well.

The activity of KGDHC is relatively low compared to that of a number of other enzymes of oxidative/energy metabolism, in skeletal[51] and heart[52] muscle and in brain.[53] In human brain, the values for KGDHC are lower than those for any other enzymes of energy metabolism, including other enzymes of the Krebs tricarboxylic acid cycle (e.g., aconitase, PDHC), enzymes of glycolysis, and electron transport complexes (TABLES 3 and 4). The values for human brain are derived largely from human brain obtained at autopsy and are therefore subject to the inherent artifacts in chemical measurements on autopsy brain.[54,55] Comparisons of activities of enzymes in large samples of brain and assayed under optimal conditions are potentially misleading if applied simplistically to analyses of brain metabolism. However, the comparison may be more interpretable when it is among "housekeeping" enzymes, which are believed to be present in all brain cells, as for the enzymes listed in TABLES 3 and 4. When the activities of such enzymes differ by orders of magnitude (e.g., KGDHC and enzymes of the electron transport chain) it is probably reasonable to assume that the activities of the enzymes with lower activity under forcing conditions are also lower under physiological conditions in the tissue being examined.

KGDHC and other enzymes of energy metabolism are not, however, distributed uniformly among cell types in the nervous system or even among different groups of neurons.[53,56–58] Cholinergic cells appear to be particularly rich in KGDHC[58] and PDHC.[56] Results for human brain are similar to but less extensive than those in rat brain (TABLE 5). As discussed below, neurons that are enriched in KGDHC may be selectively vulnerable in Alzheimer's disease (TABLE 5).

KGDHC IN NEURODEGENERATIVE DISEASES

Deficiency of KGDHC activity has been recorded in a number of conditions—for instance, antibodies to mitochondrial components in biliary cirrhosis can include antibodies to KGDHC.[59] However, deficiency of KGDHC is more characteristically associated with neurological syndromes. This association is not surprising, because of the second-to-second dependence of the function of the nervous system on oxidative metabolism. The neurological syndromes linked to KGDHC deficiency vary from deadly diseases of the neonatal period to chronic diseases of the elderly.[60] Generally, the more severe the KGDHC deficiency, the earlier the onset and the more severe the clinical syndrome. This relationship between severity of defect and severity of syndrome often occurs in inborn errors of metabolism.

Infantile Lactic Acidosis

Infantile lactic acidosis with severe psychomotor retardation is characteristic of severe deficiencies of KGDHC.[61–64] These unfortunate children typically die in in-

TABLE 5. KGDHC and selective neuronal loss in Alzheimer temporal cortex[a]

	Controls	Alzheimer's
Layer III		
NSE	184 ± 11	52 ± 6^b
KGDHC	161 ± 17	44 ± 6^b
Layer V		
NSE	133 ± 16	33 ± 4^b
KGDHC	118 ± 9	$12 \pm 1^{b,c}$

[a]Values are mean number of intensely staining cells per mm^2 for 10 fields counted per patient. NSE = neuron specific enolase, a conventional marker for neurons.
[b]$p < 0.001$ vs. control.
[c]$p < 0.001$ vs. number of NSE positive cells in AD cortex.
(Ko, Sheu, Thaler, Markesbery & Blass, in preparation; see also Ref. 86.)

fancy or even *in utero*. This syndrome can also be associated with other profound defects in the Krebs tricarboxylic acid cycle or other pathways of energy metabolism.[65] The best described of these unfortunate children have a profound deficiency of E3, with a resultant combined deficiency of KGDHC, PDHC, and BCDHC.[61–63] These children can be homozygous for mutations in *DLD*[62,63] or can be compound heterozygotes, inheriting different defective *DLD* genes from each of their parents.[61] For instance, one compound heterozygote had a TAC→TAAC mutation in one gene and an R460G mutation in the other.[61] Two Israeli patients with severe neonatal E3 deficiency had both a G229C and an insertion 105insA (Y35X) mutation.[64] Complete or nearly complete E3 deficiency may be incompatible with life even *in utero*. Homozygous *DLD* knockout in transgenic mice leads to perigastrulation lethality.[66]

Psychomotor Retardation

Psychomotor retardation in early childhood has been associated with somewhat milder deficiency of KGDHC.[67,68] Enzyme assays have implicated the core, E2k component, but studies at the gene level have not been reported.

Intermittent Psychomotor Symptoms

Intermittent psychomotor symptoms occur in Ashkenazi Jews who have a relatively mild deficiency in E3.[64] These children have intermittent attention deficit disorder with mild ataxia, incoordination, and hypotonic weakness. They have a single G229C mutation in *DLD*, without the additional insertion mutation in the Israeli patients with infantile lactic acidosis described above.[64]

Friedreich's Ataxia

Friedreich's ataxia[69] is associated with deficiencies of KGDHC.[70] These patients develop ataxia and signs of damage to the long tracts of the spinal cord, with significant clinical manifestations typically beginning in adolescence or early adult life.[70] The primary genetic defect is in the *FRDA* gene and is usually a GAA repeat.[69] *FRDA* encodes the protein frataxin, which is involved in free radical (ROS) metabolism within the mitochondria. The Friedreich mutations can lead to inactivation of a number of mitochondrial enzymes including aconitase.[71] Deficiency of the E3 component of KGDHC in Friedreich's ataxia was proposed in 1976, on the basis of a defect in both KGDHC and PDHC activity in cultured Friedreich fibroblasts.[70] E3 deficiency has been directly confirmed in Friedreich brain.[72]

Other Spinocerebellar Ataxias

A number of other spinocerebellar ataxias (SCAs) are known.[73] They involve varying mixtures of signs and symptoms associated with dysfunction of the spinal cord and cerebellum and sometimes other parts of the brains. The clinical syndromes (phenotypes) overlap among the known genotypes. The genetic abnormalities are pathologically elongated stretches of CAG repeats (CAG_n) encoding elongated polyglutamine (Q_n) sequences in proteins. The precise gene and encoded protein differ among the six spinocerebellar ataxias characterized so far.[73] A variety of lines of evidence suggest that the CAG_n/Q_n diseases are associated with a "pathological gain of function" associated with the Q_n expansions.[74,75] In SCA type I, E2 was found to be decreased in cerebellar and frontal cortices.[72]

Studies several years ago suggested that loss of mitochondrial enzyme activities including PDHC and KGDHC is commonly found in SCAs,[76,77] but extensive studies in patients whose defects have been defined at the gene level have not been done. The "toxin gain of function" of the Q_n repeats may be mediated at least in part by aberrant transglutaminase-catalyzed reactions.[74,75] Purified KGDHC is inactivated on incubation with transglutaminase and proteins containing pathologically long, but not normal length, Q_n repeats.[75]

Parkinson's Disease

Parkinson's disease (PD) is a common disease of later life that impairs secondary motor function and often goes on to impair intelligence as well. The best known defect in energy metabolism is in complex I of the electron transport chain, which may be caused in some PD patients by a defect in mitochondrial DNA (mtDNA).[78] Deficiency of KGDHC has been found in Japanese patients with PD, based on immunocytochemical studies.[79] A genetic abnormality in the *DLST* gene, which encodes the core E2k component of KGDHC, has also been reported in Japanese patients.[80] Attempted replications of these studies of KGDHC and *DLST* in PD in other populations have not been reported. The Parkinson syndrome can be induced by poisoning with the compound MPTP, which is converted in mitochondria to MPP^+. MPP^+ inhibits complex I of the electron transport chain but also inhibits KGDHC. The re-

TABLE 6. Association of the G19,11//C19,183 Allele of DLST with AD

DLST Genotype	Alzheimer	Non-Alzheimer	p
G,C/G,C	21/43 (49%)	1/13 (8%)	0.019^a
G,C/x	45/98 (46%)	6/13 (46%)	0.779
Non-G,C	16/48 (33%)	2/10 (20%)	0.708

^{a}Odds Ratio = 11.5; confidence intervals = 1.8 – 70.1. All patients in this series were positive for the ε-4 allele of the *APOE* gene. The presence or absence of Alzheimer's disease was confirmed at autopsy in these subjects. See Sheu *et al*.[15] for original data.

ported activity of KGDHC in human brain is about 10% of that of complex I (TABLES 3 and 4), suggesting that MPTP inhibition of KGDHC might be functionally more important than inhibition of complex I.

Alzheimer's Disease

Alzheimer's disease accounts for over 80% of dementia in older people.[81] The classic pathological lesions visible under the light microscope are amyloid plaques and neurofibrillary tangles.[82] Among the characteristic biochemical lesions are oxidative stress[83] and deficiency of KGDHC.[54,55,84,85] Reduction of KGDHC activity is as robust as the other lesions of AD brain, having been reported from at least four laboratories with no contravening reports.[53–55,84,85] The reduction in KGDHC activity cannot be attributed hypoxia or to postmortem change.[54,55] KGDHC is reduced in AD brain by -30% to -90%. Decreases in KGDHC activity of this magnitude can be expected to impair oxidative/energy metabolism significantly, since KGDHC appears to have the lowest activity of any enzyme in the major pathways of oxidative/energy metabolism, including the components of the electron transport chain (TABLES 3 and 4). Furthermore, neurons enriched in KGDHC appear to be selectively vulnerable in AD temporal cortex.[53] Ko *et al.*[86] showed that the selectively vulnerable neurons of layers III and V are enriched in KGDHC. In layer V, the loss of KGDHC neurons in AD was proportionally greater than the total loss of (NSE-staining) neurons, indicating selective loss of KGDHC-enriched cells within this layer (TABLE 5). These results support the assumption that mitochondrial damage in general and damage to KGDHC in human brain in particular may contribute to selective cell death.[53,86]

At least two mechanisms may contribute to the reduction of KGDHC activity in AD. A *genetic* component is implied by the finding that KGDHC deficiency persists in cultured skin fibroblasts from many but not all patients with AD.[87] More direct evidence is that polymorphisms of the *DLST* gene, which encodes the core E2k protein of KGDHC, have been associated with AD in Ashkenazi Jewish,[15] mixed white American,[14] Swedish,[18] and one[16] of two[17] Japanese series studied (TABLE 6). The Swedish series include a population sample and a sample enriched in familial AD.[18] However, the specific polymoprphisms associated with AD differ among the American, Swedish, and Japanese series. A plausibly pathogenetic mutation has not been defined, at least as yet. *Secondary damage* to KGDHC also appears to occur in AD, due to other mutations and particularly to *oxidative stress*. KGDHC is reduced in the brains of patients whose original lesion is the "Swedish" mutation in the *APP* gene.[88] As discussed above, KGDHC is sensitive to reactive oxygen species, particularly to hydroxynonenal.[42–47] Alzheimer brain is under oxidative stress, which appears to be an important part of the AD disease process.[83] Oxidative stress may contribute to the reduction of KGDHC activity in AD.

The genetic and nongenetic mechanisms are not mutually exclusive. A polymorphism (or mutation) in a gene encoding a KGDHC component may sensitize KGDHC to damage by free radicals. Indeed, other evidence indicates that *DLST* and *APOE4* may interact in the causation of AD. In the molecular genetic studies in the USA and Sweden described above,[14,15,18] the association of *DLST* with AD was significant only in the patients who were also *APOE4* positive. Very recent data[89] indicate that KGDHC deficiency correlates significantly better with clinical disability than do plaques or tangles in patients who are *APOE4* positive but not in those who are *APOE4* negative. One may speculate that a "normal polymorphism" of KGDHC might contribute to the development of AD analogously to the way that the ε-4 allele of the *APOE* gene does.[90] *APOE4* has been proposed to be deleterious due to free radical mechanisms,[90] and such mechanisms can also affect KGDHC.

CONCLUSIONS

KGDHC is a critical component of the Krebs tricarboxylic acid cycle and of glutamate metabolism. It has been well studied at both the genetic and biochemical level. In human brain, KGDHC appears to have lower activity that do other enzymes of energy metabolism. KGDHC activity is an order of magnitude lower than that of complex I and over two orders of magnitude less than that of complex IV (TABLE 3). These data suggest that KGDHC is likely to have a high metabolic control coefficient for overall oxidative/energy metabolism—that is, to be a potentially "rate-limiting" enzyme. Conditions that impair KGDHC activity are therefore likely to impair oxidative/energy metabolism in human brain. Since the brain has a second-to-second dependence on flexible regulation of energy metabolism to maintain its function, impairments of KGDHC are likely to lead to brain dysfunction that manifests itself as disease of the brain. In fact, deficiency of KGDHC has been associated with a variety of neurological diseases, including disorders that have been shown to be associated with mutations in a gene encoding a component of KGDHC. In other disorders, deficiency of KGDHC appears to be part of the disease process. Because of its critical role in energy/oxidative metabolism, impairment of KGDHC activity in the human brain is unlikely to be innocuous. Even mild deficiency of KGDHC is likely to impair the ability of brain cells to mobilize energy to repair damage or otherwise respond to the physiological stresses inherent in living. The relatively mild deficiency of KGDHC associated with AD and perhaps with PD may result from a limitation on the ability of the affected brain cells to maintain homeostasis in the face of stressors accumulating over a lifetime.

Further studies of KGDHC at the genetic, protein, and cellular levels appear warranted, particularly in relation to diseases of the nervous system.

ACKNOWLEDGMENTS

This work was supported in part by grants from the Overbrook Foundation, the Winifred Masterson Burke Relief Foundation, and the NIA (AG 09014, AG 14930).

REFERENCES

1. REED, L.J. 1988. From lipoic acid to multi-enzyme complexes. Protein Sci. **7:** 220–224.
2. REED, L.J. & R.M. Oliver. 1982. Structure-function relationships in pyruvate and α-ketoglutarate dehydrogenase complexes. Adv. Exp. Med. Biol. **148:** 231–241.
3. PETTIT, F.H., S.J. YEAMAN & L.J. REED. 1978. Purification and characterization of branched chain α-keto acid dehydrogenase complex of bovine kidney. Proc. Natl. Acad. Sci. USA **75:** 4881–4885.

4. REED, L.J. & R.M. OLIVER. 1982. Structure-function relationships in pyruvate and α-ketoglutarate dehydrogenase complexes. Adv. Exp. Med. Biol. **148:** 231–241.
5. TANAKA, N., K. KOIKE, M. HAMADA, K.-I. OTSUKA, T. SUEMATSU, M. KOIKE. 1972. Mammalian α-keto acid dehydrogenase complexes. J. Biol. Chem. **247:** 4043–4049.
6. KOIKE, K. 1998. Cloning, structure, chromosomal location and promoter analysis of human 2-oxoglutarate dehydrogenase gene. Biochim. Biophys. Acta**1385:** 373–384.
7. KOIKE, K. & S. MATSUO. 1997. Functional characterization of the 5'-flanking region of the gene encoding human 2-oxoglutarate dehydrogenase. Gene **186:** 45–53.
8. KOIKE, K. et al. 1992. Cloning and nucleotide sequence of the cDNA encoding human 2oxoglutarate dehydrogenase (lipoamide). Proc. Natl. Acad. Sci. USA **89**(5): 1963–1967.
9. PEKOVICH, S.R., P.R. MARTIN & C.K. SINGLETON. 1998. Thiamine deficiency decreases steady-state transketolase and pyruvate dehydrogenase but not α-ketoglutarate dehydrogenase levels in three human cell types. J. Nutrition **128:** 683–687.
10. NAKANO, K., C. TAKASE, T. SAKOMOTO, S. NAKAGAWA, J. INAZAWA, S. OHTA & S. MATUDA. 1994. Isolation, characterization and structural organization of the gene and pseudogene for the dihydrolipoamide succinyltransferase component of the human 2-oxoglutarate dehydrogenase complex. Eur. J. Biochem. **224:** 179–189.
11. ALI, G., W. WASCO, X. CAI, P. SZABO, K.F. SHEU, A.J. COOPER, S.M. GASTON, J.F. GUSELLA, R. TANZI & J.P. BLASS. 1994. Isolation, cloning and localization of the gene for the E2k component of the human α-ketoglutarate dehydrogenase complex. Somatic Cell Mol. Genet. **20:** 99–105.
12. NAKANO, K., S. MATUDA, T. SAKAMOTO, C. TAKASE, S. NAKAGAWA, S. OHTA, T. ARIYAMA, J. INAZAWA, T. ABE & T. MIYATA. 1993. Human dihydrolipoamide succinyltransferase. cDNA cloning and localization on chromosome 14q24.2-q24.3. Biochim. Biophys. Acta **1216:** 360–368.
13. ALI, G., W. WASCO, X. CAI, P. SZABO, K.F. SHEU, A.J. COOPER, S.M. GASTON, J.F. GUSELLA, R. TANZI & J.P. BLASS. 1994. Isolation, cloning and localization of the gene for the E2k component of the human α-ketoglutarate dehydrogenase complex. Somatic Cell Mol. Genet. 1994; **20:** 99–105.
14. SHEU, K-F.R., A.M. BROWN, B.S. KRISTAL, R.N. KALARIA, L. LILIUS, L. LANNFELT & J. P. BLASS. 1999. A *DLST* genotype associated with reduced risk for Alzheimer's Disease. Neurology **52:** 1505–1507.
15. SHEU, K-F.R., A.M. BROWN, V. HAROUTUNIAN, B.S. KRISTAL, H. THALER, M. LESSER, R.N. KALARIA, N.R. RELKIN, R.C. MOHS, L. LILIUS, L. LANNFELT & J.P. BLASS. 1999. Modulation by DLST of the genetic risk of Alzheimer's disease in a very elderly population. Ann. Neurol. **45:** 48–53.
16. NAKANO, K., S. OHTA, K. NISHIMAKI, T. MIKI & S. MATUDA. 1997. Alzheimer's disease and *DLST* genotype. Lancet **350:** 1367–1368.
17. KUNUGI, H., S. NANKO, A. UEKI, K. ISSE & H. HIRASAWA. 1998. *DLST* gene and Alzheimer's disease. Lancet **351:** 1584–1585.
18. LILIUS, L., K-F.R. SHEU, L. LANNFELT & J.P. BLASS. 1998. Association of *DLST* polymorphisms with familial Alzheimer's Disease. Neurosci. Abstr. **24:** 255.
19. CAI, X., P. SZABO, G. ALI, R.F. TANZI & J.P. BLASS. 1994. A pseudogene of dihydrolipoyl succinyltransferase (E2k) found by PCR amplification and direct sequencing of rodent-human cell hybrid DNAs. Somatic Cell Mol. Genet. **20:** 339–343.
20. SHEU, K-F.R., A.J.L. COOPER, J.G. LINDSAY & J.P. BLASS. 1994. Abnormality of the α-ketoglutarate dehydrogenase complex in fibroblasts from familial Alzheimer's disease. Ann. Neurol. **35:** 312–318.
21. SANDERSON, S.J., S.S. KHAN, G. MCCARTNEY, C. MILLER & J.G. LINDSAY. 1996. Reconstitution of mammalian pyruvate dehydrogenase and 2-oxoglutarate dehydrogenase complexes: analysis of protein X involvement and interaction of homologous and heterologous dihydrolipoamide dehydrogenases. Biochem. J. **1319:** 109–116.

22. FEIGENBAUM, A.S. & B.H. ROBINSON. 1993. The structure of the human dihydrolipoamide dehydrogenase gene (*DLD*) and its upstream elements. Genomics **17**: 376–381.
23. CAROTHERS, D.J., G. PONS & M.S. PATEL. Dihydrolipoamide dehydrogenase: functional similarities and divergent evolution of the pyridine nucleotide-disulfide oxidoreductases. Arch. Biochem. Biophys. **268**: 409–425.
24. OTULAKOWSKI, G. & B.H. ROBINSON. 1987. Isolation and sequence determination of cDNA clones for porcine and human lipoamide dehydrogenase. J. Biol. Chem. **262**: 17313–17318.
25. PONS, G., C. RAEFSKY-ESTRIN, D.J. CAROTHERS, R.A. PEPIN, A.A. JAVED, W.B. JESSE M.K. GANAPATHI, D. SAMOLA & M.S. PATEL. 1988. Cloning and cDNA sequence of the dihydrolipoamide dehydrogenase component of human α-ketodehydrogenase complexes. Proc. Natl. Acad. Sci. USA **85**:1422–1426.
26. JOHANNING, G.L., J.L. MORRIS, K.T. MADHUSUDHAN, D. SAMOLS & M.S. PATEL. 1992. Characterization of the transcriptional regulatory region of the human dihydrolipoamide dehydrogenase gene. Proc. Natl. Acad. Sci. USA **89**: 10964–10968.
27. BOURGIGNON, J., D. MACHAREL, M. NEUBURGER & R. DOUCE. 1992. Isolation, characterization, and sequence analysis of a cDNA encoding L-protein, the dihydrolipoamide dehydrogenase component of the glycine cleavage system from pea-leaf mitochondria. Eur. J. Biochem. **204**: 865–873.
28. REED, L.J. & M.L. HACKERT. 1990. Structure-function relationships in dihydrolipoamide acyltransferases. J. Biol. Chem. **265**(16): 8971–8974.
29. HACKERT, M.L., R.M. OLIVER & L.J. REED. 1983. Evidence for a multiple random coupling mechanism in the α-ketoglutarate dehydrogenase multienzyme complex of *Escherichia coli*: a computer model analysis. Proc. Natl. Acad. Sci. USA **80**: 2226–2230.
30. KNAPP, J.E., D.T. MITCHELL, M.A. YAZDI, S.R. ERNST, L.J. REED & M.L. HACKERT. 1998. Crystal structure of the truncated cubic core component of the Escherichia coli 2-oxoglutarate dehydrogenase multienzyme complex. J. Mol. Biol. **280**: 655–658
31. RICAUD, P.M., M.J. HOWARD, E.L. ROBERTS, R.W. BROADHURST & R.N. PERHAM. 1996. Three-dimensional structure of the lipoyl domain from the dihydrolipoyl succinyltransferase component of the 2-oxoglutarate dehydrogenase multienzyme complex of Escherichia coli. J. Mol. Biol. **264**: 179–190.
32. MCCARTNEY, R.G., J.E. RICE, S.J. SANDERSON, V. BUNIK, H. LINDSAY & J.G. LINDSAY. 1998. Subunit interactions in the mammalian α-ketoglutarate dehydrogenase complex. Evidence for direct association of the α-ketoglutarate dehydrogenase and dihydrolipoamide dehydrogenase components. J. Biol. Chem. **273**: 24158–24164.
33. LIU, T.C., Y.S. HONG, L.G. KOROTCHKINA, N.N. VETTAKKORUMANKANKAV & M.S. PATEL. 1999. Site-directed mutagenesis of dihydrolipoamide dehydrogenase: role of lysine-54 and glutamate-192 in stabilizing the thiolate-FAD intermediate. Protein Expression Purif. **16**: 27–39.
34. RADDATZ, G. & H. BISSWANGER. 1997. Receptor site and sterospecificity of dihydrolipoamide dehydrogenase for R- and S-lipoamide: a molecular modelling study. J. Biotech. **58**: 89–100
35. XU, D.P. & W.W. WELLS. 1996. α-Lipoic acid dependent regeneration of ascorbic acid from dehydroascorbic acid in rat liver mitochondria. J. Bioenerg. Biomembr. **28**: 77–85.
36. OLSSON, J.M., L. XIA, L.C. ERIKSSON & M. BJORNSTEDT. 1999. Ubiquinone is reduced by lipoamide dehydrogenase and this reaction is potently stimulated by zinc. FEBS Lett. **448**: 190–192.
37. KEE, K., L. NIU & E. HENDERSON. 1998. A tetrahymena thermophila G4-DNA binding protein with dihydrolipoamide dehydrogenase activity. Biochemistry **37**: 4224–4334.

38. TOYODA, T., K. SUZUKI, T. SEKIGUCHI, L.J. REED & TAKENAKA. 1998. Crystal structure of eucaryotic E3, lipoamide dehydrogenase from yeast. J. Biochem. **123:** 668–674.
39. NICHOLS, B.J. & R.M. DENTON. 1995. Towards the molecular basis for the regulation of mitochondrial dehydrogenases by calcium ions. Mol. Cell. Biochem. Aug.-Sep.: 149–150; 203–212.
40. MORENO-SANCHEZ, R., S. RODRIGUEZ-ENRIQUEZ, A. CUELLAR & N. CORONA. 1995. Modulation of 2-oxoglutarate dehydrogenase and oxidative phosphorylation by Ca^{2+} in pancreas and adrenal cortex mitochondria. Arch. Biochem. Biophys. **319:** 432–444.
41. HANSFORD, R.G. & D. ZOROV. 1998. Role of mitochondrial calcium transport in the control of substrate oxidation. Mol. Cell. Biochem. **184:** 359–369.
42. HUMPHRIES, K.M., Y. YOO & L.I. SZEWDA. 1998. Inhibition of NADH-linked mitochondrial respiration by 4-hydroxy-2-nonenal. Biochemistry **37:** 552–557.
43. HUMPHRIES, K.M. & L.I. SWEDA. 1998. Selective inactivation of α-ketoglutarate dehydrogenase and pyruvate dehydrogenase: reaction of lipoic acid with 4-hydroxy-2-nonenal. Biochemistry **37:** 15835–15841.
44. ANDERSSON, U., B. LEIGHTON, M.E. YOUNG, E. BLOMSTRAND & E.A. NEWSHOLME. 1998. Inactivation of aconitase and oxoglutarate dehydrogenase in skeletal muscle in vitro by superoxide anions and/or nitric oxide. Biochem. Biophys. Res. Commun. **249:** 512–516.
45. PARK. L.C., H. ZHANG, K.F. SHEU, N.Y. CALINGASAN, B.S. KRISTAL, J.G. LINDSAY & G.E. GIBSON. 1999. Metabolic impairment induces oxidative stress, compromises inflammatory responses, and inactivates a key mitochondrial enzyme in microglia. J. Neurochem. **72:** 1948–1958.
46. CHINOPOULOS, C., L. TRETTER & V. ADAM-VIZI. 1999. Depolarization of in situ mitochondria due to hydrogen peroxide-induced oxidative stress in nerve terminals: inhibition of α-ketoglutarate dehydrogenase. J. Neurochem. **73:** 220–228.
47. LUCAS, D.T. & L.I. SZWEDA. 1999. Declines in mitochondrial respiration during cardiac reperfusion: age-dependent inactivation of α-ketoglutarate dehydrogenase. Proc. Natl. Acad. Sci. USA **96:** 6689–6693.
48. PARK, L.C.H., G.E. GIBSON, V. BUNIK & A.J.L. COOPER. 1999. Inhibition of select mitochondrial enzymes in PC12 cells exposed to S-(1,1,2,2-tetrafluoroethyl)-L-Cysteine. J. Neurochem. In press.
49. GUTTIEREZ-CORREA, J. & A.O. STOPPANI. 1999. Inactivation of myocardial dihydrolipoamide dehydrogenase by myeloperoxidase systems: effects of halides, nitrite and thiol compounds. Free Radical Res. **30:** 105–117.
50. CAETANO, N.N., A.P. CAMPELLO, E.G. CARNIERI, M.L. KLUPPEL & M.B. OLIVIERA. 1997. Effect of methotrexate (MTX) on $NAD(P)^+$ dehydrogenases of HeLa cells: malic enzyme, 2-oxoglutarate and isocitrate dehydrogenases. Cell Biochemistry & Funct. **15:** 259–264.
51. BLOMSTRAND, E., G. RADEGRAN & B. SALTIN. 1997. Maximum rate of oxygen uptake by human skeletal muscle in relation to maximal activities of enzymes in the Krebs cycle. J. Physiol. **501:** 455–460.
52. O'DONNEL, J.M., C. DOUMEN, K.F. LANOUE, L.T. WHITE, X. YU, N.M. ALPERT & E.D. LEWANDOWSKI. 1998. Dehydrogenase regulation of metabolite oxidation and efflux from mitochondria in intact hearts. Am. J. Physiol. **274:** H467–476.
53. BLASS, J.P. Metabolic alterations common to neural and non-neural cells in Alzheimer's disease. Hippocampus **13:** 45–54.
54. GIBSON, G.E., K-F.R. SHEU, J.P. BLASS, A. BAKER, K.C. CARLSON, B. HARDING & P. PERINO. 1988. Reduced activities of thiamine-dependent enzymes in the brains and peripheral tissues of patients with Alzheimer's disease. Arch. Neurol. **45:** 841–845.
55. TERWEL, D., J. BOTHMER, E. WOLF, F. MENG & J. JOLLES. 1998. Affected enzyme activities in Alzheimer's disease are sensitive to antemortem hypoxia. J. Neurol. Sci. **161:** 47–56.

56. MILNER, T.A., C. AOKI, K.F.R. SHEU, J.P. BLASS & V.M. PICKEL. 1987. Light microscopic immunocytochemical localization of pyruvate dehydrogenase in rat brain: topographical distribution and relation to cholinergic and catecholaminergic nuclei. J. Neurosci. **7:** 3171–3190.
57. AOKI, C., T.A. MILNER, K.F.R. SHEU, J.P. BLASS & V.M. PICKEL. 1987. Regional distribution of astrocytes with intense immunoreactivity for glutamate dehydrogenase in rat brain: implications for neuron-glia interactions in glutamate transmission. J. Neurosci. **7:** 2214–2231.
58. CALINGASAN. N., H. BAKER, K.F. SHEU & G.E. GIBSON. 1994. Distribution of the α-ketoglutarate dehydrogenase complex in rat brain. J. Comp. Neurol. **346:** 461–479.
59. KOIKE, K., H. ISHIBASHI & M. KOIKE. 1998. Immunoreactivity of porcine heart lipoamide acetyl- and succinyl-transferases (PDC-E2, OGDC-E2) with primary biliary cirrhosis sera: characterization of the autoantigenic region and effects of enzymateic delipoylation and relipoylation. Hepatology **27:** 1467–1474.
60. BLASS, J.P., G.E. GIBSON & S. HOYER. 1997. Metabolism of the aging brain. *In* Advances in Cell Aging and Gerontology, Vol 2: The Aging Brain. M.P. Mattson, J.W. Gedes, P. Tamiras & E.E. Bittar, Eds.: 109–128. JAI Press. London.
61. HONG, Y.S., D.S. KERR, W.J. CRAIGEN, J. TAN, Y. PAN, M. LUSK & M.S. PATEL. 1996. Identification of two mutations in a compound heterozygous child with dihydrolioamide dehydrogenase deficiency. Hum. Mol. Genet. **5:** 1925–1930.
62. ROBINSON, B.H., K. CHUN, N. MACKAY, G. OTULAKOWSKI, R. PETROVA-BENEDICT & H. WILLARD. 1989. Isolated and combined deficiencies of the alpha-keto acid dehydrogenase complexes. Ann. N.Y. Acad. Sci. **573:** 337–346.
63. MUNNICH, A., J.M. SAUDUBRAY, J. TAYLOR, C. CHARPENTIER, C. MARSAC, F. ROCCHICCIOLI, O. AMEDEE-MANESME, F. X. COUDE, J. FREZAL & B.H. ROBINSON. 1982. Congenital lactic acidosis,alpha-ketoglutaric aciduria and variant form of maple syrup urine disease due to a single enzyme defect: dihydrolipoyl dehydrogenase deficiency. Acta Paediatr. Scand. **71:** 167–71.
64. SHAAG, A., A. SAADA, I. BERGER, H. MANDEL, A. FEIGENBAUM & O.N. ELPELEG. 1999. Molecular basis of lipoamide dehydrogenase deficiency in Ashkenazi Jews. Am. J. Med. Genet . **82:** 177–182.
65. BOURGERON, R.P., B. PARFAIT, D. CHRETIEN, A. MUNNICH & A. ROTIG. 1997. Inborn errors of the Krebs cycle: a group of unusual mitochondrial diseases in human. Biochim. Biophys. Acta. **1361:** 185–197.
66. JOHNSON, M.T., H.S. YANG, T. MAGNUSON & M.S. PATEL. 1997. Targeted disruption of the murine dihydrolipoamide dehydrogenase gene (Dld) results in perigastrulation lethality. Proc. Nat. Acad. Sci. USA **94:** 14512–14517.
67. COLLOMBET, J.M., P. DIVRY, M. MATHIEU & P. GUIBAUD. 1993. 2-Ketoglutarate dehydrogenase deficiency, a rare cause of primary hyperlactataemia: report of a new case. J. Inher. Metab. Dis. **16:** 821–825.
68. BONNEFONT, J.P., D. CHRETIEN, P. RUSTIN, B. ROBINSON, A. VASSAULT, J. AUPETIT, C. CHARPENTIER, D. RABIER, J.M. SAUDUBRAY & A. MUNNICH. 1992. Alpha-ketoglutarate dehydrogenase deficiency presenting as congenital lactic acidosis. J. Pediat. **121:** 255–228.
69. WONG, A., J. YANG, P. CAVADINI, C. GELLERA, B. LONNERDAL, F. TARONI & G. CORTOPASSI. 1999. The Friedreich's ataxia mutation confers cellular sensitivity to oxidant stress which is rescued by chelators of iron and calcium and inhibitors of apoptosis. Hum. Mol. Genet. **8**(3): 425–30.
70. BLASS, J.P., R.A.P. KARK, N. MENON & S.H. HARRIS. 1976. Decreased activities of the pyruvate and ketoglutarate dehydrogenase complexes in fibroblasts from five patients with Fredreich's ataxia. N. Engl. J. Med. **295:** 62–66.

71. ROTIG, A., P. DE LONLAY, D. CHRETIEN, F. FOURY, M. KOENIG, D. SIDI, A. MUNNICH & P. RUSTIN. 1997. Aconitase and mitochondrial iron-sulphur protein deficiency in Friedreich ataxia. Nature Genet. **17**(2): 215–217.
72. MASTROGIACOMO, F., J. LAMARCHE, S. DOZIC, G. LINDSAY, L. BETTENDORF, Y. ROBITAILLE, L. SCHUT & S.J. KISH. 1996. Immunoreactive levels of α-ketoglutarate dehydrogenase subunits in Friedreich's ataxia and spinocerebellar ataxia type I. Neurodegeneration **5**: 27–33.
73. PANDOLFO, M. & L. MONTERMINI. 1998. Molecular genetics of the hereditary ataxias. Adv. Genet. **38**: 31–68.
74. COOPER, A.J.L., F.R. SHEU, J.R. BURKE, W.J. STRITTMATTER, V. GENTILE, G. PELUSO & J.P. BLASS. 1999. Pathogenesis of inclusion bodies in $(CAG)_n/Q_n$ -expansion diseases with special reference to the role of tissue transglutaminase and to SELECTIVE VULNERABILITY. J. NEUROCHEM. **72**: 889–899.
75. COOPER, A.J.L., K.-F.R. SHEU, J.R. BURKE, O. ONODERA, W.J. STRITTMATTER, A.D. ROSES & J.P. BLASS. 1997. Glutathione-S-transferase constructs containing medium to large polyglutamine inserts are substrates of guinea pig liver transglutaminase: does transglutaminase play a role in the pathogenesis of expanded CAG/Poly Q neurodegenerative diseases? Proc. Natl. Acad. Sci. USA **94**: 12604–12609.
76. SHEU, R.K.-F., J.P. BLASS, J.M. CEDARBAUM, Y.T. KIM, B.K. HARDING & J. DECICCO. 1988. Mitochondrial enzymes in hereditary ataxias. Metab. Brain Dis. **3**: 151–160.
77. SORBI, S., S. PIACENTINI, C. FANI, S. TONINI, P. MARINI & L. AMADUCCI. 1989. Abnormalities of mitochondrial enzymes in hereditary ataxias. Acta Neurol. Scand. **80**: 103–110.
78. SCHAPIRA, A.H., M. GU, J.W. TAANMAN, S.J. TABRIZI, T. SEATON, M. CLEETER & J.M. COOPER. 1998. Mitochondria in the etiology and pathogenesis of Parkinson's disease. Ann. Neurol. **44**(Suppl. 1): S89–98.
79. MIZUNO, Y., S. MATUDA & H. YOSHINO. 1994. An immunohistochemical study on α-ketoglutarate dehydrogenase complex in Parkinson's disease. Ann. Neurol. **35**: 204–210.
80. KOBAYASHI, T., H. MATSUMINE, S. MATUDA & Y. MISUNO. 1998. Association between the gene encoding the E2 subunit of α-ketoglutarate dehydrogenase complex and Parkinson's disease. Ann. Neurol. **43**: 120–123.
81. NOLAN, K.A., M.M. LINO, A.W. SELIGMAN & J.P. BLASS. 1998. Absence of vascular dementia in an autopsy series from a dementia clinic. J. Am. Geriatr. Soc. **46**: 597–604.
82. GEDDES, J.W., T.L. TEKIRIAN, N.S. SOULTANIAN, J.W. ASHFORD, D.G. DAVIS & W.R. MARKESBERY. 1997. Comparison of neuropathologic criteria for the diagnosis of Alzheimer's disease. Neurobiol. Aging **18**(Suppl.): S99–105.
83. MARKESBERY, W.R. 1997. Oxidative stress hypothesis in Alzheimer's disease. Free Radical Biol. Med. **123**: 134–147.
84. MASTROGIACOMA, F., J.G. LINDSAY, L. BETTENDORFF, J. RICE & S.J. KISH. 1996. Brain protein and α-ketoglutarate dehydrogenase complex activity in Alzheimer's disease. Ann. Neurol. **39**: 592–598.
85. BUTTERWORTH, R.F. & A.M. BESNARD. 1990. Thiamine-dependent enzyme changes in temporal cortex of patients with Alzheimer's disease. Metab. Brain Dis. **5**: 179–184.
86. KO, L., K-F.R. SHEU & J.P. BLASS. 1993. Chemical neuroanatomy of energy metabolism: immunohistochemical studies in relation to selective vulnerability. J. Neurochem. **61**: S70.

87. BLASS, J.P., K.F. SHEU, S. PIACENTINI & S. SORBI. 1997. Inherent abnormalities in oxidative metabolism n Alzheimer's disease: interaction with vascular abnormalities. Ann. N.Y. Acad. Sci. **826:** 382–385.
88. GIBSON, G.E., H. ZHANG, K.F. SHEU, N. BOGDANOVICH, J.G. LINDSAY, L. LANNFELT, M. VESTLING & R.F. COWBURN. 1998. Alpha-ketoglutarate dehydrogenase in Alzheimer brains bearing the APP670/671 mutation. Ann. Neurol. **44:** 676–681.
89. GIBSON, G.E., V. HAROUTUNIAN, L.C.H. PARK, H. ZHANG, R. MOHS R.K-F. SHEU & J.P. BLASS. 1999. Reductions in a key mitochondrial enzyme in brains from Alzheimer's Disease patients correlate with a clinical dementia rating. J. Neurochem. **73:** S23; and submitted.
90. RAMASSAMY, C., D. AVERILL, U. BEFFERT, S. BASTIANETTO, L. THEROUX, S. LUSSIER-CACAN, J. COHN, Y. CHRISTEN, J. DAVIGNON, R. QUIRION & J. POIRIER. 1999. Oxidative damage and protection by antioxidants in the frontal cortex of Alzheimer's Disease is related to apolipoprotein E genotype. Free Radical Biol. Med. **27:** 544–553.
91. IWANGOFF, P., R. ARMBRUSTER, A. ENZ & W. MEIER-RUGE. 1980. Glycolytic enzymes from human autoptic brain cortex: normal aged and demented cases. Mech. Ageing Deve.l **14:** 203–209.
92. KISH, S.J., I. LOPES-CENDES, M. GUTTMAN, Y. FURUKAWA, M. PANDOLFO, G.A. ROULEAU, B.M. ROSS, M. NANCE, L. SCHUT, L. ANG & L. DISTEFANO. 1998. Brain glyceraldehyde-3-phosphate dehydrogenase activity in human trinucleotide repeat disorders. Arch. Neurol. **55:** 1299–1304.
93. BROWNE, S.E., A.C. BOWLING, U. MACGARVEY, M.J. BAIK, S.C. BERGER, M.M.K. MUQIT, E.D. BIRD & M.F. BEAL. 1997. Oxidative damage and metabolic dysfunction in Huntington's Disease: selective vulnerability of the basal ganaglia. Ann. Neurol. **41:** 646–653.
94. PATEL, M.S., C.A. JOHNSON, R. RAJAN & O.E. OWEN. 1975. The metabolism of ketone bodies in developing human brain: development of ketone-body–utilizing enzymes and ketone bodies as precursors for lipid synthesis. J. Neurochem. **25:** 905–908.
95. JANETZKY, B., S. HAUCK, M.B. YOUDIM, P. RIEDERER, K. JELLINGER, F. PANTUCEK, R. ZOCHLING, K.W. BOISSL & H. REICHMAN. 1994. Unaltered aconitase activity but decreased complex I activity in substantia nigra pars compacta of patients with Parkinson's disease. Neurosci. Lett. **169:** 126–128.
96. SORBI, S., E.D. BIRD & J.P. BLASS. 1983. Decreased pyruvate dehydrogenase complex activity in Huntington and Alzheimer brain. Ann. Neurol. **13:** 72–78.
97. GROSSMAN, A., R.N. ROSENBERG & L. WARMOTH. 1987. Glutamate and malate dehydrogenase activities in Joseph disease and olivopontocerebellar atrophy. Neurology **37:** 106–111.

Oxidative Stress and a Key Metabolic Enzyme in Alzheimer Brains, Cultured Cells, and an Animal Model of Chronic Oxidative Deficits

GARY E. GIBSON,[a,c] LARRY C. H. PARK,[a] HUI ZHANG,[a] SANDRO SORBI,[b] AND NOEL Y. CALINGASAN[a]

[a]*Department of Neurology and Neuroscience, Weill Medical College of Cornell University at Burke Medical Research Institute, 785 Mamaroneck Avenue, White Plains, New York 10605, USA*

[b]*Department of Neurology, University of Florence, Florence, Italy*

> ABSTRACT: Oxidative stress and diminished metabolism occur in several neurodegenerative disorders. Brains from Alzheimer's disease (AD) patients exhibit several indicators of oxidative stress and have reduced activities of the α-ketoglutarate dehydrogenase complex (KGDHC), a key mitochondrial enzyme. Whether these abnormalities are secondary to neurodegenerative processes or are inherent properties of the cells cannot be determined in autopsy brain. Studies in cultured fibroblasts suggest that AD-related differences in oxidative stress and KGDHC reflect inherent properties of AD cells. KGDHC is sensitive to oxidative stress whether the enzyme is studied in cells, in purified mitochondria, or as an isolated protein. Reductions of brain KGDHC in living rodents lead to oxidative stress and selective cell death. The results suggest that KGDHC participates in a deleterious cascade of events related to oxidative stress that are critical in selective neuronal loss in neurodegenerative diseases.

Deficits in oxidative metabolism and increases in oxidative stress may be central in the pathophysiology of Alzheimer's disease (AD). For this to be true, the oxidative deficits must occur, and they must be plausibly linked to important characteristics of AD including the known pathology, the established genetic causes, and the clinical outcome. In addition, a mechanistic explanation of the disease must address why the disease is age related. To test the involvement of oxidative stress and impaired metabolism necessitates multiple approaches including studies of brains from Alzheimer patients, cultured cells from AD patients, and animal models.

OVERVIEW OF ALZHEIMER'S DISEASE

A mechanistic comprehension of AD requires an understanding of the formation of senile plaques and neurofibrillary tangles, the pathological hallmarks of AD.

---

[c]Corresponding author. Phone: 914-597-2291; fax: 914-597-2757.
e-mail: ggibson@med.cornell.edu

Plaques and tangles were first described by Alois Alzheimer in 1907.[1] The plaques from AD brains contain primarily amyloid β-peptide, but include other proteins such as cholinesterases, hemeoxygenase, and intercellular adhesion molecules. Tangles are primarily composed of hyperphosphorylated tau proteins. These structures are not restricted to AD. This lack of uniqueness to AD suggests that additional factors are critical in the production of clinical symptoms of AD.

Both genetics and aging are important variables that influence the course of AD. When Alzheimer first described the disease,[1] it was regarded as a rare disorder. However, as the population of the world has aged, the disease is commonly recognized, since it is an age-related disorder. Estimates suggest that 10% of the people over 65 and 50% of those over 85 years old have the disease.[2–4] The patient that Alzheimer described was relatively young by today's standards. Individuals who get AD at a young age often have a genetic form of the disease (i.e., familial AD). An important advance in understanding AD has been the identification of families that have either a genetic predisposition toward the development of AD, such as apoE4 genotype, or a genetic abnormality that leads directly to AD. Altogether, clearly familial AD may account for about 10% of all AD patients. The remainder of the patients are referred to as sporadic or nonfamilial AD. Within the general population, the presence of genetic risk factors such as the presence of ε-4 allele of the apolipoprotein gene can also alter the occurrence of the disease. One of the genetic AD families that has been particularly well studied has a genetic mutation in the processing of amyloid precursor protein (APP) at the 670/671 site. APP is a transmembrane protein the function of which is unknown. The mutation at APP 670/671 leads to the overproduction of amyloid-β-peptide. This mutation appears to have 100% penetrance (i.e., all individuals who have the abnormality eventually get the disease). Studies related to this family provide the best evidence that amyloid-β-peptide can initiate AD. Another group of genetic mutations that lead to AD are the presenilin-1 (PS-1) and presenilin-2 mutations. These mutations may account for about 25–50% of all familial patients. About 43 different mutations have been reported in PS-1. Considerable evidence suggests that PS-1, which has seven transmembrane spanning regions, is localized to the endoplasmic reticulum. The pathophysiological mechanisms by which these mutations lead to AD are still controversial (see review in Ref. 3).

Many laboratories are now investigating the link between the pathological markers, the genetic mutations, and the clinical manifestations of AD. AD patients have severe memory deficits and difficulties with cognitive processing, as well as a variety of behavioral abnormalities including agitation, wandering, aggression, and depression. The following discussion supports the hypothesis that abnormal oxidative processes are a critical link in the cascade of events that lead to AD.

OVERVIEW OF BRAIN OXIDATIVE METABOLISM

The brain depends heavily upon glucose and oxygen metabolism. Although the brain represents only 2% of the body mass, it uses 20% of the glucose and oxygen. FIGURE 1 shows a simplified version of brain metabolism that highlights those features relevant to this discussion. Glycolysis converts the glucose to pyruvate, which is transported into the mitochondria. The two enzyme complexes that are shown, the

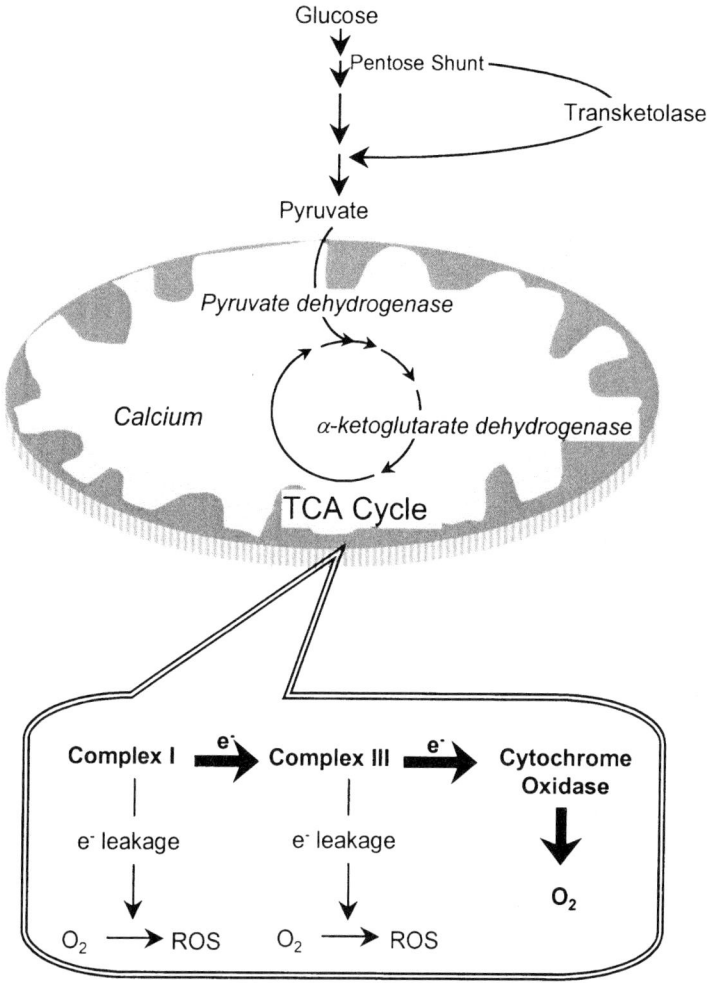

FIGURE 1. An overview of brain metabolism including the production of reactive oxygen species. Complex I and complex III show the sites in the electron transport chain that lead to electron leak.

pyruvate dehydrogenase complex (PDHC) and the α-ketoglutarate dehydrogenase complex (KGDHC), are highly regulated thiamine-dependent enzyme complexes. Each can become the rate-limiting step of the tricarboxylate cycle and is regulated by calcium. KGDHC and PDHC consist of three components, E1, E2, and E3. Each complex has unique E1 and E2 components, while the E3 component is similar for both. The reducing equivalents (i.e., NADH) that are generated in the tricarboxylic acid cycle are transferred down the electron transport chain and eventually react with molecular oxygen. Some electrons "leak" and are an important source of free radi-

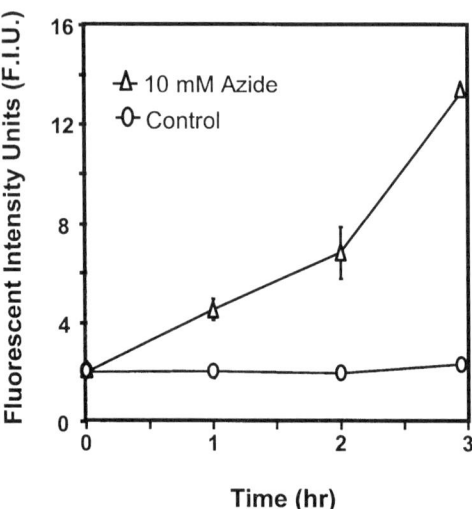

FIGURE 2. Increased ROS production in response to inhibition of oxidative metabolism in purified microglia.[5] Microglia were treated with azide to inhibit complex IV. The increase on the vertical axis reflects increased production of ROS.

cals, or reactive oxygen species (ROS). This leakage occurs at complex I and complex III of the electron transport chain and can induce oxidative stress. These free radicals may be signaling molecules, but, in excess, they can also lead to extensive damage. Blocking cytochrome oxidase (complex IV) can lead to increased production of ROS. For example, addition of azide, an inhibitor of cytochrome oxidase, to microglia elevates ROS production (FIG. 2).[5] Thus, disrupting the coupling of electron flow to oxygen increases the production of ROS and oxidative stress.

The brain has multiple sources of free radicals (or ROS) in addition to those generated by the coupling to the electron transport chain. Nitric oxide synthase (NOS) produces the free radical–containing NO• by converting L-arginine to L-citrulline. NO• can interact with proteins by nitrosylating sulfhydryl groups. The combination of NO• with superoxide radicals leads to the formation of peroxynitrite. This highly toxic oxidant readily nitrates tyrosine residues on proteins, which frequently inactivates the enzyme. ROS can also interact with unsaturated fatty acids to cause lipid peroxidation and the eventual production of reactive aldehydes. Two common aldehydes that have been implicated in brain pathology are 4-hydroxynonenal and acrolein. These reactive aldehydes react directly with the proteins, and often inactivate them.

COMPROMISED METABOLISM AND OXIDATIVE STRESS OCCUR IN BRAINS FROM ALZHEIMER'S DISEASE PATIENTS

As has been demonstrated by several methods over the course of several decades, glucose metabolism is reduced in AD brains. As each new technique has been developed for studying brain metabolism, AD patients have generally been among the earliest patients studied. A recent study used the latest PET techniques to demonstrate that glucose utilization in the AD brain is reduced by about 50%.[6] This is true even after the measures of glucose metabolism in AD brain are corrected for the partial volume-averaging effects due to the shrinkage of the AD brain. Some reports indicate that the changes in glucose metabolism precede rather than follow the clinical manifestations of AD.[7,8]

Several enzymes of energy metabolism are diminished in autopsy brains. Reductions occur in the enzymes of the tricarboxylic acid cycle and the electron transport chain. For example, reductions in cytochrome c have been reported by several groups, but the finding remains controversial.[9] Since the reduction occurs primarily in areas that are degenerating, it is difficult to know whether or not the change is secondary to the activity of that part of the brain. Although this chapter will focus on KGDHC, much of the rationale that will be discussed about whether KGDHC is important in AD will also apply to determining whether the deficit in cytochrome oxidase is important.

Striking reductions in KGDHC activity in autopsy brains from AD patients have been reported by several groups. The initial studies showed dramatic reductions in frontal cortex (−87%), caudate (−75%), occipital cortex (−92%), and mid-temporal cortex (−100%).[10] These changes in activity occur even though the activities are measured in the presence of saturating concentrations of thiamine pyrophosphate, the thiamine-containing cofactor. The activities of other mitochondrial enzymes, such as glutamate dehydrogenase, are not reduced. KGDHC activity decreases strikingly in both pathologically affected and unaffected brain regions. This result suggests that the reductions in KGDHC do not merely reflect diminished metabolism or neurodegeneration, but are an inherent property of AD cells. The reduction in KGDHC is unlikely to be related to postmortem effects. The decline in KGDHC activities in controls postmortem is small compared to the reduction with AD. Control and AD autopsy samples have been controlled for the postmortem interval. Accurate controls for agonal effects are difficult, but no differences between AD and controls were apparent. In spite of these difficulties, several groups have independently shown diminished KGDHC activities in AD brain,[11–14] and importantly no contravening reports exist. Our recent studies show that KGDHC activities are even reduced about 55% in the brains of patients with the APP670/671 mutation.[15] The changes in KGDHC activities are as replicable a finding as the changes in choline acetyltransferase, tangles, or plaques in AD. These results suggest that KGDHC activities may have a clinically relevant role in the pathophysiology of AD.

The response of the individual protein components of KGDHC to AD varies between "sporadic" and familial AD. Extensive Western blot analyses of brains from "sporadic" AD patients demonstrate that the protein levels of the individual components of the complex do not decline nearly as much as the activity.[12] From a mechanistic point of view, the results show that the enzyme protein is present but is inactive. This suggests that the protein has been damaged so that it does not associate into a functional complex. Activation of this preexisting protein may be clinically useful. On the other hand, the individual protein components of KGDHC are altered in the brains from patients bearing the APP670/671 mutation. These brains reveal dramatic reductions in E1k and E2k, but no change in E3.[15] In both the sporadic brains and the brains bearing the APP670/671 mutation KGDHC activities are reduced, but the mechanism of inactivation may be different.

Overwhelming evidence demonstrates that oxidative stress occurs in the brains of AD patients. Increased nitrotyrosine, ferritin, reactive iron, hemeoxygenase-1 (HO-1), and activated microglia are all markers of oxidative stress that occur in AD brains. Our recent studies have concentrated on measuring the reactive aldehydes, 4-hydroxynonenal and acrolein. Acrolein colocalizes with paired helical filament-1 (PHF-1) in tangles. Most tangles are stained with acrolein. In addition, all hydroxynonenal-immunoreactive structures stain with acrolein.[16] Thus, oxidative stress occurs in AD brain. Whether this oxidative stress results from or causes diminished metabolism in AD has not been established.

COMPROMISED METABOLISM AND OXIDATIVE STRESS OCCUR IN PERIPHERAL CELLS FROM ALZHEIMER'S DISEASE SUBJECTS

Although abnormal glucose oxidation has been known to accompany AD for a long time, whether this change is a cause or an aftermath of the neuronal degeneration is still unknown. This is an example of the classical chicken/egg argument that plagues neuropathology. Even under optimal conditions, when the compromises due to agonal and postmortem changes are minimized, analyzing autopsy material to assess whether the defect or the disease comes first is difficult. Dynamic processes such as calcium or ROS formation cannot be studied in postmortem materials. Investigators have explored the use of peripheral tissues to address these issues. In these tissues, changes cannot be secondary to neurodegeneration. Measures of fluids (e.g., CSF and serum) and peripheral cells (e.g., platelets, lymphocytes, and red cells) have been used. Another approach is to obtain cells from AD patients and grow them in culture. A common example is to culture fibroblasts from a skin biopsy. Cells such as lymphocytes from patients can be transformed and then studied in culture. Alternatively, mitochondrial DNA from AD patients can be transferred into neuroblastoma cells, which are then studied in culture. All of these approaches suggest that mitochondria from AD patients are different from mitochondria from controls.

Cultured fibroblasts have been used extensively in the past for determining underlying mechanisms of neurological diseases in children. For example, the use of fibroblasts was vital in determining the molecular basis of Lesch-Nyan disease (purine metabolism), Tay Sachs disease (hexosaminidase), diseases of energy metabolism (pyruvate dehydrogenase), and Refsum's disease (phytanic acid). There are numerous advantages of using fibroblasts. Any drugs or diets that the patient may be on are diluted millions of times, since fibroblasts are passed numerous times in culture. These cells are relatively easy to obtain and maintain under rigid conditions. Comparisons of multiple members of many families are possible. The effects of a genetic mutation can be measured in cells with the patient's own genetic background. This is important because the effects of genetic mutations may be altered by the genetic background of the cells.[17] Finally, the growth patterns of fibroblasts from AD and control subjects are similar. Numerous signaling abnormalities have been reported between control and AD fibroblasts. Changes include alterations in both the mitochondrial and calcium compartments, cyclic AMP regulation, G-proteins, and glucose oxidation. These changes have been extensively reviewed.[18–20]

KGDHC has been studied in fibroblasts from AD patients. Early studies reported either small, but nonsignificant, reductions of KGDHC in AD fibroblasts compared

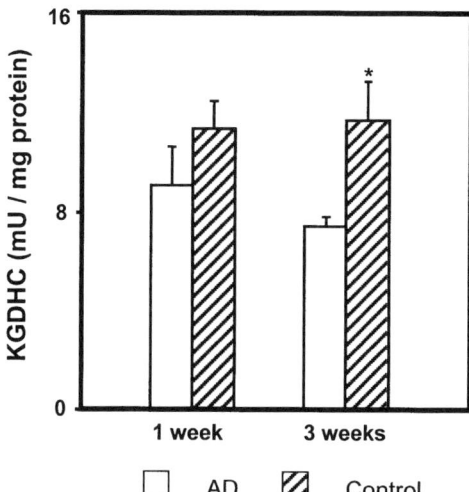

FIGURE 3. Diminished KGDHC activities in response to oxidative stress in fibroblasts. Cells bearing the PS-1 M146L mutation were exposed to a mild stress (i.e., the media was not changed for several weeks after the cells were seeded), and their KGDHC activities were monitored. Values at three weeks are means ± SEM of fibroblasts from three controls subjects and four AD subjects bearing the PS-1 M146L mutation.

to controls;[11] or slightly larger and significant decreases.[21] The early studies predated the identification of any of the known genetic mutations that lead to AD. Identification of families has helped us to better understand the role of KGDHC in AD. Studies by Sorbi and his coworkers[22] in a large series of patients show that KGDHC activity (mU/mg protein) in sporadic AD (6.4 ± 0.8) patients is much less than in cells from patients bearing PS-1 mutation AD (12.2 ± 1.9) and from controls (11.1 ± 1.0). KGDHC activity in fibroblasts from patients bearing the APP670/671 mutation do differ from controls,[20] although they do exhibit alterations in amyloid β-peptide[23] and in particular calcium compartments.[20] The results suggest that KGDHC activity is altered in fibroblasts from sporadic AD patients, but not from familial AD subjects.

Several investigators have suggested that AD-inducing genetic mutations alter the ability of cells to handle stress. Cells bearing the PS-1 M146L mutation were exposed to a mild stress (i.e., the media was not changed for several weeks after the cells were seeded), and their KGDHC activities were monitored. As shown in FIGURE 3, activities were similar after one week, but by three weeks the activity in the AD cells was significantly less than controls. The cells bearing the PS-1 mutation have an altered ability to handle stress, and KGDHC is very sensitive to this stress.

This change in KGDHC in response to stress prompted us to examine directly the ability of the AD and normal fibroblasts to handle an oxidative insult. AD and control cells were exposed to H_2O_2, and ROS production was determined with the fluorescent probe dichlorofluoroscein. The AD cells had less of a response to H_2O_2 than controls. The interpretation of these findings is complicated. One possibility is that a greater production of ROS in the AD cells may have induced their antioxidant capacity. Considerable data suggest that prior exposure to an oxidant challenge protects cells against subsequent challenges. An alternative explanation is that the membranes in the AD cells are more reactive, so that the H_2O_2 does not penetrate to the compartment of the cells that contains the probe. Nevertheless, these results are consistent with the suggestion that abnormalities in ROS metabolism are an inherent property of AD cells and not just a reflection of dying cells in AD brains.[24]

ENHANCED OXIDATIVE STRESS LINKS ALZHEIMER'S DISEASE-RELATED GENETIC ABNORMALITIES TO METABOLISM

The changes in oxidative stress that are discussed above may link the genetic changes that lead to AD to altered metabolism and to subsequent neurodegeneration. The results raise the possibility that altered metabolism is required for neurodegeneration. Evidence suggests that oxidative metabolism may be particularly sensitive to factors such as amyloid-β-peptide that generate ROS (FIG. 4). Addition of 10 nM amyloid-β-peptide diminishes the reduction of the general marker of cellular dehydrogenase activity, MTT, by nearly 50%. The concentrations required are two to three orders of magnitude less than is commonly used to kill cells. Similar findings have been reported by others,[25] showing that metabolism is very sensitive to amyloid-β-peptide.

Presenilin mutations lead to oxidative stress, to abnormalities in calcium compartments, and to a diminished ability to handle a reduction in oxidative metabolism. Inhibitors of metabolism at concentrations that do not kill normal cells destroy cells

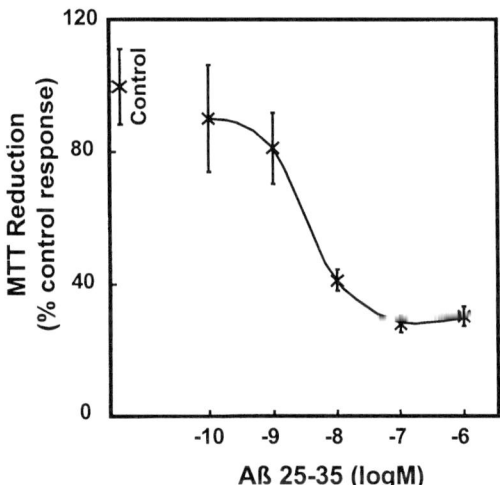

FIGURE 4. Diminished MTT [3-(4,5-dimethylthiazol-2-yl)-2,5-diphenyltetrazolium bromide] reduction in response to amyloid-β-peptide. MTT reduction is a general marker of cellular dehydrogenase activity. The concentrations required are two to three orders of magnitude less than is commonly used to kill cells.

that have been transfected with PS-1 mutant proteins.[26] 3-Nitroproprionic acid and malonate were employed as inhibitors because of their actions on succinate dehydrogenase, but at the concentrations that were used both are effective inhibitors of KGDHC (Park and Gibson, unpublished results). The enhanced sensitivity of cells bearing PS-1 mutations to metabolic inhibitors may be that enhanced ROS production in the PS-1 lines leads to inactivation of KGDHC.

KGDHC may provide a critical link between ROS, inhibition of metabolism, and cell death. In COS cells, KGDHC is the most sensitive of several mitochondrial enzymes to hyperoxia. Two days of hyperoxia does not affect glycerol phosphate dehydrogenase, but reduces NADH dehydrogenase (~50%), succinate dehydrogenase (~75%), and KGDHC (~100%).[27] The sensitivity of cellular KGDHC to ROS also occurs in cells derived from the brain, namely microglia. Inhibition of complex IV (cytochrome oxidase) by azide in microglia leads to increased ROS production (FIG. 2). Activation of microglia with lipopolysaccharide (LPS) increases NO·. Activated microglia that have been treated with an inhibitor of complex IV likely produce peroxynitrite. This combination leads to inactivation of KGDHC by about 50% (see FIG. 5).

KGDHC is also sensitive to oxidative stress in cell-free systems. When the reactive aldehyde 4-hydroxynonenal is added to isolated mitochondria, KGDHC is more sensitive than numerous other enzymes of energy metabolism including complexes I–IV of the respiratory chain, glutamate dehydrogenase, malate dehydrogenase, and pyruvate dehydrogenase.[28,29] In isolated liver mitochondria, KGDHC is inhibited by the neurotoxin 1-methyl-4-phenylpyridine (MPP^+) and the NO generator S-nitroso-N-acetyl-penicillamine (SNAP), and both effects are reversed by the radical quencher cysteine.[30] Peroxynitrite or the NO donor sodium nitroprusside inactivates

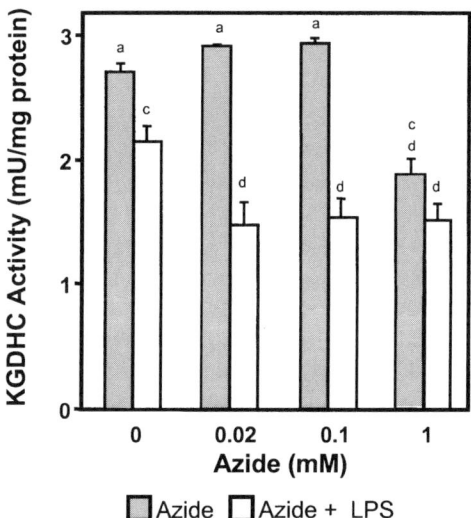

FIGURE 5. Reduced KGDHC activities following inhibition of metabolism in activated microglia. Purified microglia were treated with azide (as in FIG. 2) to block respiration and increase ROS, and with LPS, which increases NO·.

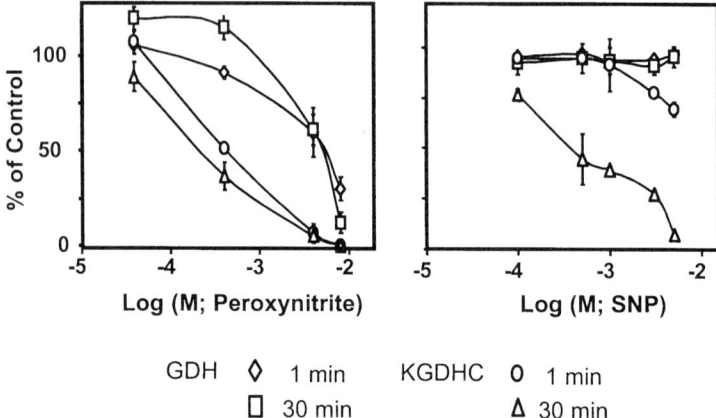

FIGURE 6. Inactivation of KGDHC by peroxynitrite and the NO donor sodium nitroprusside.[5] Microglia were incubated with the indicated concentrations of these compounds, and the activities of glutamate dehydrogenase and KGDHC were determined.

purified KGDHC (FIG. 6). With the isolated enzyme, the effects of these treatments on the individual protein components were determined. The results show that peroxynitrite increases protein nitration and reduces E1k and E2k immunoreactivity, whereas it does not affect E3 immunoreactivity. This peroxynitrite-induced change in KGDHC immunoreactivities is similar to that observed in brains from patients bearing the APP670/671 mutation. When sodium nitroprusside inactivates KGDHC, nitrotyrosine is not increased, and E1k or E2k immunoreactivities are unchanged. Sodium nitroprusside donates NO·, which has the ability to nitrosylate sulfhydryl groups such as cysteine residues in KGDHC.[5] Thus, the changes in KGDHC induced by sodium nitroprusside resemble the alterations in KGDHC in brains from sporadic AD patients. Together, the results demonstrate that KGDHC may provide a link between ROS and reduced metabolism in AD brain.

ACUTE IMPAIRMENT OF OXIDATION ALTERS THE PROTEINS THAT ARE ABNORMAL IN BRAINS FROM AD PATIENTS

Changes in the activities of key proteins of oxidative metabolism can lead to the classical hallmarks of AD—namely, the plaques and tangles. Plaques and tangles are not unique to AD. Analysis of their occurrence in other diseases suggests that interfering with oxidative metabolism can promote plaque and tangle formation. Wernicke-Korsakoff patients are a select group of alcoholics that develop severe memory disorders. These patients have reductions in thiamine-dependent enzymes in their brains, including KGDHC.[31] Increased tangles can be demonstrated by a variety of criteria in these brains. The tangles do not occur in alcoholics who do not have Wernicke-Korsadoff syndrome.[32] Plaque formation can also be stimulated by impairment of oxidative processes. For example, Bielschowsky staining revealed plaques and tangles in a 53-year-old Japanese woman with mitochondrial myopathy, encephalopathy, lactic acidosis, and stroke-like episodes.[33] Cell culture experiments support the suggestion that impairing oxidation can lead to plaque and tangle formation. For example, azide promotes accumulation of amyloidogenic fragments within cells.[34] Elevated intracellular calcium, which increases with compromised metabolism, promotes formation of amyloidogenic fragments.[35] Tau immunoreactivity in neurons increases following a challenge to energy metabolism or with glutamate excititoxicity,[36,37] and 4-hydroxynonenal can alter the phosphorylation of the tau proteins.[38] Immunoreactivity for tau proteins increases in cultured fibroblasts that are treated with the mitochondrial uncoupler, FCCP.[39] All of these data are consistent with the idea that mitochondrial deficits could lead to pathology reminiscent of AD.

ACUTE IMPAIRMENT OF OXIDATION IMPAIRS BRAIN FUNCTION IN A MANNER THAT RESEMBLES THE EFFECTS OF AGING, AND THE ALTERATIONS ARE EXAGGERATED BY AGING

Considerable data also suggest that acute impairment of oxidation alters brain function. The metabolic encephalopathies are a group of disorders in which impaired metabolism leads to diminished mental function. Disorders such as thiamine deficiency, hypoxia, and hypoglycemia are all metabolic encephalopathies.[40] Hypoxia is a prototype of these disorders. Even mild reductions in oxygen alter mental function. Among the events most sensitive to reduced oxygen are diminished short-term memory, impaired short-term memory, and loss of critical judgment. These effects resemble those associated with aging and are exaggerated by aging.[41] The alterations in mental function are at least, in part, related to reduced cholinergic function, which is a hallmark of AD. Cholinergic function is exquisitely sensitive to hypoxia. Behavioral tests in animals show that the effects of hypoxia on performance can be reversed with acetylcholinesterase inhibitors, indicating that the changes in cholinergic function are physiologically relevant. The effects of hypoxia on cholinergic function are exaggerated by normal aging, a risk factor for AD. Hypoxia also increases the release of dopamine. The same degree of hypoxia that diminishes acetylcholine (ACh) formation increases the extracellular concentrations of dopamine and glutamate, both of which can cause neuronal damage. Reducing oxygen availability reduces one neurotransmitter that diminishes processes such as memory, whereas the other neurotransmitters that cause cell damage are increased.[42]

The reductions in ACh release and synthesis with aging and hypoxia are coupled to altered calcium compartments. In both conditions, calcium uptake into the nerve terminal declines in parallel with the reductions in ACh release. The deficit in ACh and behavior in both aging and hypoxia can be ameliorated by stimulating calcium flux into the nerve terminal. These changes in calcium vary between cellular compartments (such as mitochondria and cytosol) and between different parts of the cell. If a differentiated PC12 cell is made hypoxic, cytosolic free calcium increases in the cell body and growth cone; but the calcium in these two cell regions is selectively sensitive to different calcium channel blockers.[43]

THIAMINE DEFICIENCY IS A MODEL OF CHRONIC INTERRUPTION OF OXIDATIVE METABOLISM

Thiamine deficiency (TD) is a useful model to study the effects of chronic interruption of oxidative metabolism on brain function. TD models those aspects of AD related to diminished KGDHC activities. As in AD, TD reduces KGDHC activities, and the decline is exaggerated in aged animals. As in the sporadic AD patients, the levels of the three components of KGDHC are not altered by TD. As in AD, TD induces a cholinergic lesion. Reductions in ACh turnover and reversal of TD-induced behavioral deficits by cholinesterase inhibitors both demonstrate that TD produces a cholinergic insufficiency.[44] Acetylcholinesterase inhibitors are the only drugs approved for treating AD. Furthermore, neurons are particularly sensitive to the metabolic insult produced by TD. As in AD, neurons are more sensitive to TD than other cell types. In culture, TD kills neurons, but not microglia, astrocytes, or endothelial cells.[45] Similarly, during AD neurons die, while other cell types do not.

THIAMINE DEFICIENCY IS A MODEL OF SELECTIVE NEURONAL VULNERABILITY

Selective neuronal vulnerability (i.e., death of select neuronal populations) is characteristic of all neurodegenerative disorders. The mechanisms underlying this selectivity are largely unknown. TD is a well-established model of selective cell death. The thiamine-dependent enzymes are heterogeneously distributed, but their distribution does not account for selective vulnerability. The thalamus, an area particularly vulnerable to TD, is neither particularly enriched, nor low in KGDHC. Nor are areas in which there is considerable overlap between choline acetyltransferase and KGDHC particularly vulnerable to TD. The reduction in KGDHC immunoreactivity by TD does not parallel selective vulnerability. This latter finding is of rather limited significance because the immunoreactivity does not parallel the activity measures. The results suggest that the reduction in KGDHC activity predisposes the brain to other insults.[46]

To determine the basis of selective vulnerability in TD, the earliest changes that reveal a selective change that can be related to cell death have been investigated. Just as in AD, different cell types showed a selective response; but all demonstrate oxidative stress (FIG. 7). The studies focused on the thalamus, because it is one of the selectively vulnerable regions. Neuronal loss in thalamus reliably begins in one small region, the submedial thalamic nucleus, and then spreads to include the whole thalamus. Increased hemeoxygenase-1 (HO-1) parallels a loss of neurons in this region. The increase in HO-1 is primarily in microglia and not in neurons.[47] HO-1 breaks down to heme to release iron and CO. On the subsequent days of TD, ferritin and redox active iron increase in microglia. NOS increases in both microglia and in endothelial cells. Evidence of oxidative stress also occurs in neurons. Both 4-hydroxynonenal and nitrotyrosine, a product of peroxynitrite-mediated protein modification, increase in neurons.[48] At later stages of TD, the permeability of the blood-brain barrier to IgG increases.[49] The subsequent stages in the neurodegenerative processes are accompanied by the appearance of neuritic clusters that are positive for

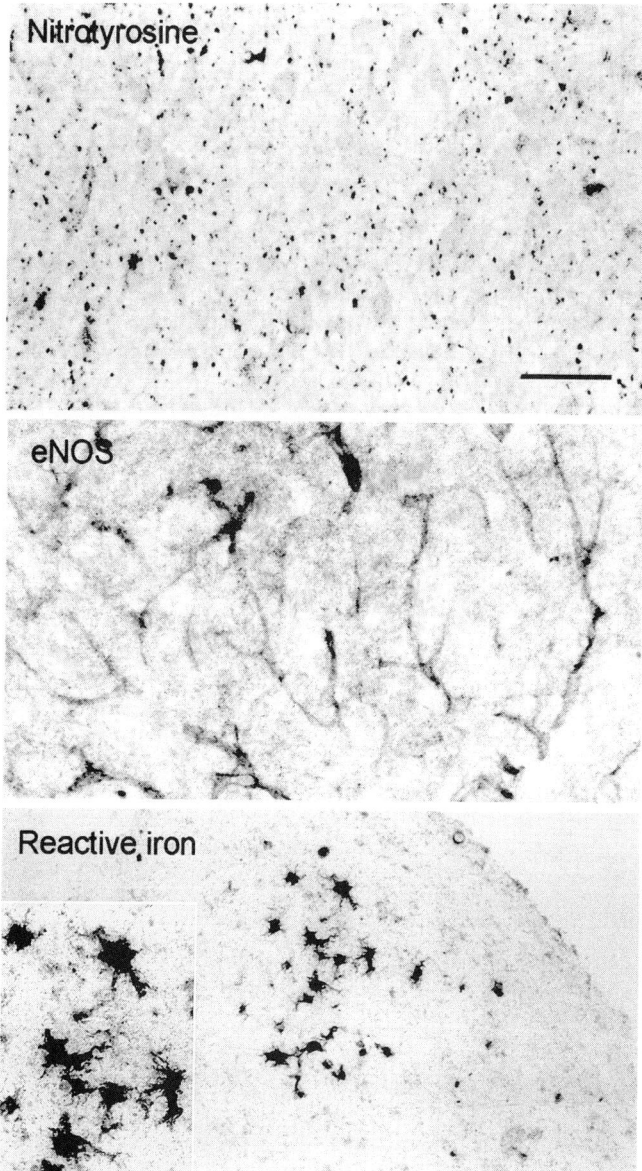

FIGURE 7. Increases in nitrotyrosine in neurons, eNOS immunoreactivities in endothelial cells, and reactive iron microglia within the thalamus following TD. Note the prominent nitrotyrosine immunostaining in axons and eNOS labeling in microvessel walls. High-magnification (*inset*) photomicrograph depicts the iron-laden microglia with plump cytoplasm and ramified processes. *Scale bar* = 25 μm (top and bottom), 100 μm (middle).

APP/APLP antibodies and morphologically resemble neuritic components of plaques in the AD brains.[50]

We postulate that the reduction in KGDHC activity predisposes vulnerable regions to other insults such as vascular changes and leads to changes in endothelial cells including induction of NOS. This alters the permeability of the BBB to allow entry of blood proteins, and affects microglia and other cells to release cytokines or deleterious compounds. The action of these compounds on metabolically compromised neurons leads to their death.

The measures of oxidative stress that occur as part of a cell death cascade in TD also occur in AD brain, and the TD model allows us to determine the mechanism underlying each of the steps. In the TD model, a temporal relation of these changes can be established to test which are critical to the induction of cell death. These data indicate that oxidative stress is an early event following compromise of oxidative metabolism. Although the neurons are the cells that die, markers of oxidative stress indicate changes in vascular and glial cells. TD is a useful model to reveal the interaction between vascular factors, HO-1 induction, lipid peroxidation, NO, and neurodegeneration in diseases associated with oxidative stress.

CONCLUSION

The experiments that are discussed in this chapter make it clear that deficits in metabolism and oxidative stress are central to the pathophysiology of AD. Studies of autopsied brain and cultured cells from patients, as well as studies of animal models and isolated enzymes demonstrate that the changes can be plausibly linked to the genetic causes of AD, to the known pathology, and to the clinical outcome. The enzyme complex KGDHC is particularly sensitive to these changes.

ACKNOWLEDGMENTS

This work was supported by NIH Grants AG14600, AG11921, and AG14930.

REFERENCES

1. TERRY, R.D. & P. DAVIES. 1980. Dementia of the Alzheimer type. Ann. Rev. Neurosci. **3:** 77–95.
2. BACHMAN, D.L. *et al.* 1993. Incidence of dementia and probable Alzheimer's disease in a general population: the Framingham Study. Neurol. 43: 515–519.
3. CARR, D.B. *et al.* 1997. Current concepts in the pathogenesis of Alzheimer's disease. Am. J. Med. **103:** 3S–10S.
4. EVANS, D.A. *et al.* 1989. Prevalence of Alzheimer's disease in a community population of older persons. Higher than previously reported. J. Am. Med. Assoc. **262:** 2551–2556.
5. PARK, L.C.H. *et al.* 1999. Metabolic impairment induces oxidative stress, compromises inflammatory responses, and inactivates a key mitochondrial enzyme in microglia. J. Neurochem. **72:** 1948–1958.
6. IBANEZ, V. *et al.* 1998. Regional glucose metabolic abnormalities are not the result of atrophy in Alzheimer's disease. Neurology **50:** 1585–1593.

7. REIMAN, E.M. *et al.* 1996 Preclinical evidence of Alzheimer's disease in persons homozygous for the epsilon 4 allele for apolipoprotein E. N. Engl. J. Med. **334:** 752–758.
8. SMALL, G.W. *et al.* 1996. Early detection of Alzheimer's disease by combining apolipoprotein E and neuroimaging. Ann. N.Y. Acad. Sci. **802:** 70–78.
9. KISH, S.J. 1997. Brain energy metabolizing enzymes in Alzheimer's disease: α-ketoglutarate dehydrogenase complex and cytochrome oxidase. Ann. N.Y. Acad. Sci. **826:** 218–228.
10. GIBSON, G.E. *et al.* 1988. Reduced activities of thiamine-dependent enzymes in the brains and peripheral tissues of patients with Alzheimer's disease. Arch. Neurol. **45:** 836–840.
11. BUTTERWORTH, R.F. *et al.* 1990. Thiamine-dependent enzyme changes in temporal cortex of patients with Alzheimer's disease. Metab. Brain Dis. **5:** 179–184.
12. MASTROGIACOMA, F. *et al.* 1996. Brain protein and alpha-ketoglutarate dehydrogenase complex activity in Alzheimer's disease. Ann. Neurol. **39:** 592–598.
13. TERWEL, D. *et al.* 1998. Affected enzyme activities in Alzheimer's disease are sensitive to antemortem hypoxia. J. Neurol. Sci. **161:** 47–56.
14. YATES, C.M. *et al.* 1990. Enzyme activities in relation to pH and lactate in postmortem brain in Alzheimer-type and other dementias. J. Neurochem. **55:** 1624–1630.
15. GIBSON, G.E. *et al.* 1998. α-Ketoglutarate dehydrogenase in Alzheimer brains bearing the APP670/671 mutation. Ann. Neurol. **44:** 676–681.
16. CALINGASAN, N.Y. *et al.* 1999. Protein-bound acrolein; a novel marker of oxidative stress in Alzheimer's disease. J. Neurochem. **72:** 751–756.
17. MELOV, S. *et al.* 1999. Mitochondrial disease in superoxide dismutase 2 mutant mice. Proc. Natl. Acad. Sci. USA **96:** 846–851.
18. GIBSON, G. *et al.* 1996. Altered oxidation and signal transduction systems in fibroblasts from Alzheimer patients. Life Sci. **59:** 477–490.
19. HUANG, H.M. *et al.* 1994. The use of cultured fibroblasts in elucidating the pathophysiology and diagnosis of Alzheimer's disease. Ann. N. Y. Acad. Sci. **747:** 225–244.
20. GIBSON, G.E. *et al.* 1997. Abnormalities in Alzheimer's disease fibroblasts bearing the APP670/671 mutation. Neurobiol. Aging **18:** 573–580.
21. SHEU, K.F. *et al.* 1994. Abnormality of the alpha-ketoglutarate dehydrogenase complex in fibroblasts from familial Alzheimer's disease. Ann. Neurol. **35:** 312–318.
22. BLASS, J.P. *et al.* 1997. Inherent abnormalities in oxidative metabolism in Alzheimer's disease: interaction with vascular abnormalities. Ann. N.Y. Acad. Sci. **826:** 382–385.
23. CITRON, M. *et al.* 1994. Excessive production of amyloid beta-protein by peripheral cells of symptomatic and presymptomatic patients carrying the Swedish familial Alzheimer disease mutation. Proc. Natl. Acad. Sci. USA **91:** 11993–11997.
24. ZHANG, H. *et al.* 1998. Increased resistance to oxidative insults in Alzheimer cells . Soc. Neurosci. (Abstr.) **28:** 771.16.
25. SHEARMAN, M.S. *et al.* 1994. Inhibition of PC12 cell redox activity is a specific, early indicator of the mechanism of beta-amyloid–mediated cell death. Proc. Natl. Acad. Sci. USA **91:** 1470–1474.
26. KELLER, J.N. *et al.* 1998. Increased sensitivity to mitochondrial toxin-induced apoptosis in neural cells expressing mutant presenilin-1 is linked to perturbed calcium homeostasis and enhanced oxyradical production. J. Neurosci. **18:** 4439–4450.
27. SCHOONEN, W.G. *et al.* 1991. Characterization of oxygen-resistant Chinese hamster ovary cells. III. Relative resistance of succinate and α-ketoglutarate dehydrogenases to hyperoxic inactivation. Free Radical Biol. Med. **10:** 111–118.
28. HUMPHRIES, K.M. *et al.* 1998. Selective inactivation of alpha-ketoglutarate dehydrogenase and pyruvate dehydrogenase: reaction of lipoic acid with 4-hydroxy-2-nonenal. Biochemistry **37:** 15835–15841.
29. HUMPHRIES, K.M. *et al.* 1998. Inhibition of NADH-linked mitochondrial respiration by 4-hydroxy-2-nonenal. Biochemistry **37:** 552–557.

30. JOFFE, G.T. *et al.* 1998. Secondary inhibition of 2-ketoglutarate dehydrogenase complex by MPTP. Neuroreport **9:** 2781–2783.
31. BUTTERWORTH, R.F. *et al.* 1993. Thiamine-dependent enzyme changes in the brains of alcoholics: relationship to the Wernicke-Korsakoff syndrome. Alcohol. Clin. Exp. Res. **17:** 1084–1088.
32. CULLEN, K.M. *et al.* 1995. Neurofibrillary tangles in chronic alcoholics. Neuropathol. Appl. Neurobiol. **21:** 312–318.
33. KAIDO, M. *et al.* 1996. Alzheimer-type pathology in a patient with mitochondrial myopathy, encephalopathy, lactic acidosis and stroke-like episodes(MELAS). Acta Neuropathol. **92:** 312–318.
34. GABUZDA, D. *et al.* 1994 Inhibition of energy metabolism alters the processing of amyloid precursor protein and induces a potentially amyloidogenic derivative. J. Biol. Chem. **269:** 13623–13628.
35. QUERFURTH, H.W. *et al.* 1994. Calcium ionophore increases amyloid beta peptide production by cultured cells. Biochemistry **33:** 4550–4561.
36. CHENG, B. *et al.* 1992. Glucose deprivation elicits neurofibrillary tangle-like antigenic changes in hippocampal neurons: prevention by NGF and bFGF. Exp. Neurol. **117:** 114–123.
37. MATTSON, M.P. *et al.* 1991. Effects of elevated intracellular calcium levels on the cytoskeleton and tau in cultured human cortical neurons. Mol. Chem. Neuropathol. **15:** 117–142.
38. MATTSON, M.P. *et al.* 1997. 4-Hydroxynonenal, a product of lipid peroxidation, inhibits dephosphorylation of the microtubule-associated protein tau. Neuroreport **8:** 2275–2281.
39. BLASS, J.P. *et al.* 1990. Induction of Alzheimer antigens by an uncoupler of oxidative phosphorylation. Arch. Neurol. **47:** 864–869.
40. PLUM, F. *et al.* 1980. The Diagnosis of Stupor and Coma. F.A. Davis. Philadelphia.
41. GIBSON, G.E. *et al.* 1991. The cellular basis of delirium and its relevance to age-related disorders including Alzheimer's disease. Int. Psychogeriatr. **3:** 373–395.
42. FREEMAN, G.B. *et al.* 1988. Dopamine, acetylcholine, and glutamate interactions in aging. Behavioral and neurochemical correlates. Ann. N.Y. Acad. Sci. **515:** 191–202.
43. GIBSON, G. *et al.* 1997. Selective changes in cell bodies and growth cones of nerve growth factor–differentiated PC12 cells induced by chemical hypoxia. J. Neurochem. **69:** 603–611.
44. GIBSON, G.E. *et al.* 1982. The role of the cholinergic system in thiamin deficiency. Ann. N.Y. Acad. Sci. **378:** 382–403.
45. PARK, L.C.H. *et al.* 1998. Metabolic impairment elicits brain cell-type selective changes in oxidative stress and cell death. Soc. Neurosci. (Abstr.). **28:** 576.9.
46. SHEU, K.F. *et al.* 1998. Immunochemical characterization of the deficiency of the α-ketoglutarate dehydrogenase complex in thiamine-deficient rat brain. J. Neurochem. **70:** 1143–1150.
47. CALINGASAN, N.Y. *et al.* 1999. Oxidative stress is associated with region-specific neuronal death during thiamine deficiency. J. Neuropathol. Exp. Neurol. In press.
48. CALINGASAN, N.Y. *et al.* 1998. Induction of nitric oxide synthase and microglial responses precede selective cell death induced by chronic impairment of oxidative metabolism. Am. J. Pathol. **153:** 599–610.
49. CALINGASAN, N.Y. *et al.* 1995. Blood-brain barrier abnormalities in vulnerable brain regions during thiamine deficiency. Exp. Neurol. **134:** 64–72.
50. CALINGASAN, N.Y. *et al.* 1996. Novel neuritic clusters with accumulations of amyloid precursor protein and amyloid precursor-like protein 2 immunoreactivity in brain regions damaged by thiamine deficiency. Am. J. Pathol. **149:** 1063–1071.

The Use of Transgenic and Mutant Mice to Study Oxygen Free Radical Metabolism

TING-TING HUANG, ELAINE J. CARLSON, INES RAINERI, ANNE MARIE GILLESPIE, HEATHER KOZY, AND CHARLES J. EPSTEIN[a]

Department of Pediatrics, University of California, San Francisco, California 94143-0748, USA

ABSTRACT: To distinguish the role of Mn superoxide dismutase (MnSOD) from that of cytoplasmic CuZn superoxide dismutase (CuZnSOD), the mouse Mn-SOD gene (*Sod2*) was inactivated by homologous recombination. *Sod2* –/– mice on a CD1 (outbred) genetic background die within the first 10 days of life (mean, 5.4 days) with a complex phenotype that includes dilated cardiomyopathy, accumulation of lipid in liver and skeletal muscle, metabolic acidosis and ketosis, and a severe reduction in succinate dehydrogenase (complex II) and aconitase (a TCA cycle enzyme) activities in the heart and, to a lesser extent, in other organs. These findings indicate that MnSOD is required to maintain the integrity of mitochondrial enzymes susceptible to direct inactivation by superoxide. On the other hand, Lebovitz *et al.* reported an independently derived MnSod null mouse (*Sod2tmlLeb*) on a mixed C57BL/6 and 129Sv background with a different phenotype. Because a difference in genetic background is the most likely explanation for the phenotypic differences, the two mutant lines were crossed into different genetic backgrounds for further analyses. To study the phenotype of *Sod2tmlLeb* mice CD1 background, the *Sod2tmlLeb* mice were crossed to CD1 for two generations before the –/+ mice were intercrossed to generate –/– mice. The life span distribution of CD1<*Sod2*–/–>Leb was shifted to the left, indicating a shortened life span on the CD1 background. Furthermore, the CD1<*Sod2*–/–>Leb mice develop metabolic acidosis at an early stage as was observed with CD1<*Sod2*–/–>Cje. When *Sod2tmlCje* was placed on C57BL/6J (B6) background, the –/– mice were found to die either during midgestation or within the first 4 days after birth. However, when the B6<*Sod2*–/+>Cje were crossed with DBA/2J (D2) for the generation of B6D2F2<*Sod2*–/–>Cje mice, an entirely different phenotype, similar to that described by Lebovitz *et al.*, was observed. The F2 *Sod*–/– mice were able to survive up to 18 days, and the animals that lived for more than 15 days displayed severe neurological abnormalities including ataxia and seizures. Their hearts were not as severely affected as were those of the CD1 mice, and neurological degeneration rather than heart defect appears to be the cause of death.

INTRODUCTION

Oxygen free radicals have been implicated in a wide variety of pathological processes, including aging and neurodegeneration. These radicals are generated by a

---

[a]Corresponding author: Charles J. Epstein, M.D., Department of Pediatrics, University of California, 531 Parnassus, U585L, San Francisco, California 94143-0748. Phone: 415-476-2981; fax: 415-476-9976.
e-mail: cepst@itsa.ucsf.edu

TABLE 1. Classification of mammalian superoxide dismutases (SODs)

	CuZn Superoxide Dismutase	Mn Superoxide Dismutase	Extracellular Superoxide Dismutase
Protein designation	CuZnSOD	MnSOD	EC-SOD
Gene designation[a]	*Sod1/SOD1*	*Sod2/SOD2*	*Sod3/SOD3*
Localization	cytoplasm	mitochondria	extracellular fluid and endothelial cell surface
Inducibility	noninducible	inducible	noninducible
Native protein	dimer	tetramer	tetramer
Metal cofactor(s)	Cu and Zn	Mn	Cu and Zn

[a]Gene designation for mouse and human, respectively.

large number of normal and abnormal events, including oxidative phosphorylation; exposure to radiation, drugs, and chemicals; and phagocytosis. The approach that we have used to study the effects of free radicals is to perturb the enzymatic machinery responsible for their metabolism. We have focused on the superoxide dismutases (SODs), which, in mammals, constitute a group of three genetically and geographically distinct enzymes (TABLE 1). The SODs convert superoxide radicals (O_2^-) to hydrogen peroxide, with hydrogen peroxide then being metabolized to water by glutathione peroxidase or catalase. We have concentrated our attention on two of the SODs: CuZnSOD, a cytoplasmic enzyme that is produced constitutively and is present in all cells; and MnSOD, the production of which, by contrast, is inducible by oxidative insults and cytokines and is present only within mitochondria. During the past 12 years we have prepared a series of transgenic and mutant mice in which the activities of these two enzymes have been either increased or decreased,[1–3] and we have investigated the consequences of these changes in SOD activity. The results we have obtained indicate that alterations in enzyme activity can have profound effects on the response of the organism to oxidative stress and that the roles of CuZnSOD and MnSOD are quite distinct.

This report will focus principally on recent findings related to mice lacking MnSOD. To put the findings in perspective, a brief summary of results obtained with CuZnSOD transgenics and knockouts will be presented.

CuZnSOD TRANSGENIC MICE

Transgenic mice carrying human *SOD1* transgenes and expressing 3- to 5-fold wild-type levels of CuZnSOD activity are grossly normal and have normal longevities (except for some suggestion that mean life span may be slightly decreased in the animals with the highest level of expression).[4] These results are at variance with what has been observed in *Drosophila,* in which transgenic flies with increased CuZnSOD activity have been found to have increased life spans.[5] When subjected to a number of treatments that produce acute oxidative stress, the transgenic mice show, in virtually all cases, decreased damage following the stressful treatment. A list of several stressors relevant to the nervous system that we have investigated is

TABLE 2. Effects of increased CuZnSOD activity in mice on the response of the central nervous system to acute oxidative stress

Treatment	Effect	References
Acute injury		
Cold injury	decreased injury	Chan et al.[6]
Blunt trauma (contusion)	decreased injury	Mikawa et al.[7]
Transient ischemia/reperfusion	decreased injury	Kinouchi et al.[8]; Yang et al.[9]; Murakami et al.[10]
	prolonged hsp70 and c-fos mRNA response	Kamii et al.[11,12]; Mikawa et al.[13]; Kondo et al.[14]
Permanent ischemia	no effect	Chan et al.[15]
Neonatal hypoxia-ischemia	increased injury	Ditelberg et al.[16]
Chemical agents		
MPTP	decreased toxicity	Przedborski et al.[17]
2'-NH_2-MPTP	decreased toxicity	Andrews et al.[18]
Methamphetamine and methamphetamine derivatives	decreased toxicity	Cadet et al.[19–21]; Hirata et al.[22,23]
3-Nitropropionic acid	decreased toxicity	Beal et al.[24]
Kainic acid	decreased toxicity	Kondo et al.[25]
β-Amyloid protein (i.c.[a])	decreased toxicity	Friedlich et al.[26]

[a]Intracerebral (by injection).

given in TABLE 2. Whether the insult was trauma, ischemia/reperfusion, or exposure to a toxic drug or chemical, the transgenic animals displayed less tissue damage and/or decreased mortality. When the mice were injected intracerebrally with the HIV antigen gp120, there was diminished activation of the adrenal axis.[27] Dopaminergic neurons derived from transgenic brains had much greater viability *in vitro* and a greater ability to populate a recipient brain after transplantation.[28–31] The one exception to a protective effect of *SOD1* transgenes was observed in neonatal mice exposed to hypoxia-ischemia.[16] In this situation, the transgenes had a deleterious effect that appears to be attributable to the fact that there was insufficient catalase or glutathione peroxidase to metabolize the increased hydrogen peroxide that was generated.[32]

Because of the difficulties in performing detailed physiological and global ischemia studies in mice, CuZnSOD transgenic rats (using the human *SOD1* gene) have been prepared.[33] As with the transgenic mice, these animals again showed enhanced resistance to the acute oxidative stress associated with global ischemia and reperfusion.

It is not as clear whether increased CuZnSOD activity has deleterious or beneficial effects in the chronic state in the absence of acute oxidative insults. The opioid response system has been found to be affected, with increased concentrations of μ-opioid receptors in the brain and enhancement of the ability of morphine to act as a reinforcer of behavior.[34–36] Reports by other investigators have suggested that there may be abnormalities in the structure of the neuromuscular junction.[37,38]

CuZnSOD MUTANT MICE

Mutant or knockout mice were produced by homologous recombination.[3,39] Homozygous mutant mice (*Sod1* −/−) totally devoid of CuZnSOD were normal grossly and, although formal aging studies have not been reported, have lived well into adulthood. These results indicated that, contrary to earlier expectations, CuZnSOD is not essential for viability. As might have been expected, heterozygous (*Sod1* −/+) animals, with 50% of wild-type CuZnSOD activity, were more sensitive to acute oxidative stress such as is induced by transient ischemia and reperfusion than were their wild-type littermates.[40] Similarly, cultured *Sod1* −/+ fibroblasts were more sensitive to paraquat than were *Sod1* +/+ fibroblasts.[3] Homozygous animals and cells were extremely sensitive to oxidative stress.

MnSOD TRANSGENIC MICE

Several strains of mice carrying *Sod2* transgenes with either a tissue-specific or endogenous promoter have been made by others and ourselves. Higher levels of MnSOD have been shown to confer increased resistance to oxygen-induced lung injury,[41] ischemia-reperfusion injury of the heart,[42] adriamycin-induced cardiomyopathy,[43] and MPTP toxicity.[44] We have found that transgenic mice with extremely high levels of MnSOD (greater than 10-fold normal) show a reduction in survival and male fertility on certain genetic backgrounds.

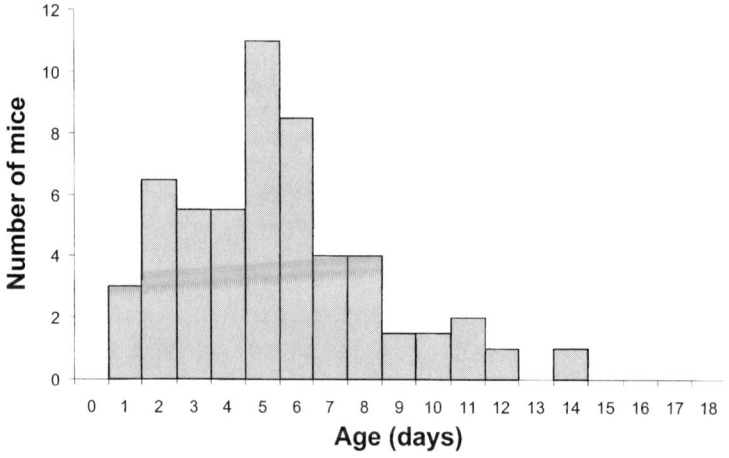

FIGURE 1. Life span distribution of CD1<*Sod2* −/−>Cje mice. A total of 55 −/− mice were monitored daily from birth. The mean life span is 5.6 days.

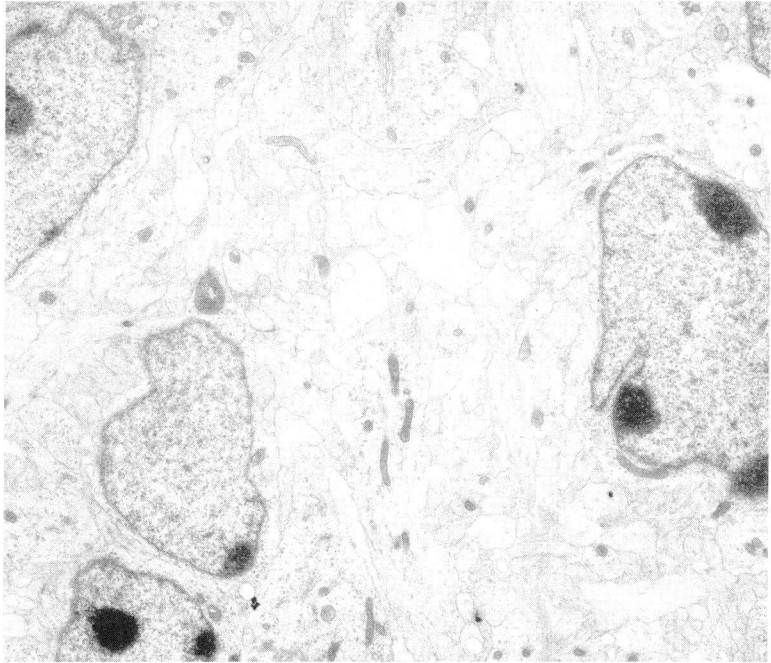

FIGURE 2. Representative brain electron micrograph of a 5-day-old CD1<*Sod2* −/−> Cje mouse. The outer cortical region is shown. No swelling and vacuolization of the mitochondria is observed. The cristae and mitochondrial membrane are intact.

MnSOD MUTANT MICE

Sod2 −/−

As an approach to determining the specific role of MnSOD in the metabolism of superoxide, *Sod2* mutant or knockout mice were generated by homologous recombination.[2] Our initial studies with homozygous animals (*Sod2* −/−) were carried out with mice generated on an outbred CD1 background. These animals could be recognized at birth—even prenatally—as being abnormal because of their paleness, small size, hyperventilation, and cool temperature. The mutant animals did poorly and generally died within the first week of life. Aside from profound weakness, there was no evidence of neurological abnormality. Formal studies of life span revealed mean and maximum life spans of 5.4 and 14 days, respectively (FIG. 1). At autopsy, the salient features were an enlarged and dilated heart and a pale fatty liver; the heart disease was considered to be the cause of death, probably by arrhythmia. Microscopic examination of the heart demonstrated all of the hallmarks of dilated cardiomyopathy, including mural thrombus formation and subendocardial fibrosis. On electron microscopy of the heart, there was a blurring of the Z bands of the sarcomeres and separation of the myofibers by large collections of mitochondria that appeared normal in structure. Skeletal muscle did not show the same Z band abnormality, but did

TABLE 3. Physical studies of 5-day old CD1<*Sod2*–/–>Cje mice

	Sod2+/+	*Sod2*–/+	*Sod2*–/–
Body weight (g)	4.3 ± 0.3 ($n = 31$)	4.0 ± 0.4 ($n = 53$)	2.6 ± 0.1 ($n = 19$)[a]
LV wall (mm)	1.37 ± 0.10 ($n = 8$)	1.27 ± 0.05 ($n = 8$)	0.95 ± 0.08 ($n = 7$)[a]
LV cavity (mm)	0.33 ± 0.09 ($n = 8$)	0.23 ± 0.04 ($n = 8$)	1.08 ± 0.18 ($n = 7$)[b]
Body temperature (°C)	32.9 ± 0.4 ($n = 5$)	32.7 ± 0.2 ($n = 10$)	30.3 ± 0.7 ($n = 4$)[a]
Organ weight (% body weight)	($n = 12$)	($n = 13$)	($n = 6$)
Liver	2.2 ± 0.1	2.4 ± 0.1	3.9 ± 0.2[c]
Brain	6.3 ± 0.2	5.9 ± 0.1	6.5 ± 0.2
Lung	2.3 ± 0.1	2.3 ± 0.1	2.2 ± 0.1
Heart	0.65 ± 0.02	0.64 ± 0.02	0.85 ± 0.04[c]
Organ water content (% wet weight)	($n = 12$)	($n = 13$)	($n = 6$)
Liver	74.1 ± 0.4	74.6 ± 0.4	64.2 ± 1.4[c]
Brain	87.1 ± 0.1	87.1 ± 0.1	86.9 ± 0.1
Lung	81.9 ± 0.2	81.9 ± 0.2	82.1 ± 0.3
Heart	81.8 ± 0.3	81.2 ± 0.4	82.6 ± 0.3

NOTE: All values are mean ± SEM.
[a]Significantly different from that of –/+ and +/+ controls, $p < 0.005$; [b] $p < 0.001$; [c] $p < 0.0001$.

display a large amount of fat deposition. On the other hand, the outer cortical region of the brain did not show overt differences from that of controls (FIG. 2). There was no swelling or vacuolization of the mitochondria, and the cristae and mitochondrial membrane were intact. Microscopic examination of the liver revealed the presence of large quantities of oil red O reactive lipid.

Physical and blood biochemistry studies of CD1<*Sod2* –/–>Cje mice are summarized in TABLES 3, 4, and 5. The salient abnormalities were reduced body weight, cardiac dilation with no edematous changes in tissues (as indicated by normal tissue dry- to wet-weight ratio), and profound metabolic acidosis with compensatory respiratory alkalosis, ketosis, and hypoglycemia. Lactic acidosis was not present, but the urine contained large amounts of 3-methylglutaconic, 3-OH-3-methylglutaric, 3-OH-isovaleric, and 2-OH-glutaric acids.[45] The acidosis was the first abnormality to be detected, and the chronology of the evolution of the acidosis, deposition of lipid, and cardiomyopathy is shown in TABLE 6 and FIGURE 3.

Several studies were undertaken to determine how the absence of MnSOD and the presumed resultant increase in superoxide levels within mitochondria might result in the observed pathology.[2,45] The normal appearance of the mitochondria on electron microscopy suggested that the elevation in superoxide was not grossly compromising membrane integrity and structure. Similarly, long-range PCR indicated that gross deletions were not present in the mitochondrial DNA, but 2–3 fold increases in 8-OH-guanine, 8-OH-adenine, and 5-OH-cytosine in total heart DNA and of 5-OH-cytosine in total brain DNA were found. The latter findings indicate that oxida-

TABLE 4. Blood chemical studies of CD1<*Sod2–/–*>Cje mice

	Sod2+/+	*Sod2–/+*	*Sod2–/–*
Na^+	133.6 ± 0.5 ($n = 3$)	133.3 ± 0.4 ($n = 7$)	137.6 ± 0.3 ($n = 3$)[a]
K^+	8.9 ± 0.8 ($n = 3$)	8.5 ± 0.3 ($n = 7$)	7.9 ± 0.9 ($n = 3$)
Cl^-	94.3 ± 0.2 ($n = 3$)	92.6 ± 0.6 ($n = 7$)	104.2 ± 4.4 ($n = 3$)[a]
Glucose	126 ± 8 ($n = 4$)	117 ± 5 ($n = 10$)	85 ± 19 ($n = 4$)
Ketones	0.2 ± 0.1 ($n = 19$)	0.3 ± 0.1 ($n = 27$)	2.2 ± 0.2 ($n = 17$)
Lactic acid	2.5 ± 0.3 ($n = 7$)	2.0 ± 0.2 ($n = 8$)	1.3 ± 0.2 ($n = 4$)[a]
Hematocrit	33.5 ± 0.6 ($n = 14$)	32.0 ± 0.6 ($n = 22$)	28.2 ± 0.7 ($n = 9$)[a]
HCO_3^a	25.7 ± 1.5 ($n = 4$)	26.2 ± 0.7 ($n = 6$)	16.0 ± 1.4 ($n = 3$)[a]
AST	237 ± 62 ($n = 4$)	263 ± 49 ($n = 6$)	414 ± 206 ($n = 3$)
Total bilirubin	0.7 ± 0.1 ($n = 4$)	0.6 ± 0.1 ($n = 6$)	1.4 ± 0.5 ($n = 3$)[a]
Triglyceride	237 ± 10 ($n = 4$)	186 ± 24 ($n = 6$)	353 ± 99 ($n = 3$)

NOTE: All values are mean ± SEM.
[a]Significantly different from that of –/+ and +/+ controls (t test, $p < 0.05$).

tive DNA damage was occurring. Since superoxide does not pass through the mitochondrial membrane, it is assumed that the damage was to mitochondrial DNA. Whether this DNA damage was sufficient to interfere with mitochondrial function is unknown.

In view of the fact that the development of cardiomyopathy suggested that energy production was impaired, a detailed analysis of oxidative phosphorylation was undertaken. These studies revealed that the activity of complexes I (NADH-dehydro-

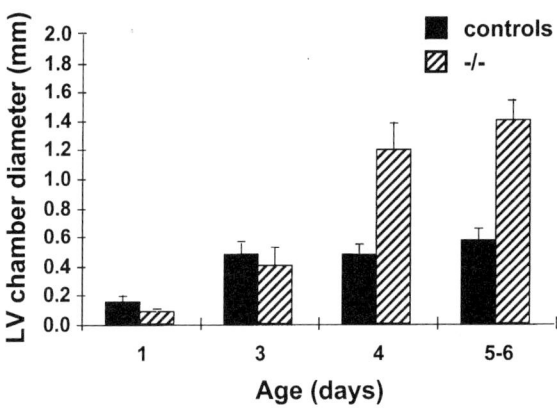

FIGURE 3. Progression of dilated cardiomyopathy in CD1<*Sod2* –/–>Cje mice. Left ventricular chamber diameter is measured as an indicator of cardiac dilatation. Hearts were removed at indicated age after euthanasia, immersed in cold PBS (Ca^{++}, Mg^{++}-free) for 1 minute and fixed in 10% neutral buffered formalin. The fixed hearts were then sectioned transversely, and the left ventricles were measured under a dissecting microscope equipped with a micrometer.

TABLE 5. Blood gas analysis of 5-day-old CD1<*Sod2*–/–>Cje mice

	Controls (+/+ and –/+) (n = 7)	*Sod2*–/– (n = 6)
pH	7.41 ± 0.01	7.38 ± 0.03^a
PCO_2 (torr)	49.3 ± 2.3	34.4 ± 2.1^a
PO_2 (torr)	44.4 ± 4.0	68.6 ± 10.2^a
BE (mM)	5.3 ± 0.9	–5.0 ± 1.2^a
HCO_3 (mM)	30.2 ± 0.9	19.9 ± 1.0^a

NOTE: All values are mean ± SEM.
^{a}Significantly different from that of controls (*t* test, $p < 0.05$).

genase) and II (succinic dehydrogenase) were significantly decreased, by 41% and 68%, respectively, in heart mitochondria.[45] A decrease in succinic dehydrogenase activity in the brain was demonstrated histochemically.[2] These complexes share the property of having [4Fe-4S] clusters in their active sites, and such clusters are known to be highly sensitive to the inhibition by superoxide.[46] The same is also true of aconitase, the first enzyme of the TCA cycle, which converts citrate to isocitrate.[47] This enzyme has been shown to be exquisitely sensitive to superoxide, which binds to and oxidizes the Fe in the cluster that serves as the substrate binding site. As a result of its oxidation, the Fe falls out of the cluster, and the enzyme is inactivated. In the heart and brain of CD1<*Sod2* –/–>Cje mice, mitochondrial aconitase activity was reduced by 89% and 67–76%, respectively.[45]

These biochemical findings provided a plausible explanation for the development of the dilated cardiomyopathy and accumulation of lipid in liver and muscle. As a result of the absence of MnSOD, superoxide levels in the mitochondria rise, and the superoxide reacts with available [4Fe-4S] clusters. Inactivation of these enzymes impairs energy production and, in the case of aconitase, the metabolism of acetyl CoA. The latter block is presumed to result in the diversion of acetyl CoA to fatty acid production. Inhibition of [4Fe-4S] cluster enzymes does not explain the generation of the organic acids found in the urine. These acids are those that are generated when the enzyme 3-OH-3-methylglutaryl CoA lyase in the leucine catabolic pathway is inhibited, and activity of this enzyme is indeed reduced by 36%.[45] Whether inhibition of this enzyme results from oxidation of a vulnerable amino acid in the active site, such as a cysteine or histidine,[48,49] remains to be determined.

Sod2 –/+

In contrast to the null animals, heterozygous animals (*Sod2* –/+), with 50% of wild-type MnSOD activity, were grossly normal; but they did have demonstrable biochemical abnormalities.[50,51] On an inbred C57BL/6 (B6) background, there was a 30% reduction in complex I and aconitase activities in liver mitochondria and a 44% decrease in state 3 respiration with duroquinol as substrate. There were also increased levels of protein carbonyls and of 8-OH-deoxyguanosine in mitochondrial proteins and DNA, respectively. The rate of induction of the mitochondrial permeability transition by Ca^{++} and *t*-butylhydroperoxide was also increased. Reduced glutathione concentrations were decreased in brain, lung, and muscle mitochondria. Taken together, these data indicate that decreasing MnSOD activity to half does have

TABLE 6. Progression of the phenotype in CD1<Sod2–/–>Cje mice

Postnatal Age	Blood pH	Liver	Heart
Day 1	acidotic	normal	normal
Day 2	acidotic	lipid deposit	N.D.
Day 3	acidotic	lipid deposit	normal
Day 4	acidotic	N.D.	dilated
Day 5–6	acidotic	lipid deposit	dilated

TABLE 7. Possible explanations for the differences in phenotype observed between Sod2<tm1Cje> and Sod2<tm1Leb>

	Sod2<tm1Cje>	Sod2<tm1Leb>
Targeting construct	remove exon 3 positive selection — neo negative selection — tk	remove exons 1 and 2 positive selection — HPRT
ES cell line	B6/Rb32Rb2H F1	129/Sv (hprt⁻)
Mouse strain (genetic background)	CD1	129/Sv and C57BL/6

TABLE 8. Effect of genetic background on *Efgr* mutant phenotype[55]

Mouse Strain (Genetic Background)	Phenotype
CF-1	periimplantation lethality
129/Sv	midgestation lethality
CD1	perinatal lethality (survive up to 3 weeks)

TABLE 9. Effect of genetic background on *Cftr* mutant phenotype[56]

Genetic Crosses	Survival Time	
(F1 × F1)	<10 Days	>6 Weeks
129/Sv; C57BL/6		24.2%
129/Sv; BALB/c		31.0%
129/Sv; CD1		29.7%
129/Sv; DBA	91.7%	

adverse effects on mitochondria, even in the absence of any specific oxidative stress. And, in the presence of increased oxidative stress resulting from permanent focal cerebral ischemia, $Sod2$ –/+ mice, this time on a CD1 background, again showed greater damage than controls.[52] They had larger infarcts, increased cerebral edema, and accelerated mitochondrial damage. In addition, these animals presented evidence of greater cerebral hydrogen peroxide production, even in the absence of ischemia. Cultured cortical neurons from fetal $Sod2$ –/+ brains were more sensitive than wild-type neurons to the toxic effects of glutamate, but not to other glutamate receptor agonists.[53]

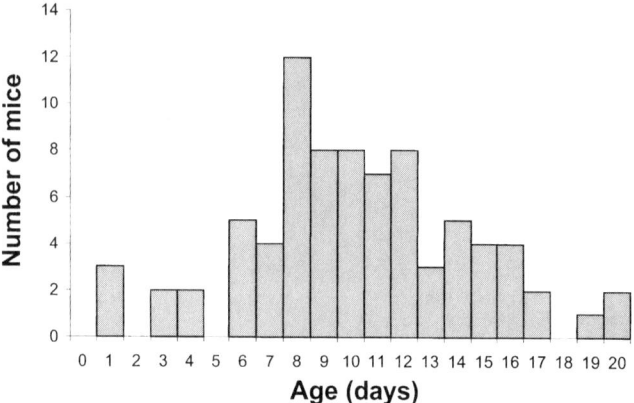

FIGURE 4. Lifespan distribution of CD1<*Sod2* –/–>Leb mice. The original Baylor mice from Dr. Lebovitz were crossed to CD1 for 2 generations before the –/+ mice were intercrossed to generate *Sod2* –/– mice for the study. A total of 80 –/– mice were monitored daily from birth. The mean life span is 10 days.

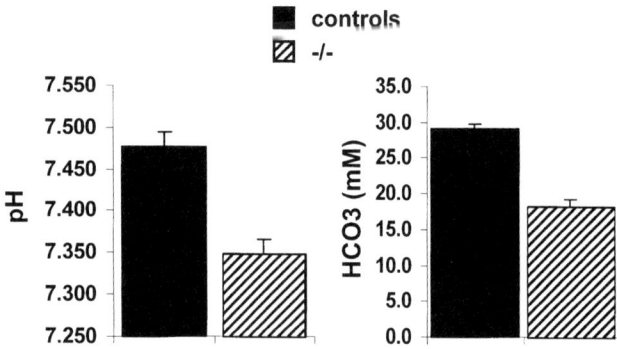

FIGURE 5. Blood gas analysis of 5-day-old Baylor *Sod2* –/– mice on CD1 background [CD1<*Sod2* –/–>Leb]. Eight –/– and 8 control (4 –/+ and 4 +/+) mice were analyzed.

Effect of Genetic Background on Sod2 –/– Phenotype

Shortly after the findings with the *Sod2* –/– mice were reported, Lebovitz et al. from Baylor described a MnSOD null mouse with a different phenotype.[54] Their animals, also produced by homologous recombination, had a much longer survival—up to 2.5 weeks—and only mild heart disease. They developed severe anemia and degeneration of the brain, with early cellular degeneration observed in the basal ganglia and adjacent subcortical gray matter beginning about 10 days after birth. In trying to understand the differences in the phenotypes of the two knockouts, we considered the three differences between the two that are summarized in TABLE 7—differences in the molecular constructs used to generate the deletions, in the stem

TABLE 10. Phenotype of *Sod2-/-* mice on different genetic backgrounds

	C57BL/6	CD1	B6D2 F2
Mean life span	1.5 days	5.4 days	11.1 days
Dilated cardiomyopathy	unknown	in all mice older than 4 days	not observed in mice older than 13 days
Metabolic acidosis	unknown	severe	mild
Ataxia	not observed	not observed	observed in mice older than 10 days
Seizures	not observed	not observed	observed in mice older than 13 days

cells that were transfected, and in the genetic backgrounds on which the mutations were being bred. Precedents from other mutant mice, such as those with an epidermal growth factor receptor (*Egfr*) or cystic fibrosis transmembrane conductance regulator (*Cftr*) mutation[55,56] (TABLES 8 and 9), suggest that a difference in strain background was the most likely explanation for the major phenotypic differences. We carried out two series of investigations. The first involved breeding the Baylor mouse onto a CD1 background; the results are shown in FIGURES 4 and 5. The life span distribution of CD1<*Sod2* –/–>Leb (Baylor) mice was shifted to the left, although a small portion of the –/– mice still survived to 20 days. This may be due to some remaining genetic influence from the original 129/Sv background. In addition, CD1<*Sod2* –/–>Leb mice develop metabolic acidosis at an early stage, as was observed in the CD1<*Sod2* –/–>Cje animals (FIG. 5).

The second approach to investigating the effect of strain background on phenotype was to breed our mutation onto the inbred C57BL/6 background; the result was quite unexpected.[57] On a B6 background, *Sod2* –/– (null) mice were found to die either during gestation (about 50%) or, if live born, within the first 4 days after birth

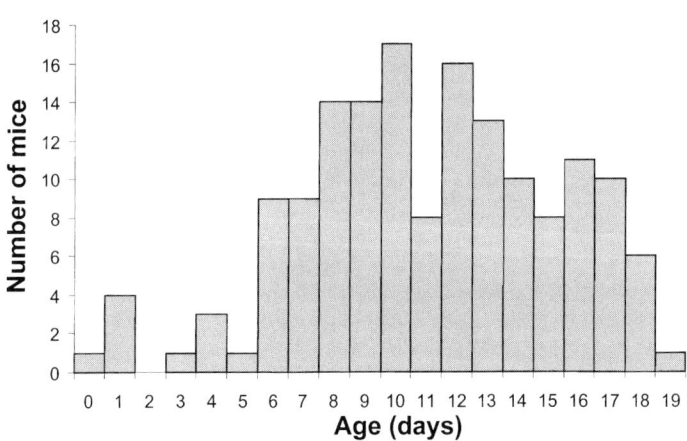

FIGURE 6. Life span distribution of B6D2 F2<*Sod2* –/–> mice. A total of 155 B6D2 F2<*Sod2* –/–> mice were followed daily from birth. The mean life span is 11.1 days.

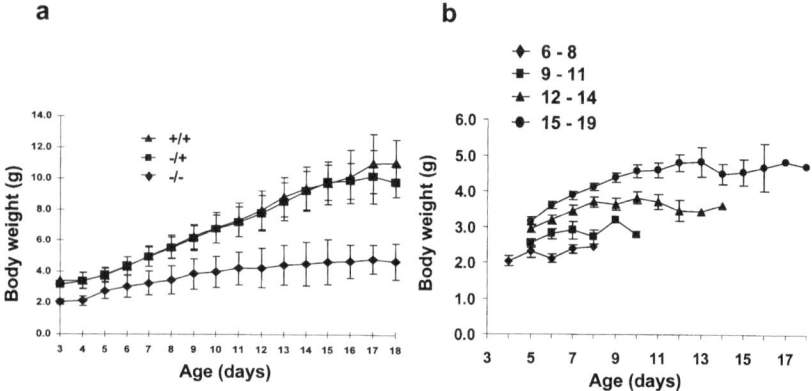

FIGURE 7. Body weight changes of B6D2 F2 mice. **a.** Body weights of the mice were recorded as soon as the mice could be marked by toe clipping. On the average, 83 +/+, 149 −/+ and 65 −/− mice were monitored daily for weight gain. **b.** Body weight changes of −/− mice grouped by life expectancy. On the average, 9 mice from the 6–8-day, 13 from the 9–11-day, 18 from the 12–14-day, and 20 from the 15–19-day group were monitored.

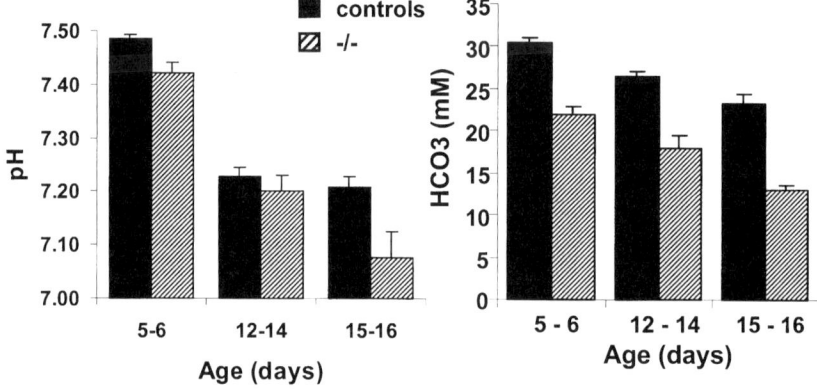

FIGURE 8. Blood gas analysis of B6D2 F2<*Sod2* −/−> mice. Blood pH and HCO_3 levels, which indicate the extent of metabolic acidosis and respiratory compensation, are shown. Eight−/− and 9 controls (5 −/+ and 4 +/+) from the 5–6-day group, 9 −/− and 22 controls (14 −/+ and 8 +/+) from the 12–14-day group, and 5 −/− and 5 controls (3 −/+ and 2 +/+) from the 15–16-day group were used for the analysis.

(mean life span 1.5 days). The cause of death has not been determined. When B6 heterozygotes (*Sod2* −/+) were crossed with inbred DBA/2 (D2) mice to generate B6D2 F1s and the F1s were then intercrossed to generate B6D2 F2s, an entirely different phenotype was observed (TABLE 10). The F2 mice survived for up to 18 days, with a mean life span of 11.1 days (FIG. 6). The animals that lived for more than 15 days displayed neurological abnormalities including ataxia and seizures. Their hearts were not as severely affected as were those of the CD1 mice; neurological degeneration rather than heart disease appeared to be the cause of death.

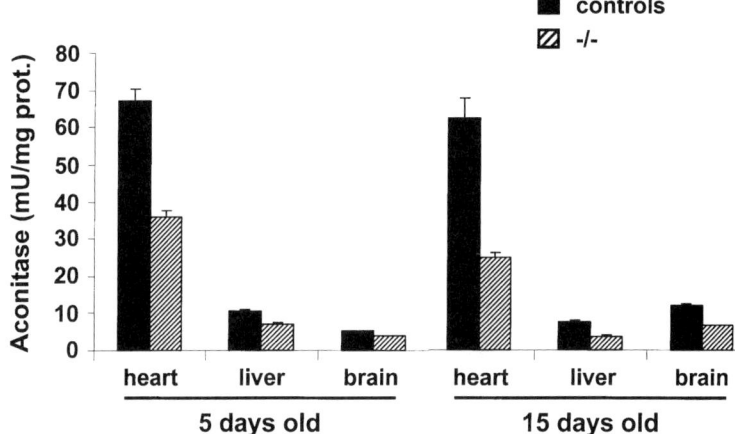

FIGURE 9. Total tissue aconitase levels of B6D2 F2<*Sod2* –/–> mice. Five –/– and 6 +/+ 5-day-old mice and 6 –/– and 5 +/+ 15-day-old mice were used for the analysis.

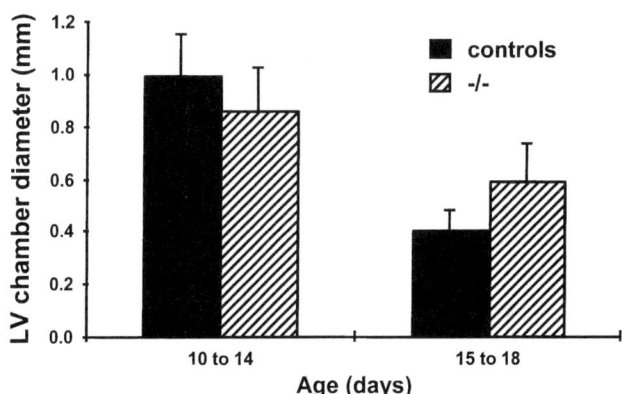

FIGURE 10. Heart measurements of B6D2 F2<*Sod2* –/–> mice. Left ventricular chamber diameter is shown here to indicate the extent of cardiac dilation. Measurements were done as described in legend to FIGURE 3.

At birth, B6D2 F2<*Sod2* –/–> mice appear paler and grow much more slowly than their –/+ and +/+ littermates. The body weight of B6D2 F2<*Sod2* –/–> mice is about 80% that of the controls at 3 days of age and drops down to about 45% that of controls by 18 days of age (FIG. 7a). There was a positive correlation between weight gain and life expectancy—mice with more rapid weight gain in general had a longer life span than those with slower weight gain (FIG. 7b). To understand the progression of *Sod2* –/– phenotype on the B6D2 F2 background, blood gas analyses, aconitase assays, and heart measurements were carried out with different age groups of –/– mice. Blood gas analyses revealed metabolic acidosis (indicated by lower levels of HCO_3) at all three age groups studied. The mice were able to compensate and maintained a normal pH until 12–14 days of age (FIG. 8). Total aconitase levels were reduced by 46%, 31%, and 26%, respectively, in heart, liver and brain of 5-day-old and by 61%, 52%, and 46% in 15-day-old −/− mice (FIG. 9). The most pronounced component of the phenotype, other than life span, that sets these mice apart from CD1<$Sod2$ −/−> mice is the lack of development of dilated cardiomyopathy. Measurement of the diameter of the left ventricular cavities showed no evidence of heart dilation up to 14 days of age (FIG. 10).

The results we have obtained with the mutant animals on inbred and F2 backgrounds indicate that genetic background can profoundly affect the phenotype and imply the existence of genetic modifiers. How these modifiers might be working to alter the null phenotype is currently the focus of our investigative efforts. Their existence does demonstrate that the metabolism of oxygen free radicals may be under more complex genetic control than we might have initially imagined.

ACKNOWLEDGMENTS

This work was supported by National Institute of Health Grants AG-08938, AG-15151, and AG-16998

We thank Rhodora Gacayan and Marguerite Doan for animal care and Alan Chong, Meagan Jones, and Carlos Adams for laboratory assistance.

REFERENCES

1. EPSTEIN, C.J., K.B. AVRAHAM, M. LOVETT, S. SMITH, O. ELROY-STEIN, G. ROTMAN, C. BRY & Y. GRONER. 1987. Transgenic mice with increased CuZn-superoxide dismutase activity: an animal model of dosage effects in Down syndrome. Proc. Natl. Acad. Sci. USA **84:** 8044–8048.
2. LI, Y., T.-T. HUANG, E.J. CARLSON, S. MELOV, P.C. URSELL, J.L. OLSON, L.J. NOBLE, M.P. YOSHIMURA, C. BERGER, P.H. CHAN, D.C. WALLACE & C.J. EPSTEIN. 1995. Dilated cardiomyopathy and neonatal lethality in mutant mice lacking manganese superoxide dismutase. Nature Genet. **11:** 376–381.
3. HUANG, T.-T., M. YASUNAMI, E.J. CARLSON, A.M. GILLESPIE, A.G. REAUME, E.K. HOFFMAN, P.H. CHAN, R.W. SCOTT & C.J. EPSTEIN. 1997. Superoxide-mediated cytotoxicity in superoxide dismutase–deficient fetal fibroblasts. Arch. Biochem. Biophys. **344:** 424–432.
4. HUANG, T.-T., E.J. CARLSON, A.M. GILLESPIE, Y. SHI & C.J. EPSTEIN. 1999. Ubiquitous overexpression of CuZn superoxide dismutase does not extend life span in mice. J. Gerontol. Biol. Sci. In press.
5. SUN, J. & J. TOWER. 1999. FLP recombinase-mediated induction of Cu/Zn-superoxide dismutase transgene expression can extend the life span of adult $Drosophila$ $melanogaster$ flies. Mol. Cell Biol. **9:** 216–228.
6. CHAN, P.H., G.Y. YANG, S.F. CHEN, E. CARLSON & C.J. EPSTEIN. 1991. Cold-induced brain edema and infarction are reduced in transgenic mice overexpressing CuZn-superoxide dismutase. Ann. Neurol. **29:** 482–486.
7. MIKAWA, S., H. KINOUCHI, H. KAMII, G.T. GOBBEL, S.F. CHEN, E. CARLSON, C.J. EPSTEIN & P.H. CHAN. 1996. Attenuation of acute and chronic damage following traumatic brain injury in CuZn-superoxide dismutase transgenic mice. J. Neurosurg. **85:** 885–891.
8. KINOUCHI, H., C.J. EPSTEIN, T. MIZUI, E. CARLSON, S. CHEN & P.H. CHAN. 1991. Attenuation of focal cerebral ischemia in transgenic mice overexpressing CuZn-superoxide dismutase. Proc. Natl. Acad. Sci. USA **88:** 11158–11162.

9. YANG, G., P.H. CHAN, J. CHEN, E. CARLSON, S.F. CHEN, P. WEINSTEIN, C.J. EPSTEIN & H. KAMII. 1994. Human copper-zinc superoxide dismutase transgenic mice are highly resistant to reperfusion injury after focal cerebral ischemia. Stroke **25:** 165–170.
10. MURAKAMI, K., T. KONDO, C.J. EPSTEIN & P.H. CHAN. 1997. Overexpression of CuZn-superoxide dismutase reduces hippocampal injury following ischemia in transgenic mice. Stroke **28:** 1797–1804.
11. KAMII, H., H. KINOUCHI, F.R. SHARP, J. KOISTINAHO, C.J. EPSTEIN & P.H. CHAN. 1994. Prolonged expression of hsp70 mRNA following transient focal cerebral ischemia in transgenic mice overexpressing CuZn-superoxide dismutase. J. Cereb. Blood Flow Metab. **14:** 478–486.
12. KAMII, H., H. KINOUCHI, F.R. SHARP, C.J. EPSTEIN, S.M. SAGAR & P.H. CHAN. 1994. C-fos mRNA expression after a mild focal brain ischemia in SOD-1 transgenic mice. Brain Res. **662:** 240–244.
13. MIKAWA, S., F.R. SHARP, H. KAMII, H. KINOUCHI, C.J. EPSTEIN & P.H. CHAN. 1995. Expression of hsp70 and c-fos mRNA after traumatic brain injury in transgenic mice overexpressing CuZn-superoxide dismutase. Mol. Brain Res. **33:** 288–294.
14. KONDO, T., K. MURAKAMI, J. HONKANIEMI, F.R. SHARP, C.J. EPSTEIN & P.H. CHAN. 1996. Expression of *hsp70* mRNA is induced in the brain of transgenic mice overexpressing human CuZn-superoxide dismutase following transient global cerebral ischemia. Brain Res. **737:** 321–326.
15. CHAN, P.H., H. KAMII, G. YANG, J. GAFNI, C.J. EPSTEIN, E. CARLSON & L. REOLA. 1993. Brain infarction is not reduced in SOD-1 transgenic mice after a permanent focal cerebral ischemia. NeuroReport **5:** 293–296.
16. DITELBERG, J.S., R.A. SHELDON, C.J. EPSTEIN & D.M. FERRIERO. 1996. Brain injury after perinatal hypoxia-ischemia is exacerbated in copper/zinc superoxide dismutase transgenic mice. Ped. Res. **39:** 204–208.
17. PRZEDBORSKI, S., V. KOSTIC, V. JACKSON-LEWIS, A.B. NAINI, S. FAHN, E. CARLSON, C.J. EPSTEIN & J.L. CADET. 1992. Transgenic mice with increased Cu/Zn-superoxide dismutase activity are resistant to N-methyl-4-phenyl-1,2,3,6-tetrahydropyridine-induced neurotoxicity. J. Neurosci. **12:** 1658–1667.
18. ANDREWS, A.M., B. LADENHEIM, C.J. EPSTEIN, J.L. CADET & D.L. MURPHY. 1996. Transgenic mice with high levels of superoxide dismutase activity are protected from the neurotoxic effects of 2'-NH_2-MPTP on serotonergic and noradrenergic nerve terminals. Mol. Pharmacol. **50:** 1511–1519.
19. CADET, J.L., P. SHENG, S. ALI, R. ROTHMAN, E. CARLSON & C.J. EPSTEIN. 1994. Attenuation of methamphetamine-induced neurotoxicity in copper/zinc superoxide dismutase transgenic mice. J. Neurochem. **62:** 380–383.
20. CADET, J.L., B. LADENHEIM, I. BAUM, E. CARLSON & C.J. EPSTEIN. 1994. CuZn-superoxide dismutase (CuZn SOD) transgenic mice show resistance of the lethal effects of methylenedioxyamphetamine (MDA) and of methylenedioxymethamphetamine (MDMA). Brain Res. **655:** 259–262.
21. CADET, J.L., B. LADENHEIM, R.B. ROTHMAN, E. CARLSON, C.J. EPSTEIN & T.H. MORAN. 1995. Superoxide radicals mediate the neurotoxicity of methylenedioxymethamphetamine (MDMA): evidence from using CuZn-superoxide dismutase transgenic mice. Synapse **21:** 169–176.
22. HIRATA, H., B. LADENHEIM, R. B. ROTHMAN, C.J. EPSTEIN & J.L. CADET. 1995. Methamphetamine-induced serotonin neurotoxicity is mediated by superoxide radicals. Brain Res. **677:** 345–347.
23. HIRATA, H., B. LADENHEIM, E. CARLSON, C. EPSTEIN & J.L. CADET. 1996. Autoradiographic evidence for methamphetamine-induced striatal dopaminergic loss in mouse brain—attenuation in CuZn-superoxide dismutase transgenic mice. Brain Res. **714:** 95–103.

24. BEAL, M.F., R.J. FERRANTE, R. HENSHAW, R.T. MATTHEWS, P.H. CHAN, N.W. KORVALL, C.J. EPSTEIN & J.B. SCHULZ. 1995. 3-Nitropropionic acid neurotoxicity is attenuated in copper/zinc superoxide dismutase transgenic mice. J. Neurochem. **65:** 919–922.
25. KONDO, T., F.R. SHARP, J. HONKANIEMI, S. MIKAWA, C.J. EPSTEIN & P.H. CHAN. 1997. DNA fragmentation and prolonged expression of c-*fos*, c-*jun* and *hsp*70 in kainic acid–induced neuronal cell death in transgenic mice overexpressing human CuZn-superoxide dismutase. J. Cereb. Blood Flow Metab. **17:** 241–256.
26. FRIEDLICH, A.L., T.W. FARRIS, E. CARLSON, C.J. EPSTEIN & L.L. BUTCHER. 1993. β-amyloid protein-induced neurotoxicity in wild type mice and in transgenic mice with elevated CuZn superoxide dismutase activity. Soc. Neurosci. Abstr. **19:** 1036.
27. RABER, J., S.M. TOGGAS, S. LEE, F.E. BLOOM, C. EPSTEIN & L. MUCKE. 1996. Central nervous system expression of HIV-1 Gp120 activates the hypothalamic-pituitary-adrenal axis: evidence for involvement of NMDA receptors and nitric oxide synthase. Virology **226:** 362–373.
28. NAKAO, N., E.M. FRODL, H. WIDNER, E. CARLSON, S.K. CHO, C.J. EPSTEIN & P. BRUNDIN. 1995. Overexpressing Cu/Zn superoxide dismutase enhances survival of transplanted neurons in a rat model of Parkinson's disease. Nature Med. **1:** 226–231.
29. PRZEDBORSKI, S., U. KHAN, V. KOSTIC, E. CARLSON, C. EPSTEIN & D. SULZER. 1996. Increased superoxide dismutase activity improves survival of cultured postnatal midbrain neurons. J. Neurochem. **67:** 1383–1392.
30. SANCHEZ-RAMOS, J.R., S. SONG, A. FACCA, A. BASIT & C.J. EPSTEIN. 1997. Transgenic murine dopaminergic neurons expressing human Cu/Zn superoxide dismutase exhibit increased density in culture, but no resistance to methylphenylpyridium-induced degeneration. J. Neurochem. **68:** 58–67.
31. MENA, M.A., U. KHAN, D.M. TOGASAKI, D. SULZER, C.J. EPSTEIN & S. PRZEDBORSKI. 1997. Effects of wild-type and mutated copper/zinc superoxide dismutase on neuronal survival and L-DOPA–induced toxicity in postnatal midbrain culture. J. Neurochem. **69:** 21–33.
32. FULLERTON, H.J., J.S. DITELBERG, S.F. CHEN, D.P. SARCO, P.H. CHAN, C.J. EPSTEIN & D.M. FERRIERO. 1998. Copper/zinc superoxide dismutase transgenic brain accumulates hydrogen peroxide after perinatal hypoxia ischemia. Ann. Neurol. **44:** 357–364.
33. CHAN, P.H., M. KAWASE, K. MURAKAMI, S.F. CHEN, Y. LI, B. CALAGUI, L. REOLA, E. CARLSON & C.J. EPSTEIN. 1998. Overexpression of SOD1 in rats protects vulnerable neurons against ischemic damage after global cerebral ischemia and reperfusion. J. Neurosci. **18:** 8292–8299.
34. KUJIRAI, K., V. JACKSON-LEWIS, E. CARLSON, C.J. EPSTEIN & J.L. CADET. 1994. Autoradiographic distribution of μ opioid receptors in the brains of Cu/Zn-superoxide dismutase mice. Synapse **17:** 76–83.
35. ELMER, G.I., J.L. EVANS, B. LADENHEIM, C.J. EPSTEIN & J.L. CADET. 1995. Transgenic superoxide dismutase mice differ in opioid-induced analgesia. Eur. J. Pharmacol. **283:** 227–232.
36. ELMER, G.I., J.L. EVANS, S.R. GOLDBERG, C.J. EPSTEIN, C.J. & J.L. CADET. 1996. Transgenic superoxide dismutase mice: increased mesolimbic μ-opioid receptors result in greater opioid-induced stimulation and opioid-reinforced behavior. Behav. Pharmacol. **7:** 628–638.
37. AVRAHAM, K.B., M. SCHICKLER, D. SAPOZNIKOV, R. YAROM & Y. GRONER. 1988. Down's syndrome: abnormal neuromuscular junction in tongue of transgenic mice with elevated levels of human Cu/Zn-superoxide dismutase. Cell **54:** 823–829.

38. AVRAHAM, K.B., H. SUGARMAN, S. ROTSHENKER & Y.GRONER. 1991. Down's syndrome: morphological remodelling and increased complexity in the neuromuscular junction of transgenic CuZn-superoxide dismutase mice. J. Neurocytol. **20:** 208–215.
39. REAUME, A.G., J.L. ELLIOTT, E.K. HOFFMAN, N.W. KOWALL, R.J. FERRANTE, D.F. SIWEK, I.M. WILCOX, D.G. FLOOD, M.F. BEAL, R.H.J. BROWN, R.W. SCOTT & W.D. SNIDER. 1996. Motor neurons in Cu/Zn superoxide-dismutase–deficient mice develop normally but exhibit enhanced cell death after axonal injury. Nature Genet. **13:** 43–47.
40. KONDO, T., A.G. REAUME, T.-T. HUANG, E. CARLSON, K. MURAKAMI, S.F. CHEN, E.K. HOFFMAN, R.W. SCOTT, C.J. EPSTEIN & P.H. CHAN. 1997. Reduction of CuZn-superoxide dismutase activity exacerbates neuronal cell injury and edema formation following transient focal cerebral ischemia. J. Neurosci. **17:** 4180–4189.
41. HO, Y.S., R. VINCENT, M.S. DEY, J.W. SLOT & J.D. CRAPO. 1998. Transgenic models for the study of lung antioxidant defense: enhanced manganese-containing superoxide dismutase activity gives partial protection to B6C3 hybrid mice exposed to hyperoxia. Am. J. Respir. Cell Mol. Biol. **18:** 538–547.
42. CHEN, Z., B. SIU, Y.S. HO, R. VINCENT, C.C. CHUA, R.C. HAMDY & B.H. CHUA. 1998. Overexpression of MnSOD protects against myocardial ischemia/reperfusion injury in transgenic mice. J. Mol. Cell. Cardiol. **30:** 2281–2289.
43. YEN, H.C., T.D. OBERLEY, C.G. GAIROLA, L.I. SZWEDA & D.K. ST. CLAIR. 1999. Manganese superoxide dismutase protects mitochondrial complex I against adriamycin-induced cardiomyopathy in transgenic mice. Arch. Biochem. Biophys. **362:** 59–66.
44. KLIVENYI, P., D. ST. CLAIR, M. WERMER, H.C. YEN, T. OBERLEY, L. YANG & M.F. BEAL. 1998. Manganese superoxide dismutase overexpression attenuates MPTP toxicity. Neurobiol. Dis . **5:** 253–258.
45. MELOV, S., P. COSKUN, M. PATEL, R. TUINSTRA, B. COTTRELL, A.S. JUN, T.H. ZASTAWNY, M. DIZDAROGLU, S.I. GOODMAN, T.-T. HUANG, H. MIZIORKO, C.J. EPSTEIN & D.C. WALLACE. 1999. Mitochondrial disease in superoxide dismutase 2 mutant mice. Proc. Natl. Acad. Sci. USA **96:** 846–851.
46. ZHANG, Y., O. MARCILLAT, C. GIULIVI, L. ERNSTER & K.J. DAVIES. 1990. The oxidative inactivation of mitochondrial electron transport chain components and ATPase. J. Biol. Chem. **265:** 16330–16336.
47. GARDNER, P.R., D.D. NGUYEN & C.W. WHITE. 1994. Aconitase is a sensitive and critical target of oxygen poisoning in cultured mammalian cells and in rat lungs. Proc. Natl. Acad. Sci. USA **91:** 12248–12252.
48. ROBERTS J.R., C. NARASINHAM, P.W. HRUZ, G.A. MITCHELL & H.M. MIZIORKO. 1994. Evaluation of cysteine 266 of human 3-hydroxy-3-methylglutaryl-CoA lyase as a catalytic residue. J. Biol. Chem. **269:** 17841–17846.
49. ROBERTS J.R., G.A. MITCHELL & H.M. MIZIORKO. 1994. Modeling of a mutation responsible for human 3-hydroxy-3-methylglutaryl-CoA lyase deficiency implicates histidine 233 as an active site residue. J. Biol. Chem. **271:** 24604–24609.
50. WILLIAMS, M.D., H. VAN REMMEN, C.C. CONRAD, T.T. HUANG, C.J. EPSTEIN & A. RICHARDSON. 1998. Reduced Mn-superoxide dismutase activity in heterozygous manganese superoxide dismutase mice is correlated to altered mitochondrial function and increased oxidative damage. J. Biol. Chem. **273:** 28510–28515.
51. VAN REMMEN, H., C. SALVADOR, H. YANG, T.T. HUANG, C.J. EPSTEIN & A. RICHARDSON. 1999. Characterization of the antioxidant status of the heterozygous manganese superoxide dismutase mouse. Arch. Biochem. Biophys. **363:** 91–97.
52. MURAKAMI, K., T. KONDO, M. KAWASE, Y. LI, S. SATO, S.F. CHEN & P.H. CHAN. 1998. Mitochondrial susceptibility to oxidative stress exacerbates cerebral infarc-

tion that follows permanent focal cerebral ischemia in mutant mice with manganese superoxide dismutase deficiency. J. Neurosci. **18:** 205–213.
53. LI, Y., J.-C. COPIN, L.F. REOLA, B. CALAGUI, G.T. GOBBEL, S.F. CHEN, S. SATO, C.J. EPSTEIN & P.H. CHAN. 1998. Reduced mitochondrial manganese-superoxide dismutase activity exacerbates glutamate toxicity in cultured mouse cortical neurons. Brain Res. **814:** 164–170.
54. LEBOVITZ, R.M., H. ZHANG, H. VOGEL, J. CARTWRIGHT, JR., L. DIONNE, N. LU, S. HUANG & M.M. MAZTUK. 1996. Neurodegeneration, myocardial injury, and perinatal death in mitochondrial superoxide dismutase–deficient mice. Proc. Natl. Acad. Sci. USA **93:** 9782–9787.
55. THREADGILL, D.W. *et al.* 1995. Targeted disruption of mouse EGF receptor: effect of genetic background on mutant phenotype. Science **269:** 230–234.
56. ROZMAHEL, R., M. WILSCHANSKI, A. MATIN, S. PLYTE, M. OLIVER, W. AUERBACH, A. MOORE, J. FOSTNER, P. DURIE, J. NADEAU, C. BEAR & L.C. TSUI. 1996. Modulation of disease severity in cystic fibrosis transmembrane conductance regulator–deficient mice by a secondary genetic factor. Nature Genet. **12:** 280–287.
57. HUANG, T.-T., E.J. CARLSON, A.M. GILLESPIE & C.J. EPSTEIN. 1998. Genetic modification of the dilated cardiomyopathy and neonatal lethality phenotype of mice lacking manganese superoxide dismutase. Age **21:** 39–40.

The Blood-Brain Barrier and Cerebrovascular Pathology in Alzheimer's Disease

RAJ N. KALARIA[a]

Institute for Health of the Elderly, Newcastle General Hospital, Westgate Road, and Department of Psychiatry, University of Newcastle, Newcastle upon Tyne NE4 6BE, United Kingdom

ABSTRACT: The pathology of Alzheimer's disease (AD) is not limited to amyloid plaques and neurofibrillary tangles. Recent evidence suggests that more than 30% of AD cases exhibit cerebrovascular pathology, which involves the cellular elements that represent the blood-brain barrier. Certain vascular lesions such as microvascular degeneration affecting the cerebral endothelium, cerebral amyloid angiopathy and periventricular white matter lesions are evident in virtually all cases of AD. Furthermore, clinical studies have demonstrated blood-brain barrier dysfunction in AD patients who exhibit peripheral vascular abnormalities such as hypertension, cardiovascular disease and diabetes. Whether these vascular lesions along with perivascular denervation are coincidental or causal in the pathogenetic processes of AD remains to be defined. In this chapter, I review biochemical and morphological evidence in context with the variable but distinct cerebrovascular pathology described in AD. I also consider genetic influences such as apolipoprotein E in relation to cerebrovascular lesions that may shed light on the pathophysiology of the cerebral vasculature. The compelling vascular pathology associated with AD suggests that transient and focal breach of the blood-brain barrier occurs in late onset AD and may involve an interaction of several factors, which include perivascular mediators as well as peripheral circulation derived factors that perturb the endothelium. These vascular abnormalities are likely to worsen cognitive disability in AD.

INTRODUCTION

Alzheimer's disease (AD) is the most common cause of dementia in the elderly. The underlying processes that lead to dementia in this disorder are not understood but late onset AD is very likely acquired by the interaction of both hereditary and environmental factors. AD is pathologically defined by the presence of senile plaques and neurofibrillary tangles. The presence of other pathologies including vascular lesions is usually ignored or regarded as insignificant. Brains of subjects with AD often bear cerebrovascular pathology consisting of degenerative microangiopathy, cerebral amyloid angiopathy (CAA), cerebral infarcts and intracerebral hemorrhages. "Pure AD" is typically considered as plaque and tangle pathology, but does

---

[a]Address for correspondence and reprint requests: CBV Path Group, MRC Unit, Newcastle General Hospital, Westgate Road, Newcastle upon Tyne NE4 6BE, UK. Phone: 0191-273-5251; fax: 0191-272-5291.
e-mail: r.n.kalaria@ncl.ac.uk

TABLE 1. The variety and percentage of distribution of cerebrovascular lesions in AD[a]

Vascular Lesions	Specific feature(s) or markers involved
Degeneration of cerebral microvessels (100%)	Loss of endothelial markers, CD34, GLUT1 Thickening of basement membrane, collagen IV
Localization of serum proteins (80%)	P component, complement, ApoE in AD lesions
Presence of cerebral amyloid angiopathy (CAA) (98%)	Aβ peptides, Cystatin C protein, inflammatory markers
Presence of lobar and intracerebral hemorrhages (10%)	CAA-related intracerebral hemorrhages
Large cerebral infarcts and cortical micro-infarcts (36%)	Variably distributed and sized
Diffuse white matter disease (35%)	Periventricular and deep white matter lesions

[a]Data (%) derived from series published previously (Premkumar et al., 1996)[5] and unpublished results (Kalaria et al.) on 300 cases.

such clear-cut pathology exist in the elderly? Interestingly, even Alzheimer in his original report[1] describing the pathology in the brain of Auguste D had written that besides "one or several fibrils in otherwise normal cells", and "numerous small miliary foci ... and ... storage of peculiar material in the cortex, one sees endothelial proliferation and also occasionally neovascularisation." Is it likely that Alzheimer described degeneration of the cortical microvessels as the seat of the blood-brain barrier (BBB) evident by modern methods rather than angiogenesis or was it that Auguste D had suffered cerebral infarcts and the "endothelial proliferation" was a consequence of the infarction? Both of these possibilities and the fact that there was moderate athersosclerosis in the basal brain arteries of Auguste D provide a basis for thinking that vascular pathology was also evident in the original case of Alzheimer, which we use today to define AD. Nevertheless, it remains to be known whether the brain vascular pathology found in AD is coincident with the disease process or whether peripheral vascular abnormalities including those that alter cerebral perfusion are causal factors in AD. Indeed, BBB abnormalities may also be caused by vascular factors linked to cardiovascular disease. These include hypertension, atrial fibrillation, and aortic and carotid atherosclerosis that may decrease cerebral perfusion and increase risk of stroke or transient ischemic attacks in AD.

THE BLOOD-BRAIN BARRIER AND CEREBROVASCULAR PATHOLOGY IN AD

At least a third of the patients with AD may exhibit a variety of brain vascular lesions (TABLE 1). These are often ignored as coincidental findings at autopsy. The cerebrovascular pathology of AD encompasses a variety of lesions including changes in endothelial and vascular smooth muscle cells, macroscopic and micro-infarction, hemorrhage and white matter changes related to small vessel disease.[2,3] In addition, amyloid β protein is involved in the degeneration of both the larger perforating arterial vessels as well as the cerebral capillaries that represent the BBB. While these microvascular changes imply that the integrity of the cerebral vasculature is impaired in AD as a result of the progressive cortical pathology, they also entail the long-term influence of peripheral vascular factors. These include longstanding hypertension, atrial fibrillation, coronary or carotid artery disease, and diabetes: conditions that may promote cerebral hypoperfusion during aging. Whether, vascular and neurodegenerative pathologies are additive in the way in which they influence clinical presentation or progression of dementia[4,5] is the subject of some debate. It also remains to be known whether each of these lesions are simple manifestations of brain aging or intrinsic to the pathogenesis of AD and the cause of dementia.[6,7]

BBB-ASSOCIATED PROTEINS IN AD

Selective biochemical changes, not necessarily related to aging, are evident in the cerebral microvasculature of AD and Down's syndrome subjects (TABLE 2). We previously reported impairment of the BBB-associated glucose transporter (GLUT1) in isolated brain microvessels and in cortical membrane fractions obtained from subjects with AD and age-matched controls.[8] The rationale for this study was based on evidence from positron emission tomography studies showing that in AD the brain has a low metabolic rate for glucose, especially in those regions which are most affected pathologically.[9,10] This may implicate impaired GLUT1 in AD although under normal conditions, glucose transport across the BBB is not rate limiting. However, under pathophysiological conditions such as in seizures and perhaps aging, transport of glucose can become a limiting factor for brain oxidative metabolism. Using the binding properties of [^{3}H]cytochalasin B and immunochemical methods,[8] we also confirmed that many collagen IV positive capillaries were absent in GLUT1 immunoreactivity.[11,12] These findings, along with unchanged binding in the cerebellum and putamen indicated somewhat selective abnormality of the GLUT1 in AD, particularly at the BBB. The reduction of the GLUT1 protein in AD subjects could not be attributed to either aging or postmortem factors[8] and suggested abnormalities in the post-translational modification of the protein. Whether amyloid β deposition in cerebral microvessels in AD impairs synthesis or increases degradation of the GLUT1 protein remains to be shown. Our recent preliminary studies in microvessels suggest that GLUT1 mRNA is also decreased in AD. It is more likely that the decreased GLUT1 in AD is related to the morphological alterations in the cerebral endothelium as described above. Since the expression of GLUT1 is limited to the endothelium-bearing tight junctions[13] these results imply increased permeability of the BBB with consequent down regulation of the GLUT1. Our observations were recently corroborated by positron emission tomography studies showing that the transport of glucose (kinetic parameter k_1) into brains of AD subjects is diminished.[14] It is also of interest that Marcus *et al.*[15] observed decreased uptake of 2-deoxy-D-glucose uptake implying reduced hexokinase activity in isolated cerebral microvessels of AD subjects. This decreased activity directly relates to our reduced GLUT1 findings and suggests that microvessels deprived of glucose may be using other sources of metabolic fuel.

Further studies were pursued to determine if different proteins associated with BBB transport functions were affected in AD.[13] We assessed γ-glutamyl transferase along with other enzymes known to be associated with the endothelium in isolated

TABLE 2. Features of the BBB and markers in AD[a]

Cellular Feature	Specific Markers Found to be Affected
Cerebral endothelium	Loss of glucose transporter, Na^+/K^+ ATPase
Endothelial membranes/ microvascular endfeet	Moderate loss of AlkP, GGT, AChE, BChE
Basement membrane	Increase in collagen proteins and perlecans
Endothelial mitochondria	Loss of carnitine aceytltransferase
Cerebral endothelium (oxidative stress)	Increased glucose-6-phosphatase
Vascular smooth muscle cells	Loss of alpha actin and accumulation of amyloid β

[a]Data derived from series published previously (Kalaria, 1992)[2] and unpublished results (Kalaria *et al.*) on 300 cases.

ABBREVIATIONS: AChE, acetylcholinesterase; AlkP, alkaline phosphatase; BuChE, butrylcholinesterase; GGT, gamma glutamyl transpeptidase.

brain microvessels (TABLE 2). There were no statistically significant changes in the activities of plasma membrane-associated enzymes including angiotensin-converting enzyme and alkaline phosphatase between AD and age-matched controls, but the activities were consistently lower in vascular fractions from AD subjects. Previous morphological evidence has shown that the cerebral endothelium is enriched in mitochondria[16] and that cerebral microvessels of a number of species are enriched in monoamine oxidase localized to the outer membrane of mitochondria. We measured monoamine oxidase activities in cerebral microvessels from AD subjects and age-matched controls. We found no significant changes in total monoamine oxidase in microvessels although carnitine acetyltransferase activity, an enzyme also localized to the mitochondria but not necessarily restricted to them, was significantly reduced in microvessels from AD subjects. Cholinesterases previously known to be localized in the cerebral microvasculature were also assessed. We found that activities of both acetylcholinesterase and butrylcholinesterase were significantly reduced in cerebral microvessels from AD subjects compared to age-matched controls (TABLE 2). These differential alterations cannot be readily explained but may relate to the deposition of amyloid β in the vasculature since cholinesterases interact with amyloid and associated proteins. However, these findings support the notion that selective or focal BBB changes may occur in the neocortex in AD.

Previous studies have suggested that the cerebral microvasculature is immunologically activated in AD. Cell adhesion molecules such as the intercellular adhesion molecule (ICAM-1) induced during inflammatory processes are upregulated in response with amyloidotic pathology[17–19] and are increased in brain capillaries and perivascular cells in AD subjects.[20] We reported that ICAM-1 immunoreactivity within capillary profiles associated with amyloid β plaques and soluble ICAM-1 determined by immunoassay were significantly increased in frontal and temporal cortex of AD subjects compared to age-matched controls.[21] Similarly, the vascular cellular adhesion molecule was increased in both capillaries and neocortical extracts from brains of AD subjects (Kalaria *et al.*, unpublished observations). These findings support the activation of the cerebral endothelium that may lead to increased permeability to circulating cells and therefore alterations in the BBB.

THE BBB AND SERUM PROTEINS IN AD LESIONS

The localization of proteins originating in or extravasated from the circulation is considered an index of BBB breakdown and provide support for microvascular abnormalities in AD. This is particularly important if the factors or proteins derived from the circulation are not produced by the brain cellular elements. Both amyloid β deposits and neurofibrillary pathology in AD acquire several specific proteins whose role is unclear in these lesions. One of the first studies[22] reported positive albumin and IgG immunoreactivity in amyloid plaques and tangles contained in brains from AD patients. More recent studies showed that these diffuse deposits of serum proteins were also evident in aging controls and concluded that such localization was not proof of BBB breakdown in AD.[23] However, antemortem factors may explain the lack of clear differences between aging controls and AD subjects.[6,13] To elucidate this issue of BBB permeability in AD, we showed that circulating proteins that particularly localize with amyloidotic lesions may accumulate over a protracted period during focal and transient leaks in the BBB. The abundance of pentraxins such as P component and C-reactive protein immunoreactivities in cerebral lesions in AD but lack of their mRNA in brain suggested that these relatively large circulation-derived proteins and possibly others, as yet uncharacterized, originate from the liver during the pathogenetic process. The specific binding of P component and amyloid β could explain the localization of some but not the more common serum proteins such as albumin and immunoglobulins in cerebral amyloid β deposits.[24] It is feasible that proteins, which do not characteristically interact with amyloid β, would either be sequestered or readily removed via the brain drainage systems. Nonetheless, these observations provide indirect evidence for an immunological link between the brain parenchyma and the circulation.

While the BBB is accepted to be unconditionally impaired in vascular dementia, it has been proposed that BBB dysfunction is involved in the aetiology and pathogenesis of AD.[13,25,26] The cerebrospinal fluid (CSF):serum albumin ratio is a generally accepted method of assessing BBB function in living subjects. Increased CSF:serum ratios have been reported in AD patients, particularly in those exhibiting peripheral vascular disease,[13,27,28] but are not apparent in others.[29,30] However, Skoog et al.[31] reported that 85-year-olds with AD had a higher CSF:serum albumin ratio than nondemented individuals, and that there were indications of a disturbed BBB function even before onset of the disease in a population-based study. It is noteworthy that chronic hypertension, considered a strong risk factor for AD, is another factor which could cause increased vascular permeability with protein extravasation. A relative BBB dysfunction may increase the possibility that substances from serum penetrate the BBB and reach the brain, where they may interact with neurons, perhaps initiating a cascade with amyloid accumulation and Alzheimer encephalopathy.

ENDOTHELIAL DEGENERATION AND BASEMENT MEMBRANE PATHOLOGY

In addition to CAA and associated intracerebral hemorrhage, AD subjects exhibit profound changes in cerebral microvessels often independent of amyloid deposition.[2] Several elegant studies using morphological and biochemical methods have demonstrated abnormalities in various cellular elements of cerebral microvessels or capillaries that relate to BBB function. These include degeneration of vascular smooth muscle cells,[32,33] focal constrictions and degenerative changes in smooth muscle cells,[34] degeneration and focal necrotic changes of the endothelium,[11] vascular basement membrane alterations accompanied by accumulation of collagen,[26,35] loss of perivascular nerve plexus,[36] decreased mitochondrial content and increased pinocytotic vesicles,[37] and loss of tight junctions.[16] Using differential immunocytochenical methods, we[11,38] have further defined the convolutional abnormalities and "collapsed" or attenuated capillaries in cortical lobes of AD subjects.[39] The differential labeling was characterized by selective degeneration of the endothelium in capillary profiles, which yet retained their basement lamina, as evidenced by markers such as collagen IV. This phenomenon was observed in virtually all amyloid β–laden cortical lobes of more than 95% of the AD as well as Down's syndrome subjects.[11] Both the length and number of degenerated microvessel profiles were significantly correlated with neocortical amyloid β deposits, but there was no apparent relationship between the degenerated microvessels and neurofibrillary tangles or existing pyramidal neurons. This vascular phenomenon along with a profound microangiopathy is concomitant with amyloid β deposition and implies abnormalities in the patency of the brain microvasculature in AD.[2] The observations support previous conclusions on disturbances in local perfusion and oxygen tension as a consequence such that neurons furthest away from the capillaries are divested.[36] However, it should be realized that the brain is a dynamic organ and that reactive or compensatory responses are presumably not limited to neurons or glia[38] but that the cerebral endothelium must also be constantly changing even in aging.[2] It is conceivable that the BBB may be able to sustain subtle damage within certain regions and that the deposition of extracellular amyloid predisposes the endothelium to further degeneration. Although, these vascular abnormalities combined with the vascular amyloid deposition imply breach of the BBB in AD, clear functional evidence to support this is not apparent from non-invasive imaging and permeability studies (see Kalaria, 1992[2]). However, it is possible that breakdown of the BBB occurs focally and transiently over a protracted period in association with reactive mechanisms that direct repair and growth.[38]

CEREBRAL AMYLOID ANGIOPATHY, INTRACEREBRAL HEMORRHAGE AND APOLIPOPROTEIN E

Amyloid β–associated CAA has been reported to be present in 82-97% of AD subjects and is consistently always present in Down's syndrome.[3] Our recent analysis on isolated cerebral vessels in parallel with brain tissue from a series of 300 cases of AD indicates that CAA is more frequent in AD than previously thought.[5] It involves the leptomeninges, small pial vessels, intracortical arterioles as well as brain capillaries.[39] The lesion was characterized by sporadic focal deposits in surface vessels to complete infiltration of numerous meningeal and intracortical vessels throughout all cortical lobes.[39] There can be little doubt that cerebral vascular amyloid deposition resulting in CAA compounds the aging-related microvascular abnormalities in AD.[2] However, amyloid β–associated CAA may coexist with other neurodegenerative disorders such as Creutzfeldt-Jakob disease,[40] and it may be exclusive to certain disorders such as hereditary cerebral hemorrhage or hemorrhagic stroke with amyloidosis of the Dutch type and the Flemish type. CAA was frequently prominent in the occipital lobes and more profound in the sulci compared to the gyri of the neocortex. Vascular amyloid β deposits were rare in the large cranial arteries or muscular vessels of peripheral organs even in patients with relatively high degree of cerebral amyloid β burden. CAA could result from head injury or sporadic hemorrhagic strokes causing vascular amyloid accumulation when cerebral vessels are subjected to trauma, oxidative stress, or hemodynamic stress.[41] The characteristic cerebral distribution of CAA also implies that the process may be largely limited to brain vessels associated with a tight or continuous endothelium and when exposed to molecular triggers which may include soluble amyloid β itself that may even originate in perivascular plaques.[2] An intriguing hypothesis has been proposed to explain the mechanism of CAA and accumulation of amyloid β in brain. Weller et al.[42] have suggested that the characteristic vascular deposition of amyloid is related to the lack of clearance of amyloid β via the interstitial drainage pathways. Irrespective of the mechanism of CAA, it is highly likely that the characteristic vascular deposition in AD and the other amyloid angiopathies compromise BBB function and promote chronic hypoperfusion.[43]

CAA is considered an important cause of intracerebral and lobar hemorrhages. We have estimated that up to 10% of AD subjects exhibit CAA-related intracerebral hemorrhages.[5,44] We have moreover shown that AD subjects with evidence of intracerebral hemorrhage exhibit higher proportions of the more pathogenic amyloid β(42) compared to amyloid β(40) in the vasculature (Kalaria et al., unpublished observations). Whereas intracerebral bleeds characterize the Dutch and Flemish variants of cerebral hemorrhage with amyloidosis, it can cause premature death in the elderly and AD patients. Consideration of the Dutch disease provides certain clues to a link between CAA and stroke. It is thought that the first stroke-like episode triggers multiple cerebral bleeds, which may be accompanied by white matter lesions that in turn lead to rapid decline of cognitive functions.[45] We have previously proposed that certain routinely classified AD cases with predominant microvascular lesions and vascular amyloid β deposition might in fact be CAA variants of AD.[38] This would be consistent with the pathological features of the Dutch hereditary disease where severe CAA is present in the absence of profound cortical pathology.

Previously implicated as a susceptibility factor for cardiovascular disease, the inheritance of the ε4 allele of the apolipoprotein E gene (APOE) is considered to be the most important genetic factor in nonfamilial AD. A three- to fourfold increased frequency of the APOE-ε4 allele is not only linked to late-onset AD, but also to middle-aged individuals with coronary heart disease[46] and atherosclerosis.[47] In accord with the implications that apolipoprotein E might promote pathological alterations in the vascular wall, it is intriguing that the APOE-ε4 allele is also a strong factor in the development of CAA in AD.[5,48] We examined the frequencies of APOE-ε4 alleles in age-matched controls and subgroups of 200 AD subjects exhibiting CAA and other frequently associated vascular lesions. APOE-ε4-allele frequency (48%) in AD subjects with moderate to severe CAA was six times higher than those who exhibited mild CAA. In the subjects with severe CAA, the occurrence of an ε4 allele was increased by a factor of more than 15. This was despite the fact that neocortical amyloid β plaque densities in the advanced and mild CAA groups were similar and that all the subjects had met the accepted neuropathological criteria for AD. More remarkably, the ε4-allele frequency was highly associated with AD subjects exhibiting lobar or intracerebral hemorrhage, all of whom had advanced CAA. These findings suggested that the *APOE*-ε4 allele is a significant factor in the development of CAA in AD and revealed the possibility that *APOE* is a factor in CAA and other vascular abnormalities associated with AD. Our observations on the relationship between *APOE*-ε4 allele and CAA-related intracerebral hemorrhage were confirmed by others,[49–52] but it was later demonstrated that the ε4 allele does not appear to be an independent risk factor for CAA-related hemorrhage. Surprisingly, recent studies implicate *APOE*-ε2 allele as the strong factor for intracerebral hemorrhage in amyloid-laden vessels that may cause rupture of the vessel walls by inducing specific cellular changes.[50,53,54]

Recent observations suggest *APOE*-ε4 allele frequency also increases the risk of dementia in stroke survivors[55] and that the allele frequency is similarly increased in vascular dementia.[56] Bronge *et al*.[57] have recently reported that *APOE*-ε4 homozygotes have extensive white matter lesions seen upon magnetic resonance imaging in the deep white matter than those with the ε3/ε3 genotypes. These reports are in accord with our postmortem studies showing that the ε4 allele frequency in 36% of the AD subjects, who exhibited concomitant cerebrovascular pathology resulting from single infarcts, multiple microinfarcts, ischemic white matter lesions, or petechial hemorraghes, was significantly higher than in those without such pathology. That additive effects of peripheral vascular disease and *APOE* may also be important in AD has been implied by some studies.[58–60] We found that 77% of the patients with AD had cardiovascular disease defined by the presence of variable arteriosclerosis of the aortic blood vessels at autopsy. We also reported that *APOE* frequencies in patients who exhibited cardiovascular disease were significantly different from those who did not and over 60% of the AD subjects with arteriosclerosis carried at least one *APOE*-ε4 allele. Interestingly, those who carried the *APOE*-ε4 allele were almost three times likely to have had both AD and cardiovascular disease. These findings are consistent with those of Hofman *et al*.[58] implicating an interaction between carotid artery thickening and *APOE* ε4 in the progression of AD.

APOE is considered to have more widespread effects than any other genetic factor implicated in AD, but the mechanism(s) by which it exerts its effect remain largely unknown. However, several scenarios involving molecular events in brain pathology and physiological actions on the cardiovascular system have been proposed. An increased reutilization of apolipoprotein E-lipid complexes may explain our recent finding that apolipoprotein E in CSF is decreased both in AD and vascular dementia.[61,62] Both hypertension and hyperlipidemia are major risk factors for atherosclerosis.[63] Since high serum cholesterol level during middle age was associated with an increased risk for AD in old age, it is conceivable that a part of the effect of the *APOE*-ε4 allele on the risk for AD is mediated through high serum cholesterol concentrations. Alternatively, it can have direct effects on the endothelium rendering it leaky and leading to compromise of BBB functions.[63] In view of these developments, it remains to be seen whether cholesterol-lowering drugs will be useful to prevent or slowdown progression of AD. Indeed, the outcome of such an approach may help to define the nature of the interaction between apolipoprotein E, atherosclerosis,[58] and ischemic white matter lesions[61,64] in the etiology of AD.

OTHER PATHOLOGY IN AD: WHITE MATTER LESIONS AND CEREBRAL INFARCTION

Although white matter lesions may influence the course of AD there is no clear consensus to suggest that volume or localization of the lesions is predictive of AD. Ischemic white matter lesions associated with lipohyalinosis and narrowing of the lumen of the small perforating arteries as well as arterioles which nourish the deep white matter, have also been amply described in AD.[64–66] Neuropathological correlative studies comparing magnetic resonance imaging (MR) findings with post-mortem neuropathological examination have determined that the hyperintense deep white matter lesions, identified in more than 60% of AD patients as well as dementia with Lewy bodies,[67] consist mainly of demyelination, reactive gliosis and arteriosclerosis.[68] In late-onset AD subjects, white matter lesions are differently distributed than in early-onset AD[66] and in vascular dementia patients[69] and that they have more severe leukoariosis.[70] In a recent study of nondemented individuals with extensive neuropathology of AD, de la Monte[71] suggested that the white matter degeneration precedes the cortical atrophy in AD. It is known that long-standing hypertension causes lipohyalinosis and thickening of the vessel walls.[72] In this context, it is intriguing that AD-type pathology may be precipitated by hypertensive insults. For example, Sparks et al.[73] found an increased amount of neurofibrillary tangles and senile plaques in the brains of nondemented individuals with hypertension. However, knowledge about risk factors for the development of ischemic white matter lesions, their progress over time, influence on the clinical course of AD and their relationship with other vascular pathologies remains to be still clarified.

Individuals with AD may be at increased risk for stroke and cerebral infarction, the main diagnostic features in vascular dementia. Current findings suggest that almost 35% of AD subjects bear evidence of cerebral infarction at autopsy.[5,44] Cortical lobar infarcts as well as microinfarcts, which generally tend to be more frequent and invariably localized in the temporal and occipital lobes, occur in AD (TABLE 1). Both remote and recent infarcts have been evident at autopsy. The microinfarcts may or may not be associated with petechial hemorrhages. Multiple microinfarcts may occur as a result of endothelial damage or luminal blockage or thrombotic events induced in microvessels. CAA may also increase the tendency of cerebral infarction in AD. Neither the strategic location nor the volume of such infarcts has been precisely determined in AD and correlated with measures of cognitive decline. It is not understood whether the isolated infarcts are a consequence of or responsible for triggering the classical pathological lesions in AD. However, certain features or conditions may rapidly precipitate strokes in AD patients. This is corroborated by circumstantial evidence suggesting that rapid development of cerebral infarctions in some AD patients occurs because of a strong interaction between severity of cerebral amyloid angiopathy—i.e., amyloid infiltration of the media and adventitia of cerebral vessels, and hypertension.[5,44] Conversely, recent longitudinal clinical and epidemiological studies suggest that stroke episodes may lead to characteristic degenerative changes of AD demonstrated by the progressive onset and course of dementia.[74,75] Kokmen et al.[75] suggested that stroke may account for as cause of AD in as many as 50% of demented cases in older age groups. Such studies have also emphasized that AD is three times likely to precipitate in the elderly after a stroke episode or a transient ischemic attack. While the substantial presence of strokes is likely to influence the processes that cause dementia in AD, prospective follow-up studies are necessary to evaluate their exact role in AD, especially where preexisting stroke episodes or cardiovascular disease are not considered as an exclusion criteria for the diagnosis of AD.

CONCLUSIONS

Cerebrovascular abnormalities underscore the role for the blood-brain barrier in the etiopathogenesis of AD. There are profound morphological and biochemical changes such as that in the glucose transporter of the cortical microvasculature in subjects with late onset AD. In addition, amyloid β protein is involved in the degeneration of both the larger perforating arterial vessels as well as the cerebral capillaries that represent the BBB. These vascular changes not only imply that the integrity of the cerebral microvasculature is impaired in AD but may relate to long-term peripheral influences associated with cardiovascular disease or peripheral vascular disease. Genetic factors such as *APOE* and the ε4 allele may modify or attenuate cerebrovascular function during aging and increase the susceptibility to pathogenetic processes in AD. I suggest the microvascular disease causes transient and focal breach of the BBB in late onset AD and it results from an interaction of several factors, which include perivascular mediators and circulation derived damage to the endothelium. These may collectively contribute to the deterioration of cognitive functions in AD.

ACKNOWLEDGMENTS

Research reported here was supported by grants from MRC (UK), NIH (NINDS) and a Zenith award from the National Alzheimer's Association (Chicago, USA).

REFERENCES

1. ALZHEIMER, A. 1907. Uber eine eigenartig Erkrankung der Hirnrinde. Allg. Z. Psychiatrie Psych. Ger. Med. **64:** 146–148.
2. KALARIA, R.N. 1996. Cerebral vessels in ageing and Alzheimer's disease. Pharm. Therap. **72:** 193–214.
3. VINTERS, H.V. 1987. Cerebral amyloid angiopathy: a critical review. Stroke **18:** 311–324.
4. HEYMAN, A. *et al.* 1998. Cerebral infarcts in patients with autopsy-proven Alzheimer's disease. CERAD Part XVIII. Neurology **51:** 159–162.
5. PREMKUMAR, D.R.D. *et al.* 1996. Apolipoprotein E ε4 alleles in cerebral amyloid angiopathy and cerebrovascular pathology in Alzheimer's disease. Am. J Pathol. **148:** 2083–2095.
6. HACHINSKI, V. & D. MUNOZ. 1997. Cerebrovascular pathology in Alzheimer's disease: cause, effect or epiphenomeneon. Ann. N.Y. Acad. Sci. **826:** 1–6.
7. PASQUIER, F. *et al.* 1998. The influence of coincidental vascular pathology on symptomatology and course of Alzheimer's disease. J. Neural Transm. Suppl. **54:** 117–127.
8. KALARIA, R.N. & HARIK, S.I. 1989. Reduced glucose transporter at the blood-brain barrier and cerebral cortex in Alzheimer's disease. J. Neurochem. **53:** 1083–1088.
9. FREY, K.A. 1998. Neurochemical imaging of Alzheimer's disease and other degenerative dementias. Q. J. Nucl. Med. **42:** 166–178.
10. MIELKE, R. & W.D. HEISS. 1998. Positron emission tomography for diagnosis of Alzheimer's disease and vascular dementia. J. Neural Transm. Suppl. **53:** 237–250.

11. KALARIA, R.N. & P. HEDERA. 1995. Differential degeneration of the endothelium and basement membrane of capillaries in Alzheimer's disease. Neuroreport **6:** 477–480.
12. KAWAI, M. *et al.* 1990. The relationship of amyloid plaques to cerebral capillaries in Alzheimer's disease. Am. J. Pathol. **37:** 1435–1446.
13. KALARIA, R.N. 1992. The blood-brain barrier and cerebral microcirculation in Alzheimer's disease. Cerebrovascular Brain Metab. **4:** 226–260.
14. JAGUST, W.J. *et al.* 1991. Diminished glucose transport in Alzheimer's disease. Cereb. Blood Flow Metab. **11:** 323–330.
15. MARCUS, D.L. & M.L. FREEDMAN. 1997. Decreased brain glucose metabolism in microvessels from patients with Alzheimer's disease. Ann. N.Y. Acad. Sci. **826:** 248–253.
16. STEWART, P.A. *et al.* 1992. A mophometric study of the blood brain barrier in Alzheimer's disease. Lab. Invest. **67:** 734–742.
17. AKIYAMA, H. *et al.* 1993. Expression of intercellular adhesion moleculae (ICAM)-1 by a subset of astrocytes in Alzheimer disease and some other degenerative neurological disorders. Acta Neuropathol. **85:** 628–634.
18. FROHMAN, E.M. 1991. Expression of intercellular adhesion molecule (ICAM-1) in Alzheimer's disease. J. Neurol. Sci. **106:** 105–111.
19. VERBEEK, M.M. *et al.* 1994. Accumulation of intercellular adhesion molecule-1 in senile plaques in brain tissue of patients with Alzheimer's disease. Am. J. Pathol. **144:** 104–116.
20. KALARIA, R.N. 1993. The immunopathology of Alzheimer's disease. Brain Pathol. **3:** 333–347.
21. LERNER, A. *et al.* 1993. Cell adhesion molecules in the vascular and neocortical pathology of Alzheimer's disease. Soc. Neurosci. Abstr. **19:** 623.
22. WISNIEWSKI, H.M. & P.B. KOZLOWSKI. 1982. Evidence for bloodbrain barrier changes in senile dementia of the Alzheimer type (SDAT). Ann. N. Y. Acad. Sci. **396:** 119–131.
23. MUNOZ, D.G. *et al.* 1997. Serum protein leakage in Alzheimer's disease revisited. Ann. N. Y. Acad. Sci. **826:** 173–189.
24. KALARIA, R.N. *et al.* 1991. Serum amyloid P in Alzheimer's disease. Implications for dysfunction of the blood-brain barrier. Ann. N. Y. Acad. Sci. **640:** 145–148.
25. HARDY, J. *et al.* 1986. An integrative hypothesis concerning the pathogenesis and progression of Alzheimer's disease. Neurobiol. Aging **7:** 489–502.
26. PERLMUTTER, L.S. *et al.* 1994. Vascular basement membrane components and the lesions of Alzheimer's disease: Light and electron microscopic analyses. Microscopy Res. Tech. **28:** 204–215.
27. BLENNOW, K. *et al.* 1990. Blood-brain barrier disturbance in patients with Alzheimer's disease is related to vascular factors. Acta Neurol. Scand. **81:** 349–351.
28. HAMPEL, H. *et al.* 1995. Evidence of blood-cerebrospinal fluid-barrier impairment in a subgroup of patients with dementia of the Alzheimer type and major depression: a possible indicator for immunoactivation. Dementia **6:** 348–354.
29. KAY, A. *et al.* 1987. CSF and serum concentration of albumin and IgG in Alzheimer's disease. Neurobiol. Aging **8:** 2125.
30. MECOCCI, P. *et al.* 1991. Blood.brain barrier in a geriatric population: barrier function in degenerative and vascular dementia. Acta Neurol. Scand. **84:** 210–213.
31. SKOOG, I. *et al.* 1998. A population-study on blood-brain barrier function in 85-year-olds. Relation to Alzheimer's disease and vascular dementia. Neurology **50:** 966–971.
32. KAWAI, M. *et al.* 1993. Degeneration of amyloid precursor protein-containing smooth muscle cells in cerebral amyloid angiopathy. Brain Res. **623:** 142–146.
33. PERRY, G. *et al.* 1998. Cerebrovascular muscle atrophy is a feature of Alzheimer's disease. Brain Res . **791:** 63–66.

34. MIYAKAWA, T. 1997. Electron microscopy of amyloid fibrils and microvessels. Ann. N. Y. Acad. Sci. **826:** 25–34.
35. KALARIA, R.N. & A.B. PAX. 1995. Increased collagen content of cerebral microvessels in Alzheimer's disease. Brain Res. **705:** 349–352.
36. SCHEIBEL, A.B. *et al.* 1987. Denervation microangiopathy in senile dementia, Alzheimer type. Alz. Dis. Assoc. Dis. **1:** 19–37.
37. CLAUDIO, L. 1996. Utrastructural features of the blood-brain barrier in biopsy tissue from Alzheimer's disease patients. Acta Neuropathol. **91:**6–14.
38. KALARIA, R.N. *et al.* 1998. Vascular endothelial growth factor in Alzheimer's disease and experimental cerebral ischemia. Mol. Brain Res. **62:** 101–105.
39. KALARIA, R.N. *et al.* 1996. Production and increased detection of amyloid β-protein and amyloidogenic fragments in brain microvessels, meningeal vessels and choroid plexus in Alzheimer disease. Mol. Brain Res. **35:** 58–68.
40. GRAY, F. 1994. Creutzfeldt-Jakob disease and cerebral amyloid angiopathy. Acta Neuropathol. **88:** 106–111.
41. DE LA TORRE, J.C. 1997. Hemodynamic consequences of deformed microvessels in the brain in Alzheimer's disease. Ann. N. Y. Acad. Sci. **826:** 75–91.
42. WELLER, R.O. *et al.* 1998. Cerebral amyloid angiopathy: amyloid beta accumulates in putative interstitial fluid drainage pathways in Alzheimer's disease. Am. J. Pathol. **153:** 725–733.
43. DE JONG, G.I *et al.* 1997. Cerebrovascular hypoperfusion: a risk factor for Alzheimer's disease? Ann. N. Y. Acad. Sci. **826:** 56–74.
44. OLICHNEY, J.M. *et al.* 1995. Cerebral infarction in Alzheimer's disease is associated with severe amyloid angiopathy and hypertension. Arch. Neurol. **52:** 702–708.
45. HAAN, J. *et al.* 1990. Dementia in hereditary cerebral hemorrhage with amyloidosis-Dutch type. Arch. Neurol. **47:** 965–968.
46. WILSON, P.W.F. *et al.* 1994. Apolipoprotein E alleles, dyslipidemia, and coronary heart disease: the Framingham offspring study. JAMA **272:** 1666–1671.
47. DAVIGNON, J. *et al.* 1988. Apolipoprotein E polymorphism and atherosclerosis. Arteriosclerosis **8:** 1–21.
48. KALARIA, R.N. & D.R.D. PREMKUMAR. 1995. Apolipoprotein E genotype and cerebral amyloid angiopathy in Alzheimer's disease. Lancet **346:**1424.
49. GREENBERG, S.M. *et al.* 1995. Apolipoprotein E epsilon 4 and cerebral hemorrhage associated with amyloid angiopathy. Ann. Neurol. **38:** 254–259.
50. NICOLL, J.A. *et al.* 1997. High frequency of apolipoprotein epsilon 2 allele in hemorrhage due to cerebral amyloid angiopathy. Ann. Neurol. **41:** 716–721.
51. OLICHNEY, J.M. *et al.* 1996. Apolipoprotein epsilon 4 allele is associated with increased neuritic plaques and cerebral amyloid angiopathy in Alzheimer's disease and Lewy body variant. Neurology **47:** 190–196.
52. ZAROW, C. *et al.* 1999. Cerebral amyloid angiopathy in Alzheimer's disease is associated with apolipoprotein E4 and cortical neuron loss. Alz. Dis. Assoc. Dis. **13:** 1–8
53. GREENBERG, S.M. *et al.* 1998. Association of apolipoprotein E epsilon 2 and vasculopathy in cerebral amyloid angiopathy. Neurology **50:** 961–965.
54. MCCARRON, M.O. & J.A. NICOLL. 1998. High frequency of apolipoprotein E epsilon 2 allele is specific for patients with cerebral amyloid angiopathy-related haemorrhage. Neurosci. Lett. **247:** 45–48.
55. MARGAGLIONE, M. *et al.* 1998. Prevalence of apolipoprotein E alleles in healthy subjects and survivors of ischemic stroke: an Italian Case-Control Study. Stroke **29:** 399–403.
56. FRISONI, G.B. *et al.* 1994. Apolipoprotein E ε4 allele in Alzheimer's disease and vascular dementia. Dementia **5:** 240–242.

57. BRONGE, L. et al. 1999. White matter lesions in Alzheimer patients are influenced by apolipoprotein E genotype. Dement. Geriatr. Cogn. Disord. **10:** 89–96.
58. HOFMAN, A. et al. 1997. Atherosclerosis, apolipoprotein E, and the prevalence of dementia and Alzheimer's disease in the Rotterdam Study. Lancet **349:** 151–154.
59. KALARIA, R.N. 1997. Apolipoprotein E, arteriosclerosis and Alzheimer's disease. Lancet **349:** 1174–1175.
60. KOSUNEN, O. et al. 1995. Relation of coronary atherosclerosis and apolipoprotein E genotypes in Alzheimer patients. Stroke **26:** 743–748.
61. SKOOG, I. et al. 1997. Apolipoprotein E in cerebrospinal fluid in 85-year-olds. Relation to dementia, apolipoprotein E polymorphism, cerebral atrophy, and white-matter lesions. Arch. Neurol. **54:**267–272.
62. SKOOG, I. et al. 1998b. A population study of Apo E genotype at the age of 85: relation to dementia, cerebrovascular disease and mortality. J. Neurol. Neurosur. Psychiatr. **64:** 37–43.
63. NOTKOLA, I-L. et al. 1998. Serum total cholesterol, apolipoprotein E ε4 allele, and Alzheimer's disease. Neuroepidemiology **17:** 14–20.
64. WALLIN, A. 1998. The overlap between Alzheimer's disease and vascular dementia: the role of white matter changes. Dement. Geriatr. Cogn. Disord. **9**(Suppl. 1)**:** 30–35.
65. ENGLUND, E. 1998. Neuropathology of white matter changes in Alzheimer's disease and vascular dementia. Dement. Geriatr. Cogn. Disord. **9**(Suppl. 1)**:** 6-12.
66. SCHELTENS, P. et al. 1992. White matter lesions on magnetic resonance imaging in clinically diagnosed Alzheimer's disease. Evidence for a heterogeneity. Brain **115:** 735–748.
67. BARBER, R. et al. 1999. White matter lesions on magnetic resonance imaging in dementia with lewy bodies, Alzheimer's disease, vascular dementia, and normal aging. J. Neurol. Neurosurg. Psychiatry **67:** 66–72.
68. VAN GIJN, G. 1998. Leukoariosis and vascular dementia. Neurology **51**(Suppl. 3)**:** S3–S8.
69. WAHLUND, L.O. et al. 1994. White matter hyperintensities in dementia: does it matter? Magn. Reson. Imaging **12:** 387–394.
70. LEYS, D. et al. 1991. Could Wallerian degeneration contribute to "leuko-araiosis" in subjects free of any vascular disorder? J. Neurol. Neurosurg. Psychiatr. **54:** 46–50.
71. DE LA MONTE, S.M. 1989. Quantitation of cerebral atrophy in preclinical and end-stage Alzheimer's disease. Ann. Neurol. **25:** 450–459.
72. SKOOG, I. 1997. The relationship between blood pressure and dementia: a review. Biomed. Pharmacother. **51:** 367–375.
73. SPARKS, D.L. et al. 1990. Cortical senile plaques in coronary artery disease, aging and Alzheimer's disease. Neurobiol. Aging **11:** 601–607.
74. HENON, H. et al. 1997. Pre-existing stroke in patients: baseline frequency, associated factors and outcome. Stroke **28:** 2429–2436.
75. KOKMEN, E. et al. 1996. Dementia after ischaemic stroke: a population based study in Rochester, Minnesota (1960–1984). Neurology **46:** 154–159.

Neurological Changes Induced by Stress in Streptozotocin Diabetic Rats

LAWRENCE P. REAGAN, ANA MARIA MAGARIÑOS, AND BRUCE S. McEWEN[a]

The Harold and Margaret Milliken Hatch Laboratory of Neuroendocrinology, The Rockefeller University, New York, New York 10021, USA

ABSTRACT: Previous studies from our laboratory demonstrated that chronic stress produces molecular, morphological, and ultrastructural changes in the rat hippocampus that are accompanied by cognitive deficits. Glucocorticoid impairment of glucose utilization is proposed as a causative factor involved in stress-induced changes. Current studies have examined the neurological changes induced by stress in rats with a preexisting strain upon their homeostatic load—namely, in streptozotocin (stz)–diabetic rats. Administration of stz (70 mg/kg, iv) produced diabetic symptoms such as weight loss, polyuria, polydipsia, hyperglycemia, and neuroendocrine dysfunction. Morphological analysis of hippocampal neurons revealed that diabetes alone produced dendritic atrophy of CA3 pyramidal neurons, an effect potentiated by 7 days of restraint stress. Analysis of genes critical to neuronal homeostasis revealed that glucose transporter 3 (GLUT3) mRNA and protein levels were specifically increased in the hippocampus of diabetic rats, while stress had no effect upon GLUT3 expression. Insulin-like growth factor (IGF) receptor expression was also increased in the hippocampus of diabetic rats subjected to stress. In spite of the activation of these adaptive mechanisms, diabetic rats subjected to stress also had signs of neuronal damage and oxidative damage. Collectively, these results suggest that the hippocampus of diabetic rats is extremely susceptible to additional stressful events, which in turn can lead to irreversible hippocampal damage.

INTRODUCTION

Diabetes mellitus is an endocrine disorder of carbohydrate metabolism resulting primarily from inadequate insulin release (type 1 diabetes) or insulin insensitivity coupled with inadequate compensatory insulin release (type 2 diabetes). According to current projections, over 200 million people worldwide will be diagnosed with diabetes by the year 2010.[1] While the symptoms of diabetes can be controlled by insulin replacement, drugs, diet and exercise, the long-term complications associated with diabetes can lead to cardiovascular disease, as well as renal and ocular disorders. In addition, diabetes leads to the development of peripheral neuropathies, including motor and sensory polyneuropathies, as well as autonomic neuropathy.[2] The neurological consequences of diabetes mellitus in the central nervous system (CNS) have most recently been receiving greater attention. For example, hyperglycemia po-

---

[a]Corresponding author: Bruce S. McEwen, Ph.D., The Alfred E. Mirsky Professor, Head, The Harold and Margaret Milliken Hatch Laboratory of Neuroendocrinology, The Rockefeller University, 1230 York Avenue, Box 165, New York, New York 10021. Phone: 212-327-8624; fax: 212-327-8634.
e-mail: mcewen@rockvax.rockefeller.edu

tentiates neurological damage in stroke and ischemia, contributes to cerebrovascular complications, disruption of the blood-brain barrier (BBB), and cerebral edema.[3] Cognitive impairments are also recognized as one of the long-term neurological complications associated with diabetes.[4] Cognitive performance has been positively correlated with metabolic control in diabetic patients,[5] and improved glycemic control has been shown to ameliorate cognitive dysfunction in elderly type 2 patients.[6,7] Collectively, these studies demonstrate that diabetes produces molecular, cellular, morphological, and behavioral changes in the CNS and also illustrate that the neurological complications of diabetes are not limited to peripheral neuropathies.

Neuroendocrine dysfunction is also observed in diabetes. For example, cortisol levels are increased in type 1 diabetic patients,[8] suggesting that corticosteroids may participate in the neurological complications associated with diabetes. Closer examination of the actions of the corticosteroids in the brain reveals an apparent paradox in that glucocorticoids (GCs) and mineralocorticoids (MCs) are involved in central functions that are essential for neuronal plasticity, as well as neuronal death. As a result, corticosteroids appear to have a biphasic effect upon neuronal development, function, and survival. The paradoxical actions of corticosteroids are most evident in the hippocampus, a brain region proposed to serve as a critical integration center for memory formation and retrieval.[9] For example, 21-day administration of stress or stress levels of GCs produces selective atrophy of the apical dendrites of CA3 pyramidal neurons,[10,11] morphological changes that are associated with memory impairments.[12] Conversely, exposure to stress levels of GCs for prolonged periods produces not only reversible dendritic remodeling, but also neuronal damage and neuronal death in rats[13,14] and primates.[15,16] In pathophysiological settings, GCs potentiate neuronal damage observed following ischemia and stroke.[17] The GC potentiation of ischemia- and excitotoxicity-mediated neuronal damage can be prevented by energy supplementation,[9] suggesting that GCs create an energy-compromised environment that makes hippocampal neurons more vulnerable to neurotoxic events.

GCs have been shown to decrease glucose utilization in both clinical and experimental settings.[18–20] Under such conditions, GCs may increase neuronal vulnerability by decreasing glucose utilization,[9] thereby increasing neuronal allostatic load.[21] In an attempt to identify the factors that may be involved in the transition from stress-mediated neuronal plasticity towards stress-mediated neuronal death, our more recent studies have examined the effects of stress in an experimental model of increased allostatic load. We have investigated the neurological changes mediated by stress in streptozotocin (stz)–treated rats, an experimental model of type 1 diabetes used extensively to examine the physiological and pathophysiological consequences of diabetes.[3,4] Our studies to date indicate that compensatory mechanisms are activated in the hippocampus of diabetic rats subjected to stress. Nonetheless, stz-diabetic rats subjected to stress appear to have increased neuronal vulnerability in the hippocampus that may in turn produce irreversible neuronal damage.

DIABETES-INDUCED MORPHOLOGICAL CHANGES IN THE CNS

While the development of peripheral neuropathies is well documented in diabetic patients,[2,22] the development of central nervous system complications in diabetes has previously been attributed to degenerative vascular disease.[23] Seminal investigations by Reske-Nielsen and co-workers introduced the concept of diabetic encephalopathy independent of vascular disease.[24,25] These investigators reported numerous morphologic abnormalities in type 1 diabetic patients, including neuronal degeneration in the cortex, cerebellum, and brain stem. Subsequent studies have confirmed and extended these initial findings.[26] For example, type 1 diabetic patients have significant increases in lesion size and lesion number when compared to aged-matched controls.[27] Experimental models of type 1 diabetes have provided similar results in that stz-diabetic rats have both functional and morphological alterations in the hypothalamus.[28,29] The long-term effects of stz-induced diabetes also give rise to reductions in brain weight, as well as reductions in neocortical volume and in the density of neocortical neurons.[30] Electron microscopic analysis of the gracile nucleus in the stz-diabetic rat has shown morphological abnormalities, including degenerating dendritic and axonal terminals, mitochondrial swelling, and vesicular clustering.[31] Collectively, the results from both the experimental and clinical settings show that diabetic encephalopathy is a complication associated with diabetes and suggests that such morphologic abnormalities may contribute to cognitive impairments seen in elderly diabetic patients.

Morphological Analysis of Hippocampal Neurons in Diabetic Rats Subjected to Stress

To determine if increases in allostatic load would accelerate the actions of GCs in the hippocampus, we initiated studies to examine the morphological changes in diabetic rats and diabetic rats subjected to seven days of restraint stress.[32] These studies showed that approximately 10 days following stz administration, diabetes reduced the number of apical branch points, as well as the total dendritic length of the apical dendrites of CA3 pyramidal neurons. Diabetic rats subjected to seven days of restraint stress had more pronounced dendritic remodeling. These results demonstrate that increases in neuronal homeostatic load accelerate dendritic remodeling in the hippocampus. In addition, these results suggest that diabetes may make hippocampal neurons more vulnerable to subsequent stressors.

BRAIN GLUCOSE TRANSPORTERS AND DIABETES

The family of facilitative glucose transporter (GLUT) proteins are responsible for the entry of glucose into cells throughout the periphery and the brain.[33] The expression, regulation and activity of glucose transporters play an essential role in neuronal homeostasis, since glucose represents the primary energy source for the brain.[34,35] While many GLUT isoforms have been identified in the brain, four of these transporters have been extensively studied and are believed to be responsible for the majority of glucose transport and utilization in the CNS. In particular, GLUT1 is highly expressed in capillaries at the blood-brain barrier and is responsible for the transport of circulating glucose into the CNS, while GLUT5 is expressed in microglial cells. GLUT4, the insulin-sensitive glucose transporter, is expressed in the cerebellum and cortex, as well as the hypothalamus.[36] GLUT3, the neuron-specific glucose transporter, is solely responsible for the essential function of glucose delivery into neurons. GLUT3 mRNA and protein have a widespread distribution in the brain, including the hippocampus.[37,38]

The regulation of glucose transporter expression in the brain during diabetes remains controversial. For example, some investigators have reported that GLUT1 mRNA expression is increased in experimental models of diabetes,[39–41] while other studies reported that GLUT1 expression was not altered following chronic hyperglycemia.[42,43] Investigations into the regulation of GLUT3 by diabetes have also failed to reach a consensus. Chronic hyperglycemia induced by stz does not alter GLUT3 protein expression as determined by immunoblot analysis.[42] GLUT3 mRNA expression has region-specific regulation during development in *db/db* mice,[41] but has been reported to be unaffected by stz administration.[43]

Regulation of GLUT3 Expression in the Hippocampus of Diabetic Rats Subjected to Stress

In view of the discrepancies in these studies, we examined the regulation of GLUT3 in the hippocampus of diabetic rats subjected to stress.[44] *In situ* hybridization histochemistry showed that diabetes, but not 7 days of restraint stress, increases GLUT3 mRNA expression in the hippocampus. Diabetes-induced increases in GLUT3 mRNA expression observed in our study were specific for the hippocampus, in that diabetes did not modulate the expression of GLUT3 mRNA expression in the cortex of diabetic rats. These results suggest that some of the discrepancies observed in the diabetes-mediated regulation of glucose transporter expression may be the result of regionally specific changes. We also examined the regulation of GLUT3 protein expression in the hippocampus of diabetic rats subjected to stress using a radioimmunocytochemical approach. The results of this study revealed that GLUT3 protein expression was increased in the hippocampus of diabetic rats subjected to stress; conversely, stress alone did not affect GLUT3 protein expression. These results suggest that stz-induced increases in GLUT3 mRNA and protein expression in the hippocampus may represent a compensatory mechanism to increase glucose utilization during diabetes. In addition, the results of this study suggest that the impairment of glucose utilization by stress levels of GCs does not occur at the transcriptional or translational levels for GLUT3.

DIABETES, OXIDATIVE STRESS, AND MITOCHONDRIAL FUNCTION

Oxidative stress, lipid peroxidation, and increased production of reactive oxygen species are known to be increased in diabetes and by stress.[45-47] For instance, increases in superoxide production are observed in the serum of type 1 diabetic patients, increases that are reduced with improved glycemic control.[48] Oxidative damage in rat brain is increased by experimentally induced hyperglycemia.[49] Lipid peroxidation products are also increased in the brains of type 2 diabetic mice, while the activity of antioxidant enzymes such as catalase and superoxide dismutase (SOD) are decreased.[50,51] Increases in lipid peroxidation products, such as 4-hydroxynonenal (HNE), reduce the activity of a variety of enzymes that are critical to neuronal homeostasis, including glucose transporters.[52] HNE has also been shown to impair mitochondrial function in brain synaptosomes.[53] More recent studies showed that exposure of intact cardiac mitochondria to HNE resulted in significant decreases in mitochondrial respiration.[54] Subsequent studies found that α-ketoglutarate dehydrogenase and pyruvate dehydrogenase serve as specific mitochondrial enzyme targets for HNE protein conjugation, resulting in decreases in enzyme activity.[55] Similar decreases in pyruvate dehydrogenase activity are observed in the brains of type 2 diabetic mice, suggesting that mitochondrial enzymes may also be targets of oxidative stress in diabetes.[56] Collectively, these results suggest that diabetes-mediated increases in oxidative stress, decreases in antioxidant enzyme activity, as well as diabetes-induced mitochondrial dysfunction, may make the hippocampus more vulnerable to subsequent pathophysiological events.

Oxidative Stress Is Increased in the Hippocampus of Diabetic Rats Subjected to Stress

In an attempt to determine if lipid peroxidation is increased in the hippocampus of diabetic rats subjected to stress, radioimmunocytochemistry was performed using previously characterized polyclonal antisera that recognize HNE-conjugated proteins[57] and ^{35}S-labeled secondary antisera. The results of this study showed that HNE protein conjugation is specifically increased in all subregions of the hippocampus in diabetic rats subjected to stress.[58] Conversely, HNE protein conjugation was not increased in the cortex of diabetic rats subjected to stress compared to control rats. These results show that oxidative stress is specifically increased in the hippocampus of diabetic rats subjected to stress and suggests that the hippocampus may be particularly sensitive to increases in neuronal homeostatic load.

STRESS, DIABETES, AND THE INSULIN-LIKE GROWTH FACTOR SYSTEM

The insulin-like growth factors (IGFs) are proposed to mediate a variety of functions in the CNS to promote neuronal development, differentiation, and function.[59] The peripheral IGF system is modulated during diabetes in both the experimental and clinical settings.[60] However, few studies have examined the diabetes-induced regulation of the central IGF system, although IGF-I and IGF-II levels have been reported to be decreased in the brains of diabetic animals.[61–63]

The IGF system has also been shown to be regulated by glucocorticoids (GCs). For example, previous studies have demonstrated that GCs represent an important component in the regulation of the IGF system in experimental models of diabetes.[64–66] In addition, the IGF system is modulated by GCs or stress under normal glycemic conditions in both animals and man.[67–70]

Stress/Diabetes Regulation of Hippocampal IGF Receptor Expression

Since previous studies have determined that the peripheral IGF system is modulated by diabetes and have suggested that the IGF system may mediate responses to challenges in neuronal homeostasis, we examined the regulation of IGF-Ir and IGF-IIr mRNA expression in the hippocampus of diabetic rats subjected to stress. Streptozotocin-induced diabetes did not modulate IGF-Ir levels in hippocampal pyrami-

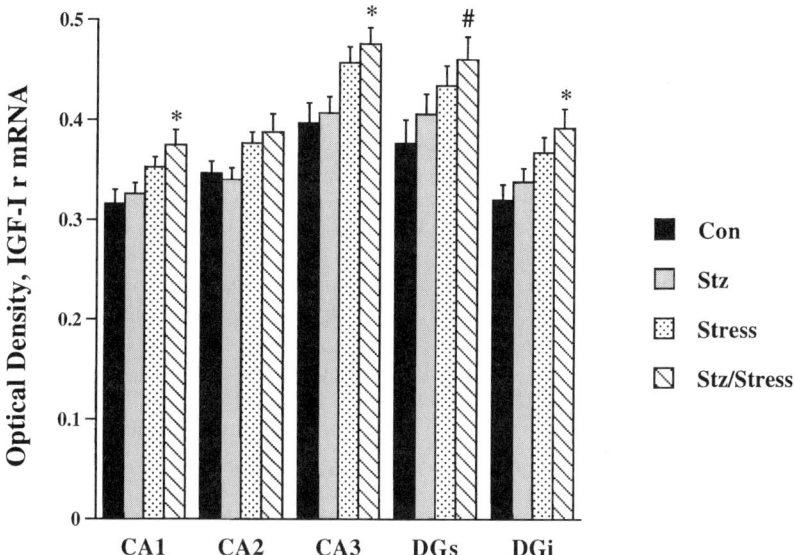

FIGURE 1. Autoradiographic image analysis of hippocampal IGF-I receptor (IGF-Ir) mRNA expression in control animals (Con), diabetic rats (Stz), rats subjected to 7 days of restraint stress (Stress), and diabetic rats subjected to 7 days of restraint stress (Stz/Stress). Statistical analysis revealed that IGF-Ir mRNA levels were significantly increased in diabetic rats subjected to restraint stress in CA1 pyramidal neurons [$F_{3,28} = 4.33$] and CA3 pyramidal neurons [$F_{3,28} = 4.956$] compared to controls. IGF-Ir mRNA expression was also increased in the superior blade of the dentate gyrus (DGs) [$F_{3,28} = 2.762$] and the inferior blade of the dentate gyrus (DGi) [$F_{3,28} = 3.956$] in diabetic rats subjected to stress compared to controls. *, $p \leq 0.01$; #, $p \leq 0.05$.

dal neurons and granule neurons of the dentate gyrus (FIG. 1) compared to vehicle-treated, nonstressed controls. Similarly, 7 days of restraint stress alone did not significantly modulate IGF-Ir mRNA expression. Diabetic rats subjected to 7 days of restraint stress had statistically significant increases in IGF-Ir mRNA expression in CA1 pyramidal neurons, CA3 pyramidal neurons, granule neurons in the superior blade of the dentate gyrus (DGs), and granule neurons of the inferior blade of the dentate gyrus (DGi) (FIG. 1). CA2 pyramidal neurons showed a trend towards increased mRNA levels in diabetic rats subjected to stress, although these increases did not achieve statistical significance. These results suggest that diabetes may provide modest disinhibition of IGF-Ir mRNA expression, while stress enhances IGF-Ir mRNA expression. The combined effects of diabetes and stress produce significant increases in IGF-Ir mRNA levels in the hippocampus, suggesting that IGF-Ir expression is modulated by both hyperglycemia and stress. In support of this hypothesis, GCs have previously been shown to be an important element in the diabetes-induced regulation of the IGF system.[64–66] Emulsion autoradiographic analysis of hippocampal nonprincipal neurons in CA1-oriens, CA1-radiatum, CA3, and the hilus of diabetic rats, rats subjected to restraint stress, and diabetic rats subjected to restraint stress did not reveal any differences in IGF-Ir mRNA expression when compared to vehicle controls (data not shown).

Unlike IGF-Ir mRNA levels in principal neurons, IGF-IIr levels were not modulated in pyramidal neurons of the hippocampus and granule neurons of the dentate gyrus. In particular, diabetic rats, rats subjected to restraint stress as well as diabetic rats subjected to restraint stress did not have changes in IGF-IIr mRNA expression in principal neurons of the hippocampus compared to nonstressed, vehicle controls (data not shown). Similarly, IGF-IIr mRNA levels in hippocampal nonprincipal neurons were unaffected by stress or diabetes when compared to vehicle controls. Diabetic rats subjected to 7 days of restraint stress had significant increases in IGF-IIr mRNA levels compared to diabetic, nonstressed rats in nonprincipal neurons of the hilus, CA1-oriens and CA1-radiatum. and CA3 (FIG. 2). These results demonstrate that under normal glycemic conditions stress has no effect upon IGF-IIr expression in hippocampal nonprincipal neurons, while under hyperglycemic conditions stress increases IGF-IIr mRNA expression. Interestingly, the GC modulation of IGFBP expression has previously been shown to be most evident during insulin deficiency.[64–66]

SUMMARY AND PERSPECTIVES

Collectively, the results of the current studies have expanded our understanding of the molecular and cellular alterations observed in an experimental model of in-

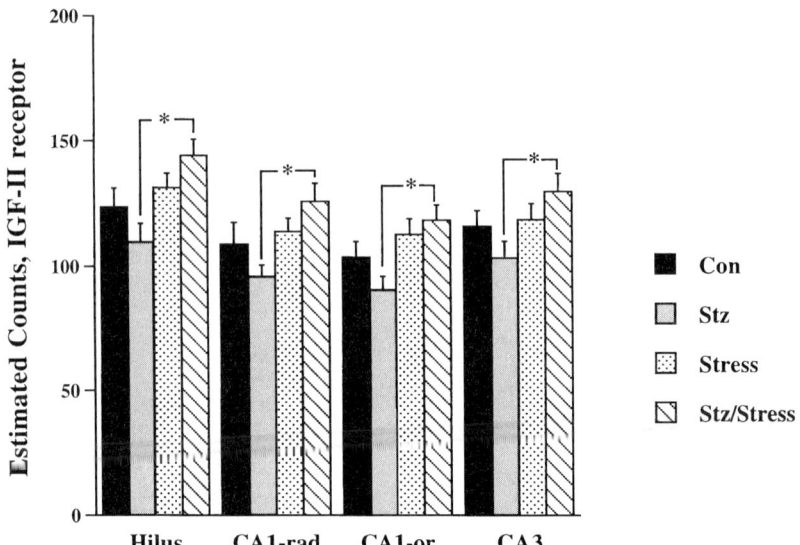

FIGURE 2. Statistical analysis of emulsion autoradiography for IGF-II receptor (IGF-IIr) mRNA expression in nonprincipal neurons of the rat hippocampus. Diabetic rats subjected to 7 days of restraint stress (Stz/Stress) exhibited significant increases in IGF-IIr mRNA expression compared to nonstressed diabetic rats (Stz) in nonprincipal neurons of the hilus [F3,28 = 4.242], CA1-radiatum [F3,28 = 3.37], CA1-oriens [F3,28 = 4.03], and CA3 [F3,28 = 2.746]. *, $p \leq 0.01$.

creased allostatic load, and more fully characterize the morphological and histochemical indices of hippocampal neuronal plasticity produced by diabetes and stress. Many comparisons can be drawn between the chronic effects of diabetes and the neurological impairments observed in Alzheimer's disease (AD). For example, AD patients have increased cognitive function in response to glucose administration[71] and insulin therapy,[72] presumably by increasing hippocampal glucose utilization. Neuronal glucose transport and utilization have been shown to be decreased in AD.[73] Cognitive deficits are also observed in diabetic patients,[5] and cognitive performance can be enhanced with improved glycemic control.[6,7] Previous studies from our laboratory have demonstrated that cognitive impairments induced by chronic stress are associated with atrophy of the rat hippocampus.[74] Similarly, AD patients also exhibit hippocampal atrophy,[75] and this morphological deficit has been proposed as an indicator of susceptibility to AD.[76] Diabetes has also been shown to promote morphological changes in the brain, including neuronal degeneration.[3,4] The results of our current studies indicate that diabetic rats subjected to stress develop neuronal atrophy, as well as indices of irreversible neuronal damage.[32] In both clinical and experimental settings, morphological changes in the hippocampus are associated with cognitive deficits in AD, diabetes, and chronic stress. The mechanisms that produce the neurological complications in AD, diabetes, and stress show striking similarities. For instance, several studies indicate that neuronal damage observed in AD patients is mediated by β-amyloid peptides.[77] In experimental models of hyperglycemia, β-amyloid toxicity is significantly potentiated in the hippocampus of stz-treated rats.[78] Similarly, our preliminary studies have demonstrated that HNE protein conjugation, which is proposed to mediate β-amyloid toxicity, is increased in the hippocampus of diabetic rats subjected to stress. Diabetes and AD are associated with impairments in mitochondrial function. Several critical enzymes involved in mitochondrial respiration—namely, α-ketoglutatarate dehydrogenase and pyruvate dehydrogenase— have been identified as targets of HNE protein conjugation.[54,55] Since AD and diabetes are both associated with impaired glucose utilization and deficits in mitochondrial activity, metabolic dysfunction may be considered an important component of both of these diseases. The similarities between AD and the neurological consequences of diabetes have lead to the proposal that the life-long effects of hyperglycemia may predispose diabetic patients to AD,[79] although this hypothesis remains to be confirmed.[80] The similarities observed between diabetes and AD provide an exciting opportunity to achieve a better understanding of the neurological consequences of these disorders, and may assist in the development of pharmacological strategies to assist in the treatment of the morphological and cognitive impairments associated with Alzheimer's disease and diabetes.

REFERENCES

1. MANDRUP-POULSEN, T. 1998. Diabetes. Br. Med. J. **316:** 1221–1225.
2. BROWN, M.J. & A.K. ASBURY. 1983. Diabetic neuropathy. Ann. Neurol. **15:** 2–12.
3. MOORADIAN, A.D. 1988. Diabetic complications of the central nervous system. Endocr. Rev. **9:** 346–356.
4. MCCALL, A.L. 1992. The impact of diabetes on the CNS. Diabetes **41:** 557–570.
5. RYAN, C.M. 1988. Neurobehavioral complications of type I diabetes. Examination of possible risk factors. Diabetes Care **11:** 86–93.

6. GRADMAN, T.J., A. LAWS, L.W. THOMPSON & G.M. REAVEN. 1993. Verbal learning and/or memory improves with glycemic control in older subjects with non-insulin-dependent diabetes mellitus. J. Am. Geriatr. Soc. **41:** 1305–1312.
7. MENEILLY, G.S., E. CHEUNG, D. TESSIER, C. YAKURA & H. TUOKKO. 1993. The effect of improved glycemic control on cognitive functions in the elderly patient with diabetes. J. Gerontol. **48:** M117–M121.
8. COUCH, R.M. 1992. Dissociation of cortisol and adrenal androgen secretion in poorly controlled insulin-dependent diabetes mellitus. Acta Endocrinol. **127:** 115–117.
9. MCEWEN, B.S. & R.M. SAPOLSKY. 1995. Stress and cognitive function. Curr. Opin. Neurobiol. **5:** 205–216.
10. WOOLLEY, C.S., E. GOULD & B.S. MCEWEN. 1990. Exposure to excess glucocorticoids alters dendritic morphology of adult hippocampal pyramidal neurons. Brain Res. **531:** 225–231.
11. WATANABE, Y., E. GOULD & B.S. MCEWEN. 1992. Stress induces atrophy of apical dendrites of hippocampal CA3 pyramidal neurons. Brain Res. **588:** 341–345.
12. CONRAD, C.D., L.A. GALEA, Y. KURODA & B.S. MCEWEN. 1996. Chronic stress impairs rat spatial memory on the Y maze, and this effect is blocked by tianeptine pretreatment. Behav. Neurosci. **110:** 1321–1334.
13. SAPOLSKY, R.M., L.C. KREY & B.S. MCEWEN. 1985. Prolonged glucocorticoid exposure reduces hippocampal neuron number: implications for aging. J. Neurosci. **5:** 1222–1227.
14. KERR, D.S., L.W. CAMPBELL, M.D. APPLEGATE, A. BRODISH & P.W. LANDFIELD. 1991. Chronic stress-induced acceleration of electrophysiologic and morphometric biomarkers of hippocampal aging. J. Neurosci. **11:** 1316–1324.
15. UNO, H., R. TARARA, J.G. ELSE, M.A. SULEMAN & R.M. SAPOLSKY. 1989. Hippocampal damage associated with prolonged and fatal stress in primates. J. Neurosci. **9:** 1705–1711.
16. SAPOLSKY, R.M., H. UNO, C.S. REBERT & C.E. FINCH. 1990. Hippocampal damage associated with prolonged glucocorticoid exposure in primates. J. Neurosci. **10:** 2897–2902.
17. REAGAN, L.P. & B.S. MCEWEN. 1997. Controversies surrounding glucocorticoid-mediated cell death in the hippocampus. J. Chem. Neuroanat. **13:** 149–167.
18. KADEKARO, M., M. ITO & P.M. GROSS. 1988. Local cerebral glucose utilization is increased in acutely adrenalectomized rats. Neuroendocrinology **47:** 329–334.
19. HORNER, H.C., D.R. PACKAN & R.M. SAPOLSKY. 1990. Glucocorticoids inhibit glucose transport in cultured hippocampal neurons and glia. Neuroendocrinology **52:** 57–64.
20. DE LEON, M.J., T. MCRAE, H. RUSINEK, A. CONVIT, S. DE SANTI, C. TARSHISH, J. GOLOMB, N. VOLKOW, K. DAISLEY, N. ORENTREICH & B.S. MCEWEN. 1997. Cortisol reduces hippocampal glucose metabolism in normal elderly, but not in Alzheimer's disease. J. Clin. Endocrinol. Metab. **82:** 3251–3259.
21. MCEWEN, B.S. 1998. Protective and damaging effects of stress mediators. N. Engl. J. Med. **338:** 171–179.
22. NIAKAN, E., Y. HARATI & J.P. COMSTOCK 1986. Diabetic autonomic neuropathy. Metabolism **35:** 224–234.
23. DEJONG, R.N. 1977. CNS manifestations of diabetes mellitus. Postgrad. Med. **61:** 101–107.
24. RESKE-NIELSEN, E. & K. LUNDBAEK. 1963. Diabetic encephalopathy: diffuse and focal lesions of the brain in long-term diabetes. Acta Neurol. Scand. **39:** 273–290.
25. RESKE-NIELSEN, E., K. LUNDBAEK & O.U. RAFAELSEN. 1965. Pathological changes in the central and peripheral nervous systems of young long-term diabetics. 1. Diabetic encephalopathy. Diabetologia **1:** 233–241.
26. OLSSON, Y., J. SAVE-SODERBERGH, P. SOURANDER & L. ANGERVALL. 1968. A patho-anatomical study of the central and peripheral nervous system in diabetes of early onset and long duration. Pathol. Eur. **3:** 62–79.

27. DEJGAARD, A., A. GADE, H. LARSSON, V. BALLE, A. PARVING & H.-H. PARVING. 1990. Evidence for diabetic encephalopathy. Diabetic Med. **8:** 162–167.
28. BESTETTI, G.E., M.J. REYMOND, C.E. BOUJON, T. LEMARCHAND-BÉRAUD & G.L. ROSSI. 1989. Functional and morphological aspects of impaired TRH release by mediobasal hypothalamus of STZ-induced diabetic rats. Diabetes **38:** 1351–1356.
29. BESTETTI, G., V. LOCATELLI, F. TIRONE, G.L. ROSSI & E.E. MÜLLER. 1985. One month of streptozotocin-diabetes induces different neuroendocrine and morphological alterations in the pituitary axis of male and female rats. Endocrinology **117:** 208–216.
30. JAKOBSEN, J., P. SIDENIUS, H.J.G. GUNDERSEN & R. OSTERBY. 1987. Quantitative changes of cerebral neocortical structure in insulin-treated long-term streptozotocin-induced diabetes in rats. Diabetes **36:** 597–601.
31. TAY, S.S.W. & W.C. WONG. 1991. Gracile nucleus of streptozotocin-induced diabetic rats. J. Neurocytol. **20:** 356–364.
32. MAGARIÑOS, A.M. & B.S. MCEWEN. 1998. The hippocampal morphology of diabetic rats shows an increased vulnerability to repeated stress. Soc. Neurosci. Abstr. **24:** 1921.
33. MAHER, F., S.J. VANNUCCI & I.A. SIMPSON. 1994. Glucose transporter proteins in brain. FASEB J. **8:** 1003–1011.
34. LUND-ANDEREN, H. 1979. Transport of glucose from blood to brain. Physiol. Rev. **59:** 305–310.
35. PARDRIDGE, W.M. 1983. Brain metabolism: a perspective from the blood-brain barrier. Physiol. Rev. **63:** 1481–1535.
36. LELOUP, C., M. ARLUISON, N. KASSIS, N. LEPETIT, N. CARTIER, P. FERRÉ & L. PÉNICAUD. 1996. Discrete brain areas express the insulin-responsive glucose transporter GLUT4. Mol. Brain Res. **38:** 45–53.
37. NAGAMATSU, S., J.M. KORNHAUSER, C.F. BURANT, S. SEINO, K.E. MAYO & G.I. BELL. 1992. Glucose transporter expression in brain. J. Biol. Chem. **267:** 467–472.
38. MCCALL, A.L., A.M. VAN BUEREN, M. MOHOLT-SIEBERT, N.J. CHERRY & W.R. WOODWARD. 1994. Immunohistochemical localization of the neuron-specific glucose transporter (GLUT3) to neuropil in adult rat brain. Brain Res. **659:** 292–297.
39. CHOI, T.B., R.J. BOADO & W.M. PARDRIDGE. 1989. Blood-brain barrier glucose transporter mRNA is increased in experimental diabetes mellitus. Biochem. Biophys. Res. Commun. **164:** 375–380.
40. LUTZ, A.J. & W.M. PARDRIDGE. 1993. Insulin therapy normalizes GLUT1 glucose transporter mRNA but not immunoreactive transporter protein in streptozotocin-diabetic rats. Metabolism **42:** 939–944.
41. VANNUCCI, S.J., E.M. GIBBS & I.A. SIMPSON. 1997. Glucose utilization and glucose transporter proteins GLUT-1 and GLUT-3 in brains of diabetic (*db/db*) mice. Am. J. Physiol. **272:** E267–E274.
42. KAINULAINEN, H., A. SCHURMANN, P. VILJA & H.G. JOOST. 1993. *In-vivo* glucose uptake and glucose transporter proteins GLUT1 and GLUT3 in brain tissue from streptozotocin-diabetic rats. Acta Physiol. Scand. **149:** 221–225.
43. NAGAMATSU, S., H. SAWA, N. INOUE, Y. NAKAMICHI, H. TAKESHIMA & T. HOSHINO. 1994. Gene expression of GLUT3 glucose transporter regulated by glucose *in vivo* in mouse brain and *in vitro* in neuronal cell cultures from rat embryos. Biochem. J. **300:** 125–131.
44. REAGAN, L.P., A.M. MAGARIÑOS, L.R. LUCAS, A. VAN BUEREN, A.L. MCCALL & B.S. MCEWEN. 1999. Regulation of GLUT3 glucose transporter in the hippocampus of diabetic rats subjected to stress. Am. J. Physiol. **276:** E879–E886.
45. BAYNES, J.W. 1991. Role of oxidative stress in development of complications in diabetes. Diabetes **40:** 405–412.
46. WOLFF, S.P. 1993. Diabetes mellitus and free radicals. Brit. Med. Bull. **49:** 642–652.

47. LIU, J., X. WANG, M.K. SHIGENAGA, H.C. YEO, A. MORI & B.N. AMES. 1996. Immobilization stress causes oxidative damage to lipid, protein, and DNA in the brain of rats. FASEB J. **10:** 1532–1538.
48. CERIELLO, A., D. GIUGLIANO, A. QUATRARO, P. DELLO RUSSO & P.J. LEFÈBVRE. 1991. Metabolic control may influence the increased superoxide generation in diabetic serum. Diabetic Med. **8:** 540–542.
49. ARAGNO, M., E. BRIGNARDELLO, E. TAMAGNO, O. DANNI & G. BOCCUZZI. 1997. Dehydroepiandrosterone administration prevents the oxidative damage induced by acute hyperglycemia in rats. J. Endocrinol. **155:** 233–240.
50. KUMAR, J.S.S. & V.P. MENON. 1993. Effect of diabetes on levels of lipid peroxides and glycolipids in rat brain. Metabolism **42:** 1435–1439.
51. MAKAR, T.K., K. RIMPEL-LAMHAOUAR, D.G. ABRAHAM, V.S. GOKHALE & A.J.L. COOPER. 1995. Antioxidant defense systems in the brains of type II diabetic mice. J. Neurochem. **65:** 287–291.
52. MATTSON, M.P. 1998. Modification of ion homeostasis by lipid peroxidation: roles in neuronal degeneration and adaptive plasticity. Trends Neurosci. **21:** 53–57.
53. KELLER, J.N., Z. PANG, J.W. GEDDES, J.G. BEGLEY, A. GERMEYER, G. WAEG & M.P. MATTSON. 1997. Impairment of glucose and glutamate transport and induction of mitochondrial oxidative stress and dysfunction in synaptosomes by amyloid β-peptide: role of the lipid peroxidation product 4-hydroxynonenal. J. Neurochem. **69:** 273–284.
54. HUMPHRIES, K.M., Y. YOO & L.I. SZWEDA. 1998. Inhibition of NADH-linked mitochondrial respiration by 4-hydroxy-2-nonenal. Biochemistry **37:** 552–557.
55. HUMPHRIES, K.M. & L.I. SZWEDA. 1998. Selective inactivation of α-ketoglutarate dehydrogenase and pyruvate dehydrogenase: reaction of lipoic acid with 4-hydroxy-2-nonenal. Biochemistry **37:** 15835–15841.
56. MAKAR, T.K., B.L. HUNGUND, G.A. COOK, K. KASHFI & A.J.L. COOPER. 1995. Lipid metabolism and membrane composition are altered in the brains of type II diabetic mice. J. Neurochem. **64:** 2159–2168.
57. UCHIDA, K., L.I. SZWEDA, H.-Z. CHAE & E.R. STADTMAN. 1993. Immunochemical detection of 4-hydroxynonenal protein adducts in oxidized hepatocytes. Proc. Natl. Acad. Sci. USA **90:** 8742–8746.
58. REAGAN, L.P., A.M. MAGARIÑOS & B.S. MCEWEN. 1999. Oxidative stress is specifically increased in the hippocampus of diabetic rats subjected to stress. Soc. Neurosci. Abstr. In press.
59. JONES, J.I. & D.R. CLEMMONS. 1995. Insulin-like growth factors and their binding proteins: biological actions. Endocr. Rev. **16:** 3–34.
60. LEROITH, D., M. ADAMO, H. WERNER & ROBERTS, JR. 1991. Insulinlike growth factors and their receptors as growth regulators in normal physiology and pathologic states. Trends Endocrinol. Metab. **2:** 134–139.
61. WUARIN, L., R. NAMDEV, J.G. BURNS, Z. FEI & D.N. ISHII. 1996. Brain insulin-like growth factor-II mRNA content is reduced in insulin-dependent and non-insulin-dependent diabetes mellitus. J. Neurochem. **67:** 742–751.
62. BUSIGUINA, S., J.A. CHOWEN, J. ARGENTE & I. TORRES-ALEMAN. 1996. Specific alterations of the insulin-like growth factor I system in the cerebellum of diabetic rats. Endocrinology **137:** 4980–4987.
63. ZHUANG, H.-X., L. WUARIN, Z. FEI & D.N. ISHII. 1997. Insulin-like growth factor (IGF) gene expression is reduced in neural tissues and liver from rats with non-insulin-dependent diabetes mellitus, and IGF treatment ameliorates diabetic neuropathy. J. Pharmacol. Exp. Ther. **283:** 366–374.
64. UNTERMAN, T.G., J.J. JENTEL, D.T. OEHLER, R.G. LACSON & J.F. HOFERT. 1993. Effects of glucocorticoids on circulating levels and hepatic expression of insulin-

like growth factor (IGF)–binding proteins and IGF-I in the adrenalectomized streptozotocin-diabetic rat. Endocrinology **133**: 2531–2539.
65. RODGERS, B.D., A.M. STRACK, M.F. DALLMAN, L. HWA & C.S. NICHOLL. 1995. Corticosterone regulation of insulin-like growth factor I, IGF-binding proteins, and growth in streptozotocin-induced diabetic rats. Diabetes **44**: 1420–1425.
66. RODGERS, B.D., R.M. BAUTISTA & C.S. NICHOLL. 1995. Regulation of insulin-like growth factor binding proteins in rats with insulin-dependent diabetes mellitus. Proc. Soc. Exp. Biol. Med. **210**: 234–241.
67. ADAMO, M., H. WERNER, W. FARNSWORTH, ROBERTS, JR., M. RAIZADA & D. LEROITH. 1988. Dexamethasone reduces steady state insulin-like growth factor I messenger ribonucleic acid levels in rat neuronal and glial cells in primary culture. Endocrinology **123**: 2526–2570.
68. OHYAMA, T., M. SATO, M. NIIMI, N. HIZUKA & J. TAKAHARA. 1997. Effects of short- and long-term dexamethasone treatment on growth and growth hormone (GH)–releasing hormone (GRH)-GH–insulin-like growth factor–I axis in conscious rats. Endocr. J. **44**: 827–835.
69. BERNTON, E., D. HOOVER, R. GALLOWAY & K. POPP. 1995. Adaptation to chronic stress in military trainees. Adrenal androgens, testosterone, glucocorticoids, IGF-I, and immune function. Ann. N. Y. Acad. Sci. **774**: 217–231.
70. ISLAM, A., C. AYER-LELIEVRE, C. HEIGENSKÖLD, N. BOGANOVIC, B. WINBLAD & A. ADEM. 1998. Changes in IGF-1 receptors in the hippocampus of adult rats after long-term adrenalectomy: receptor autoradiography and in situ hybridization histochemistry. Brain Res. **797**: 342–346.
71. CRAFT, S., S.E. DAGOGO-JACK, B.V. WIETHOP, C. MURPHY, R.T. NEVINS, S. FLEISCHMAN, V. RICE, J.W. NEWCOMER & P.E. CRYER. 1993. Effects of hyperglycemia on memory and hormone levels in dementia of the Alzheimer type: a longitudinal study. Behav. Neurosci. **6**: 926–940.
72. CRAFT, S., J. NEWCOMER, S. KANNE, S.E. DAGOGO-JACK, P. CRYER, Y. SHELINE, J. LUBY, A. DAGOGO-JACK & A. ANDERSON. 1996. Memory improvement following induced hyperinsulinaemia in Alzheimer's Disease. Neurobiol. Aging **17**: 123–130.
73. PIERT, M., R.A. KOEPPE, B. GIORDANI, S. BERENT & D.E. KUHL. 1996. Diminished glucose transport and phosphorylation in Alzheimer's Disease determined by dynamic FDG-PET. J. Nucl. Med. **37**: 201–208.
74. CONRAD, C.D., A.M. MAGARIÑOS, J.E. LEDOUX & B.S. MCEWEN. 1999. Repeated restraint stress facilitates fear conditioning independently of causing hippocampal CA3 dendritic atrophy. Behav. Neurosci. In press.
75. DE LEON, M.J., A.E. GEORGE, J. GOLOMB, C. TARSHISH, A. CONVIT, A. KLUGER, S. DE SANTI, T. MCRAE, S.H. FERRIS, B. REISBERG, C. INCE, H. RUSINEK, M. BOBINSKI, B. QUINN, D.C. MILLER & H.M. WISNIEWSKI. 1997. Frequency of hippocampal formation atrophy in normal aging and Alzheimer's disease. Neurobiol. Aging **18**: 1–11.
76. DE LEON, M.J., A.E. GEORGE, L.A. STYLOPOULOS, G. SMITH & D.C. MILLER. 1989. Early marker for Alzheimer's disease: the atrophic hippocampus. Lancet **2**: 672–673.
77. TROJANOWSKI, J.Q., C.M. CLARK, M.L. SCHMIDT, S.E. ARNOLD & V.M.-Y. LEE. 1997. Strategies for improving the postmortem neuropathological diagnosis of Alzheimer's disease. Neurobiol. Aging **18**: S75–S79.
78. SMYTH, M.D., J.P. KESSLAK, B.J. CUMMINGS & C.W. COTMAN. 1994. Analysis of brain injury following intrahippocampal administration of beta-amyloid in streptozotocin-treated rats. Neurobiol. Aging **15**: 153–159.
79. MESSIER, C. & M. GAGNON. 1996. Glucose regulation and cognitive functions: relation to Alzheimer's disease and diabetes. Behav. Brain Res. **75**: 1–11.
80. FINCH, C.E. & D.M. COHEN. 1997. Aging, metabolism, and Alzheimer disease: review and hypothesis. Exp. Neurol. **143**: 82–102.

Functional Brain Imaging in the Resting State and during Activation in Alzheimer's Disease

Implications for Disease Mechanisms Involving Oxidative Phosphorylation

STANLEY I. RAPOPORT

Laboratory of Neurosciences, National Institute on Aging, National Institutes of Health, Bethesda, Maryland 20892, USA

ABSTRACT: *In vivo* brain imaging of patients with Alzheimer's disease (AD) using positron emission tomography (PET) demonstrates progressive reductions in resting-state brain glucose metabolism and blood flow in relation to dementia severity, more so in association than primary cortical regions. During cognitive or psychophysical stimulation, blood flow and metabolism in the affected regions can increase to the same extent in mildly demented AD patients as in age-matched controls, suggesting that energy delivery is not rate limiting. Activation declines with dementia severity, and is markedly reduced in severely demented patients. These results suggest that there is an initial "normal" functionally-responsive stage in AD, followed by a late less responsive stage. Studies of biopsied and postmortem brain indicate that the initial stage is accompanied by selective and potentially reversible down-regulation of the brain enzymes, including cytochrome oxidase, which mediate mitochondrial oxidative-phosphorylation.

INTRODUCTION

The mammalian brain is distinguished by a high rate of glucose consumption and by high activities of enzymes involved in mitochondrial oxidative phosphorylation (OXPHOS). OXPHOS produces ATP, which is consumed by the Na,K-ATPase pump to maintain synaptic function.[1] Because of these relations, *in vivo* brain imaging of regional cerebral metabolic rates for glucose ($rCMR_{glc}$) or of regional cerebral blood flow (rCBF) [which is coupled to glucose metabolism] has been used to quantify functional synaptic activity. In diseases like Alzheimer's disease (AD), cerebral metabolic or flow changes, which can be localized and quantified by means of positron emission tomography (PET), are correlated with cognitive and behavioral deficits and thus can be used to understand and interpret mechanisms of functional failure. In this paper, I review PET studies that suggest that two general stages of functional failure occur in the course of AD, the first in mild dementia in which brain responsiveness is largely maintained, the second in severe dementia in which it is not. Studies of biopsied and postmortem brain indicate that the initial stage is accompanied

[a]Phone: 301-496-1765; fax: 301-402-0074.
e-mail: sir@helix.nih.gov

by selective and potentially reversible down-regulation of brain enzymes, including cytochrome oxidase, which mediate mitochondrial OXPHOS.

CORRELATIONS BETWEEN RESTING-STATE BRAIN METABOLISM AND COGNITIVE DEFICITS IN ALZHEIMER'S DISEASE

Cross-sectional PET studies of subjects diagnosed as having AD demonstrate resting-state (eyes covered, ears plugged with cotton) reductions in $rCMR_{glc}$, measured with ^{18}F-fluoro-2-deoxy-D-glucose, throughout the neocortex in proportion to dementia severity.[2,3] The reductions represent intrinsic reductions in metabolism per gram brain tissue, as they remain significant after correction for brain atrophy.[4] They are not uniformly distributed. Neocortical association areas are affected earlier and more severely than are primary visual, auditory and somatosensory areas. Thalamic and basal ganglia nuclei are relatively spared.[2,3] Large pyramidal neurons with long intracortical and intercortical axons are most affected and correlated functional interactions between them are reduced,[5] leading to the interpretation that the cognitive changes in AD represent a "cortical disconnection" syndrome.[6]

Within the neocortex, regional resting-state PET metabolic reductions correlate with regional densities of neurofibrillary tangles (NFTs) but not of senile (neuritic) plaques, two neuropathological hallmarks of AD.[7–9] The regions which are affected earliest and most severely constitute a telencephalic "association" system. They include association neocortex, posterior hippocampus, cortico-basal nucleus of the amygdaloid complex, transentorhinal cortex, entorhinal cortex and nucleus basalis of Meynert, with the transentorhinal and entorhinal cortices likely being affected in the earliest stage of disease.[10,11] Comparative anatomic studies indicate that this association system expanded and differentiated during primate but particularly hominid evolution. AD has been hypothesized to be a phylogenic, distinct human disease in which the telencephalic system is vulnerable because of the genetic factors that promoted its evolution.[10,12]

Patterns of metabolic reductions in individual AD patients are quite heterogeneous, and correspond to patterns of cognitive and behavioral abnormalities in individual patients (FIG. 1). In one PET study of AD patients, four independent $rCMR_{glc}$ patterns could be identified using a statistical "principal components" analysis.[13] In the most common patient group, $rCMR_{glc}$ was reduced in superior and inferior parietal lobules and posterior medial temporal lobe, and the patients had depression and poor visuospatial and memory performance.

As predicted from brain lesion studies,[14] abnormal left-right hemispheric metabolic asymmetries or abnormal frontal-parietal metabolic gradients (FIG. 1) correlated with specific cognitive "discrepancies" in AD patients.[15–18] Mildly-moderately demented AD patients[b] with a lower right- than left-sided $rCMR_{glc}$ on average had worse scores on visuospatial tests (e.g., Range Drawing Test) than on language tests (e.g., Syntax Comprehension Test), and the reverse was true in patients with a relatively lower left-sided $rCMR_{glc}$. Hemispheric metabolic asymmetries in mildly de-

---

[b]Dementia severity can be stratified according to the Mini-Mental State Examination;[19] a score of 30–21 defines mild dementia, of 20–11 moderate dementia and of 10–0 severe dementia.

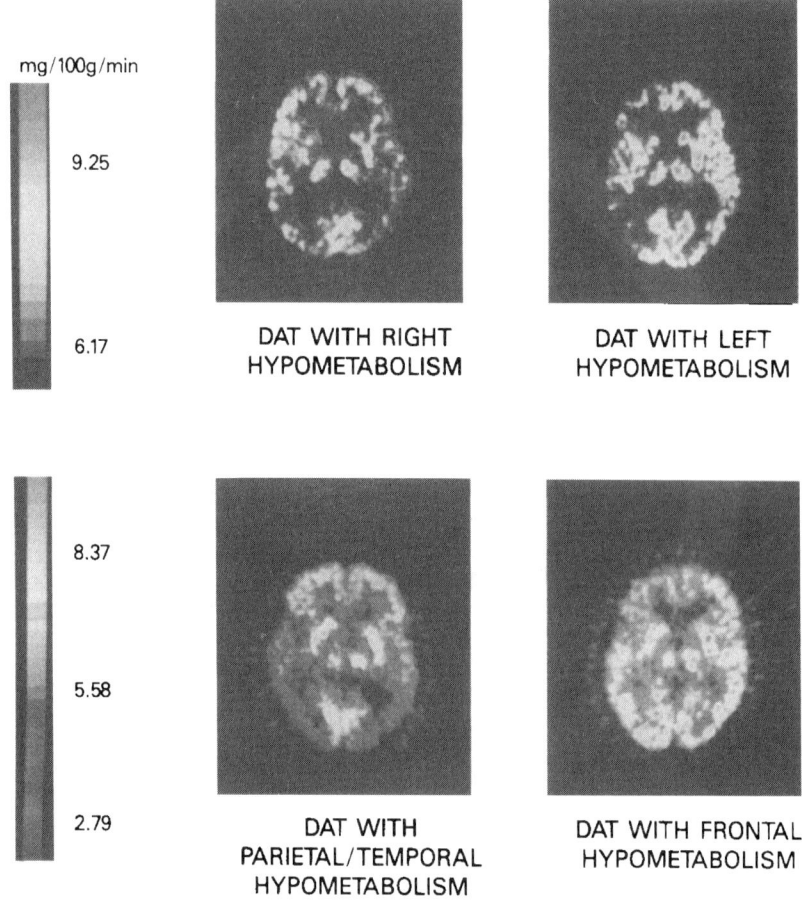

FIGURE 1. Variable patterns of resting-state brain glucose metabolism in dementia of the Alzheimer type (DAT) patients, measured using positron emission tomography in horizontal brain slices. The most common pattern is that on the *lower left*, involving parietal/temporal hypometabolism. In moderately-severely demented patients, each pattern corresponds to a specific profile of cognitive deficits, illustrating the principle that brain network function underlies cognition and behavior.[11] Abnormal right-left hemispheric asymmetries (*top*) may occur in mildly demented patients in the absence of deficits in the visuospatial or language functions. They predict the pattern of such deficits that will appear 1–3 years later, thus PET measures of metabolism are more sensitive markers of the AD process than are current cognitive tests (see text).

mented AD patients with only a memory deficit predicted the visuospatial-language "discrepancies" (differences in rank ordered cognitive test scores) that appeared 1–3 years later. Subjects with lower right- than left-sided metabolic rates later had worse visuospatial than language abilities, and vice versa.

Reliable predictions of neocortically mediated cognitive deficits from early metabolic cortical asymmetries imply that resting-state metabolic measurements are more sensitive markers of neocortical dysfunction than are current psychometric tests. Once an asymmetry is established, its direction will usually be maintained throughout the course of disease. If right-sided metabolism initially is less than left-sided metabolism in an individual patient, it is less on follow-up, and vice versa.[16,17] A time invariant direction of asymmetry implies that rates of metabolic decline are approximately the same in the two hemispheres of a given patient. The rates may thus be related to the AD brain lipids having a reduced "critical temperature" (temperature at which brain lipids form a stable bilayer in solution), as then membrane breakdown would be proportional to the difference between the observed critical temperature and its normal value of 37°C.[20,21] The reduced critical temperature has been ascribed to an abnormal plasmalogen content of brain lipids.[22]

There is considerable variation in PET metabolic patterns among AD patients (FIG. 1), as well as considerable overlap of metabolic rates between mildly demented patients and controls. This limits our ability to use PET scanning to reliably identify a subject early on as having AD. Discriminant analysis combined with multiple regression can be used to overcome this limitation. The discriminant procedure constructs a linear combination of observed variables to best describe group differences and to classify group membership of any individual.[23] It provides a probablistic statement regarding the likelihood of an individual PET or MRI scan being similar to scans from a disease or an age-matched control group.

A discriminant function, derived using $rCMR_{glc}$ PET data from identified AD patients and controls, could classify subjects in either category with 87% accuracy. When later applied to brain metabolic data from an unaffected at-risk familial-AD (FAD) subject, the function indicated that the subject had a significant AD pattern. A "probable" clinical diagnosis of AD was confirmed one year later, when parietal lobe $rCMR_{glc}$ was seen to be reduced in a second scan when the patient had a clear dementia.[25] The same discriminant function could identify AD patterns in nondemented older Down syndrome subjects who later went on to develop AD-type dementia.[26]

Subtle but statistically significant mean PET abnormalities have been reported in at-risk FAD subjects and indicate that measurable metabolic dysfunction in AD can precede even a "possible" clinical diagnosis.[c] In one such study, FAD subjects with only memory impairment were divided according to their apolipoprotein e4-allele status: e4 heterozygotes and noncarriers of this allele.[27] Compared with the noncarriers, the heterozygotes had reduced mean $rCMR_{glc}$ in the parietal cortex; both groups had less parietal $rCMR_{glc}$ than affected AD patients. In another study, middle-aged cognitively normal e4 homozygotes had an AD-like group PET pattern on cortical projection maps of $rCMR_{glc}$.[28] A third study showed that FAD subjects at risk because of a chromosome-14 linkage, an APP mutation, or other genetic factors had reduced mean global and parietotemporal $rCMR_{glc}$ compared with age-matched controls.[29] In each of the above studies, there was considerable overlap of brain metabolic values between the experimental FAD group and controls. Discriminant analysis (see above) and follow-up will be necessary to establish the power of an individual metabolic pattern to predict dementia in as yet cognitively normal subjects.

[c]A "possible" AD diagnosis is made on the basis of history of memory decline and measurable memory deficit, whereas a "probable" diagnosis requires an additional cognitive deficit.[24]

STRESSING THE ALZHEIMER'S BRAIN BY FUNCTIONAL STIMULATION

It remains unclear as to whether resting-state brain metabolic and flow reductions early in AD are due to rate-limiting ATP availability causing neuronal dysfunction, or are due to neuronal dysfunction causing reduced energy demand and metabolism. Evidence that $rCMR_{glc}$ and rCBF can be markedly increased by activation in early AD suggests that the first case is more likely.

We used $H_2^{15}O$ with PET to quantify rCBF in occipitotemporal visual association regions that subserve object recognition, while AD patients or healthy controls per-

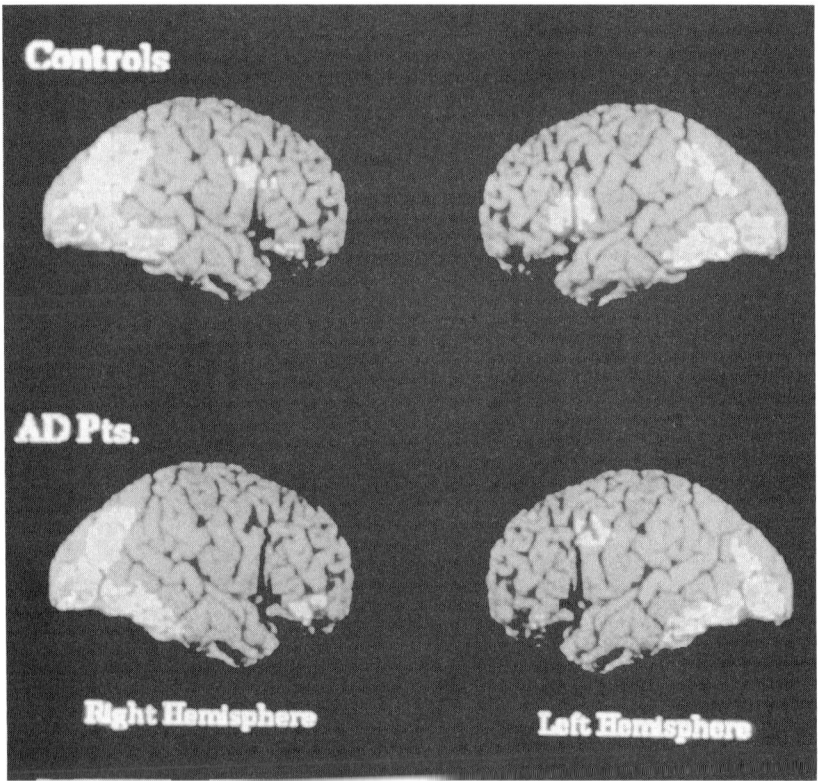

FIGURE 2. Significant increments in ΔrCBF in lateral images of brain in control and Alzheimer disease patients, using statistical parametric mapping. Difference image identifies pixels in which flow was increased by more than 30% above baseline, intensity of color the extent of significance above $p < 0.01$. Note involvement of two visual pathways, occipital temporal and occipital parietal, as well as some regions in frontal lobe. There are no major differences in distribution between subject groups. Mean (±S.D.) accuracy for controls and AD subjects was 92 ± 4 % and 88 ± 4 % (n.s.), whereas mean reaction time was 1.97 ± 0.54 s and 3.39 ± 1.29 s; variance differed significantly ($p < 0.05$). Areas that were activated, primary striate cortex and occipital-temporal and occipital parietal visual cortex, demonstrated reduced resting-state rCBF (TABLE 1). (Reprinted from Grady et al.[30] by permission.)

TABLE 1. rCBF in occipitotemporal visual association cortex (Brodmann areas 19 and 37) in Alzheimer's disease patients and normal volunteers performing a control of face matching task[a]

Task	Control Subjects ($n = 13$)	Alzheimer Disease Patients ($n = 11$)
	rCBF (ml/100g/min)	
Baseline control task	48.4 ± 0.9	40.1 ± 1.1*
Face matching task	53.1 ± 0.9	44.1 ± 1.2*
Difference ΔrCBF, face matching–control task	4.7 ± 0.2	4.0 ± 0.4

[a]From Rapoport and Grady.[31]
*Mean ± SE differs significantly from control mean, $p < 0.0001$.

formed a control or a face-matching task.[30,31] rCBF during the control task (a button was pressed alternately with right and left thumbs in response to a neutral visual stimulus) was subtracted from rCBF during the face-matching task (the button was pressed with the appropriate thumb after deciding whether the right or left face was to be matched) to produce a "difference" image of ΔrCBF values (FIG. 2). Mildly-moderately demented AD patients were chosen who could perform the face-matching task as accurately (85 + 8 (SD)% correct choices) as controls (92 + 5%, respectively), although their reaction times during the task were more variable than in the controls, 3.34 + 1.46 s compared with 2.07 + 0.54 s. As illustrated in lateral views of the brain (FIG. 2), statistically significant rCBF increments occurred generally in the identical regions, mainly occipital-temporal visual areas, in AD patients and controls.

As illustrated in TABLE 1, "control task" rCBF was lower in occipitotemporal regions (Brodmann areas 17, 19, and 37) of AD than of control subjects. During the face-matching task, the mean flow increment ΔrCBF (ml/100g/min) did not differ between patients and controls. The affected AD brain areas, which in the resting state showed reduced rCBF, could be "normally" activated during the face recognition task.[31]

Like the rCBF response, the gross $rCMR_{glc}$ response to stimulation may be within normal limits in mildly demented AD patients.[32,33] $rCMR_{glc}$ was measured in the resting-state (eyes covered/ears blocked) in AD and control subjects, and in the same PET session during audiovisual stimulation by a movie not requiring subject compliance. In the controls, audiovisual stimulation significantly increased $rCMR_{glc}$ in visual and auditory cortical areas. In mildly demented patients, $rCMR_{glc}$ increments in these regions were within two standard deviations (95 percentile limits) of the mean control response. The increments fell below these limits in moderately demented patients and were further reduced in severely demented patients. In a parallel fashion, brain responsiveness to the anti-cholinesterase drug, Donepezil, also appears to decline with disease progression. Chronic administration of this drug produces significant and positive mean "cognitive" enhancement in mildly but not moderately demented AD patients.[34,35]

Graded activation paradigms, in which a parameter such as stimulus intensity or task difficulty is varied discretely while rCBF is measured, provide an opportunity

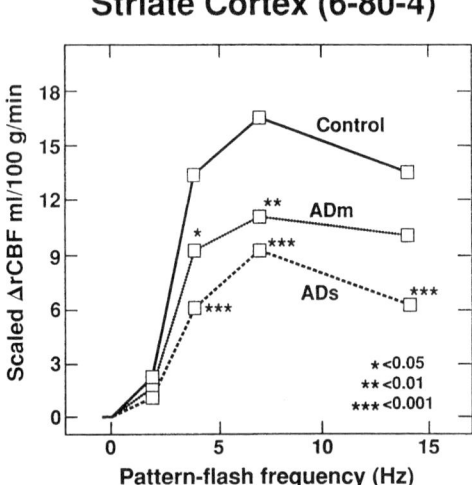

FIGURE 3. Mean stimulus-rCBF response curves in striate cortex during patterned flash stimulation at different frequencies, in control subjects and AD patients with mild dementia (m) and moderate-to-severe dementia (s). This is an example of "parametric" (varying) stimulation to explore function deficits in AD (see text), giving "dose-response" type curves which can be interpreted in terms of synaptic efficacy and drug modulation.[31] The decline in the control flow curve at 14 Hz has been ascribed to failed responses of synapses in the parvocellular visual system (see text). Scaled ΔrCBF corrects for global flow differences between groups. *Asterisks* indicate significant difference from control means. Coordinates in Talairach space. (Reprinted from Mentis *et al.*[39] by permission.)

to construct stimulus-rCBF response curves (like pharmacological dose-response curves) which can be interpreted in terms of synaptic efficacy and drug modulation.[31,36] Using such a graded approach, patterned alternating flashes of red light grids were presented by means of goggles to control and AD subjects, and rCBF was measured at each frequency with $H_2^{15}O$ and PET (FIG. 3).[37–39] In the controls, striate cortex ΔrCBF increased linearly between stimulus frequencies of 0 and 8 Hz, then fell at higher frequencies. In mildly demented AD patients, ΔrCBF at frequencies of 4 Hz were generally less than in the controls; differences were greater and more significant in moderately-severely demented patients, but the brain nevertheless could be activated in these patients.

The control biphasic stimulus-rCBF response curve has been ascribed to drop-out of synaptic responses of the parvocellular visual system at frequencies above 8 Hz, while high frequency–responding synapses in the magnocellular visual system continue to be activated.[40,41] Both visual systems are found within the striate cortex. A middle temporal region (V5/MT) exists which contains only magnocellular neurons. This region could be activated at 1 Hz (when movement was apparent) in the controls but not in the AD patients. High-frequency responding and presumably more metabolically dependent synapses are more vulnerable in AD than are low-frequency responding synapses. In Brodmann visual area 18 of AD patients, PET abnormalities at frequencies above 8 Hz could be partially reversed by acute administration of the anticholinesterase drug, physostigmine.[42]

SELECTIVE DOWN-REGULATION OF OXIDATIVE PHOSPHORYLATION (OXPHOS) IN ALZHEIMER'S DISEASE

TABLE 2 illustrates that enzyme activity of cytochrome oxidase (COX), the rate-limiting enzyme complex of OXPHOS within mitochondria, and mRNA levels for complex I–V subunits of the electron transport chain, coded by mitochondrial DNA (mtDNA) or by nuclear DNA (nDNA), are all reduced in midtemporal cortex of the postmortem AD brain. Similar changes are not found in unaffected motor cortex. Thus, levels of mRNA for the mtDNA-encoded ND1 and ND4 subunits of complex

TABLE 2. Mitochondrial and nuclear DNA markers in midtemporal cortex of Alzheimer's disease brain, and in lateral geniculate nucleus of monkey brain seven days after injecting tetrodotoxin into vitreous humor of one eye

Marker	% Decrease in Alzheimer's Disease vs. Control Temporal Cortex	% Decrease in LGN TTX-Treated vs. Control Monkey[a]
COX enzyme activity	20–25[b]	23 ± 1[c]
COX protein	n.d.	23 ± 2
mtDNA	n.s.[d]	26 ± 4
COX I mRNA (mtDNA)[e]	58 ± 3[f]	49 ± 3
COX III mRNA (mtDNA)	54 ± 5[f]	n.d.
NADH dehydrogenase enzyme activity	40[g]	n.d.
ND1 mRNA (mtDNA)	50–60[f]	n.d.
ND4 mRNA (mtDNA)	60 ± 8[b]	n.d.
12S rRNA (mtDNA)	n.s.[f]	n.d.
28S rRNA (nDNA)	n.s[f]	n.d.
COX IV mRNA (nDNA)	40 ± 8[f,i]	18 ± 3
COX VIII mRNA (nDNA)	n.d.	29 ± 3
ATPsyn.β mRNA (nDNA)	50–60[f]	n.d.
β-actin mRNA (nDNA)	n.s.[f]	n.d.
LDH-B mRNA (nDNA)	n.s.[f]	n.d.

ABBREVIATIONS: ATPsyn.β, ATP synthase subunit β; COX, cytochrome oxidase; COX I, III, IV, VIII, cytochrome oxidase subunits; LDH-B, lactate dehydrogenase subunit B; LGN, lateral geniculate nucleus; NADH, reduced nicotinamide adenine dinucleotide; ND1, ND4, subunits of NADH dehydrogenase; TTX, tetradotoxin.[1] Hevner and Wong-Riley;[48] [2]Hatanpää et al.;[43] [3]M. Tamataini, K. Chandrasekaran, and C. R. Filburn (unpublished observations); [4]Chandrasekaran et al.;[47] [5]Fukuyama et al.;[44] [6]Chandrasekaran et al.;[75] [7]Parker et al.[76]

[a]Mean ± SEM; n.s., not significant; n.d., not determined.
[b]Parenthesis identifies whether mRNA is encoded by mitochondrial DNA (mtDNA) or nuclear DNA (nDNA).

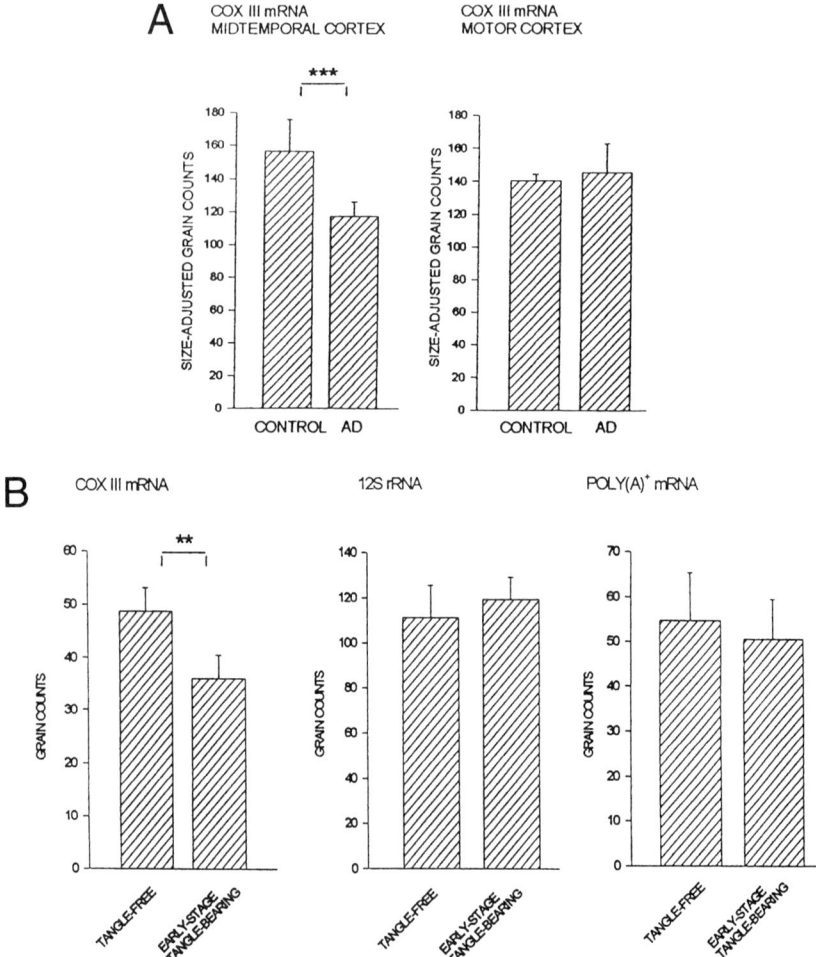

FIGURE 4. Correlations between neurofibrillary pathology and cytochrome oxidase subunit III (COX III) gene expression in pyramidal neurons Alzheimer's disease brain. **A:** COX III mRNA is reduced in tangle-free neurons of midtemporal but not motor cortex ($n = 3$–4) of AD brain. **B:** COX III mRNA declines progressively as tangles fill < 50% of cell cytoplasm (early stage) or > 50% (late stage) in pyramidal cells of AD midtemporal cortex, whereas non-OXPHOS mitochondrial 12S RNA and total poly(A)+ mRNA are reduced significantly only in late-stage tangle-bearing neurons. Grain counts give mRNA levels. Statistical significance: ***$p < 0.001$, **$p < 0.01$, *$p < 0.05$. (Reprinted from Hatanpää et al.[43] by permission.)

I, and for COX I and III subunits of complex IV, are significantly decreased. Levels of the nDNA-encoded mRNA for COX IV and for the beta subunit of ATP synthase (ATPsyn.β) of complex V are reduced as well. The molecular changes are specific to OXPHOS. They are unrelated to mitochondrial drop-out or reduced transcription of non-OXPHOS-related mtDNA, as levels of mtDNA-encoded 12S rRNA and total mtDNA are normal. Nor are they related to a general reduction in nuclear transcription, as nDNA-encoded mRNA for β-actin, lactic acid dehydrogenase-B, and 28S rRNA are unchanged.[43–47]

TABLE 2 also illustrates that a comparable pattern of reduced COX enzyme activity, and reduced mRNA levels for COX subunits coded for by nDNA as well as mtDNA, was found in the monkey lateral geniculate nucleus after 3–7 days of reduced retinal electrical activity due to chronic retinal exposure to tetrodotoxin through intravitreal injection.[48] As these reductions were reversed after the tetrodotoxin was removed,[49] they likely reflected physiological and fully reversible down-regulation of COX subunit gene expression in response to reduced neuronal energy demand. Mechanisms involved may include changes in the precursor pool for mitochondrial peptides or proteins[50] or in transcriptional or post-transcriptional factors produced by the nuclear genome.[45,51]

The topography of COX enzyme activity and of mRNA levels for some COX subunits within mammalian brain argues for a direct role of COX in maintaining synaptic function. COX activity and mitochondrial density are highest in terminal fields that receive major afferent pathways, and COX mRNA levels are highest in the cell bodies of these terminal fields, where transcription requiring the nuclear factors occurs.[49] For example, in the primate neocortex, COX enzyme activity is most intense in layers I and IV, and COX II mRNA is localized mainly in pyramidal neuron cell bodies in layers III and V–VI.[52–54] That selective reductions of OXPHOS subunit gene expression similar to those in the AD brain were found in the tetrodotoxin-treated monkeys and could be reversed, suggests that the reductions in AD represent potentially reversible down-regulation due to reduced energy demand by synapses. Parallel declines in synaptic markers and in COX enzyme activity are reported in association cortex of the AD brain.[55]

OXPHOS enzyme subunit expression appears to be down-regulated by stages in pyramidal neurons of the AD brain in relation to intracellular accumulation of NFTs. This was demonstrated by using *in situ* hybridization to measure levels of mtDNA-encoded COX III mRNA and 12S rRNA and of poly(A)$^+$ mRNA coded by mtDNA plus nDNA in pyramidal neurons of midtemporal AD cortex (FIG. 3).[43] Intracellular NFT density, evidenced by tau protein abnormally phosphorylated at its 396 and 404 serine sites, also was quantified in the cytoplasm of these neurons using an appropriate antibody. Pyramidal neurons without NFTs showed reduced COX III mRNA levels compared with NFT-free neurons in control brain, but mRNA for mtDNA-encoded 12S rRNA was normal. When NFTs filled less than 50% of the cell cytoplasm, COX III mRNA was further reduced while poly(A)$^+$ mRNA and 12S rRNA remained unchanged. When more than 50% of the neuronal cytoplasm was filled with NFTs, poly(A)$^+$ mRNA and 12S rRNA levels were reduced as well, indicative of non-transcribing, dying neurons.

Because down-regulation of OXPHOS gene expression was not accompanied by reduced mRNA levels of non-OXPHOS genes in the early stages of NFT accumulation, ATP production would not be expected to be rate limiting, and the down-regulation should be potentially reversible. Maintained responsiveness of OXPHOS agrees with evidence that mitochondrial respiration can be stimulated to the same extent by uncoupling agents in biopsied AD and control brain.[56] In late stage disease, when brain activation *in vivo* is markedly depressed (see above), depression of OXPHOS is likely rate limiting as neurons are dying and their general transcription is depressed. Cells with high NFT content may have less unphosphorylated tau available to react with tubulin to form fibrils necessary for fast axonal transport, thus depriving dendrites and axon terminals of mitochondria themselves.[57] Their proteins also demonstrate severe oxidative damage.[58]

The discrepancy between selective down-regulation of OXPHOS in postmortem samples of AD cortex (TABLE 2) and loss of general transcriptional ability in cortical pyramidal neurons whose cytoplasm is more than 50% filled with NFTs,[43] could be explained if the NFT-filled pyramidal cells in the postmortem cortical sample were incapable of significant transcription. Transcription by the remaining "viable cells" then would account for the observed selective down-regulation of OXPHOS in the samples.

In contrast to the significant correlation between NFT density and COX III expression in individual pyramidal neurons of the AD brain, COX expression is not correlated with neuronal proximity to senile (neuritic) plaques.[59] This, as well as evidence that regional NFT density but not plaque density correlates with regional reductions in $rCMR_{glc}$ in life,[9] argues that senile plaques do not contribute to decreased brain energy metabolism in AD.

DISCUSSION

In vivo brain imaging of patients with AD using PET demonstrates progressive reductions in resting-state brain glucose metabolism and blood flow in relation to dementia severity, more so in association than primary cortical regions. During cognitive or psychophysical stimulation, blood flow and glucose metabolism in affected regions can increase to the same extent in mildly demented AD patients as in age-matched controls, suggesting that energy availability is not rate-limiting. Responses to stimulation decline however with dementia severity; they are markedly reduced but nevertheless present to some extent in severely demented patients.

To the extent that brain metabolism and flow are markers of synaptic integrity,[1] the *in vivo* stages of brain responsiveness to stimulation likely reflect staging in synaptic loss and dysfunction in the brain itself. Biopsy and postmortem studies suggest that, of the pathological hallmarks of AD, synaptic loss best correlates with dementia severity in life.[31,43,46,55,60–62] Like blood flow and OXPHOS changes, synaptic changes in AD appear to be staged. In the first stage there is loss of presynaptic terminals with compensatory increments in the size of the remaining terminals; the net area of apposition between presynaptic and postsynaptic elements is maintained. In the second, additional presynaptic terminals are lost and the area of apposition is significantly reduced.[60,61]

A significant activation of brain in mildly demented AD patients is consistent with a physiologically and potentially reversible down-regulation of OXPHOS, the cause of which likely is reduced energy demand by dysfunctional synapses (see above).[63] *In vivo* regulation of OXPHOS enzyme expression appears coordinated with expression of nuclear genes involved in ATP production in a number of mitochondrial disorders,[64,65] and with the activity of Na,K-ATPase.[66] It is not unreasonable to think that in AD down-regulation of OXPHOS is coordinated with down-regulation of glucose delivery and ATP consumption, possibly through common transcriptional or post-translational mechanisms.[66–68] In affected AD brain regions, protein levels have been reported to be decreased for the GLUT1 glucose transporter in capillaries, the GLUT3 glucose transporter in neurons and Na,K-ATPase. Additionally, the neuron-specific α3 subunit of brain Na,K-ATPase is reduced.[69–72] The α-ketoglutarate dehydrogenase enzyme complex, which catalyzes the conversion of α-ketoglutarate to succinyl-coenzyme A in the tricarboxylic acid cycle within mitochondria, has been found to be reduced in FAD involving a mutation causing overproduction of amyloid β peptide, but in both affected and unaffected brain regions.[73] Its reduction in the unaffected regions may be related to increased expression of this peptide.

Taken together, data suggest that stimulation can increase $rCMR_{glc}$ and rCBF markedly, by approximately normal amounts, in the early stage of AD, and that mechanisms regulating glucose incorporation, OXPHOS and energy consumption are coordinately down-regulated at this stage of disease. This coordinated down-regulation likely is a response to reduced energy demand at the synapse. Even a "physiological" down-regulation of OXPHOS may increase brain vulnerability to reactive oxygen species during conditions of ischemia, excitotoxicity or even normal brain excitation, leading to mtDNA damage. The net effect would be an irreversible degenerative process involving mitochondrial dysfunction in the AD brain.[45,46,65,74]

REFERENCES

1. SOKOLOFF, L. 1991. Relationship between functional activity and energy metabolism in the nervous system: whether, where and why? *In* Brain Work and Mental Activity. Quantitative Studies with Radioactive Tracers. Alfred Benzon Symposium VIII. N.A. Lassen *et al.*, Eds. Munksgaard. Copenhagen. pp. 52–67.
2. KUMAR, A. *et al.* 1991. High-resolution PET studies in Alzheimer's disease. Neuropsychopharmacology **4:** 35–46.
3. RAPOPORT, S.I. 1991. Positron emission tomography in Alzheimer's disease in relation to disease pathogenesis: A critical review. Cerebrovasc. Brain Metab. Rev. **3:** 297–335.
4. IBÁÑEZ, V. *et al.* 1998. Regional glucose metabolic abnormalities are not the result of atrophy in Alzheimer's disease. Neurology **50:** 1585–1593.
5. HORWITZ, B. *et al.* 1987. Intercorrelations of regional cerebral glucose metabolic rates in Alzheimer's disease. Brain Res. **407:** 294–306.
6. MORRISON, J.H. *et al.* 1990. Cellular pathology in Alzheimer's disease: implications for corticocortical disconnection and differential vulnerability. *In* Imaging, Cerebral topography and Alzheimer's disease, Research and Perspectives in Alzheimer's disease. S.I. Rapoport *et al.*, Eds. Fondation Ipsen, Springer-Verlag. Berlin. pp. 19–40.
7. KHACHATURIAN, Z.S. 1985. Diagnosis of Alzheimer's disease. Arch. Neurol. **42:** 1097–1105.
8. LEWIS, D.A. *et al.* 1987. Laminar and regional distributions of neurofibrillary tangles and neuritic plaques in Alzheimer's disease: A quantitative study of visual and auditory cortices. J. Neurosci. **7:** 1799–1808.
9. DECARLI, C.S. *et al.* 1992. Post-mortem regional neurofibrillary tangle densities but not senile plaque densities are related to regional cerebral metabolic rates for glucose during life in Alzheimer's disease patients. Neurodegeneration **1:** 113–121.
10. RAPOPORT, S.I. 1988. Brain evolution and Alzheimer's disease. Rev. Neurol. (Paris) **144:** 79–90.
11. BRAAK, H., E. BRAAK & J. BOHL. 1993. Staging of Alzheimer-related cortical destruction. Eur. Neurol. **33:** 403–408.

12. RAPOPORT, S.I. 1990. Integrated phylogeny of the primate brain, with special reference to humans and their diseases. Brain Res. Rev. **15:** 267–294.
13. GRADY, C.L. et al. 1990. Subgroups in dementia of the Alzheimer type identified using positron emission tomography. J. Neuropsychiatry Clin. Neurosci. **2:** 373–384.
14. BENTON, A. 1985. Visuoperceptual, visuospatial, and visuoconstructive disorders. In Clinical Neuropsychology, 2nd edition. K.M. Heilman & E. Valenstein, Eds. Oxford University Press. New York. pp. 151–185.
15. HAXBY, J.V. et al. 1986. Neocortical metabolic abnormalities precede nonmemory cognitive deficits in early Alzheimer's-type dementia. Arch. Neurol. **43:** 882–885.
16. GRADY, C.L. et al. 1988. Longitudinal study of the early neuropsychological and cerebral metabolic changes in dementia of the Alzheimer type. J. Clin. Exp. Neuropsychol. **10:** 576–596.
17. GRADY, C.L. et al. 1988. Cerebral metabolic asymmetries predict decline in language performance in dementia of the Alzheimer type (DAT). J. Clin. Exp. Neuropsychol. **10:** 39.
18. HAXBY, J.V. et al. 1990. Longitudinal study of cerebral metabolic asymmetries and associated neuropsychological patterns in early dementia of the Alzheimer type. Arch. Neurol. **47:** 753–760.
19. FOLSTEIN, M.F., S.E. FOLSTEIN & P.R. MCHUGH. 1975. Mini Mental State. A practical method for grading the cognitive state of patients for the clinician. J. Psychiatr. Res. **12:** 189–198.
20. GINSBERG, L. et al. 1993. Regional specificity of membrane instability in Alzheimer's disease brain. Brain Res. **615:** 355–357.
21. GERSHFELD, N.L. 1989. Spontaneous assembly of a phospholipid bilayer as a critical phenomenon: influence of temperature, composition, and physical state. J. Phys. Chem. **93:** 5256–5261.
22. FAROOQUI, A.A., S.I. RAPOPORT & L.A. HORROCKS. 1997. Membrane phospholipid alterations in Alzheimer disease: Deficiency of ethanolamine plasmalogens. Neurochem. Res. **22:** 523–527.
23. CLARK, C.M. et al. 1991. The FDG/PET methodology for early detection of disease onset: A statistical model. J. Cereb. Blood Flow Metab. **11:** A96–A102.
24. MCKHANN, G. et al. 1984. Clinical diagnosis of Alzheimer's disease: report of the NINCDS-ADRDA work group under the auspices of Department of Health and Human Services task force on Alzheimer's disease. Neurology **34:** 939–944.
25. AZARI, N.P. et al. 1993. Early detection of Alzheimer's disease: A statistical approach using positron emission tomographic data. J. Cereb. Blood Flow Metab. **13:** 438–447.
26. AZARI, N.P. et al. 1994. Detection of an Alzheimer disease pattern of cerebral metabolism in Down syndrome. Dementia **5:** 69–78.
27. SMALL, G.W. et al. 1995. Apolipoprotein E type 4 allele and cerebral glucose metabolism in relatives at risk for familial Alzheimer disease. JAMA **273:** 942–947.
28. REIMAN, E.M. et al. 1996. Preclinical evidence of Alzheimer's disease in persons homozygous for the ε4 allele for apolipoprotein E. N. Engl. J. Med. **334:** 752–758.
29. KENNEDY, A.M. et al. 1995. Deficits in cerebral glucose metabolism demonstrated by positron emission tomography in individuals at risk of familial Alzheimer's disease. Neurosci. Lett. **186:** 17–20.
30. GRADY, C.L. et al. 1993. Activation of cerebral blood flow during a visuoperceptual task in patients with Alzheimer-type dementia. Neurobiol. Aging **14:** 35–44.
31. RAPOPORT, S.I. & C.L. GRADY. 1993. Parametric in vivo brain imaging during activation to examine pathological mechanisms of functional failure in Alzheimer disease. Int. J. Neurosci. **70:** 39–56.

32. PIETRINI, P. et al. 1999. Association between brain functional failure and dementia severity in Alzheimer's disease: Resting versus stimulation PET study. Am. J. Psychiatry **158:** 470–473.
33. PIETRINI, P. et al. Cerebral metabolic response to passive audiovisual stimulation in Alzheimer disease patients and healthy controls: a PET "stress test" for dementia. Submitted.
34. ROGERS, S.L. & L.T. FRIEDHOFF. 1998. Long-term efficacy and safety of donepezil in the treatment of Alzheimer's disease: An interim analysis of the results of a US multicentre open label extension study. Eur. Neuropsychopharmacol. **8:** 67–75.
35. LÖW, G. 1998. Personal communication. Director Eisai-Pfizer, Germany
36. VANMETER, J.W. et al. 1995. Parametric analysis of functional neuroimages: application to a variable-rate motor task. Neuroimage **2:** 272–383.
37. MENTIS, M.J. et al. 1996. Visual cortical dysfunction in Alzheimer's disease evaluated with a temporally graded "stress test" during PET. Am. J. Psychiatry **153:** 32–40.
38. MENTIS, M.J. et al. 1997. Frequency variation of a pattern-flash visual stimulus during PET differentially activates brain from striate through frontal cortex. Neuroimage **5:** 116–128.
39. MENTIS, M.J. et al. 1998. Increasing required neural response to expose abnormal brain function in mild versus moderate or severe Alzheimer's disease: PET study using parametric visual stimulation. Am. J. Psychiatry **155:** 785–794.
40. FOX, P.T. & M.E. RAICHLE. 1985. Stimulus rate determines regional brain blood flow in striate cortex. Ann. Neurol. **17:** 303–305.
41. LIVINGSTONE, M. & D. HUBEL. 1988. Segregation of form, color, movement, and depth: anatomy, physiology, and perception. Science **240:** 740–749.
42. MENTIS, M.J. et al. 1997. Acetylcholine (ACh) projection tracts modulating graded visual stimuli are differentially impaired in Alzheimer disease (AD) as assessed by physostigmine during PET. Soc. Neurosci. Abstr. **23:** 28.
43. HATANPÄÄ, K. et al. 1996. Neuronal activity and early neurofibrillary tangles in Alzheimer's disease. Ann. Neurol. **40:** 411–420.
44. FUKUYAMA, R. et al. 1996. Gene expression of ND4, a subunit of complex I of oxidative phosphorylation in mitochondria, is decreased in temporal cortex of brains of Alzheimer's disease patients. Brain Res. **713:** 290–293.
45. CHANDRASEKARAN, K. et al. 1996. Evidence for physiological down-regulation of brain oxidative phosphorylation in Alzheimer's disease. Exp. Neurol. **142:** 80–88.
46. RAPOPORT, S.I. et al. 1996. Brain energy metabolism, cognitive function and down-regulated oxidative phosphorylation in Alzheimer disease. Neurodegeneration **5:** 473–476.
47. CHANDRASEKARAN, K. et al. 1997. Decreased expression of nuclear and mitochondrial DNA-encoded genes of oxidative phosphorylation in association neocortex of Alzheimer disease. Brain Res. Mol. Brain Res. **44:** 99–104.
48. HEVNER, R.F. & M.T.T. WONG-RILEY. 1993. Mitochondrial and nuclear gene expression for cytochrome oxidase subunits are disproportionately regulated by functional activity in neurons. J. Neurosci. **13:** 1805–1819.
49. WONG-RILEY, M.T.T. 1989. Cytochrome oxidase: an endogenous metabolic marker for neuronal activity. Trends Neuroscience **12:** 94–101.
50. LIU, S. & M. WONG-RILEY. 1994. Nuclear-encoded mitochondrial precursor protein: intramitochondrial delivery to dendrites and axon terminals of neurons and regulation of neuronal activity. J. Neurosci. **14:** 5338–5351.
51. SCARPULLA, R.C. 1996. Nuclear respiratory factors and the pathways of nuclear-mitochondrial interaction. Trends Cardiovasc. Med. **6:** 39–45.

52. HEVNER, R.F. & M.T.T. WONG-RILEY. 1991. Neuronal expression of nuclear and mitochondrial genes for cytochrome oxidase (CO) subunits analyzed by in situ hybridization: comparison with CO activity and protein. J. Neurosci. **11:** 1942–1958.
53. CHANDRASEKARAN, K. *et al.* 1992. Localization of cytochrome oxidase (COX) activity and COX mRNA in the hippocampus and entorhinal cortex in the monkey brain: correlation with specific neuronal pathways. Brain Res. **579:** 333–336.
54. CHANDRASEKARAN, K. *et al.* 1993. Localization of cytochrome oxidase (COX) activity and COX mRNA in the perirhinal and superior temporal sulci of the monkey brain. Brain Res. **606**: 213–219.
55. HATANPÄÄ, K. *et al.* 1999. Loss of brain synaptic proteins regulating plasticity in human aging and Alzheimer's disease. J. Neuropathol. Exp. Neurol. **58:** 637–643.
56. SIMS, N.R. *et al.* 1987. Mitochondrial function in brain tissue in primary degenerative dementia. Brain Res. **436:** 30–38.
57. YAFFE, M.P. 1999. The machinery of mitochondrial inheritance and behavior. Science **283:** 1493–1497.
58. CALINGASAN, N.Y., K. UCHIDA & G.E. GIBSON. 1999. Protein-bound acrolein: a novel marker of oxidative stress in Alzheimer's disease. J. Neurochem. **72:** 751–756.
59. HATANPÄÄ, K. *et al.* 1998. No association between Alzheimer plaques and decreased levels of cytochrome oxidase subunit mRNA, a marker of neuronal energy metabolism. Brain Res. Mol. Brain Res. **59:** 13–21.
60. SCHEFF, S.W., S.T. DEKOSKY & D.A. PRICE. 1990. Quantitative assessment of cortical synaptic density in Alzheimer's disease. Neurobiol. Aging **11:** 29–37.
61. DEKOSKY, S.T. & S.W. SCHEFF. 1990. Synapse loss in frontal cortex biopsies in Alzheimer's disease: Correlation with cognitive severity. Ann. Neurol. **27:** 457–464.
62. TERRY, R.D. *et al.* 1991. Physical basis of cognitive alterations in Alzheimer's disease: Synapse loss is the major correlate of cognitive impairment. Ann. Neurol. **30:** 572–580.
63. RAPOPORT, S.I. 1997. Deux stades, réversible et irréversible, de l'insuffisance fonctionnelle dans le cerveau Alzheimerien. *In* De la Neurophysiologie à la Maladie D'Alzheimer. Symposium en Hommage à Yvon Lamour. Y. Christien, Ed. Solal. Marseille. pp. 165–172.
64. HEDDI, A. *et al.* 1993. Mitochondrial DNA expression in mitochondrial myopathies and coordinated expression of nuclear genes involved in ATP production. J. Biol. Chem. **268:** 12156–12163.
65. WALLACE, D.C. 1999. Mitochondrial diseases in man and mouse. Science **283:** 1482–1488.
66. HEVNER, R.F., R.S. DUFF & M.T. WONG-RILEY. 1992. Coordination of ATP production and consumption in brain: Parallel regulation of cytochrome oxidase and Na^+,K^+-ATPase. Neurosci. Lett. **138:** 188–192.
67. BOADO, R.J. & W.M. PARDRIDGE. 1993. Glucose deprivation causes posttranscriptional enhancement of brain capillary endothelial glucose transporter gene expression via GLUT1 mRNA stabilization. J. Neurochem. **60:** 2290–2296.
68. KUMAGAI, A.K. *et al.* 1995. Upregulation of blood-brain barrier GLUT1 transporter protein and mRNA in experimental chronic hypoglycemia. Diabetes **44:** 1399–1404.
69. HARIK, S.I., M.J. MITCHELL & R.N. KALARIA. 1989. Ouabain binding in the human brain. Effects of Alzheimer's disease and aging. Arch. Neurol. **46:** 951–954.
70. KALARIA, R.N. & S.I. HARIK. 1989. Reduced glucose transporter at the blood-brain barrier and in cerebral cortex in Alzheimer disease. J. Neurochem. **53:** 1083–1088.
71. SIMPSON, I.A. *et al.* 1994. Decreased concentrations of GLUT1 and GLUT3 glucose transporters in the brains of patients with Alzheimer's disease. Ann. Neurol. **35:** 546–551.

72. CHAUHAN, N.B. & G.J. SIEGEL. 1996. In situ analysis of Na,K-ATPase α1- and α2-isoform mRNAs in aging rat hippocampus. J. Neurochem. **66:** 1742–1751.
73. GIBSON, G.E. *et al.* 1998. α-Ketoglutarate dehydrogenase in Alzheimer brains bearing the APP670/671 mutation. Ann. Neurol. **44:** 676–681.
74. BEAL, M.F. 1998. Mitochondrial dysfunction in neurodegenerative diseases. Biochim. Biophys. Acta **1366:** 211–223.
75. CHANDRASEKARAN, K. *et al.* 1994. Impairment of mitochondrial cytochrome oxidase gene expression in Alzheimer disease. Brain. Res. Mol. Brain Res. **24** (Suppl. 1)**:** 336–340.
76. PARKER JR., W.D. *et al.* 1994. Electron transport chain defects in Alzheimer's disease brain. Neurology **44:** 1090–1096.
77. LURIA, A.R. 1973. The Working Brain. An Introduction to Neuropsychology. Basic Books. New York. p. 398.

Cellular and Molecular Mechanisms Underlying Perturbed Energy Metabolism and Neuronal Degeneration in Alzheimer's and Parkinson's Diseases

MARK P. MATTSON,[a] WARD A. PEDERSEN, WENZHEN DUAN, CARSTEN CULMSEE, AND SIMONETTA CAMANDOLA

Laboratory of Neurosciences, National Institute on Aging, Baltimore, Maryland 21224, USA

ABSTRACT: Synaptic degeneration and death of nerve cells are defining features of Alzheimer's disease (AD) and Parkinson's disease (PD), the two most prevalent age-related neurodegenerative disorders. In AD, neurons in the hippocampus and basal forebrain (brain regions that subserve learning and memory functions) are selectively vulnerable. In PD dopamine-producing neurons in the substantia nigra–striatum (brain regions that control body movements) selectively degenerate. Studies of postmortem brain tissue from AD and PD patients have provided evidence for increased levels of oxidative stress, mitochondrial dysfunction and impaired glucose uptake in vulnerable neuronal populations. Studies of animal and cell culture models of AD and PD suggest that increased levels of oxidative stress (membrane lipid peroxidation, in particular) may disrupt neuronal energy metabolism and ion homeostasis, by impairing the function of membrane ion-motive ATPases and glucose and glutamate transporters. Such oxidative and metabolic compromise may thereby render neurons vulnerable to excitotoxicity and apoptosis. Studies of the pathogenic mechanisms of AD-linked mutations in amyloid precursor protein (APP) and presenilins strongly support central roles for perturbed cellular calcium homeostasis and aberrant proteolytic processing of APP as pivotal events that lead to metabolic compromise in neurons. Specific molecular "players" in the neurodegenerative processes in AD and PD are being identified and include Par-4 and caspases (bad guys) and neurotrophic factors and stress proteins (good guys). Interestingly, while studies continue to elucidate cellular and molecular events occurring in the brain in AD and PD, recent data suggest that both AD and PD can manifest systemic alterations in energy metabolism (e.g., increased insulin resistance and dysregulation of glucose metabolism). Emerging evidence that dietary restriction can forestall the development of AD and PD is consistent with a major "metabolic" component to these disorders, and provides optimism that these devastating brain disorders of aging may be largely preventable.

[a]Address for correspondence: Laboratory of Neurosciences, National Institute on Aging, GRC 4FO1, 5600 Nathan Shock Drive, Baltimore, MD 21224. Phone: 410-558-8463; fax: 410-558-8465.
e-mail: mmattson@grc.nia.nih.gov

INTRODUCTION

Other chapters in this volume describe the overwhelming evidence that brain energy metabolism is perturbed in AD and PD. Among those data are the many brain imaging studies documenting reduced radiolabeled glucose uptake into brain cells of living AD patients (see refs. 1 and 2 and chapter by Rapoport in this volume), and studies showing reduced mitochondrial function in affected brain regions and in fibroblasts from AD and PD patients.[3,4] The purpose of the present chapter is to integrate the available information concerning the cellular and molecular events leading to neuronal degeneration in AD and PD to arrive at a "best-guess" working picture of the pathogenesis of these two disorders. Although many of the same neurodegenerative cascades are likely operative in other neurodegenerative disorders (e.g., Huntington's disease and amyotrophic lateral sclerosis), we have elected to focus on AD and PD because these are the two most prominent neurodegenerative disorders, and because the evidence for metabolic disturbances in these disorders is more extensive than for other neurodegenerative disorders. Any consideration of the pathogenesis of AD and PD has to include the fact that, although genetic and environmental factors influence risk for these disorders, increasing age is the major risk factor. As is evident from many studies, increased levels of oxidative stress, mitochondrial dysfunction, and metabolic aberrancies are prominent features of aging in many different organ systems including the brain.[3,5] Critical questions in the field of neurodegenerative disorders therefore include: (1) how do disease-specific initiators promote oxidative stress and mitochondrial dysfunction? (2) how do genetic factors promote neurodegenerative cascades? (3) does altered peripheral energy metabolism (e.g., increased insulin resistance) initiate and/or predispose to neurodegenerative cascades?

In AD, synapses and neurons in brain regions that subserve learning and memory functions, including the hippocampus, entorhinal cortex, basal forebrain, and neocortical association cortices, degenerate.[6] The brains of AD victims are characterized by extensive extracellular deposits of amyloid β-peptide (Aβ) and by degenerating neurons that contain abnormal filaments composed mainly of the microtubule-associate protein tau (which is oxidatively modified and hyperphosphorylated).[7,8] Aberrant proteolytic processing of APP is implicated in initiation of the neurodegenerative cascade in AD, and may promote neurodegeneration by increasing production of neurotoxic forms of Aβ and/or by decreasing production of neuroprotective forms of secreted APP.[7] Excellent progress in understanding the pathophysiology of AD has been made as a direct result of the identification of three genes (APP, PS-1, and PS-2) in which mutations cause early-onset autosomal dominant AD.[9,10] As described below, the two general consequences of these mutations (as elucidated in studies of transfected cells and transgenic mice harboring the mutations) are aberrant APP processing, with increased neuronal vulnerability to excitotoxic and metabolic insults, and apoptosis.

PD is characterized by motor abnormalities including akinesia, tremor, and rigidity that result largely from progressive degeneration of dopaminergic neurons in the substantia nigra (SN). The axons of the SN dopaminergic neurons form synapses on target neurons in the striatum. It is believed that in PD, and in animal models of PD (see below), the dopaminergic neurons cease producing dopamine long before the cells degenerate and die, and hence the quite striking beneficial effect of L-dopa ad-

ALZHEIMER'S DISEASE	
Presence in Neurofibrillary Tangles	**Presence in Brain Tissue and/or CSF**
4-hydroxynonenal adducts	Lipid peroxidation products
Oxidized proteins	Oxidized proteins
Nitrotyrosine	Nitrotyrosine
DNA oxidation products	DNA oxidation products

PARKINSON'S DISEASE	
Presence in Lewy Bodies	**Presence in Brain Tissue**
4-hydroxynonenal adducts	Free and protein-bound 4-hydroxynonenal
Oxidized proteins	Oxidized proteins
Nitrotyrosine	Nitrotyrosine

FIGURE 1. Evidence for increased oxidative stress in neurons in AD and PD models (see text for discussion and references).

ministration in PD patients during the early stages of the disease. The causes of neuronal degeneration in PD have not been identified, but it is increasingly clear that age-related increases in levels of oxidative stress[11,12] and mitochondrial dysfunction[4] play important roles. Two prominent properties of SN dopaminergic neurons that are of potential relevance to their selective vulnerability are their high content of iron and their high levels of the pigment melanin.[12] Although the specific factors that initiate the neurodegenerative process in PD are unknown, environmental factors appear to be particularly important.[13] However, a very small percentage of cases of PD are caused by mutations in the α-synuclein gene,[14] and α-synuclein is a conspicuous component of Lewy bodies, suggesting a role for abnormalities in the metabolism/function of this protein in PD.

OXIDATIVE STRESS AND PERTURBED MITOCHONDRIAL FUNCTION IN AD AND PD

Analyses of tissue homogenates from postmortem brain tissue have provided evidence for increased levels of cellular oxidative stress in vulnerable regions of AD and PD brains compared to the same brain regions from age-matched controls and to less vulnerable brain regions from the same patients.[11,12,13–17] Immunohistochemical analyses of brain sections from AD patients reveal increased protein oxidation, protein nitration, and lipid peroxidation in neurofibrillary tangles and neuritic plaques (FIG. 1).[15,18,19] In addition, levels of the lipid peroxidation product 4-hydroxynonenal in the ventricular CSF of AD patients are increased 2–3-fold compared to age-matched controls.[20] Consistent with increased levels of cellular oxidative stress in AD are data showing alterations in levels of antioxidant enzymes in vulnerable regions of AD brain. For example, protein and activity levels of catalase were decreased, while levels of Cu/Zn-SOD and Mn-SOD were increased, in vulnerable (and to a lesser extent in nonvulnerable) regions of AD brain compared to age-matched controls (FIG. 1).[21]

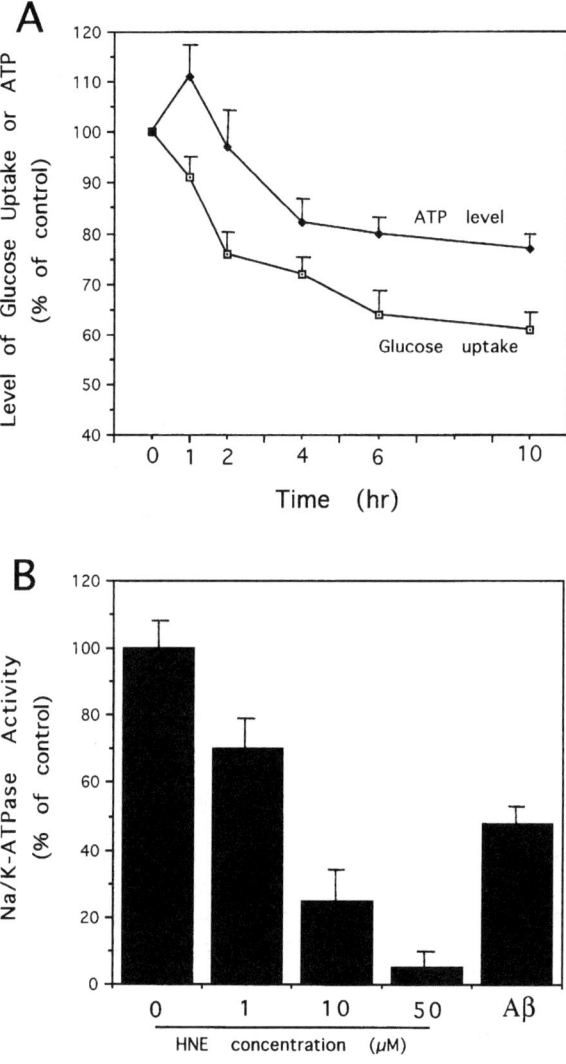

FIGURE 2. AD-relevant insults impair membrane ion-motive ATPase actiivty and glucose transport by a lipid peroxidation-mediated mechanism. **A:** Cultured rat cortical neurons were exposed to 10 μM Aβ25-35 for the indicated time periods and levels of [^{3}H]-glucose uptake and cellular ATP levels were quantified (modified from ref. 36). **B:** Cultured rat hippocampal neurons were exposed to the indicated concentrations of 4-hydroxynonenal (HNE) or 50 μM Aβ25-35 for 3 h and Na$^+$/K$^+$-ATPase activity was quantified (modified from Ref. 35).

Analyses of mitochondrial functional parameters in AD brain tissue reveal a striking decrease in activity of α-ketoglutarate dehydrogenase complex.[22] Such mitochondrial abnormalities have also been reported to occur in fibroblasts from AD patients.[23]

Data from studies of postmortem brain tissue from PD patients has provided evidence for increased levels of lipid peroxidation[11] and protein nitration[24] in substantia nigra of PD patients. In addition, there is an increase in levels of the DNA oxidation product 8-hydroxyguanine in substantia nigra,[25] and a generalized increase in protein carbonyls in the brain[26] in PD patients. Immunohistochemical analyses have shown that levels of the 4-hydroxynonenal are increased specifically in degenerating neurons in SN of PD patients.[27] A striking feature of PD is a profound decrease in the level of mitochondrial complex I activity in SN, a change that may arise from and/or contribute to increased cellular oxidative stress.[4] Interestingly, dysfunction can also be observed in mitochondria from peripheral cells from PD patients. Oxidative decarboxylation of pyruvate was impaired in fibroblasts from patients with PD,[28] and transformation of cells with mitochondria from PD patients resulted in perturbed cellular calcium homeostasis.[29]

ANIMAL AND CELL CULTURE MODELS OF AD AND PD

What are the events that lead to increased levels of oxidative stress and mitochondrial dysfunction in AD and PD? Increasing age is the major risk factor for both AD and PD, and it can therefore be presumed that the aging process itself plays an important role in promoting oxidative stress in the brain. Because they are postmitotic and have a very high metabolic rate, neurons are particularly susceptible to life-long accumulation of oxidative damage. Age-related alterations in proteolytic processing of APP may play a major role in the increased levels of oxidative stress in neurons in AD[7] and, conversely, oxidative stress and metabolic impairment may alter APP processing.[30] When in an aggregating form (as in the brains of AD patients), Aβ can induce membrane lipid peroxidation in cultured rat hippocampal neurons and human hippocampal synaptosomes.[31,32] The oxidative stress induced by Aβ may render neurons vulnerable to excitotoxicity[33] and apoptosis.[34] Interestingly, exposure of hippocampal neurons to Aβ also induces time- and dose-dependent decreases in catalase activity and increases in CuZnSOD and MnSOD activities,[21] suggesting a role for Aβ in the altered antioxidant enzyme profile in AD brain. Studies of the impact of lipid peroxidation on cultured hippocampal and cortical neurons suggest a scenario in which lipid peroxidation results in impairment of membrane ion-motive ATPases (Na$^+$/K$^+$-ATPase and Ca^{2+}-ATPase) and glucose transporters (FIG. 2[32,35,36]). ATP levels are decreased following exposure of neurons to Aβ or the lipid peroxidation product 4-hydroxynonenal (FIG. 2[32,36]). This leads to membrane depolarization, energy depletion and disruption of cellular calcium homeostasis. Membrane lipid peroxidation, as induced by Fe^{2+} or Aβ, also impairs glutamate transport in astrocytes and synaptosomes[37,38] which would be expected to further promote excitotoxic injury. Consistent with a central role for oxidative stress in the mechanism of Aβ-induced neuronal death are data showing that antioxidants (e.g., vitamin E, uric acid, propyl gallate, glutathione and estrogens) protect neurons against Aβ toxicity.[31,35,39,40] In addition to its direct actions on neurons, Aβ may also promote metabolic compromise of neurons by damaging vascular endothelial cells. We have found that exposure of vascular endothelial cells to Aβ results in impairment of glucose transport and barrier functions in these cells.[41] The latter effects of Aβ were prevent-

TABLE 1. Evidence that metabolic compromise and overactivation of glutamate receptors contributes to the neurodegenerative process in AD and PD

Evidence	Refs.
Glutamate induces neurofibrillary tangle (NFT)-like changes in cultured neurons.	33, 91
Glucose deprivation induces NFT-like changes in cultured neurons.	92
Aβ induces NFT-like changes in cultured neurons.	33, 93
Excitotoxins induce NFT-like changes in hippocampal neurons of adult rats.	94
Glucocorticoids and stress enhance NFT-like alterations *in vivo*.	94, 95
4-hydroxynonenal prevents tau dephosphorylation in cultured neurons	8

ed by antioxidants, thus implicating oxidative stress in Aβ's actions on vascular endothelial cells.

Further evidence for the involvement of oxidative stress, metabolic compromise and overactivation of glutamate receptors in the pathogenesis of AD comes from studies of the neuronal cytoskeleton in cell culture and animal models of AD (see TABLE 1 and references cited therein). Exposure of cultured rat hippocampal neurons to glutamate, Aβ and glucose deprivation results in antigenic changes in tau similar to those present in the neurofibrillary tangles in AD. Calcium influx and increased oxidative stress may be important mediators of such cytoskeletal alterations. Administration of excitotoxins to adult rats elicits antigenic changes in tau similar to those present in neurofibrillary tangles. Interestingly, stress (and consequent glucocorticoid production) exacerbate the AD-like alterations in hippocampal neurons. Neurofibrillary degenerative changes can be suppressed by treatment of neurons with neurotrophic factors and antioxidants (TABLE 1 and M.P.M., unpublished data).

Experimental data demonstrate that AD-relevant insults can impair mitochondrial function in neurons. Exposure of cultured hippocampal neurons and cortical synaptosomes to Aβ or Fe^{2+} leads to mitochondrial membrane depolarization and increased levels of mitochondrial ROS.[34,37] Such mitochondrial dysfunction in these models is secondary to increased oxidative stress because administration of antioxidants or overexpression of Mn-SOD results in preservation of mitochondrial function.[37,42] The mitochondrial dysfunction, in turn, contributes to the disruption of calcium homeostasis believed to occur in neurons in AD.[43] In addition to causing mitochondrial membrane depolarization, and increased mitochondrial oxidative stress and calcium dysregulation,[37,42,43] Aβ can cause damage to mitochondrial DNA[44] although it is not known if the latter alteration precedes or follows mitochondrial dysfunction.

Analyses of the mechanisms whereby mutations in APP and presenilin-1 (PS-1) cause AD have proven particularly informative in elucidating the events that result in increased oxidative stress and impaired mitochondrial function in neurons (FIG. 3). APP mutations have been shown to increase production of Aβ in cultured cells and transgenic mice, and the increased amyloidogenesis of Aβ under such conditions is likely to promote oxidative stress and metabolic impairment as described above (and see Ref. 7). In addition to increasing levels of Aβ, APP mutations may result in decreased levels of the secreted form of APP (sAPPα).[45,46] sAPPα has potent neuroprotective actions on hippocampal neurons in cell culture[31,47] and *in vi-*

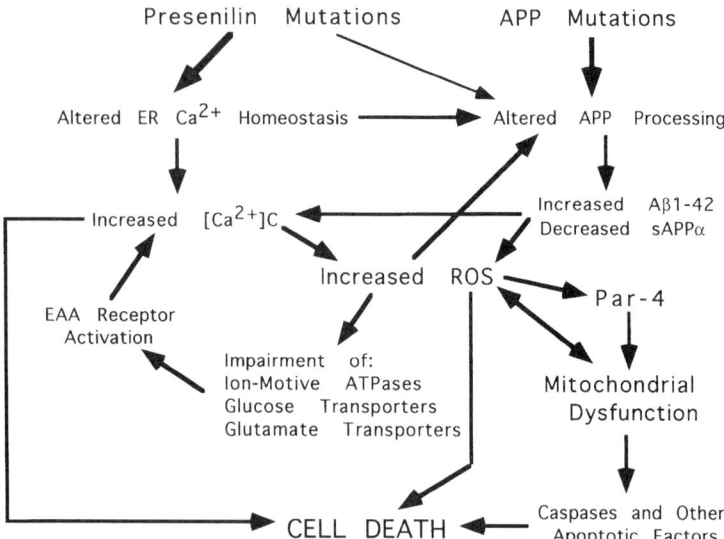

FIGURE 3. Working model for mechanisms whereby mutations in the amyloid precursor protein and presenilin-1 promote neuronal degeneration in AD (see text for discussion).

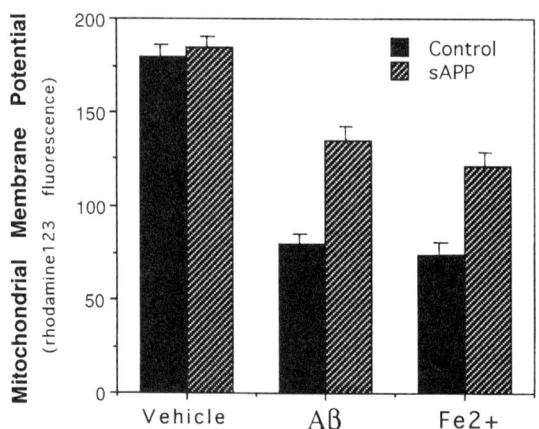

FIGURE 4. Impairment of synaptic mitochondrial function by AD-relevant insults: protective action of sAPPα. Rat cortical synaptosomes were pretreated for 1 h with 1 nM sAPPα or vehicle (Control). Synaptosomes were then exposed for 4 h to vehicle, 50 μM Aβ25-35 or 50 μM Fe^{2+} and levels of rhodamine 123 fluorescence (a measure of mitochondrial membrane potential) were quantified. Values are the mean and SEM of determinations made in at least 4 synaptosome preparations.

vo.[48] Pretreatment of neurons with sAPPα increases their resistance to oxidative injury induced by Fe^{2+} and Aβ.[31,45] The mechanism whereby sAPPα protects neurons appears to involve activation of potassium channels[49] and activation of the transcription factor NF-κB,[50] a transcription factor that was recently shown to protect neurons against apoptosis and excitotoxicity in several different culture paradigms.[51,52] In addition, we recently found that sAPPα induces an increase in the basal level of glucose transport, and attenuates oxidative impairment of glucose transport in cortical synaptosomes.[53] sAPPα can also preserve mitochondrial function in synaptosomes exposed to Aβ or Fe^{2+} (FIG. 4).

Mutations in presenilins (PS-1 and PS-2) are responsible for many cases of early-onset autosomal dominant AD.[10] Presenilins are integral membrane proteins localized predominantly in the endoplasmic reticulum. Expression of AD-linked PS-1 mutations in cultured neural cells and transgenic mice results in aberrant processing of APP[9] and to increased vulnerability of neurons to apoptosis and excitotoxicity.[54-57] Data suggest that a primary alteration in neurons expressing mutant PS-1 is perturbed endoplasmic reticulum calcium homeostasis resulting in enhanced calcium release when neurons are "challenged" with physiological (e.g., glutamate) or pathophysiological (e.g. exposure to Aβ) insults.[56,58] This perturbed calcium homeostasis appears to contribute to increased levels of cellular oxidative stress and mitochondrial dysfunction in neurons subjected to apoptotic insults. As evidence, manipulations that block calcium release from ER (dantrolene and xestospongin), block influx through plasma membrane voltage-dependent channels (nifedipine), or buffer cytoplasmic calcium (overexpression of calbindin D-28k) protect neurons expressing mutant PS-1 against Aβ-induced death.[54,55] Studies of PS-1 mutant knockin mice show that a PS-1 mutation increases vulnerability of hippocampal neurons to excitotoxicity *in vivo* and in cell culture; hippocampal neurons expressing mutant PS-1 exhibit enhanced calcium responses to glutamate.[56] The importance of oxidative stress and mitochondrial dysfunction in the cell death-enhancing actions of PS-

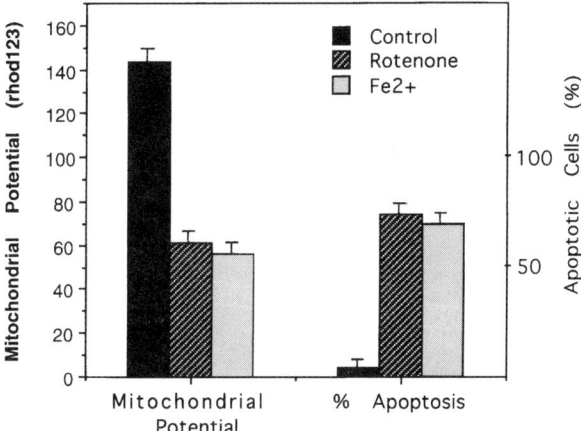

FIGURE 5. The complex I inhibitor rotenone and Fe^{2+} induce apoptosis of cultured human dopaminergic cells. Cultured SK-N-MC cells were exposed to vehicle (Control), 1 μM rotenone or 5 μM Fe^{2+} and levels of rhodamine 123 fluorescence (a measure of mitochondrial transmembrane potential) and apoptosis were quantified 12 and 24 h later, respectively. Values are the mean and SEM of determinations made in 4-6 cultures. (Modified from Ref. 61.)

1 mutations is demonstrated by the ability of the following manipulations to counteract the "endangering" actions of PS-1 mutations: (1) treatment with antioxidants including vitamin E, uric acid and 17β-estradiol[54,59]; (2) overexpression of manganese superoxide dismutase[60]; (3) treatment of neurons with cyclosporin A, an inhibitor of mitochondrial permeability transition pore formation.[42]

The factors responsible for increased oxidative stress in SN dopaminergic neurons in PD are not completely clear, but roles for Fe^{2+}, mitochondrial dysfunction, environmental toxins akin to MPTP (1-methyl-4-phenyl-1,2,3,6-tetrahydropyridine), and dopamine itself have been suggested.[4,12,13] Exposure of cultured dopaminergic cells to Fe^{2+}, MPP^+ (the toxic metabolite of MPTP), the complex I inhibitor rotenone, or dopamine can cause cell death[61] (FIG. 5). An excellent model of PD involves administration of the toxin MPTP to monkeys, which causes a clinical phenotype remarkably similar to that seen in human PD patients.[62] MPTP administration results in loss of tyrosine hydroxylase production and eventual degeneration of SN dopaminergic neurons. Levels of oxidative stress are increased, and complex I activity is decreased, in SN dopaminergic neurons prior to their degeneration in this model. The role of α-synuclein in the neurodegenerative process in PD remains to be established. It is of considerable interest that α-synuclein forms aggregates in the cytoplasm of degenerating neurons (Lewy bodies) and that expression of mutant α-synuclein in cultured cells may promote their degeneration.[63] If and how such protein aggregates induce oxidative stress and mitochondrial dysfunction remains to be determined.

Recent studies have identified specific mechanism whereby membrane lipid peroxidation, which appears to be a key event in the neurodegenerative process in both AD and PD, impairs the function of membrane transport proteins. Lipid peroxidation liberates, from membrane fatty acids, several aldehydes of varying carbon chain lengths. Among such aldehydes, 4-hydroxynonenal (HNE) appears to play a central role in impairment of protein function.[64] HNE covalently modifies proteins on cysteine, lysine and histidine residues. Immunoprecipitation-western blot analyses using antibodies that recognize HNE-modified proteins, in combination with antibodies against specific transport proteins, have demonstrated direct covalent modification of glucose transporter[36] and glutamate transporter[37] proteins following exposure of cultured neurons and synaptosomes to insults (e.g., Fe^{2+} and Aβ) that induce membrane lipid peroxidation. Presumably the covalent modification of these transporter proteins by HNE promotes protein crosslinking and impairs protein function.

THE CONCEPT OF "SYNAPTIC APOPTOSIS" AND ITS POSSIBLE ROLES IN AD AND PD

Synaptic compartments (presynaptic terminals and postsynaptic dendritic spines) are regions of neurons that are exposed to high levels of oxidative and metabolic stress. This is the case largely because glutamate receptors and calcium channels are concentrated in synaptic compartments, and the membrane depolarization and calcium influx resulting from activation of these ion channels results in oxidative stress and a high energy (ATP) demand. There are many sound reasons to believe that syn-

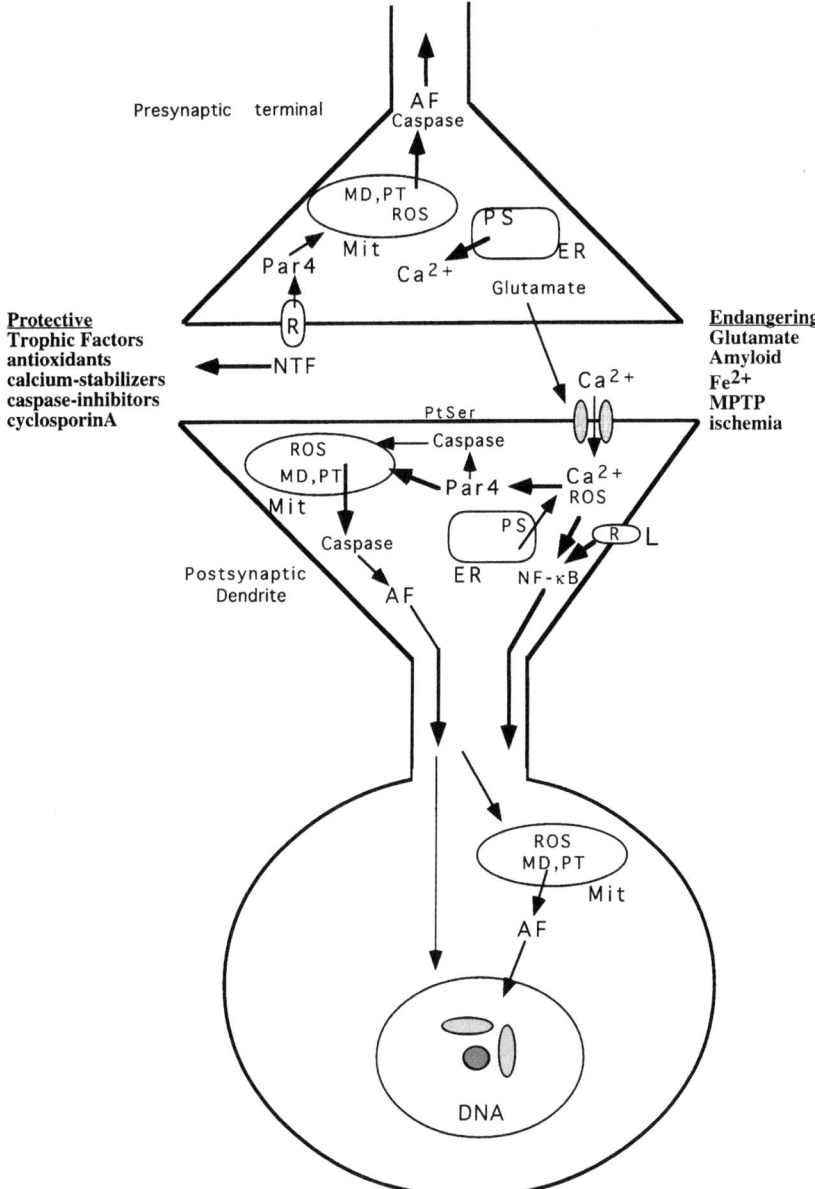

FIGURE 6. Working model of apoptotic biochemical cascades and their roles in synaptic degeneration and neuronal death. See text for discussion. (Modified from Ref. 67.)

apses are the sites where the neurodegenerative process begins in AD and PD (as well as in other neurodegenerative conditions). First, the extent of synapse loss is tightly correlated with cognitive deficits in AD patients,[6] and synapse loss in stria-

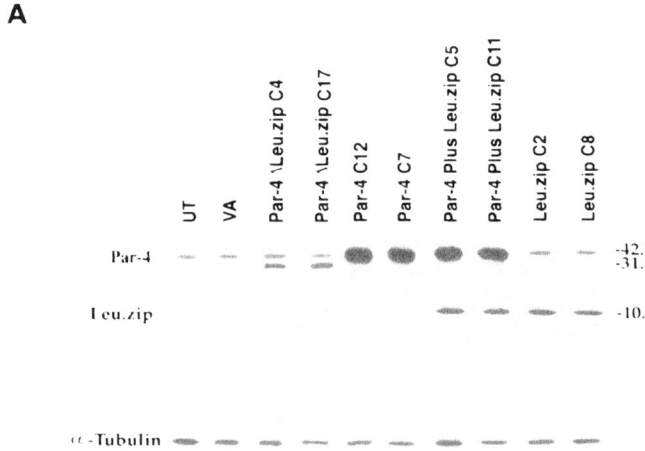
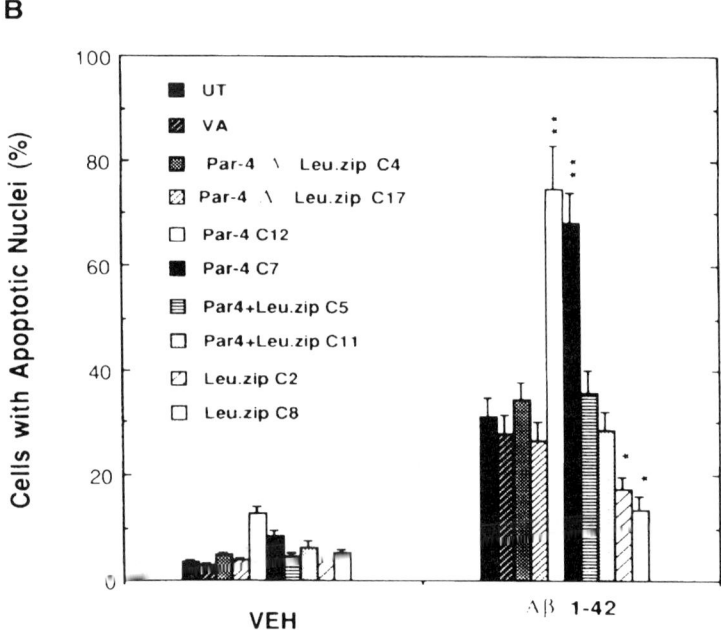

FIGURE 7. Evidence that Par-4 suppresses NF-κB activation and promotes neuronal apoptosis. **A:** Western blot analysis of Par-4 protein levels in selected PC12 cell clones: UT, untransfected; VA, clones transfected with vector alone; Par-4ΔLeu.zip, clones expressing Par-4 lacking the leucine zipper domain; Par-4, clones overexpressing full-length Par-4; Leu.zip, clones expressing the leucine zipper domain of Par-4. **B:** Cultures of the indicated PC12 cell clones were exposed for 24 h to vehicle (VEH) or 50 μM Aβ1-42 and the percentage of cells with apoptotic nuclei quantified (modified from Ref. 71).

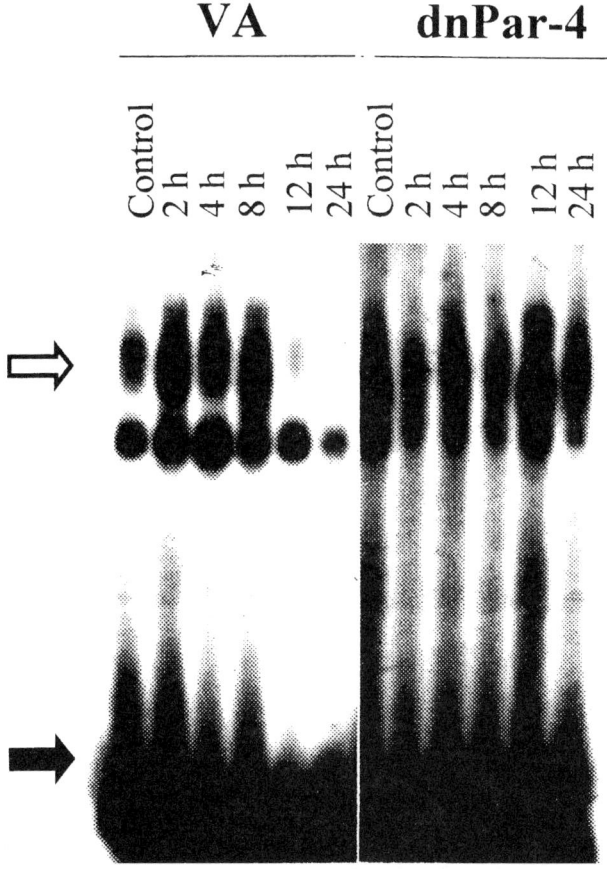

FIGURE 7 (continued). C: Gel-shift analysis of activated NF-κB prior to (Control) and at the indicated time points following trophic factor withdrawal in vector-transfected PC12 cells (VA) and PC12 cells overexpressing the dominant negative Par-4 leucine zipper domain (dnPar-4). Each lane was loaded with 5 μg of cell protein.

tum also appears to correlate with motor dysfunction in PD patients.[65] Second, overactivation of glutamate receptors, which are localized to postsynaptic regions of neuronal dendrites, plays an important role in the neuronal death process in several different animal and cell culture models of AD[33,35,56] and PD.[66] Third, recent studies have shown that apoptotic biochemical cascades are activated in vulnerable neuronal populations in AD and PD, and can also be activated locally in synaptic compartments following exposure to insults relevant to AD (Aβ) and PD (MPP$^+$ and rotenone)[67–69] (FIG. 6). The latter studies showed that exposure of synaptosomes or intact synaptically connected neurons to disease-relevant insults results in the following apoptosis-related events: caspase activation; loss of plasma membrane phospholipid asymmetry; induction of Par-4 expression (at the translational level);

mitochondrial membrane depolarization; mitochondrial oxyradical production; mitochondrial calcium uptake; and release into the cytosol of factors capable of inducing nuclear chromatin condensation and fragmentation.

Recent findings from studies of postmortem brain tissue, and animal and cell culture models, suggest that Par-4 (prostate apoptosis response-4) may serve as a critical link in the chain of events that leads neuronal degeneration in AD and PD. Par-4 is a leucine zipper and death domain-containing protein originally identified for its role in apoptosis of prostate cells,[70] and more recently implicated as a pivotal effector of neuronal apoptosis.[71] Levels of Par-4 mRNA and protein are increased in vulnerable regions (and to a lesser extent in nonvulnerable regions) of AD brain.[71] The latter study also showed that approximately 40% of neurofibrillary tangle-bearing neurons are also Par-4 immunoreactive, thus establishing a direct relationship between increased Par-4 expression and neuronal degeneration in AD. Cell culture data further suggest a central role for Par-4 in the cell death process in AD. Par-4 levels increase wihin 1–2 hours of exposure cultured hippocampal neurons to Aβ, and treatment of the neurons with a Par-4 antisense oligonucleotide prevents neuronal apoptosis.[71] Par-4 induction occurs prior to, and is required for, mitochondrial dysfunction and caspase activation following exposure of neurons to Aβ. Overexpression of the Par-4 leucine zipper domain prevents apoptosis (FIG. 7[71]), suggesting a necessary role for Par-4 interactions with another protein in its pro-apoptotic action. Par-4 interacts with regulatory domain of PKCζ and this interaction may inhibit the enzymatic activity of PKCζ.[72] We have recently found that Par-4 induction suppresses activation of the transcription factor NF-κB in PC12 cells (FIG. 7), which may promote apoptosis because activation of NF-κB prevents neuronal death induced by a variety of oxidative and metabolic insults.[51,52]

Par-4 levels increase dramatically in midbrain dopaminergic neurons of monkeys and mice exposed to MPTP.[61] The increase in Par-4 levels occurs in both neuronal cell bodies in the substantia nigra and their axon terminals in the striatum, and precedes loss of tyrosine hydroxylase immunoreactivity and cell death. Interestingly, Par-4 levels also increase following MPTP administration in neurons in the red nucleus, lateral geniculate nucleus and cerebral cortex.[61] Par-4 may play a role in neuronal dysfunction and/or apoptosis in the latter brain regions because, previous studies have documented metabolic alterations and/or neurodegenerative changes in these regions.[73] Exposure of cultured human dopaminergic neural cells to the complex I inhibitor rotenone, or to Fe^{2+}, resulted in Par-4 induction, mitochondrial dysfunction, and subsequent apoptosis.[61] Blockade of Par-4 induction by antisense treatment prevented rotenone- and Fe^{2+}-induced mitochondrial dysfunction and apoptosis demonstrating a critical role for Par-4 in the cell death process. Collectively, the data therefore suggest that Par-4 is a mediator of neuronal apoptosis associated with the pathogenesis of PD.

DO PERIPHERAL ALTERATIONS IN ENERGY METABOLISM CONTRIBUTE TO THE PATHOGENESIS OF AD AND PD?

There have been a large number of reports documenting alterations in glucose metabolism in AD patients.[1,2] Studies of fibroblasts and platelets from AD and PD patients add further weight to the evidence for widespread metabolic alterations in these two disorders.[23,28,74] In addition, several studies have documented "diabetes-like" alterations in AD patients including increased insulin resistance and abnormal glucose tolerance.[75,76] Moreover, alterations in the hypothalamic-pituitary-adrenal axis that controls glucocorticoid production have been widely reported.[77] Despite these findings, there is as yet no definitive evidence that a generalized metabolic disturbance precedes and contributes to the neurodegenerative process. We have recently found that glucose regulation is altered in transgenic mice expressing the APP "Swedish" mutation. Specifically, we have found that the APP mutant mice exhibit an altered response in a glucose tolerance test such that levels of glucose rise to much higher levels than in wild-type control mice (FIG. 8). Basal levels of glucose and insulin were not different in the wild-type and APP mutant transgenic mice. The APP mutant mice are hypersensitive to fasting such that they die within days to weeks when subjected to an alternate day feeding regimen.[78]

Studies of PD patients have also provided evidence for altered peripheral glucose metabolism in this disorder.[79] The alterations are, in general, similar to those in AD and include abnormal glucose tolerance tests and increased insulin resistance. These findings are particularly interesting because studies in rodents indicate that chronic hyperglycemia can cause dysfunction of dopaminergic transmission.[80]

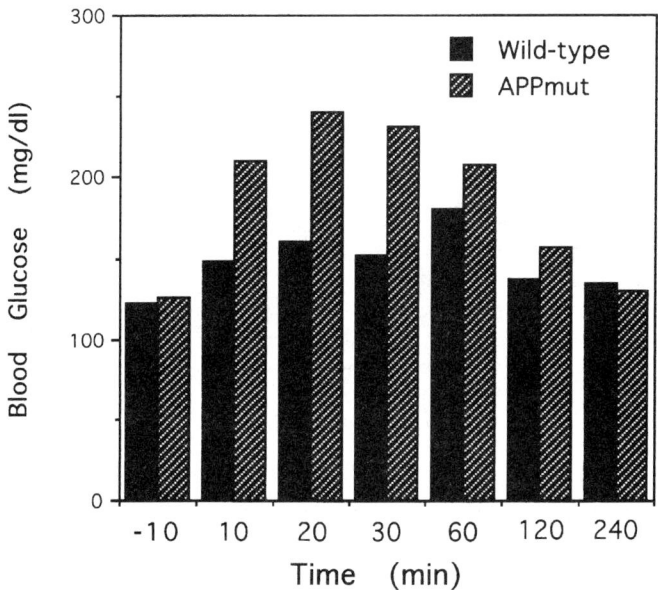

FIGURE 8. Glucose metabolism is altered in transgenic mice expressing an AD-linked APP mutation. Glucose levels were measured in blood samples from wild-type mice and APP mutant transgenic mice (96) taken 10 min prior to and at the indicated time points following administration of an intraperitoneal bolus of glucose. Values are the mean and SD ($n = 3$–4 mice).

CAN DIETARY RESTRICTION PREVENT AD AND PD?

Despite the convincing evidence that high food intake is a risk factor for age-related disorders such as cardovascular disease and diabetes, the possibility that high food intake might also increase risk for neurodegenerative disorders is largely unexplored. There are several reasons why it might be expected that reduced life-long food intake might ward off age-related neurodegenerative disorders such as AD and PD. First, dietary restriction (DR; reduced calorie intake with maintenance of miconutrient and vitamin nutrition) dramatically extends lifespan and reduces development of age-related disease in rodents[81] and monkeys.[82,83] Second, DR reduces levels of cellular oxidative stress in several different organ systems including the brain.[81,84,85] Third, epidemiological data suggest that the incidences of AD and PD are lower in countries with low per capita food consumption (e.g., China and Japan) compared to countries with high per capita food consumption (e.g., USA and Canada).[86]

We have begun to critically test the hypothesis that DR will increase resistance of neurons to AD, PD and other age-related neurodegenerative conditions. In one set of experiments rats were fed *ad libitum* (AL) or were maintained on an alternate day feeding schedule (DR) for periods of 2 weeks to 4 months. The rats were then administered either kainate (an excitotoxin that selectively damages hippocampal CA3 and CA1 neurons) or 3-nitropropionic acid (3NP, a mitochondrial toxin that selectively damages striatal neurons). Kainate-induced hippocampal damage and associated deficits in visuospatial memory were markedly reduced in rats maintained on DR for 3–4 months compared to AL-fed rats.[87] 3NP-induced damage to striatal neurons and associated motor deficits were significantly attenuated in rats maintained on the DR diet. Mice maintained on DR for 3–4 months were relatively resistant to MPTP-induced loss of tyrosine hydroxylase-positve neurons in the SN, and associated behavioral deficits, compared to AL-fed mice (FIG. 9[88]).

Reasoning that DR exerts its neuroprotective actions by a mechanism involving reduced glucose availability to neurons, we endeavored to mimic the effect of DR in AL-fed rats. To this end we determined whether administration of 2-deoxy-D-glucose (2DG), a non-metabolizable analog of glucose, to adult rats would result in increased resistance of neurons to excitotoxic, metabolic and oxidative insults. Rats administered 2DG (daily dose of 100 mg/kg body weight for 7 days) exhibited increased resistance of hippocampal neurons to kainate-induced damage[89] and increased resistance of cortical and striatal neurons to focal ischemia-reperfusion injury.[90] The vulnerability of SN dopaminergic neurons to MPTP-induced damage was decreased in mice administered 2DG.[88] In all three animal models behavioral outcome was significantly improved in animals pretreated with 2DG. The beneficial effects of 2DG on brain neurons *in vivo* are likely due to direct effects of 2DG on the neurons. As evidence, pretreatment of cultured hippocampal neurons with 2DG increases their resistance to excitotoxic, metabolic and oxidative insults.[89,90] Similarly, 2DG pretreatment protects cultured human dopaminergic cells against death induced by rotenone and Fe^{2+}.[88]

Collectively, the findings suggest that DR may protect the brain by a mechanism involving a "conditioning" response in which the DR can be viewed as a mild stress that induces upregulation of stress proteins in neurons. If DR exerts similar benefi-

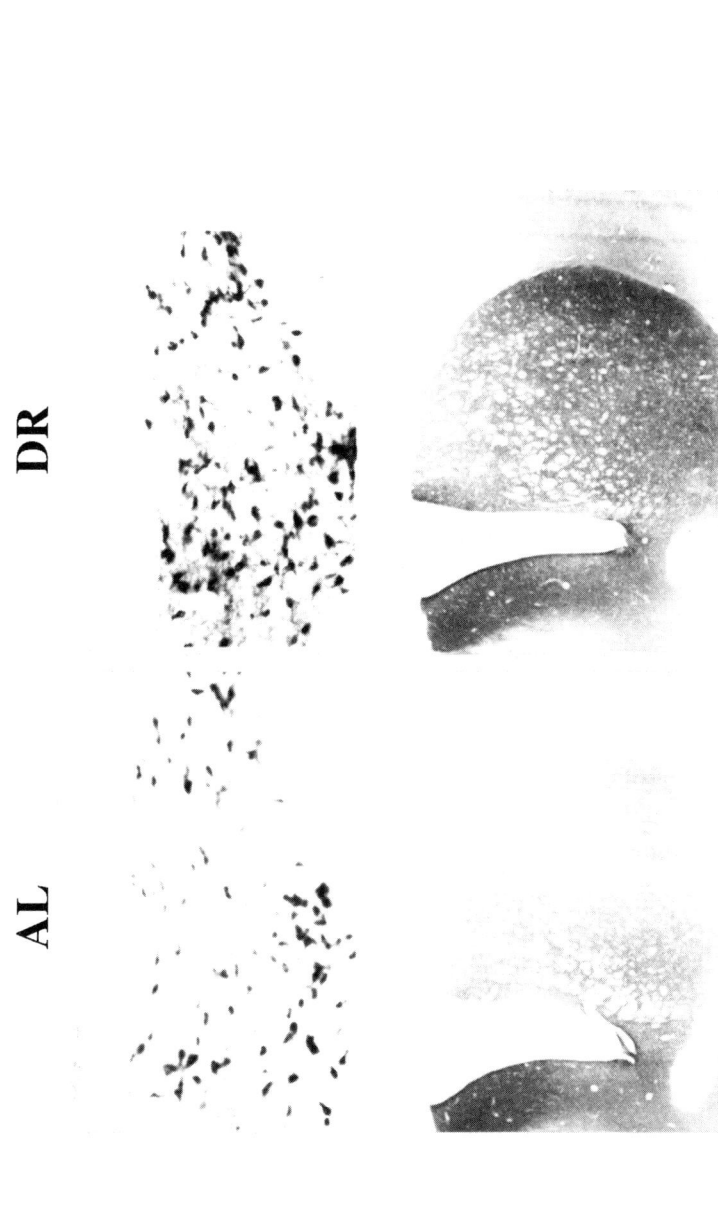

FIGURE 9. Dietary restriction increases resistance of dopaminergic nigro-striatal neurons in a mouse model of Parkinson's disease. Tyrosine hydroxylase (TH) immunoreactivity in substantia nigra (*upper panels*) and striatum (*lower panels*) 24 h following administration of MPTP in a control mouse fed *ad libitum* (AL) and a mouse maintained on dietary restriction for 3 months (DR). Note decreased loss of TH-positive neurons in the substantia nigra, and of TH-positive axons in the striatum, of the DR mouse.

cial effects in humans then DR may be a useful approach for reducing the incidence and severity of several different age-related neurodegenerative conditions.

ACKNOWLEDGMENTS

Research presented in this paper was supported by grants to M.P.M. from the NIH (NIA and NINDS).

REFERENCES

1. HOYER, S., R. NITSCH & K. OESTERREICH. 1991. Predominant abnormality in cerebral glucose utilization in late-onset dementia of the Alzheimer type: a cross-sectional comparison against advanced late-onset and incipient early-onset cases. J. Neural Transm. **3:** 1–14.
2. JAGUST, W.J., J.P. SEAB, R.H. HUESMAN, P.E. VALK, C.A. MATHIS, B.R. REED, P.G. COXSON & T.F. BUDINGER. 1991. Diminished glucose transport in Alzheimer's disease: Dynamic PET studies. J. Cereb. Blood Flow Metab. **11:** 323–330.
3. BENZI, G. & A. MORETTI. 1997. Contributions of mitochondrial alterations to brain aging. In The Aging Brain. M.P. Mattson & J.W. Geddes, Eds. JAI Press. Greenwich, CT. Adv. Cell Aging Gerontol. **2:** 129–160.
4. SCHAPIRA, A.H., M. GU, J.W. TAANMAN, S.J. TABRIZI, T. SEATON, M. CLEETER & J.M. COOPER. 1998. Mitochondria in the etiology and pathogenesis of Parkinson's disease. Ann. Neurol. **44:** S89–S98.
5. BUTTERFIELD, D.A. & E.R. STADTMAN. 1997. Protein oxidation processes in aging brain. In The Aging Brain. M.P. Mattson & J.W. Geddes, Eds. JAI Press. Greenwich, CT. Adv. Cell Aging Gerontol. **2:** 161–191.
6. DEKOSKY, S.T., S.W. SCHEFF & S.D. STYREN. 1996. Structural correlates of cognition in dementia: quantification and assessment of synapse change. Neurodegeneration **5:** 417–421.
7. MATTSON, M.P. 1997. Cellular actions of β-amyloid precursor protein, and its soluble and fibrillogenic peptide derivatives. Physiol. Rev. **77:** 1081–1132.
8. MATTSON, M.P., W. FU, G. WAEG & K. UCHIDA. 1997. 4-hydroxynonenal, a product of lipid peroxidation, inhibits dephosphorylation of the microtubule-associated protein tau. NeuroReport **8:** 2275–2281.
9. HARDY, J. 1997. Amyloid, the presenilins and Alzheimer's disease. Trends Neurosci. **20:** 154–159.
10. MATTSON, M. P. & Q. GUO. 1998. The Presenilins. Neuroscientist **5:** 112–124.
11. DEXTER, D.T., C.J. CARTER, F.R. WELLS, F. JAVOY-AGID, Y. AGID, A. LEES, P. JENNER & C.D. MARSDEN. 1989. Basal lipid peroxidation in substantia nigra is increased in Parkinson's disease. J. Neurochem. **52:** 381–389.
12. JENNER, P. & C.W. OLANOW. 1998. Understanding cell death in Parkinson's disease. Ann. Neurol. **44:** S72–S84.
13. LANGSTON, J.W. 1998. Epidemiology versus genetics in Parkinson's disease: progress in resolving an age-old debate. Ann. Neurol. **44:** S45–S52.
14. POLYMEROPOULOS, M.H. 1998. Autosomal dominant Parkinson's disease and α-synuclein. Ann. Neurol. **44:** S63–S64.
15. SMITH, C.D., J.M. CARNEY, P.E. STARKE-REED, C.N. OLIVER, E.R. STADTMAN, R.A. FLOYD & W.R. MARKESBERY. 1991. Excess brain protein oxidation and enzyme dysfunction in normal aging and in Alzheimer disease. Proc. Natl. Acad. Sci. USA **88:** 10540–10543.
16. MOCCOCI, P., M.S. MACGARVEY & M.F. BEAL. 1994. Oxidative damage to mitochondrial DNA is increased in Alzheimer's disease. Ann. Neurol. **36:** 747–751.

17. LOVELL, M.A., W.D. EHMANN, S.M. BUTLER & W.R. MARKESBERY. 1995. Elevated thiobarbituric acid-reactive substances and antioxidant enzyme activity in the brain in Alzheimer's disease. Neurology **45:** 1594–1601.
18. SMITH, M.A., P.L.R. HARRIS, L.M. SAYRE, J.S. BECKMAN & G. PERRY. 1997. Widespread peroxynitrite-mediated damage in Alzheimer's disease. J. Neurosci. **17:** 2653–2657.
19. GOOD, P.F., P. WERNER, A. HSU, C.W. OLANOW & D.P. PERL. 1996. Evidence of neuronal oxidative damage in Alzheimer's disease. Am. J. Pathol. **149:** 21–28.
20. LOVELL, M.A., W.D. EHMANN, M.P. MATTSON & W.R. MARKESBERY. 1997. Elevated 4-hydroxynonenal levels in ventricular fluid in Alzheimer's disease. Neurobiol. Aging **18:** 457–461.
21. BRUCE, A.J., S. BOSE, W. FU, C.M. BUTT, M.-E. MIRAULT, N. TANIGUCHI & M.P. MATTSON. 1997. Amyloid β-peptide alters the profile of antioxidant enzymes in hippocampal cultures in a manner similar to that observed in Alzheimer's disease. Pathogenesis **1:** 15–30.
22. GIBSON, G.E., K.F. SHEU & J.P. BLASS. 1998. Abnormalities of mitochondrial enzymes in Alzheimer disease. J. Neural. Transm. **105:** 855–870.
23. SHEU, K.F., A.J. COOPER, K. KOIKE, M. KOIKE, J.G. LINDSAY & J.P. BLASS. 1994. Abnormality of the alpha-ketoglutarate dehydrogenase complex in fibroblasts from familial Alzheimer's disease. Ann. Neurol. **35:** 312–318.
24. GOOD, P.F., A. HSU, P. WERNER, D.P. PERL & C.W. OLANOW. 1998. Protein nitration in Parkinson's disease. J. Neuropathol. Exp. Neurol. **57:** 338–342.
25. ALAM, Z.I., A. JENNER, S.E. DANIEL, A.J. LEES, N. CAIRNS, C.D. MARSDEN, P. JENNER & B. HALLIWELL. 1997. Oxidative DNA damage in the parkinsonian brain: an apparent selective increase in 8-hydroxyguanine in substantia nigra. J. Neurochem. **69:** 1196–1203.
26. ALAM, Z.I., S.E. DANIEL, A.J. LEES, D.C. MARSDEN, P. JENNER & B. HALLIWELL. 1997. A generalised increase in protein carbonyls in the brain in Parkinson's but not incidental Lewy body disease. J. Neurochem. **69:** 1326–1329.
27. YORITAKA, A., N. HATTORI, K. UCHIDA, M. TANAKA, E.R. STADTMAN & Y. MIZUNO. 1996. Immunohistochemical detection of 4-hydroxynonenal protein adducts in Parkinson disease. Proc. Natl. Acad. Sci. USA **93:** 2696–2701.
28. MYTILINEOU, C., P. WERNER, S. MOLINARI, A. DI ROCCO, G. COHEN & M. D. YAHR. 1994. Impaired oxidative decarboxylation of pyruvate in fibroblasts from patients with Parkinson's disease. J. Neural. Transm. Park. Dis. Dement. Sect. **8:** 223–228.
29. SHEEHAN, J.P., R.H. SWERDLOW, W.D. PARKER, S.W. MILLER, R.E. DAVIS & J.B. TUTTLE. 1997. Altered calcium homeostasis in cells transformed by mitochondria from individuals with Parkinson's disease. J. Neurochem. **68:** 1221–1233.
30. GABUZDA, D., J. BUSCIGLIO, L. CHEN, P. MATSUDAIRA & B. YANKNER. 1994. Inhibition of energy metabolism alters the processing of amyloid precursor protein and induces a potentially amyloidogenic derivative. J. Biol. Chem. **269:** 13623–13628.
31. GOODMAN, Y. & M.P. MATTSON. 1994. Secreted forms of β-amyloid precursor protein protect hippocampal neurons against amyloid β-peptide-induced oxidative injury. Exp. Neurol. **128:** 1–12.
32. MARK, R.J., K. HENSLEY, D.A. BUTTERFIELD & M.P. MATTSON. 1995. Amyloid β-peptide impairs ion-motive ATPase activities: evidence for a role in loss of neuronal Ca^{2+} homeostasis and cell death. J. Neurosci. **15:** 6239–6249.
33. MATTSON, M.P., B. CHENG, D. DAVIS, K. BRYANT, I. LIEBERBURG & R.E. RYDEL. 1992. β-amyloid peptides destabilize calcium homeostasis and render human cortical neurons vulnerable to excitotoxicity. J. Neurosci. **12:** 376–394.
34. KRUMAN, I., A.J. BRUCE-KELLER, D.E. BREDESEN, G. WAEG & M.P. MATTSON. 1997. Evidence that 4-hydroxynonenal mediates oxidative stress-induced neuronal apoptosis. J. Neurosci. **17:** 5089–5100.

35. MARK, R.J., M.A. LOVELL, W.R. MARKESBERY, K. UCHIDA & M.P. MATTSON. 1997. A role for 4-hydroxynonenal, an aldehydic product of lipid peroxidation, in disruption of ion homeostasis and neuronal death induced by amyloid β-peptide. J. Neurochem. **68:** 255–264.
36. MARK, R.J., Z. PANG, J.W. GEDDES, K. UCHIDA & M.P. MATTSON. 1997. Amyloid β-peptide impairs glucose uptake in hippocampal and cortical neurons: involvement of membrane lipid peroxidation. J. Neurosci. **17:** 1046–1054.
37. KELLER, J.N., Z. PANG, J.W. GEDDES, J.G. BEGLEY, A. GERMEYER, G. WAEG & M.P. MATTSON. 1997. Impairment of glucose and glutamate transport and induction of mitochondrial oxidative stress and dysfunction in synaptosomes by amyloid-β peptide: Role of the lipid peroxidation product 4-hydroxynonenal. J. Neurochem. **69:** 273–284.
38. BLANC, E.M., J.N. KELLER, S. FERNANDEZ & M.P. MATTSON. 1998. 4-hydroxynonenal, a lipid peroxidation product, inhibits glutamate transport in astrocytes. Glia **22:** 149–160.
39. GOODMAN, Y., A.J. BRUCE, B. CHENG & M.P. MATTSON. 1996. Estrogens attenuate and corticosterone exacerbates excitotoxicity, oxidative injury and amyloid β-peptide toxicity in hippocampal neurons. J. Neurochem. **66:** 1836–1844.
40. KELLER, J.N. & M.P. MATTSON. 1997. 17β-estradiol attenuates oxidative impairment of synaptic Na^+/K^+-ATPase activity, glucose transport and glutamate transport induced by amyloid β-peptide and iron. J. Neurosci. Res. **50:** 522–530.
41. BLANC, E.M., M. TOBOREK, R.J. MARK, B. HENNIG & M.P. MATTSON. 1997. Amyloid β-peptide disrupts barrier and transport functions and induces apoptosis in vascular endothelial cells. J. Neurochem. **68:** 1870–1881.
42. KELLER, J.N., M.S. KINDY, F.W. HOLTSBERG, D.K. ST CLAIR, H.-C. YEN, A. GERMEYER, S.M. STEINER, A.J. BRUCE-KELLER, J.B. HUTCHINS & M.P. MATTSON. 1998. Mitochondrial MnSOD prevents neural apoptosis and reduces ischemic brain injury: suppression of peroxynitrite production, lipid peroxidation and mitochondrial dysfunction. J. Neurosci. **18:** 687–697.
43. KRUMAN, I., Z. PANG, J.W. GEDDES & M.P. MATTSON. 1999. Pivotal role of mitochondrial calcium uptake in neural cell apoptosis and necrosis. J. Neurochem. **72:** 529-540.
44. BOZNER, P., V. GRISHKO, S.P. LEDOUX, G.L. WILSON, Y.C. CHYAN & M.A. PAPPOLLA. 1997. The amyloid beta protein induces oxidative damage of mitochondrial DNA. J. Neuropathol. Exp. Neurol. **56:** 1356–1362.
45. FURUKAWA, K., B. SOPHER, R.E. RYDEL, J.G. BEGLEY, G.M. MARTIN & M.P. MATTSON. 1996. Increased activity-regulating and neuroprotective efficacy of α-secretase-derived secreted APP is conferred by a C-terminal heparin-binding domain. J. Neurochem. **67:** 1882–1896.
46. LANNFELT, L., H. BASUN, L.O. WAHLUND, B.A. ROWE & S.L. WAGNER. 1995. Decreased α-secretase-cleaved amyloid precursor protein as a diagnostic marker for Alzheimer's disease. Nature Med. **1:** 829–832.
47. MATTSON, M.P., B. CHENG, A. CULWELL, F. ESCH, I. LIEBERBURG & R.E. RYDEL. 1993. Evidence for excitoprotective and intraneuronal calcium-regulating roles for secreted forms of β-amyloid precursor protein. Neuron **10:** 243–254.
48. SMITH-SWINTOSKY, V.L., L.C. PETTIGREW, S.D. CRADDOCK, A.R. CULWELL, R.E. RYDEL & M.P. MATTSON. 1994. Secreted forms of β-amyloid precursor protein protect against ischemic brain injury. J. Neurochem. **63:** 781–784.
49. FURUKAWA, K., S.W. BARGER, E. BLALOCK & M.P. MATTSON. 1996. Activation of K^+ channels and suppression of neuronal activity by secreted β-amyloid precursor protein. Nature **379:** 74–78.

50. BARGER, S.W. & M.P. MATTSON. 1996. Induction of neuroprotective κB-dependent transcription by secreted forms of the Alzheimer's β-amyloid precursor. Mol. Brain Res. **40:** 116–126.
51. BARGER, S.W., D. HORSTER, K. FURUKAWA, Y. GOODMAN, J. KRIEGLSTEIN & M.P. MATTSON. 1995. Tumor necrosis factors α and β protect neurons against amyloid β-peptide toxicity: evidence for involvement of a κB-binding factor and attenuation of peroxide and Ca^{2+} accumulation. Proc. Natl. Acad. Sci. USA **92:** 9328–9332.
52. MATTSON, M.P., Y. GOODMAN, H. LUO, W. FU & K. FURUKAWA. 1997. Activation of NF-κB protects hippocampal neurons against oxidative stress-induced apoptosis: evidence for induction of Mn-SOD and suppression of peroxynitrite production and protein tyrosine nitration. J. Neurosci. Res. **49:** 681–697.
53. MATTSON, M.P., Z.H. GUO & J.D. GEIGER. 1999. Secreted form of amyloid precursor protein attenuates oxidative impairment of glucose and glutamate transport in synaptosomes by a cyclic GMP-mediated mechanism. J. Neurochem. **73:** 532–537.
54. GUO, Q., B.L. SOPHER, D.G. PHAM, K. FURUKAWA, N. ROBINSON, G.M. MARTIN & M.P. MATTSON. 1997. Alzheimer's presenilin mutation sensitizes neural cells to apoptosis induced by trophic factor withdrawal and amyloid β-peptide: involvement of calcium and oxyradicals. J. Neurosci. **17:** 4212–4222.
55. GUO, Q., S. CHRISTAKOS, N. ROBINSON & M.P. MATTSON. 1998. Calbindin blocks the pro-apoptotic actions of mutant presenilin-1: reduced oxidative stress and preserved mitochondrial function. Proc. Natl. Acad. Sci. USA **95:** 3227–3232.
56. GUO, Q., W. FU, B.L. SOPHER, M.W. MILLER, C.B WARE, G.M. MARTIN & M.P. MATTSON. 1999. Increased vulnerability of hippocampal neurons to excitotoxic necrosis in presenilin-1 mutant knockin mice. Nature Med. **5:** 101–107.
57. BEGLEY, J.G., W. DUAN, K. DUFF & M.P. MATTSON. 1999. Altered calcium homeostasis and mitochondrial dysfunction in cortical synaptic compartments of presenilin-1 mutant mice. J. Neurochem. **72:** 1030–1039.
58. GUO, Q., K. FURUKAWA, B.L. SOPHER, D.G. PHAM, N. ROBINSON, G.M. MARTIN & M.P. MATTSON. 1996. Alzheimer's PS-1 mutation perturbs calcium homeostasis and sensitizes PC12 cells to death induced by amyloid β-peptide. NeuroReport **8:** 379–383.
59. MATTSON, M.P., N. ROBINSON & Q. GUO. 1997. Estrogens stabilize mitochondrial function and protect neural cells against the pro-apoptotic action of mutant presenilin-1. NeuroReport **8:** 3817–3821.
60. GUO, Q., W. FU, B.L. SOPHER, F.W. HOLTSBERG, S.M. STEINER & M.P. MATTSON. 1999. Superoxide mediates the apoptosis-enhancing action of presenilin-1 mutations. J. Neurosci. Res. **56:** 457–470.
61. DUAN, W., D.M. GASH, V. RANGNEKAR & M.P. MATTSON. 1999. Participation of Par-4 in degeneration of dopaminergic neurons in primate and rodent models of Parkinson's disease. Ann. Neurol. **46:** 587–597.
62. TIPTON, K.F. & T.P. SINGER. 1993. Advances in our understanding of the mechanisms of the neurotoxicity of MPTP and related compounds. J. Neurochem. **61:** 1191–1206.
63. EL-AGNAF, O.M., R. JAKES, M.D. CURRAN, D. MIDDLETON, R. INGENITO, E. BIANCHI, A. PESSI, D. NEILL & A. WALLACE. 1998. Aggregates from mutant and wild-type alpha-synuclein proteins and NAC peptide induce apoptotic cell death in human neuroblastoma cells by formation of beta-sheet and amyloid-like filaments. FEBS Lett. **440:** 71–75.
64. MATTSON, M.P. 1998. Modification of ion homeostasis by lipid peroxidation: roles in neuronal degeneration and adaptive plasticity. Trends Neurosci. **21:** 53–57.

65. ANGLADE, P., A. MOUATT-PRIGENT, Y. AGID & E. HIRSCH. 1996. Synaptic plasticity in the caudate nucleus of patients with Parkinson's disease. Neurodegeneration **5:** 121–128.
66. MITCHELL, I.J. & C. B. CARROLL. 1997. Reversal of parkinsonian symptoms in primates by antagonism of excitatory amino acid transmission: potential mechanisms of action. Neurosci. Biobehav. Rev. **21:** 469–475.
67. MATTSON, M.P., J.N. KELLER & J.G. BEGLEY. 1998. Evidence for synaptic apoptosis. Exp. Neurol. **153:** 35–48.
68. MATTSON, M.P., J. PARTIN & J.G. BEGLEY. 1998. Amyloid β-peptide induces apoptosis-related events in synapses and dendrites. Brain Res. **807:** 167–176.
69. DUAN, W., V. RANGNEKAR & M.P. MATTSON. 1999. Par-4 production in synaptic compartments following apoptotic and excitotoxic insults: evidence for a pivotal role in mitochondrial dysfunction and neuronal degeneration. J. Neurochem. **72:** 2312–2322.
70. SELLS, S.F., S.-S. HAN, S. MUTHUKKUMAR, N. MADDIWAR, R. JOHNSTONE, E. BOGHAERT, D. GILLIS, G. LIU, P. NAIR, S. MONNIG, P. COLLINI, M.P. MATTSON, V. P. SUKHATME, S.G. ZIMMER, D.P. WOOD, J.W. MCROBERTS, Y. SHI & V.M. RANGNEKAR. 1997. Expression and function of the leucine zipper protein Par-4 in apoptosis. Mol. Cell. Biol. **17:** 3823–3832.
71. GUO, Q., W. FU, J. XIE, H. LUO, S.F. SELLS, J.W. GEDDES, V. BONDADA, V.M. RANGNEKAR & M.P. MATTSON. 1998. Par-4 is a mediator of neuronal degeneration associated with the pathogenesis of Alzheimer's disease. Nature Med. **4:** 957–962.
72. DIAZ-MECO, M.T., M.M. MUNICIO, S. FRUTOS, P. SANCHEZ, J. LOZANO, L. SANZ & J. MOSCAT. 1996. The product of par-4, a gene induced during apoptosis, interacts selectively with the atypical isoforms of protein kinase C. Cell **86:** 777–786.
73. GNANALINGHAM, K.K., N.A. MILKOWSKI, L.A. SMITH, A.J. HUNTER, P. JENNER & C.D. MARSDEN. 1995. Short and long-term changes in cerebral [^{14}C]-2-deoxyglucose uptake in the MPTP-treated marmoset: relationship to locomotor activity. J. Neural. Transm. Gen. Sect. **101:** 65–82.
74. SORBI, S., S. PIACENTINI, S. LATORRACA, P. PIERSANTI & L. AMADUCCI. 1995. Alterations in metabolic properties in fibroblasts in Alzheimer disease. Alzheimer Dis. Assoc. Disord. **9:** 73–77.
75. MESSIER, C. & M. GAGNON. 1996. Glucose regulation and cognitive functions: relation to Alzheimer's disease and diabetes. Behav. Brain Res. **75:** 1–11.
76. VANHANEN, M. & H. SOININEN. 1998. Glucose intolerance, cognitive impairment and Alzheimer's disease. Curr. Opin. Neurol. **11:** 673–677.
77. MOLCHAN, S.E., J.L. HILL, A.M. MELLOW, B.A. LAWLOR, R. MARTINEZ & T. SUNDERLAND. 1990. The dexamethasone suppression test in Alzheimer's disease and major depression: relationship to dementia severity, depression, and CSF monoamines. Int. Psychogeriatr. **2:** 99–122.
78. PEDERSEN, W.A., C. CULMSEE, D. ZIEGLER, J.P. HERMAN & M.P. MATTSON. 1999. Aberrant stress response associated with severe hypoglycemia in a transgenic mouse model of Alzheimer's disease. J. Mol. Neurosci. **13:** 159–165.
79. SANDYK, R. 1993. The relationship between diabetes mellitus and Parkinson's disease. Int. J. Neurosci. **69:** 125–130.
80. SHIMIZU, H., Y. SHIMOMURA, M. TAKAHASHI, I. KOBAYASHI & S. KOBAYASHI. 1990. Dopamine receptor in the streptozotocin-induced diabetic rats. Exp. Clin. Endocrinol. **95:** 263–266.
81. SOHAL, R.S. & R. WEINDRUCH. 1996. Oxidative stress, caloric restriction, and aging. Science **273:** 59–63.
82. CEFALU, W.T., J.D. WAGNER, Z.Q. WANG, A.D. BELL-FARROW, J. COLLINS, D. HASKELL, R. BECHTOLD & T. MORGAN. 1997. A study of caloric restriction and car-

diovascular aging in cynomolgus monkeys (*Macaca fascicularis*): a potential model for aging research. J. Gerontol. A Biol. Sci. Med. Sci. **52:** B10–B19.
83. LANE, M.A., D.J. BAER, W.V. RUMPLER, R. WEINDRUCH, D.K. INGRAM, E.M. TILMONT, R.G. CUTLER & G.S. ROTH. 1996. Calorie restriction lowers body temperature in rhesus monkeys, consistent with a postulated anti-aging mechanism in rodents. Proc. Natl. Acad. Sci. USA **93:** 4159–4164.
84. DUBEY, A., M.J. FORSTER, H. LAL & R.S. SOHAL. 1996. Effect of age and caloric intake on protein oxidation in different brain regions and on behavioral functions of mouse. Arch. Biochem. Biophys. **333:** 189–197.
85. AKSENOVA, M.V., M.Y. AKSENOV, J.M. CARNEY & D.A. BUTTERFIELD. 1998. Protein oxidation and enzyme activity decline in old brown Norway rats are reduced by dietary restriction. Mech. Ageing Dev. **100:** 157–168.
86. GRANT, W. 1997. Dietary links to Alzheimer's disease. Alz. Dis. Rev. **2:** 42-55.
87. BRUCE-KELLER, A.J., G. UMBERGER, R. MCFALL & M.P. MATTSON. 1999. Food restriction reduces brain damage and improves behavioral outcome following excitotoxic and metabolic insults. Ann. Neurol. **45:** 8–15.
88. DUAN, W. & M.P. MATTSON. 1999. Dietary restriction and 2-deoxyglucose administration improve behavioral outcome and reduce degeneration of dopaminergic neurons in models of Parkinson's disease. J. Neurochem. Submitted.
89. LEE, J., A.J. BRUCE-KELLER, U. KRUMAN, S.L. CHAN & M.P. MATTSON. 1999. 2-deoxy-D-glucose protects hippocampal neurons against excitotoxic and oxidative injury: involvement of stress proteins. J. Neurosci. Res. In press.
90. YU, Z.F. & M.P. MATTSON. 1999. Dietary restriction and 2-deoxyglucose administration reduce focal ischemic brain damage and improve behavioral outcome: evidence for a preconditioning mechanism. J. Neurosci. Res. **57:** 830–839.
91. MATTSON, M.P. 1990. Antigenic changes similar to those seen in neurofibrillary tangles are elicited by glutamate and Ca^{2+} influx in cultured hippocampal neurons. Neuron **4:** 105–117.
92. CHENG, B. & M.P. MATTSON. 1992. Glucose deprivation elicits neurofibrillary tangle-like antigenic changes in hippocampal neurons: prevention by NGF and bFGF. Exp. Neurol. **117:** 114–123.
93. BUSCIGLIO, J., A. LORENZO, J. YEH & B.A. YANKNER. 1995. β-amyloid fibrils induce tau phosphorylation and loss of microtubule binding. Neuron **14:** 879–888.
94. ELLIOTT, E., M.P. MATTSON, P. VANDERKLISH, G. LYNCH, I. CHANG & R.M. SAPOLSKY. 1993. Corticosterone exacerbates kainate-induced alterations in hippocampal tau immunoreactivity and spectrin proteolysis in vivo. J. Neurochem. **61:** 57–67.
95. STEIN-BEHRENS, B., M.P. MATTSON, I. CHANG, M. YEH & R.M. SAPOLSKY. 1994. Stress exacerbates neuron loss and cytoskeletal pathology in the hippocampus. J. Neurosci. **1:** 5373–5380.
96. HSIAO, K., P. CHAPMAN, S. NILSEN, C. ECKMAN, Y. HARIGAYA, S. YOUNKIN, F. YANG & G. COLE. 1996. Correlative memory deficits, Aβ elevation, and amyloid plaques in transgenic mice. Science **274:** 99–102.

Use of Cytoplasmic Hybrid Cell Lines for Elucidating the Role of Mitochondrial Dysfunction in Alzheimer's Disease and Parkinson's Disease

SOUMITRA S. GHOSH,[a,c,d] RUSSELL H. SWERDLOW,[b,c] SCOTT W. MILLER,[a] BRINA SHEEMAN,[a] W. DAVIS PARKER, JR.,[b] AND ROBERT E. DAVIS[a]

[a]*MitoKor, San Diego, California 92121, USA*

[b]*Department of Neurology, University of Virginia School of Medicine, Charlottesville, Virginia 22904, USA*

ABSTRACT: There is substantial evidence of mitochondrial defects in neurodegenerative disorders such as Alzheimer's and Parkinson's diseases (AD and PD). We have probed the molecular implications of mitochondrial dysfunction in these diseases by transferring mitochondria from platelets obtained from disease and control donors into mitochondrial DNA-depleted recipient neuron–based cells (ρ^0 cells). This process creates cytoplasmic hybrid (cybrid) cells where the mitochondrial DNA (mtDNA) from the donor is expressed in the nuclear and cellular background of the host ρ^0 cell. Differences in phenotype between disease and control groups can thus be attributed to the exogenous mitochondria and mtDNA. Key methodological issues relating to this approach were addressed by demonstrating that recipient ρ^0 cells have <1 mtDNA copy/cell, and that exclusive repopulation with donor mtDNA occurs in cybrid cells. Further, we describe that sampling of heterogeneous cell populations is a valid approach for cybrid analysis. Our studies show that the focal respiratory chain defects reported in platelets of AD and PD cybrids can be recapitulated in AD and PD cybrids. In addition, both AD and PD cybrids display increased oxidative stress and perturbations in calcium homeostasis. These data suggest that the transfer of a mtDNA defect from disease donor platelets is the likely cause of the cybrid biochemical phenotype, and highlight the potential value of these cell lines as cellular disease models.

INTRODUCTION

Mitochondrial dysfunction has been implicated in the pathophysiology of several common neurodegenerative disorders that include Alzheimer's disease (AD), Parkinson's disease (PD), Huntington's disease, and amyotrophic lateral sclerosis.[1] In addition to generating ATP via respiration, mitochondria are critical for maintaining calcium homeostasis in cells and play a central role in apoptotic and necrotic cell

---

[c]These authors contributed equally in writing this chapter.

[d]Address for correspondence: Soumitra Ghosh, Ph.D., Director, Bio-Organic Chemistry, MitoKor, 11494 Sorrento Valley Road, San Diego, CA 92121. Phone: 858-509-5603; fax: 858-793-7805.

e-mail: ghoshs@mitokor.com

death pathways. It is therefore not surprising that neurons, which depend primarily on mitochondrial ATP production to meet their bioenergetic demands, are at risk when mitochondrial function is impaired. While most diseases with clear mitochondrial genetic origins are quite rare in the general population, it is becoming increasingly apparent that many common late-onset diseases exhibit mitochondrial pathology either as a primary causative event or as a secondary consequence of the disease state. Disease-associated mitochondrial DNA (mtDNA) alterations have been identified for only a few rare diseases. However, it has been difficult to convincingly demonstrate a direct link between the mutant genotype and a disease phenotype, even in rare mitochondrial diseases, because of incomplete penetrance and the influence of environmental triggers (e.g., Leber's hereditary optic neuropathy).[2,3]

The characterization of mitochondrial dysfunction at the molecular level and of its association with functional defects in cells or tissue has been hampered by the inability to manipulate the mitochondrial genome by standard molecular biology techniques. For example, it has not been feasible to engineer and express site-directed mutants of mitochondrial genes to study the effect of a mutation in isolation from the numerous polymorphic nucleotide changes in the mitochondrial genome. Attardi and coworkers have partially circumvented this problem by pioneering techniques that permit the transfer of mitochondria along with other cytoplasmic contents from enucleated donor cells or cells lacking a nucleus (e.g., platelets) into cells experimentally depleted of endogenous mtDNA (termed ρ^0 cells).[4,5] Cells containing donor mitochondria are fused to the recipient ρ^0 cells to create cytoplasmic hybrid (cybrid) cell lines that display aberrant or normal respiratory chain phenotype depending upon whether or not the transferred mtDNA harbors deleterious alterations. Thus, the cybrid methodology is a useful approach for demonstrating the functional consequences of known mtDNA mutations and for screening for mtDNA alterations in candidate diseases. In this chapter, we describe the use of the cybrid technique to probe the basis of mitochondrial dysfunction in Alzheimer's disease (AD) and Parkinson's disease (PD), and the applicability of cybrid cellular systems as models for these diseases.

ALZHEIMER'S DISEASE CYBRID STUDIES

Substantial evidence suggests that AD is associated with focal defects in energy metabolism with accompanying increases in oxidative stress. Positron emission tomography studies have reported regionally specific deficits in energy metabolism in AD brains.[6–8] AD pathology displays prominent signs of oxidative injury implicating reactive oxygen species (ROS) as seen in increased DNA, protein, and lipid oxidation.[9–14] Functional magnetic resonance spectroscopy studies indicate decreased production of ATP in AD brain as inferred from elevated inorganic phosphate to phosphocreatine ratios.[15,16] Several studies have reported increased female-to-male ratio in the parental generation of probands, suggesting maternal inheritance in sporadic AD.[17,18] The observation by Parker and others of specific defects in catalytic activity of cytochrome c oxidase (complex IV of the electron transport chain) in AD brain and platelets provides strong evidence linking mitochondrial dysfunction with metabolic deficits in AD.[19–25] The activities of other components of the electron transport chain are normal in AD brain and platelets. These findings differ from

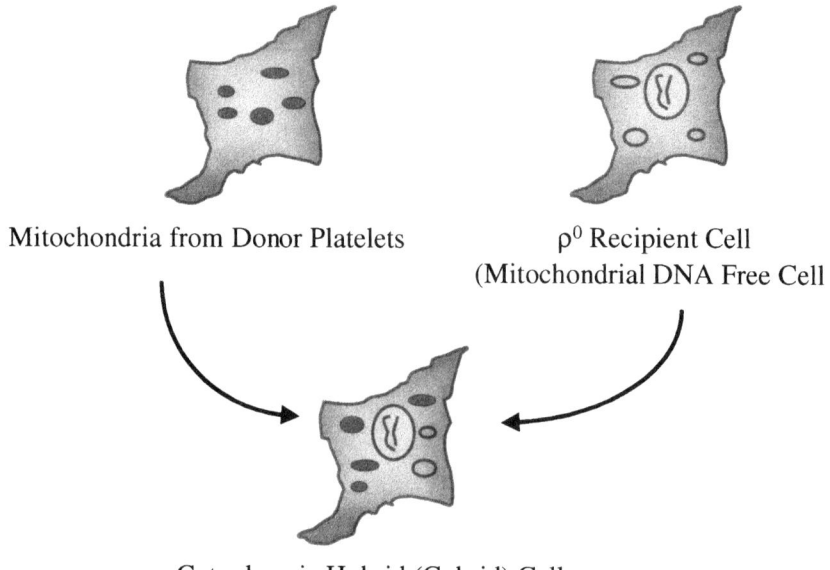

FIGURE 1. Creation of cybrid cells. Long-term treatment of cells with ethidium bromide, an inhibitor of mtDNA replication, is used to produce ρ^0 recipient cells that lack mtDNA. Exogenous mitochondrial DNA is reintroduced into these cells via polyethylene glycol–mediated fusion with platelets from disease-affected or control donors.

those demonstrating a complex I dysfunction in PD brain and platelets (discussed in section on PD) and thus suggest that the complex IV defect associated with AD is disease specific, systemic, and is not simply a nonspecific consequence of neurodegeneration, postmortem change, or aging.

The molecular implications of these observations were probed by fusion of platelets from multiple AD and control donor individuals into mtDNA-depleted SH-SY5Y neuroblastoma and Ntera/D1 (NT2) human teratocarcinoma cells (FIG. 1). The choice of these neuron-based recipient cell lines was guided by the close resemblance of their nuclear backgrounds to that of the disease-affected tissue. This procedure creates cybrids where the mtDNA from the AD or control donor is expressed in the nuclear and cellular environment of the host ρ^0 cell. After growth in culture to allow mtDNA repopulation, cybrid cells can regain the capacity to use oxidative phosphorylation and can become aerobically competent. However, if the donor mitochondria display biochemical or genetic abnormalities, the resulting cybrids may exhibit similar defects, and oxidative phosphorylation may not fully recover. The key feature of our approach is the use of age-matched normal controls for comparison. Thus, any differences in the mitochondrial phenotype (with attendant cellular ramifications) between the AD and control cybrids must arise as a consequence of the transferred mitochondria and mtDNA.

There are several criteria that should be met when creating cybrid cell lines. First, the ρ^0 cell lines should be depleted of endogenous mtDNA, such that reversion to the

TABLE 1. Mitochondrial DNA copy number/cell SH-SY5Y cells following extended treatment with ethidium bromide

Cell Type	EtBr (mg/ml)	Days of Treatment	mtDNA Copy/Cell
SH-SY5Y	0		4172 ± 623
SH-SY5Y	0.1	64	2144 ± 162
SH-SY5Y	0.5	64	3070 ± 111
SH-SY5Y	1.0	64	1848 ± 293
SH-SY5Y	5.0	64	0.48 ± 0.18
SH-SY5Y	5.0	225	0.000019

NOTE: Total cellular DNA was isolated from 4×10^6 cells by the QIAamp Tissue Kit Protocol (Qiagen Inc.). Quantitation of mtDNA was carried out using Taqman Real Time PCR in the ABI Prism 7700 Sequence Detection System (Perkin Elmer). A specific Taqman Probe/primer set combination was designed for amplification and detection of the mtDNA-encoded cytochrome c oxidase subunit 2 gene. Pure mtDNA was isolated from a mitochondrial fraction from brain tissue as decribed by Mecocci et al.[13] and was used to generate calibration curves for the assay. Values are shown as the mean ± SD from four independent determinations. EtBr = ethidium bromide.

parental mtDNA genotype does not occur. Second, exclusive repopulation of foreign mtDNA should be demonstrated once the cybrids are established in culture. We previously created and characterized ρ^0 cell lines derived from SH-SY5Y neuroblastoma cells.[26] mtDNA content was lowered to below the limits of detection in SH-SY5Y cells using slot-blot DNA hybridization procedures when the cells were treated with ethidium bromide (EtBr), a potent inhibitor of mtDNA replication. EtBr caused a dose-dependent decrease of KCN-sensitive oxygen consumption that was undetectable in cells treated with 5 μg/ml EtBr for 64 days ($\rho^0 64/5$ cells). In addition, these cells showed essentially an absence of enzymatic activity of cytochrome c oxidase, three of whose subunits are mtDNA-encoded.[26] The relative lack of impairment of nuclear-encoded citrate synthase, a mitochondrial matrix enzyme, is consistent with depletion of mtDNA and not nuclear-encoded gene products. We have recently revisited this issue utilizing real-time quantitative PCR detection to assess the content of mtDNA-encoded cytochrome c oxidase subunit 2 gene in EtBr-treated SH-SY5Y cells. TABLE 1 indicates that EtBr concentrations at or below 1 μg/ml were ineffective in depleting mtDNA content. In contrast, $\rho^0 64/5$ cells contain less than one mtDNA copy/cell, and incubation for 225 days with 5 μg/ml EtBr essentially results in a "mtDNA knockout" cell line. Thus, by this very sensitive technique SH-SY5Y cells treated with EtBr for extended periods were shown to lack detectable amounts of mtDNA and meet molecular criteria as ρ^0 cells.

NT2 cells subjected to EtBr treatment require NMDA and AMPA receptor inhibition to protect cells from glutamate toxicity during the ρ^0 creation process.[27] Whereas ρ^0 NT2 cells are auxotrophic for pyruvate and uridine, ρ^0 SH-SY5Y cells show auxotrophy only for pyruvate. Pyruvate is presumably needed to reoxidize excess NADH produced by glycolysis. These auxotrophic markers provide a convenient selection system for cybrid production.

Successful transformation of ρ^0 SH-SY5Y cells with exogenous mtDNA was confirmed by tracking the transfer of a homoplasmic single-nucleotide polymorphism at mtDNA position 7028 (C to T substitution) found in blood cells of a donor. Clonal sequencing analysis revealed that the cybrid but not the parental SH-SY5Y

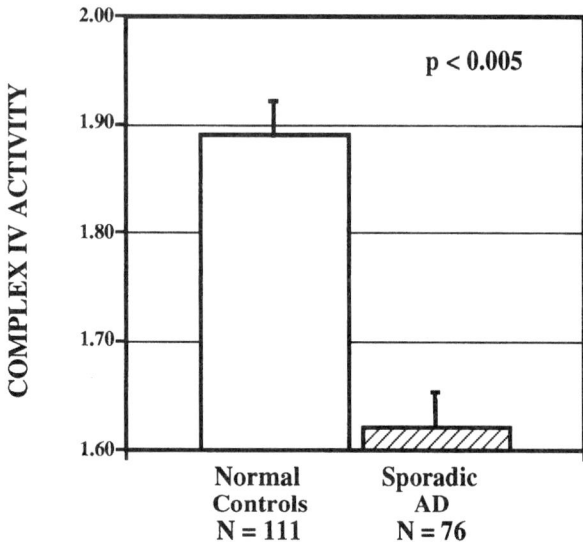

FIGURE 2. Characterization of complex IV activity of SH-SY5Y–derived AD and control cybrids. Complex IV activities were determined as described in Miller et al.[26] and are reported as $\Delta A.\min^{-1}mg^{-1}$ total cellular protein.

cell line harbored the homoplasmic polymorphism.[26] Recently, a more detailed study was undertaken that followed the transfer of an AD patient's mtDNA harboring a rare homoplasmic single-nucleotide polymorphism (G to A substitution) at mtDNA position 6366. Multiple clonal sequencing of the entire mitochondrial genome (10 clones per PCR fragment) revealed that the mtDNA of the cybrid line was identical to that isolated from the patient's blood cells.[28] These data provide compelling evidence that the residual endogenous mtDNA in ρ^0 cells does not contribute to cybrid mtDNA genotype.

SH-SY5Y-derived cybrids were made with platelet-rich buffy coat fractions of blood obtained from age-matched control and AD donors. The cell lines created from each donor were grown as mixed-population cultures and were evaluated separately for ETC enzyme activities. Complex I activity was similar in AD cybrids and control cybrids.[29] In contrast, the mean complex IV activities of AD cybrids ($n = 111$) was significantly different ($p < 0.005$, Student's t-test) from age-matched control cybrids ($n = 76$) (FIG. 2). The AD cybrids had a mean decrease in complex IV activity of 15% compared to age-matched control cybrids, which decrease is similar to that reported in blood and brain of AD patients. This pattern of a focal ETC defect was also mirrored in NT2 AD cybrids.[27] To further characterize this complex IV defect, we randomly selected a subset of mixed-population AD and mixed-population control cybrids. Multiple single-cell subclones were selected and expanded from these mixed-population cybrids for clonal analysis. The subclones exhibited variability in complex IV activity that was distributed around the mean complex IV activity of the parental mixed population from which they were derived (FIG. 3). A

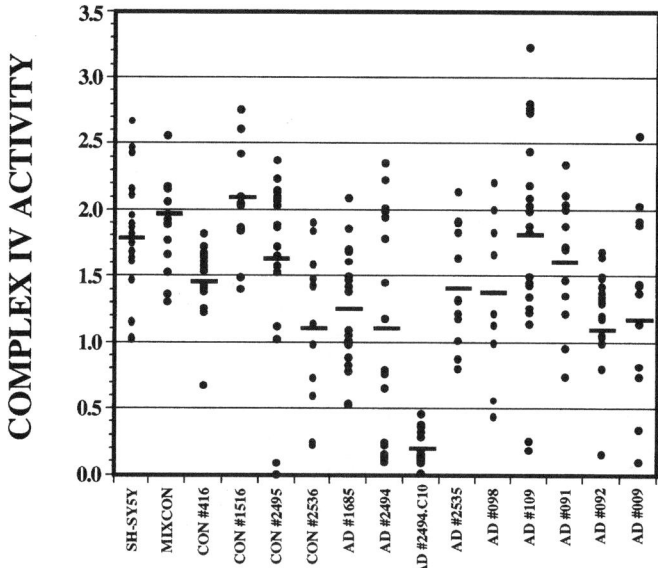

FIGURE 3. Single-cell subclones were cultured from parental SH-SY5Y–derived cybrid populations from both clinically diagnosed AD patients and nondisease age-matched controls. The complex IV activities for the parent population *(bars)* were determined at the same time as those of the subclones *(dots)*.

strong positive correlation was seen between the mean complex IV activity of subclones and the complex IV activity of the parental mixed cybrid population (FIG. 4).

Next, we selected and expanded single-cell subclones from a single AD cybrid clone. It was anticipated that subcloning from single-cell clones might reduce potential contributions of nuclear aneuploidy or unique nuclear–mitochondrial interactions to the mitochondrial dysfunction of AD cybrids. However, the complex IV activity of the subclones again varied around the complex IV activity of the original clone (FIG. 3, entry AD #2494-C10). Taken together, these data suggest that mixed-population cybrid cells reliably represent the total heterogeneous cellular population, while individual clones reflect random samples from a heterogeneous population. Slight differences in the complex IV activities of the parental cybrids and mean complex IV activity of their subclones probably reflects sampling bias in the selection of clones. The variability in complex IV activity among the subclones may reflect the heterogeneity of the mitochondria used to create the cybrid cells, a unique interaction between the nuclear genome of the ρ^0 cell and the mitochondria from the donor, experimental variability in the assessment of complex IV activity, or any combination of these factors. It is not surprising that the nuclear environment of the recipient ρ^0 cell might influence the expression of a mitochondrial defect. Indeed, Dunbar *et al.*[30] has shown that the nuclear environment can have profound effects on the mitochondrial genotype in cybrid cells.

To address the possibility that the distribution of complex IV activities in the subclones could arise from varying degrees of repopulation of mtDNA, we estimated

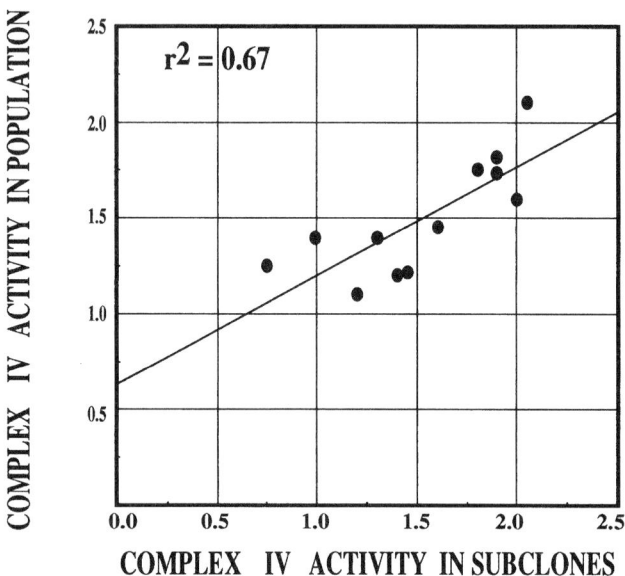

FIGURE 4. Mean of subclone complex IV activities correlates with activities of parental heterogeneous populations.

mtDNA copy number in these cell lines by making real-time quantitative PCR measurements. The data was normalized to copy number of the nuclear glucose 6-phosphate dehydrogenase gene. The lack of correlation between complex IV activity and mtDNA copy number in subclones indicates that differential repopulation of mtDNA is not the likely cause for the noted range of complex IV activities.

It is conceivable that the complex IV defect in our AD cybrids results from a consistent incorporation of a donor-derived nuclear factor or contaminant that exists in AD donor cells and not in control donor cells and that survives the transformation process and long-term cell culture conditions. Our platelet-enriched fractions contain a large excess of nonnucleated platelets relative to nucleated lymphocytes, making any meaningful contribution of nuclear contaminants unlikely. Previously, a screen for hyperdiploidy in cybrids proved negative.[26] To further exclude the possibility of donor-derived nuclear contamination of our cybrids, distinctive nuclear markers were tracked in parental SH-SY5Y cells, ρ⁰ SH-SY5Y cells, blood cells from donors, and in cybrids. The parental SH-SY5Y cell line carries an APO E ε3/3 genotype. Regardless of the genotype of our blood donors, all of the cybrids screened thus far have exclusively displayed the APO E ε3/3 genotype of the host cell line. These data imply that nuclear contamination by donor cells of our cybrids is unlikely.

A complex IV defect should lead to increased leakage of electrons from the ETC that can subsequently react with molecular oxygen to generate ROS. As anticipated, the ROS-sensitive probe dichlorofluorescin diacetate (DCF-DA)[31] detected significantly elevated production of ROS in AD cybrids as compared to control cy-

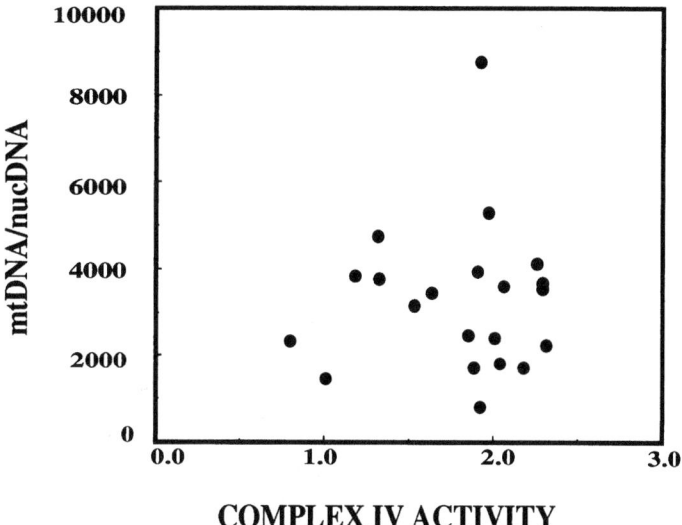

FIGURE 5. Lack of correlation between complex IV activity and mtDNA copy number in subclones. DNA was isolated from pelleted SH-SY5Y–derived cybrid cells by SDS/proteinase K treatment, followed by extraction with the Phase Lock Gel I system (5 Prime > 3′Prime). mtDNA was quantitated as described in note to TABLE 1. Nuclear DNA was quantitated by amplification of a fragment of the glucose 6-phosphate dehydrogenase gene using a specific Taqman primer/probe combination and by using nuclear DNA isolated from ρ^0 225/5 SH-SY5Y cells for generating calibration curves. mtDNA copy numbers were normalized to nuclear DNA copy number.

brids.[27,29] In response to increased oxidative stress, we have noted significant induction of radical scavenging enzymes such as glutathione reductase and glutathione peroxidase in SH-SY5Y AD cybrids. ROS production in AD cybrids was found to decline over time due to these cellular compensatory mechanisms (Ref. 28 and Davis & Miller, unpublished results). Upregulation of Cu/Zn superoxide dismutatase (SOD) and Mn SOD in NT2 AD cybrids has been observed as a compensatory response to increased ROS.[27] In addition, AD cybrids display increased basal cytosolic calcium concentration and impaired intracellular calcium homeostasis,[32] and abnormal mitochondrial morphology.[33]

While a clear molecular link has yet to be established, our studies provide strong evidence that AD cybrids prepared from two different human "neuron-like" host cell lines (SH-SY5Y and NT2) recapitulate important features of the AD phenotype. A characteristic of Alzheimer's disease is decreased activity of cytochrome *c* oxidase in platelets and brain. We have transferred a stable defect in cytochrome *c* oxidase activity from the blood of AD donors to these cell lines that suggests that the cytochrome *c* oxidase defect is carried into cybrids with mitochondria from these AD donors. It is unlikely, therefore, that the cytochrome *c* oxidase defect in AD arises from a secondary insult that is associated with disease morbidity.

PARKINSON'S DISEASE CYBRID STUDIES

Mitochondria were implicated as potential players in Parkinson's disease neurodegeneration in the 1980s. At the start of that decade, young intravenous drug users began presenting to San Francisco–area hospitals with parkinsonism. The cause of their clinical syndrome was a meperidine contaminant, 1-methyl-4-phenyl-1,2,3,6-tetrahydropyridine (MPTP).[34] The mechanisms by which MPTP exposure leads to neurodegeneration and parkinsonism unraveled over the next several years, culminating in the discovery by Nicklas *et al.* that a metabolite of MPTP, 1-methyl-4-phenylpyridine (MPP$^+$), inhibits complex I of the mitochondrial electron transport chain (ETC).[35] The relevance of this biochemical event to PD neurodegeneration was established in 1989 when several groups demonstrated that complex I dysfunction occurs in idiopathic PD.[36–38]

Parker *et al.* further proposed that complex I dysfunction in PD might arise from mutation(s) in mtDNA.[36] An mtDNA origin was hypothesized over nuclear genetic or toxic etiologies since Mendelian inheritance is rarely seen in PD and since no obvious environmental toxin appears to account for the majority of the cases.[39] Although several studies have failed to reveal an obvious point mutation or deletion of mtDNA in PD subjects,[40–45] these studies were not designed to address the relevance of single-nucleotide polymorphisms or to detect the presence of low-abundance heteroplasmy/compound heteroplasmy. Cybrid technology was therefore utilized to explore these possibilities.

The host ρ^0 cell type selected for the first PD cybrid studies was the neuron-like SH-SY5Y neuroblastoma ρ^0 cell line of Miller *et al.*[26] The mtDNA of these SH-SY5Y neuroblastoma ρ^0 cells was replenished by the transfer of platelet mitochondria derived from 24 sporadic PD and 28 control subject blood samples.[46] Platelets were chosen as the mtDNA donor source in these experiments because of the procurement advantages of this tissue and since complex I dysfunction is apparent in PD platelet mitochondria.[36, 47–50]

The resultant cybrid lines were maintained in continuous culture for about six weeks. During this time multiple cell divisions occurred, substantially diluting any nonperpetuating transferred components (nuclear encoded proteins, RNA, potential toxins). At the end of this period the cybrid cell lines were harvested, digitonin was used to solubilize cell membranes, and enriched mitochondrial fractions were prepared through centrifugation. Measurement of ETC component function revealed that complex I activity in the PD cell lines was 20% less than that of the controls. Complex IV activities were similar between the PD and control groups.[46] The mean enzyme activities from this study are shown in TABLE 2. These data demonstrate that

TABLE 2. Complex I and IV activites in Parkinson's disease and control cybrid cell lines

Cybrid	Complex I Activity (nmol/min/mg protein)	Complex IV Activity (sec^{-1}/mg protein)
Parkinson's disease (n = 24)	27.1 ± 0.7*	0.059 ± 0.003
Control (n = 28)	33.8 ± 0.7	0.066 ± 0.004

NOTE: Data are from Swerdlow *et al.*[46] Values are shown as the group mean ± SEM. *Significant at $p < 0.0001$; unpaired, two-way Student's *t*-test.

there is mtDNA aberration in at least the platelets of PD subjects, and that this genetic defect is sufficient to drive complex I dysfunction *in vitro*.

The study of Gu *et al.* using an A549 human lung ρ^0 cell line provided similar results and reached similar conclusions.[51] Complex I activity was also depressed in directly assayed platelet mitochondria. The magnitude of the cybrid complex I defect was about 25%, and was similar in magnitude to the biochemical defect measured in platelets. Shults and Miller subsequently reported a third series of sporadic PD cybrid lines. In this work, the mean complex I activity of their PD cybrid lines ($n = 4$) was 31% less than that of their control lines ($n = 6$).[52]

Several questions about the fidelity of PD cybrid data and their relevance to PD in general warrant discussion. Perhaps most importantly, the possibility that complex I dysfunction in PD cybrids arises as a consequence of some etiologic factor other than mtDNA requires consideration. Several "negative control studies" of cybrids expressing mitochondrial genes from persons with other diseases support the likelihood that cybrid ETC defects are indeed a specific consequence and sensitive indicator of underlying mtDNA abnormality. Huntington's disease (HD), nonalchoholic steatohepatitis (NASH), and some forms of dystonia are all clinical syndromes in which mitochondrial dysfunction occurs. Cybrids studies of these syndromes do not reveal the presence of perpetuable mitochondrial dysfunction.[53–55] Furthermore, an analysis of cybrid lines containing mtDNA from PD-affected members of the Contursi kindred was recently presented. This family manifests a highly penetrant, autosomal dominant form of PD that arises from mutation of the α-synuclein gene on chromosome 4.[56,57] Contursi PD cybrids do not demonstrate complex I dysfunction. This supports the contention that depressed complex I dysfunction in sporadic PD cybrids truly does arise as a consequence of mtDNA mutation.[58]

The pathophysiologic relevance of the *in vitro* PD cybrid complex I defect to PD neurodegeneration is also subject to debate. The magnitude of complex I activity reduction in PD cybrids is 20–30%. However, this may underestimate the true extent of the defect for several reasons. Mitochondrial enrichment in cybrid ETC activity assays was limited since pure mitochondrial fractions were not prepared through density centrifugation. The resulting loss of biochemical sensitivity would tend to minimize any detectable defect. Also, platelets served as the mtDNA donor source in these experiments, and may not possess as pervasive a degree of mitochondrial dysfunction as brain does in this disease.

Perhaps more relevant to this issue are the functional consequences of the mtDNA-determined complex I defect found in PD cybrids. Two types of experimental data indicate these cell lines experience increased oxidative stress. First, incubation of PD cybrids with the dye DCF-DA yields a greater fluorescent signal than does incubation of the dye with control cybrid lines.[46] This is a direct indication of increased ROS generation in the PD cybrids, since fluorescence is dependent on reaction of DCF-DA with ROS.[31] Second, antioxidant enzyme activity levels are increased in PD cybrid lines.[59] Relative to control cybrid lines, activities for total superoxide dismutase (SOD), manganese SOD, copper-zinc SOD, glutathione peroxidase, glutathione reductase, and catalase are elevated. Exposure of control cybrid lines to MPP$^+$ reproduces a similar pattern of antioxidant enzyme activity elevations. As MPP$^+$-induced inhibition of complex I is associated with ROS generation,[60] the most likely interpretation of these experiments is that antioxidant enzyme levels are secondarily elevated in PD cybrids in response to their mtDNA-determined complex I defect. Overall, these data parallel studies of PD brain that show increased oxidative stress and antioxidant enzyme activities.[61–66]

The ability of mitochondria to sequester calcium is diminished in PD cybrids. Relative to control cybrid lines, the cytosolic rise in calcium concentration that follows exposure to the ETC uncoupler carbonyl cyanide m-chlorophenylhydrazone (CCCP) is blunted in PD cybrid cells.[67] This perhaps occurs as a consequence of decreased mitochondrial membrane potential.[68] Further, the ability of PD cybrid lines to reset basal cytosolic calcium levels is retarded following calcium wave induction by the cholinomimetic molecule carbachol.[67]

One interesting aspect of PD cybrids is their apparent tendency to develop intracellular inclusions that share antigenic determinants with Lewy bodies.[69] These inclusions stain positively with antibodies to ubiquitin and α-synuclein. Other abnormal morphologic features of PD cybrids include mitochondrial enlargement and mitochondria with pale matrices/disrupted cristae.[33]

The results of the cybrid experiments discussed above do not, however, address whether or not mtDNA aberration in sporadic PD is inherited or acquired. A recent study of cybrids expressing mitochondrial genes from members of a family with maternally inherited PD is relevant to this issue. A kindred originally described by Wooten *et al.* displays PD over three generations.[70] Transmission of the disease respects matrilineal lines in this family. We generated cybrid lines containing mtDNA from maternally and paternally descended family members. Complex I activity was diminished, oxidative stress was increased, and abnormal structural deposition was increased in cybrid lines expressing mtDNA from the maternal descendents.[71] This pattern was observed even in cybrids containing mtDNA from maternal descendents in their third decade who do not currently express a PD phenotype. In this family, then, both cybrid and epidemiologic data suggest that inherited mtDNA mutation occurs and may play a role in PD development. Mutational analysis of mtDNA in this kindred is currently underway.

In summary, the 20–31% decrease in complex I activity observed in PD cybrids (relative to control cybrid lines)[46,51,52] is comparable to what is seen in assays of similarly prepared mitochondria obtained from the platelets of PD subjects.[47,48] Whatever the true magnitude of the complex I defect in PD cybrids, it appears sufficient to drive oxidative stress, perturb calcium homeostasis, and facilitate formation of intracellular inclusions *in vitro*. As these are phenomena that are either implicated in neurodegenerative pathophysiology or are believed to mark pathologic events in PD brain, it is reasonable to expect that the mtDNA-determined complex I defect of PD cybrids would result in neuronal dysfunction and cell loss were it to occur in brain. Finally, while the question of whether mtDNA aberration in PD is inherited or acquired remains unanswered, indications are that at least in some cases it is inherited.

CONCLUSIONS

The cybrid technology serves as a valuable tool for investigating the role of mtDNA in neurodegenerative processes that show deficits in energy metabolism. The salient points of the biochemical characterization of our AD and PD cybrid lines

TABLE 3. Biochemical characterization of AD and PD cybrid cell lines: summary

- Recipient ρ^0 cells have <1 mtDNA copy/cell and do not contribute to mtDNA genotype
- Exclusive repopulation with donor mtDNA in cybrid cells
- Nuclear background of cybrids may not be contaminated by nuclear factors from donor
- Mean complex IV activity of single-cell clones strongly correlates with activity of parental mixed-cell population
- No correlation of mtDNA copy number with ETC biochemical defect
- Cybrids recapitulate the same biochemical phenotype as exhibited in diseased tissue
- Negative control studies show specificity of transferable mitochondrial defect
→ Molecular analysis of cybrids at DNA level in progress

is summarized in TABLE 3. Our studies demonstrate that the focal ETC defects reported in platelets of PD and AD patients can be recapitulated in PD and AD cybrids. These data argue that the transfer of defective mtDNA is the most likely cause of the manifestation of ETC dysfunction in the cybrids. We have shown that sampling of mixed cell populations is a valid approach for cybrid analysis and that the use of control cybrids for comparison strongly implicates a transferable mtDNA defect with disease phenotype. Others have achieved similar success with mixed population cybrids in Parkinson's disease,[51] providing further corroboration of our approach. Since AD and PD cybrids reproduce key phenotypic characteristics of the diseases, they are potentially valuable as cellular disease models. Future studies will focus on molecular characterization at the DNA level and on understanding the interplay of the mitochondrial and nuclear genomes that underlie the disease phenotype.

REFERENCES

1. FISKUM, G., A.N. MURPHY & M.F. BEAL. 1999. Mitochondria in neurodegeneration: acute ischemia and chronic neurodegenerative diseases. J. Cereb. Blood Flow Metab. **19:** 351–369.
2. HOWELL, N. 1997. Leber hereditary optic neuropathy: how do mitochondrial DNA mutations cause degeneration of the optic nerve? J. Bioenerg. Biomembr. **29:** 165–173.
3. HOWELL, N. 1998. Leber hereditary optic neuropathy: respiratory chain dysfunction and ration of the optic nerve. Vision Res. **38:** 1495–1504.
4. a. KING, M.P. & G. ATTARDI. 1989. Human cells lacking mtDNA: repopulation with exogenous mitochondria by complementation. Science **246:** 500–503.
   b. KING, M.P. & G. ATTARDI. 1989. Injection of mitochondria into human cells leads to a rapid replacement of the endogenous mitochondrial DNA. Cell **52:** 811–819.
5. CHOMYN, A., S.T. LAI, R. SHAKELEY, N. BRESOLIN, G. SCARLATO & G. ATTARDI. 1994. Platelet-mediated transformation of mtDNA-less human cells: analysis of phenotypic variability among clones from normal individuals and complementation behavior of the tRNAlys mutation causing myoclonic epilepsy and ragged red fibers. Am. J. Hum. Genet. **54:** 966–974.
6. KUHL, D.E., E.J. METTER & W.H. RIEGE. 1985. Patterns of cerebral glucose utilization in depression, multiple infarct dementia, and Alzheimer's disease. Res. Publ. Assoc. Nerv. Ment. Dis. **63:** 211–226.
7. HAXBY, J.V., C.L. GRADY, E. KOSS, B. HORWITZ, L. HESTON, M. SCHAPIR, R.P. FRIEDLAND & S.I. RAPOPORT. 1990. Longitudinal study of cerebral metabolic asymmetries and associated neuropsychological patterns in early dementia of the Alzheimer type. Arch. Neurol. **47:** 753–760.
8. AZARI, N.P., K.D. PETTIGREW, M.B. SCHAPIRO, J.V. HAXBY, C.L. GRADY, P. PIETRINI, J.A. SALERNO, L.L. HESTON, S.I. RAPOPORT & B. HORWITZ. 1993. Early detection of

Alzheimer's disease: a statistical approach using positron emission tomography. J. Cereb. Blood Flow Metab. **13:** 438–447.
9. PALMER, A.M. & M.A. BURNS. 1994. Selective increase in lipid peroxidation in the inferior temporal cortex in Alzheimer's disease. Brain Res. **645:** 338–342.
10. PAPPOLLA, M.A., R.A. OMAR, K.S. KIM & N.K. ROBAKIS. 1992. Immunohistochemical evidence of antioxidant stress in Alzheimer's disease. Am. J. Pathol. **140:** 621-628.
11. JEANDEL, C., M.B. NICOLAS, F. DUBOIS, F. NABET-BELLEVILLE, F. PENIN & G. CUNY. 1989. Lipid peroxidation and free radical scavengers in Alzheimer's disease. Gerontology **35:** 275–282.
12. BALAZS, L. & M. LEON. 1994. Evidence of an oxidative challenge in the Alzheimer's brain. Neurochem. Res. **19:** 1131-1137.
13. MECOCCI, P., U. MACGARVEY & M.F. BEAL. 1994. Oxidative damage to mitochondrial DNA is increased in Alzheimer's disease. Ann. Neurol. **36:** 747–751.
14. SMITH, M.A., G. PERRY, P.L. RICHEY, L.M. SAYRE, V.M. ANDERSON, M.F. BEAL & N. KOWALL. 1996. Oxidative damage in Alzheimer's Disease. Nature **382:** 120–121.
15. PETTEGREW, J.W., W.E. KLUNK, K. PANCHALINGAM, J.N. KANFER & R.J. MCCLURE. 1994. Alterations of cerebral metabolism in probable Alzheimer's disease. Neurobiol. Aging **15:** 117–132.
16. PETTEGREW, J.W., W.E. KLUNK, E. KANAL, K. PANCHALINGAM & R.J. MCCLURE. 1995. Changes in brain membrane phospholipid and high-energy phosphate metabolism precede dementia. Neurobiol. Aging **16:** 973–975.
17. DUARA, R., R.F. LOPEZ-ALBEROLA, W.W. BARKER, D.A. LOEWENSTEIN, M. ZATINSKY, C.E. EISDORFER & G.B. WEINBERG. 1993. A comparison of familial and sporadic Alzheimer's disease. Neurology **43:** 1377–1384.
18. EDLAND, S.D., J. SILVERMAN, E.R. PESKIND, D. TSAUNG, E. WIJSMAN & J.C. MORRIS. 1996. Increased risk of dementia in mothers of Alzheimers disease cases: evidence for maternal inheritance. Neurology **47:** 254–256.
19. PARKER, W.D., C.M. FILLEY & J.K. PARKS. 1990. Cytochrome oxidase deficiency in Alzheimers disease. Neurology **40:** 1320–1303.
20. KISH, S.J., C. BERGERON, A. RAJPUT, S. DOZIC, F. MASTROGIACOMO, L.J. CHANG, J.M. WILSON, L.M. DISTEFANO & J.N. NOBREGA. 1992. Brain cytochrome oxidase in Alzheimer's disease. J. Neurochem. **59:** 776–779.
21. PARKER, W.D., J.K. PARKS, C.M. FILLEY & B.K. KLEINSCHMIDT-DEMASTERS. 1994. Electron transport chain defects in Alzheimer's disease. Neurology **44:** 1090–1096.
22. PARKER, JR., W.D., N.J. MAHR & C.M. FILLEY. 1994. Reduced platelet cytochrome oxidase activity in Alzheimers disease. Neurology **44:** 1086–1090.
23. MUTISYA, E.M., A.C. BOWLING & M.F. BEAL. 1994. Cortical cytochrome oxidase activity is reduced in Alzheimer's disease. J. Neurochem. **63:** 2179–2184.
24. CHAGNON, P., C. BETARD, Y. ROBITAILLE, A. CHOLETTE & D. GAUVREAU. 1995. Distribution of brain cytochrome oxidase activity in various neurodegenerative diseases. NeuroReport **6:** 711–715.
25. PARKER, JR., W.D. & J.K. PARKS. 1995. Cytochrome c oxidase in Alzheimer's disease brain: purification and characterization. Neurology **45:** 482–486.
26. MILLER, S.W., P.A. TRIMMER, W.D. PARKER, JR. & R.E. DAVIS. 1996. Creation and characterization of mitochondrial DNA depleted cell lines with "neuronal-like" properties. J. Neurochem. **67:** 1897–1907.
27. SWERDLOW, R.H., J.K. PARKS, B.S. CASSARINO, D.J. MAGUIRE, R.S. MAGUIRE, J.P. BENNETT, R.E. DAVIS & W.D. PARKER, JR. 1997. Cybrids in Alzheimer's disease: a cellular model of the disease? Neurology **49:** 918–925.
28. GHOSH, S.S., S.E. MILLER, C. HERRNSTADT, J.A. DYKENS, A.N. MURPHY & R.E. DAVIS. 1999. Genetic and biochemical characterization of a cybrid cell line with a stable cytochrome c oxidase defect. Soc. Neurosci. Abstr. **25(1):** 337.

29. DAVIS, R.E., S. MILLER, C. HERRNSTADT, S.S. GHOSH, E. FAHY, L.E. SHINOBU, D. GALASKO, L.J. THAL, M.F. BEAL, N. HOWELL & W.D. PARKER, JR. 1997. Mutations in mitochondrial cytochrome c oxidase genes segregate with late-onset Alzheimer disease Proc. Natl. Acad. Sci. USA **94:** 4526–4531.
30. DUNBAR, D.R., P.A. MOONIE, H.T. JACOBS & I.J. HOLT. 1995. Different cellular backgrounds confer a marked advantage to either mutant or wild-type mitochondrial genomes. Proc. Natl. Acad. Sci. USA **92:** 6562–6566.
31. ROYALL, J.A. & H. ISCHIROPOULOS. 1993. Evaluation of 2′7′-dichlorofluorescin and dihydrorhodamine 123 as fluorescent probes for intracellular H_2O_2 in cultured endothelial cells. Arch. Biochem. Biophys. **302:** 348–355.
32. SHEEHAN, J.P., R.H. SWERDLOW, S.W. MILLER, R.E. DAVIS, J.K. PARKS, W.D. PARKER & J.B. TUTTLE. 1997. Calcium homeostasis and reactive oxygen species production in cells transformed by mitochondria from individuals with sporadic Alzheimer's disease. J. Neurosci. **17:** 4612–4622.
33. TRIMMER, P.A., R.H. SWERDLOW, J.K. PARKS, S.W. MILLER, R.E. DAVIS & W.D. PARKER, JR. Abnormal mitochondrial morphology in sporadic Parkinson's and Alzheimer's disease cybrid lines. Exp. Neurol. In press.
34. LANGSTON, J.W., P.A. BALLARD, J.W. TETRUD & I. IRWIN. 1983. Chronic parkinsonism in humans due to a product of meperidine-analog synthesis. Science **219:** 979–980.
35. NICKLAS, W.J., I. VYAS & R.E. HEIKKILA. 1985. Inhibition of NADH-linked oxidation in brain mitochondria by 1-methyl-4-phenylpyridine, a metabolite of the neurotoxin, 1-methyl-4-phenyl-1,2,3,6-tetrahydropyridine. Life Sci. **36:** 2503–2508.
36. PARKER, W.D., S.J. BOYSON & J.K. PARKS. 1989. Electron transport chain abnormalities in idiopathic Parkinson's disease. Ann. Neurol. **26:** 719–723.
37. SCHAPIRA, A.H.V., J.M. COOPER, D. DEXTER, P. JENNER, J.B. CLARK & C.D. MARSDEN. 1989. Mitochondrial complex I deficiency in Parkinson's disease. Lancet **1:** 1289.
38. MIZUNO, Y., S. OHTA, M. TANAKA, S. TAKAMIYA, K. SUZUKI, T. SATO, H. OYA, T. OZAWA & Y. KAGAWA. 1989. Deficiencies in complex I subunits of the respiratory chain in Parkinson's disease. Biochem. Biophys. Res. Commun. **163:** 1450–1455.
39. PARKER, W.D. 1990. Sporadic neurologic disease and the electron transport chain: a hypothesis. *In* Proceedings of the 1989 Meeting of the American Society for Neurologic Investigation: New Developments in Neuromuscular Disease. R.M. Pascuzzi, Ed.: 59–64. Indiana University Printing Services. Bloomington.
40. SCHAPIRA, A.H.V., I.J. HOLT, M. SWEENEY, A.E. HARDING, P. JENNER & C.D. MARSDEN. 1990. Mitochondrial DNA analysis in Parkinson's disease. Movement Disorders **5:** 294–297.
41. LESTIENNE, P., J. NELSON, P. RIEDERER, K. JELLINGER & H. REICHMANN. 1990. Normal mitochondrial genome in brain from patients with Parkinson's disease and complex I defect. J. Neurochem. **55:** 1810–1812.
42. LESTIENNE, P., I. NELSON, P. REIDERER, H. REICHMANN & K. JELLINGER. 1991. Mitochondrial DNA in postmortem brain from patients with Parkinson's disease. J. Neurochem. **56:** 1819.
43. MANN, V.M., J.M. COOPER & A.H.V. SCHAPIRA. 1992. Quantitation of a mitochondrial DNA deletion in Parkinson's disease. FEBS Lett. **299:** 218–222.
44. SANDY, M.S., J.W. LANGSTON, M.T. SMITH & D.A. DI MONTE. 1993. PCR analysis of platelet mtDNA: lack of specific changes in Parkinson's disease. Movement Disorders **8:** 74–82.
45. DIDONATO, S., M. ZEVIANI, P. GIOVANNINI, N. SAVARESE, M. RIMOLDI, C. MARIOTTI, F. GIROTTI & T. CARACENI. 1993. Respiratory chain and mitochondrial DNA in muscle and brain in Parkinson's disease patients. Neurology **43:** 2262–2268.

46. SWERDLOW, R.H., J.K. PARKS, S.W. MILLER, J.B. TUTTLE, P.A. TRIMMER, J.P. SHEEHAN, J.P. BENNETT, R.E. DAVIS & W.D. PARKER. 1996. Origin and functional consequences of the complex I defect in Parkinson's disease. Ann. Neurol. **40:** 663–671.
47. KRIGE, D., M.T. CARROL, J.M. COOPER, C.D. MARSDEN & A.H.V. SCHAPIRA. 1992. Platelet mitochondrial function in Parkinson's disease. The Royal Kings and Queens Parkinson Disease Research Group. Ann. Neurol. **32:** 782–788.
48. YOSHINO, H., Y. NAKAGAWA-HATTORI, T. KONDO & Y. MIZUNO. 1992. Mitochondrial complex I and II activities of lymphocytes and platelets in Parkinson's disease. J. Neural Transm. **41:** 27–34.
49. BENECKE, R., P. STRUMPER & H. WEISS. 1993. Electron transfer complexes I and IV of platelets are abnormal in Parkinson's disease but normal in Parkinson-plus syndromes. Brain **116:** 1451–1455.
50. HAAS, R.H., F. NASIRIAN, K. NAKANO, D. WARD, M. PAY, R. HILL & C.W. SHULTS. 1995. Low platelet mitochondrial complex I and complex II/III activity in early untreated Parkinson's disease. Ann. Neurol. **37:** 714–722.
51. GU, M., J.M. COOPER, J.W. TAANMAN & A.H.V. SCHAPIRA. 1998. Mitochondrial DNA transmission of the mitochondrial defect in Parkinson's disease. Ann. Neurol. **44:** 177–186.
52. SHULTS, C.W. & S.W. MILLER. 1998. Reduced complex I activity in parkinsonian cybrids. Movement Disorders **13** (Suppl. 2): 217.
53. TABRIZI, S.J., J.M. COOPER & A.H.V. SCHAPIRA. 1998. Mitochondrial DNA in focal dystonia: a cybrid analysis. Ann. Neurol. **44:** 258–261.
54. SWERDLOW, R.H., J.K. PARKS, M.B. HARRISON & W.D. PARKER. 1999. Normal electron transport chain activities in Huntington's cybrids. Soc. Neurosci. Abstr. **24**(1): 476.
55. CALDWELL, S.H., R.H. SWERDLOW, E.M. KAHN, J.C. IEZZONI, E.E. HESPENEIDE, J.K. PARKS & W.D. PARKER, JR. 1999. Mitochondrial abnormalities in non-alcoholic steatohepatitis. J. Hepatol. **31:** 430–434.
56. GOLBE, L.I., G. DI IORIO, V. BONAVITA, D.C. MILLER & R.C. DUVOISIN. 1990. A large kindred with autosomal dominant Parkinson's disease. Ann. Neurol. **27:** 276–282.
57. POLYMEROPOULOS, M.H., C. LAVEDAN, E. LEROY, S.E. IDE, A. DEHAJIA, A. DUTRA, B. PIKE, H. ROOT, J. RUBINSTEIN, R. BOYER, E.S. STENROOS, S. CHANDRASEKHARAPPA, A. ATHANASSIADOU, T. PAPAPETROPOLOUS, W.G. JOHNSON, A.M. LAZZARINI, R.C. DUVOISIN, G. DI IORIO, L.I. GOLBE & R.L. NUSSBAUM. 1997. Mutation in the α-synuclein gene identified in families with Parkinson's disease. Science **276:** 2045–2047.
58. PARKER, W.D., L.I. GOLBE, J.K. PARKS, G. DI IORIO, G.F. WOOTEN & R.H. SWERDLOW. 1999. Cybrid analysis of Contursi kindred mtDNA. Soc. Neurosci. Abstr. **25**(2): 1336.
59. CASSARINO, D.S., C.P. FALL, R.H. SWERDLOW, T.S. SMITH, E.M. HALVORSEN, S.W. MILLER, J.K. PARKS, W.D. PARKER & J.P. BENNET. 1997. Elevated reactive oxygen species and antioxidant enzyme activities in animal and cellular models of Parkinson's disease. Biochim. Biophys. Acta **1362:** 77–86.
60. ADAMS, J.D., JR., L.K. KLAIDMAN & A. LEUNG. 1992. MPP^+ and $MPDP^+$ induced oxygen radical formation with mitochondrial enzymes. Free Radical Biol. Med. **15:** 181–186.
61. MARTILLA, R.J., H. LORENTZ & U.K. RINNE. 1988. Oxygen toxicity protecting enzymes in Parkinson's disease. Increase of superoxide dismutase-like activity in the substantia nigra and basal nucleus. J. Neurol. Sci. **86:** 321–331.
62. SAGGU, H., J. COOKSEY, D. DEXTER, F.R. WELLS, A. LEES, P. JENNER & C.D. MARSDEN. 1989. A selective increase in particulate superoxide dismutase activity in parkinsonian substantia nigra. J. Neurochem. **53:** 692–697.

63. KALRA, J., A.H. RAJPUT, S.V. MANTHA & K. PRASAD. 1992. Serum antioxidant enzyme activity in Parkinson's disease. Mol. Cell Biochem. **110:** 165–168.
64. DAMIER, P., E.C. HIRSCH, P. ZHANG, Y. AGID & F. JAVOY-AGID. 1993. Glutathione peroxidase, glial cells and Parkinson's disease. Neuroscience **52:** 1–6.
65. DEXTER, D.T., A.E. HOLLEY, W.D. FLITTER, T.F. SLATER, F.R. WELLS, S.E. DANIEL, A.J. LEES, P. JENNER & C.D. MARSDEN. 1994. Increased levels of lipid hydroperoxides in the Parkinsonian substantia nigra: an HPLC and ESR study. Movement Disorders **9:** 92–97.
66. SANCHEZ-RAMOS, J.R., E. OVERVIK & B.N. AMES. 1994. A marker of oxyradical-mediated DNA damage (8-hydroxy-2'deoxyguanosine) is increased in nigro-striatum of Parkinson's disease brain. Neurodegeneration **3:** 197–204.
67. SHEEHAN, J.P., R.H. SWERDLOW, W.D. PARKER, S.W. MILLER, R.E. DAVIS & J.B. TUTTLE. 1997. Altered calcium homeostasis in cells transformed by mitochondria from individuals with Parkinson's disease. J. Neurochem. **68:** 1221–1233.
68. CASSARINO, D.S., P.M. KENNEY, J.P. BENNETT, JR. 1999. Oxidative stress reduces mitochondrial membrane potentials in Parkinson's and Alzheimer's disease cybrids. Soc. Neurosci. Abstr. **25**(2)**:** 1336.
69. TRIMMER, P.A. 1999. PD cybrids contain inclusion bodies that share antigenic determinants with Lewy bodies. Soc. Neurosci. Abstr. **25**(2)**:** 1336.
70. WOOTEN, G.F. L.J. CURRIE, J.P. BENNETT, J.M. TRUGMAN & M.B. HARRISON. 1997. Maternal inheritance in two large kindreds with Parkinson's disease. Neurology **48:** A333.
71. SWERDLOW, R.H., J.K. PARKS, J.N. DAVIS, D.S. CASSARINO, P.A. TRIMMER, L.J. CURRIE, J. DOUGHERTY, S. BRIDGES, J.P. BENNETT, G.F. WOOTEN & W.D. PARKER. 1998. Matrilineal inheritance of complex I dysfunction in a multigenerational Parkinson's disease family. Ann. Neurol. **44:** 873–881.

Polyglutamine Domain Proteins with Expanded Repeats Bind Neurofilament, Altering the Neurofilament Network

YOSHITAKA NAGAI,[a,c] OSAMU ONODERA,[a,c] WARREN J. STRITTMATTER,[a–c] AND JAMES R. BURKE[a,c,d]

[a]Department of Medicine (Neurology), [b]Department of Neurobiology, and [c]Deane Laboratory, Duke University Medical Center, Durham, North Carolina 27710, USA

ABSTRACT: Proteins with expanded polyglutamine (polyQ) repeats cause eight inherited neurodegenerative diseases. Nuclear and cytoplasmic polyQ protein is a common feature of these diseases, but its role in cell death remains debatable. Since the neuronal intermediate filament network is composed of neurofilament (NF) and NF abnormalities occur in neurodegenerative diseases, we examined whether pathologic-length polyQ domain proteins interact with NF. We expressed polyQ-green fluorescent fusion proteins (GFP) in a neuroblast cell line, TR1. Pathologic-length polyQ-GFP fusion proteins form large cytoplasmic aggregates surrounded by neurofilament. Immunoisolation of pathologic-length polyQ proteins co-isolated 68 kD NF protein demonstrating molecular interaction. These observations suggest that polyQ interaction with NF is important in the pathogenesis of the polyglutamine repeat diseases.

INTRODUCTION

Eight inherited neurodegenerative diseases are caused by expanded CAG repeats encoding a polyglutamine domain in different disease-producing genes.[1] These diseases, Huntington's disease (HD), dentatorubral-pallidoluysian atrophy (DRPLA), spinocerebellar atrophy 1, 2, 3 (Machado-Joseph disease (MJD)), 6 and 7 (SCA 1, 2, 3, 6 and 7), and spinobulbar muscular atrophy (SBMA) share common molecular, cellular and clinical features.

The expanded polyglutamine domain plays a central role in pathogenesis, since the length of the polyglutamine repeat determines the age of symptom onset and disease phenotype, and expression of proteins containing large glutamine domains is toxic to *E. coli*, *C. elegans*, *Drosophila*, mammalian cells and transgenic mice.[2–5] The polyglutamine domain itself is pathogenic because cell death occurs whether the repeat domain is contained in a fragment, the entire native protein or fused to non-native partners such as green fluorescent protein (GFP) or hypoxanthine phosphoribosyltransferase (Hprt).[4–8]

[d]Corresponding author. Phone: 919-684-0054; fax: 919-684-6514.
e-mail: james.burke@duke.edu

A pathologic hallmark of the polyglutamine repeat diseases is intracellular protein aggregation. The nuclear aggregate is well studied, but there is little information on the composition of the cytoplasmic aggregate.[9,10] A number of proteins have been found to bind to polyglutamine including glyceraldehyde-3-phosphate dehydrogenase (GAPDH),[11,12] huntingtin associated protein (HAP-1),[13] huntingtin interacting protein (HIP-1),[14] apopain,[15] leucine-rich acidic nuclear protein (LANP),[10] cystathionine β-synthase, WW domain proteins,[17] SH3-GL3 protein and vimentin,[4] but none of these proteins is known to be involved in the cytoplasmic aggregates.

In a search for candidate proteins that may be involved in polyglutamine induced cell death we examined polyglutamine-intermediate filament interactions. We previously showed that expression of pathologic-length polyglutamine-GFP fusion protein in transfected COS7 cells causes cytoplasmic aggregation, disrupts the vimentin intermediate filament network and causes cell death,[4] but mature neurons do not contain vimentin.[19] Vimentin is homologous with the neuronal intermediate filament proteins, neurofilament (NF).[19] Pathologic evidence suggests that the NF distribution is altered in neurons from polyglutamine repeat and other neurodegenerative diseases.[20–24] There is no direct evidence linking polyglutamine repeat containing proteins to NF.

To examine whether long-polyglutamine domain proteins interact with NF and alter the NF network, we expressed varying length polyglutamine domainGFP fusion proteins in TR1 cells. TR1 cells are a clonal murine cell line derived by sequential oncogenic retroviral infection of neocortical neuronal precursors.[25] TR1 cells have neuron-like features and express NF and neuron-specific enolase, but not glial fibrillary acidic protein (GFAP) or vimentin (see RESULTS).

In this study we found that polyglutamine-GFP fusion proteins display length-dependent, perinuclear aggregation in a neuronal cell line (TR1), that NF proteins accumulate around the aggregate, and that immunoisolation of expanded polyglutamine-containing proteins co-isolates NF protein. Our data suggest that the interaction of polyglutamine with NF proteins is involved in the pathogenesis of polyglutamine repeat diseases.

RESULTS

Polyglutamine-GFP Fusion Proteins Aggregate in the Perinuclear Region of TR1 Cells in a Polyglutamine Length-Dependent Manner

Cells transfected with the various Q-GFP constructs produced approximately equal amounts of polyglutamine-GFP fusion (not shown). Fusion proteins are named Q followed by the number of uninterrupted glutamines and GFP for green fluorescent protein. TR1 cells transfected with Q19-GFP display persistent diffuse fluorescence. In contrast, cells expressing Q56-GFP or Q80-GFP initially displayed diffuse fluorescence, but within 24 hours formed single or multiple large cytoplasmic aggregates. The aggregates were primarily located in the perinuclear region and only occasionally found elsewhere in the cell.

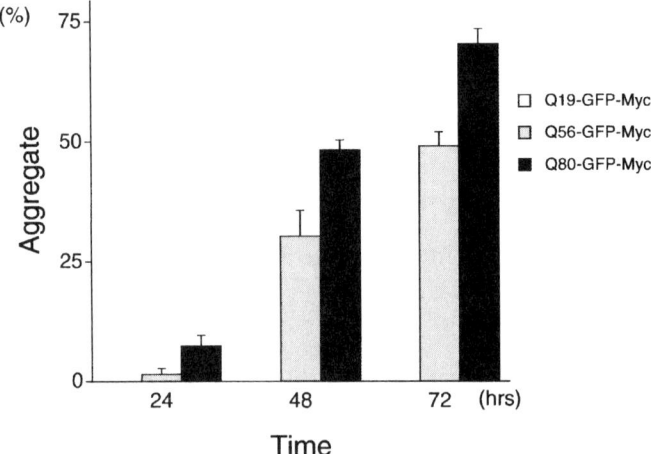

FIGURE 1. Polyglutamine-GFP aggregates in a length-dependent pattern. The percentage of cells with visible aggregates of fusion protein containing 19, 56, or 80 glutamines were determined by dividing the number of cells with aggregates by the total number of cells expressing green fluorescence. *Error bars* = standard error.

Polyglutamine-GFP Forms Aggregate in TR1 Cells in a Length-Dependent Pattern

To examine the temporal formation of aggregates, we determined the percentage of transfected cells with aggregates at 1, 2 and 3 days (FIG. 1). Aggregates were rare in cells expressing Q19-GFP, while cells expressing Q56-GFP and Q80-GFP frequently produced fluorescent protein aggregates. By 3 days after transfection, approximately half of the Q56-GFP expressing cells contained fluorescent aggregate and 70% of the Q80-GFP expressing cells did.

Neurofilament Protein Accumulates around the Expanded Polyglutamine Aggregate

The TR1 cell intermediate filament network consists of NF-L, NF-M, and NF-H protein, but not vimentin (Western blot and immunocytochemistry—data not shown). To determine whether polyglutamine-containing proteins interact with neurofilament proteins, we used confocal microscopy to examine TR1 cells expressing polyglutamine-GFP fusion. The NF network extends throughout the cytoplasm of nontransfected cells and cells expressing Q19-GFP protein (compare the non-transfected cells in FIGURE 2 to see the normal appearance of the neurofilament network with various neurofilament antibodies). In contrast, the appearance of the NF network is different in cells expressing Q56- and Q80-GFP; NF no longer extends throughout the cell, it accumulates around the polyglutamine aggregate. The NF network pattern is identical when probed with antibodies to phosphorylated NF-M and NF-H (FIG. 2 A and B), non-phosphorylated NF-M and NF-H (C and D) and NF-L (E and F). Unlike NF, the distribution of the microtubule and actin network is unaffected by the presence of polyglutamine-GFP aggregates (FIG. 2, G and H).

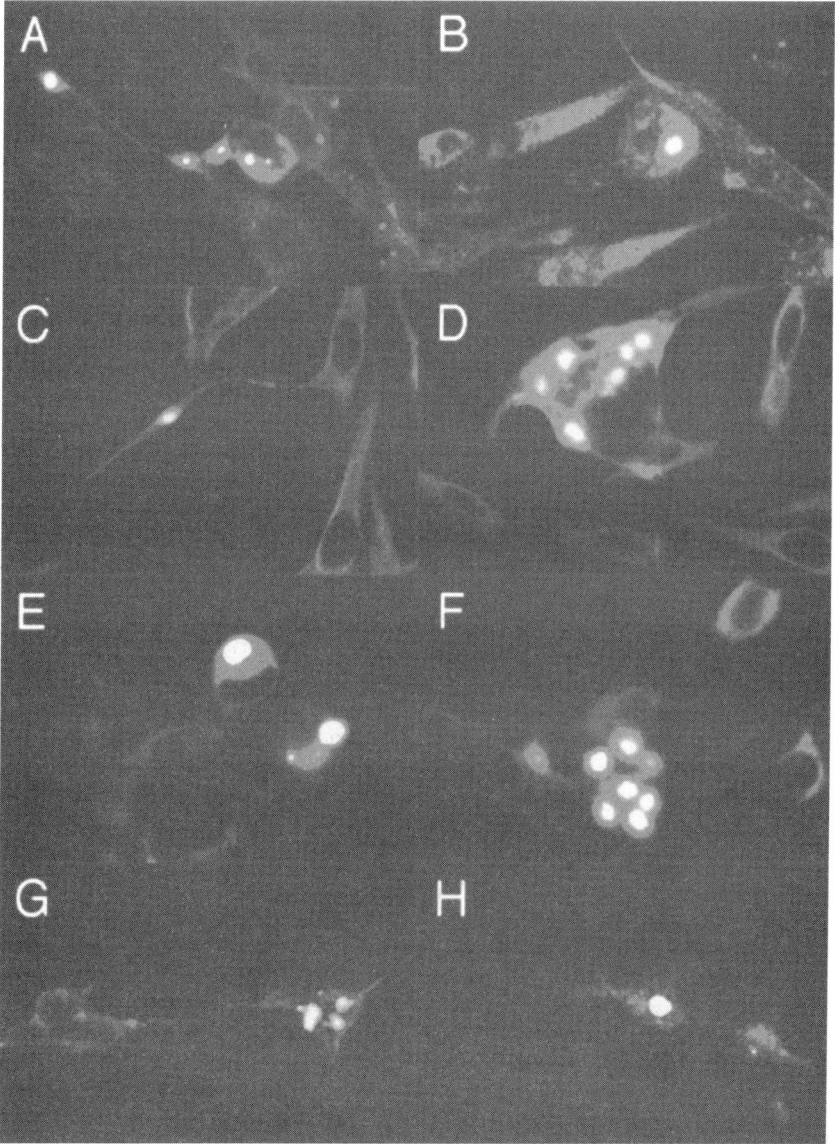

FIGURE 2. Neurofilament protein accumulates around the polyglutamine-GFP fusion protein aggregate in TR1 cells. Cells expressing Q80-GFP were fixed 24 hours after transfection and incubated with indicated antibody. Anti-phosphorylated neurofilament antibody (SMI 31) (**A** and **B**); Anti-nonphosphorylated neurofilament antibody (SMI 311) (**C** and **D**); Anti-NF-L 68 (**E and F**); Anti-β tubulin (**G**) and Anti-actin (AC40) (**H**). The cells were examined using a confocal microscope for GFP using FITC optics and for neurofilament or other cytoskeletal proteins using Texas red optics. Polyglutamine-GFP fusion protein produced green signal and red signal represents neurofilament protein. Magnification ×600.

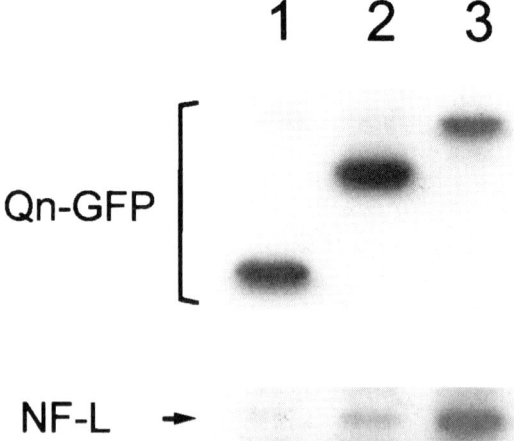

FIGURE 3. Pathologic-length polyglutamine-GFP fusion proteins co-immunoisolate neurofilament. TR1 cells expressing Q19-GFP (*lane 1*); Q56-GFP; (*lane 2*) or Q-80-GFP (*lane 3*) were lysed and immunoisolated with anti-Myc antibody. Immunoisolates were separated with magnetic beads coated with secondary antibody. Bound proteins were eluted and analyzed on Western blots. *Upper panel* shows Western blot probed with anti-GFP antibody. *Lower panel* shows Western blot probed with anti-neurofilament 68-kD antibody. Polyglutamine-GFP vector contains a myc-epitope.

NF-L Co-immunoisolates with Long Polyglutamine Fusion Proteins

The accumulation of NF protein around aggregated Q56- and Q80-GFP does not prove direct molecular interaction. To determine whether pathologic-length polyglutamine proteins bind NF protein we immunoisolated polyglutamine-GFP fusion proteins from TR1 cells expressing Q19-GFP, Q56-GFP or Q80-GFP. NF-L coimmunoisolated with the Q56- and Q80-GFP, but not with Q19-GFP (FIG. 3). In contrast, other cytoskeletal proteins, β-tubulin and actin did not co-immunoisolate (not shown). We could not determine whether NF-M and NF-H co-immunoisolated with polyglutamine-GFP because these NF proteins nonspecifically adsorbed to the magnetic beads used for the isolation procedure (data not shown). These data support the conclusion that expanded polyglutamine domain fusion proteins bind 68 kD NF.

DISCUSSION

The mechanism by which expanded polyglutamine domain proteins cause neurodegenerative disease is unknown, but expansion of this domain causes protein aggregation and fibril formation both *in vitro* and *in vivo*.[4,26–28] We demonstrated that pathologic-length polyglutamine domain proteins bind NF and alter the intracellular NF distribution. The interaction of polyglutamine and NF may be important in pathogenesis and explain cell specific vulnerability.

Neurofilament Protein and the Polyglutamine Repeat Diseases

NF proteins are members of the intermediate filament (IF) protein family and have sequence and structural homology with vimentin. The normal neurofilament structure is disrupted in polyglutamine-repeat disease neurons. In HD, the striatal neuron neurofilament network is condensed and fragmented.[22] DiFiglia et al. reported that dystrophic neurites in HD contain filamentous aggregates associated with NF positive axonal fibers.[26] Cortical neurons in HD accumulate 10 to 15 nanometer fibrils which resemble NF.[27] Neurofilament aggregates (torpedoes and spheroids and neuroaxonal dystrophy) are also found in spinocerebellar ataxias and DRPLA.[29,30]

Neurofilaments are the predominant IF protein in adult neurons and are the major determinant of neuronal size and axonal caliber.[19,31] Disruption of NF can cause neuronal atrophy and reduction of axon caliber which are pathologic features in HD, DRPLA, MJD and SBMA.[24,32–35] Iwabuchi et al. proposed that reduction of axonal diameter and cell volume is the characteristic neuropathologic feature of the CAG triplet repeat diseases.[24,32,36]

Neurofilament Proteins Form a Fibrillar Network

There are three major neurofilament proteins based on molecular weight: NF light (NF-L; 68 kD); NF medium (NF-M; 160 kD) and NF heavy (NF-H; 200 kD). NF fibrils are obligate hetero-oligomers requiring NF-L plus NF-M and/or NF-H for filament formation.[37] NF assembly can be altered by post-translational modifications including phosphorylation and glycosylation.[19] In our experiments we observed no difference in the appearance of the collapsed NF network whether the cells were probed with phosphorylated or nonphosphorylated NF antibodies. In addition to forming an elaborate three-dimensional network within the cell, NF proteins interact with other cytoskeletal proteins including intermediate filament associated proteins (IFAP).[38]

Mechanisms of NF Accumulation

NF protein accumulates around the polyglutamine-GFP aggregates in TR1 cells. The mechanism by which expanded polyglutamine fusion protein aggregates cause NF accumulation is not known, but may involve alterations in synthesis, transport or collapse of the NF network. The NF network is a stable cytoskeletal element but can be disrupted by intracytoplasmic injection of antineurofilament antibody,[39] expression of mutated neurofilament protein[40,41] or changes in the relative expression levels of NF protein.[19,42,43] IF networks can also be collapsed by expression of mutant IF-associated proteins (e.g., plectin, desmoplakin and BPAG1n).[38] Expanded polyglutamine domain proteins, therefore, might alter the NF network by changing expression or directly interacting with NF blocking transport or causing collapse. Polyglutamine containing proteins may also bind to NF indirectly via NF-associated proteins.

Johnston et al. recently presented evidence that misfolded proteins are deposited in perinuclear inclusions called aggresomes.[44] They proposed that aggresomes form when the capacity of cells to remove misfolded protein is exceeded. Interestingly, they found that these structures were surrounded by the IF protein vimentin. A similar process may be occurring in neurons.

Specificity of Neuronal Degeneration

Normal and expanded polyglutamine domain proteins are expressed in virtually every cell type and tissue, but the expanded polyglutamine domain proteins cause degeneration restricted to specific groups of neurons. The mechanism(s) responsible for cellular specificity are unknown. The absence of pathology in nonneural cells in the CAGtriplet repeat diseases may be due to several factors: lower levels of protein expression; the ability of nonneural cells to replicate and replace dying cells; molecular interactions imposed by other domains of the polyglutamine domain proteins; cell-specific processing; or to cells-pecific interacting proteins.[10,13,15,45]

NF is predominantly expressed in neurons and alteration in the NF network can cause neuron specific pathology. Transgenic mice overexpressing NF-L or NF-H develop NF aggregates in motor neurons producing pathology similar to motor neuron disease (amyotrophic lateral sclerosis).[40,43,46] Similarly, low levels of NF-M expression (2–25% of wild-type) in transgenic mice cause age-dependent accumulation of neurofilament aggregates in selected neurons including the cerebral cortex and cerebellar Purkinje cells.[47] Expression of truncated NF-H-lac-Z fusion protein in transgenic mice produces Lewy body-like inclusions in several types of neurons, but only Purkinje cell degeneration occurs, a common finding in spinocerebellar ataxia.[49] Selective binding of polyglutamine proteins to NF protein or NF associated proteins could, therefore, explain cytoplasmic aggregation and the specificity of neuronal degeneration seen in the polyglutamine repeat diseases.

MATERIALS AND METHODS

Construction of Polyglutamine Clones

Plasmid vectors expressing polyglutamineGFP fusion proteins containing a myc-epitope were synthesized as previously described. The nucleotide sequences of all constructs were confirmed by the dideoxynucleotide chain terminator method. The amino acid sequence of each clone is:

MVSTHHHHH(**Q**)**1980**HHGNSGPPR<u>ILQSTVPRARDPPVAT</u>GFP-Myc-His.

The underlined sequence represents the sequence of the multicloning site of pEGFPN1.

Cell Culture and Transfection

TR1 cells were cultured in Opti-MEM I (Gibco BRL; Gaithersburg, MD) on poly-D-lysine or Cell-Tak (Becton-Dickinson Labware, Bedford, MA) coated tissue culture plates. Transfection was performed using lipofectamine reagent according to the protocol from the supplier (Gibco BRL). The percentage of cells with visible fusion protein aggregates was determined by counting cells with visible aggregates and dividing by the total number of fluorescent cells in four independent experiments. In each series more than 500 transfected cells was counted.

Immunofluorescence and Confocal Laser Microscopy Analysis

For immunofluorescent labeling, 24 hours after transfection, cells were fixed in methanol at −20°C. Cells were incubated with primary antibodies at room temperature for 30 minutes, extensively rinsed, and exposed to secondary fluorescent antibodies at room temperature for 30 minutes. The following immunoreagents were used: monoclonal antibodies to phosphorylated and nonphosphorylated neurofilament (SMI 31 and SMI 311, respectively) (Sternberger Monoclonals, Baltimore, MD); neurofilament 68 and actin (monoclonal AC40 and polyclonal rabbit) (Sigma BioSciences, St. Louis, MO); β-tubulin (Amersham International, Buckinghamshire, England); anti-vimentin (V9) BoehringerMannheim, Indianapolis, IN); Texas redlabeled horse antimouse IgG (H + L) and horse antirabbit IgG (H + L) (Vector Laboratories, Burlingame, CA). Images were obtained using a Zeiss microscope equipped with fluorescein and rhodamine filter sets and Zeiss 32× lens (Carl Zeiss, Inc., Thornwood, NY) and the images photographed. Confocal images were obtained using BioRad MRC600 Confocal System (BioRad Labs, Richmond, CA) with an argon laser and Nikon Diaphot microscope equipped with a Nikon 60× oil immersion objective.

Co-Immunoisolation of Polyglutamine Fusion Proteins and Neurofilament

TR1 cells were transfected with Q19GFP, Q56GFP or Q80GFP expression plasmids in poly-D-lysine coated 6 well plates. 24 hours after transfection, the cells were rinsed with PBS, then incubated with 0.5 ml extraction buffer for 30 minutes at 4°C. After removing the supernatant the remaining material was incubated with cell lysis buffer[4,48] for 60 minutes at 4°C. The cells were scraped with a rubber policeman, homogenized and centrifuged. The supernatant was preincubated with Dynabeads (M-450) sheep anti-mouse IgG (Dynal, Lake Success, NY) at 4°C for one hour to remove nonspecific binding. After separating the beads using a magnetic separation unit, the supernatant was incubated with antiMyc monoclonal antibody (Invitrogen, San Diego, CA) for 16 hours. 50 µl Dynabeads (M450) sheep antimouse IgG were suspended in 0.5 ml binding buffer[4,48] and incubated with the supernatant for 2 hours at 4°C. After incubation, the beads were separated using a magnetic separation unit and washed three times with washing buffer.[4,48] Washed beads were boiled 5 min in Laemmli and the supernatant subjected to SDS-PAGE and Western blot analysis. NF protein was detected with SMI31 (Sternberger Monoclonals) at 1:1000 dilution, anti-NF160 at 1:400 dilution and anti-NF68 (Sigma) at 1:5000 dilution.

ACKNOWLEDGMENTS

We are grateful to Timothy Tucker for technical assistance, to Shoji Tsuji, M.D., Ph.D. (Niigata University, Niigata, Japan) for providing the DRPLA cDNA from which the polyglutamine constructs were made, and to J. Chun and Alleix Biopharmaceuticals (Mississaugh, Ontario) for supplying the TR1 cells. We thank Larry Hawkey for assistance with confocal microscopy and William Matthews for assistance with fluorescence microscopy. This work was supported by a Wills Foundation Post-doctoral fellowship (Y.N.) and a Beeson Physician-Faculty Scholar Award from the American Federation for Aging Research (J.R.B.).

REFERENCES

1. KOSHY, B.T. *et al.* 1997. The CAG/polyglutamine tract diseases: gene products and molecular pathogenesis. Brain Pathol. **7:** 927–942.
2. ONODERA, O. *et al.* 1996. Toxicity of expanded polyglutamine-domain proteins in *Escherichia coli*. FEBS Letters **399:** 135–139.
3. WARRICK, J.M. *et al.* 1998. Expanded polyglutamine protein forms nuclear inclusions and causes neural degeneration in *Drosophila*. Cell **93:** 939–949.
4. ONODERA, O. *et al.* 1997. Oligomerization of expanded-polyglutamine domain fluorescent fusion proteins in cultured mammalian cells. Biochem. Biophys. Res. Commun. **238:** 599–605.
5. MANGIARINI, L. *et al.* 1996. Exon 1 of the HD gene with an expanded CAG repeat is sufficient to cause a progressive neurological phenotype in transgenic mice. Cell **87:** 493–506.
6. BURRIGHT, E.N. *et al.* 1995. SCA1 transgenic mice: a model for neurodegeneration caused by an expanded CAG trinucleotide repeat. Cell **82:** 937–948.
7. IKEDA, H. *et al.* 1996. Expanded polyglutamine in the Machado-Joseph disease protein induces cell death in vitro and in vivo. Nature Genet. **13:** 196–202.
8. ORDWAY, J.M. *et al.* 1997. Ectopically expressed CAG repeats cause intranuclear inclusions and a progressive late onset neurological phenotype in the mouse. Cell **91:** 753–763.
9. DAVIES, S.W. *et al.* 1998. Are neuronal intranuclear inclusions the common neuropathology of triplet-repeat disorders with polyglutamine-repeat expansions? Lancet **351:** 131–133.
10. MATILLA, A. *et al.* 1997. The cerebellar leucine-rich acidic nuclear protein interacts with ataxin-1. Nature **389:** 974–978.
11. BURKE, J.R. *et al.* 1996. Huntingtin and DRPLA proteins selectively interact with the enzyme GAPDH. Nature Med. **2:** 347–350.
12. KOSHY, B. *et al.* 1996. Spinocerebellar ataxia type-1 and spinobulbar muscular atrophy gene products interact with glyceraldehyde-3-phosphate dehydrogenase. Hum. Molec. Genet. **5:** 1311–1318.
13. LI, X.J. *et al.* 1995. A huntingtin-associated protein enriched in brain with implications for pathology. Nature **378:** 398–402.
14. KALCHMAN, MA. *et al.* 1997. HIP1, a human homologue of S. cerevisiae Sla2p, interacts with membrane-associated huntingtin in the brain. Nature Genet. **16:** 44–53.
15. GOLDBERG, Y.P. *et al.* 1996. Cleavage of huntingtin by apopain, a proapoptotic cysteine protease, is modulated by the polyglutamine tract. Nature Genet. **13:** 442–449.
16. BOUTELL, J.M. *et al.* 1998. Huntingtin interacts with cystathionine beta-synthase. Hum. Molec. Genet. **7:** 371–378.
17. FABER, P.W. *et al.* 1998. Huntingtin interacts with a family of WW domain proteins. Hum. Molec. Genet. **7:** 1463–1474.
18. SITTLER, A. *et al.* 1998. SH3GL3 associates with the huntingtin exon 1 protein and promotes the formation of polygln-containing protein aggregates. Molec. Cell **2:** 427–436.
19. LEE, M.K. *et al.* 1996. Neuronal intermediate filaments. Ann. Rev. Neurosci. **19:** 187–217.
20. DICKSON D.W. *et al.* 1996. Cytoskeletal pathology in non-Alzheimer degenerative dementia: new lesions in diffuse Lewy body disease, Pick's disease, and corticobasal degeneration. J. Neural Transm. Suppl. **47:** 31–46.
21. O'DONNELL, D.M. *et al.* 1995. Trinucleotide repeat disorders in pediatrics. Curr. Opin. Pediat. **7:** 715-725.
22. NIHEI, K. *et al.* 1992. Neurofilament and neural cell adhesion molecule immunocytochemistry of Huntington's disease striatum. Ann. Neurol. **31:** 59–63.
23. TROJANOWSKI, J.Q. *et al.* 1993. Altered tau and neurofilament proteins in neurodegenerative diseases: diagnostic implications for Alzheimer's disease and Lewy body dementias. Brain Pathol. **3:** 45-54.

24. IWABUCHI, K. et al. 1993. A clinicopathological study on autosomal dominant hereditary dentatorubro-pallidoluysian atrophy (Naitoh-Oyanagi's disease). Adv. Neurol. Sci. **37:** 678–692.
25. CHUN, J. et al. 1996. Clonal cell lines produced by infection of neocortical neuroblasts using multiple oncogenes transduced by retroviruses. Molec. Cell. Neurosci. **7:** 304–321.
26. DIFIGLIA, M. et al. 1997. Aggregation of huntingtin in neuronal intranuclear inclusions and dystrophic neurites in brain. Science **277:** 1990–1993.
27. JACKSON, M. et al. 1995. The cortical neuritic pathology of Huntington's disease. Neuropath. Appl. Neurobiol. **21:** 18–26.
28. PAULSON, H.L. et al. 1997. Intranuclear inclusions of expanded polyglutamine protein in spinocerebellar ataxia type 3. Neuron **19:** 333–344.
29. HIRANO, A. et al. 1994. Structure of neurons in the aging nervous system. In Neurodegenerative Diseases. D.B. Caine, Ed. :3–14. Saunders. Philadelphia, PA.
30. LOWE, J. et al. 1997. Disorders of movement and systems degeneration. In Greenfield's Neuropathology. D.I. Graham & P.L. Lantos, Eds. :281-366. Arnold. London, England.
31. XU, Z. et al. 1996. Subunit composition of neurofilaments specifies axonal diameter. J. Cell Biol. **133:** 1061–1069.
32. IWABUCHI, K. et al. 1995. Clinicopathological study on autosomal dominantheredatary spastic spastic ataxia (Greenfield): its relationship to ataxo-choreoathetosis form of DRPLA, spinopontine degeneration, Machado-Joseph disease and SCA3. Adv. Neurol. Sci. **39:** 164–187.
33. LI, M. et al. 1995. Primary sensory neurons in X-linked recessive bulbospinal neuropathy: histopathology and androgen receptor gene expression. Muscle & Nerve **18:** 301–308.
34. OYANAGI, K. et al. 1989. A quantitative investigation of the substantia nigra in Huntington's disease. Ann. Neurol. **26:** 13–19.
35. FARMER, T.W. et al. 1989. Ataxia, chorea, seizures, and dementia. Pathologic features of a newly defined familial disorder. Arch. Neurol. **46:** 774–779.
36. IWABUCHI, K. 1995. Some problems on neuropathology of Huntington disease. Rinsho Shinkeigaku - Clinical Neurol. **35:** 1537–1539.
37. LEE, M.K. et al. 1993. Neurofilaments are obligate heteropolymers in vivo. J. Cell Biol. **122:** 1337–1350.
38. FUCHS, E. et al. 1998. A structural scaffolding of intermediate filaments in health and disease. Science **279:** 514–519.
39. DURHAM, H.D. 1992. An antibody against hyperphosphorylated neurofilament proteins collapses the neurofilament network in motor neurons but not in dorsal root ganglion cells. J. Neuropathol. Exp. Neurol. **51:** 287–297.
40. Lee M.K. et al. 1994. A mutant neurofilament subunit causes massive, selective motor neuron death: implications for the pathogenesis of human motor neuron disease. Neuron **13:** 975–988.
41. WONG, P.C. et al. 1990. Characterization of dominant and recessive assembly-defective mutations in mouse neurofilament NF-M. J. Cell Biol. **111:** 1987–2003.
42. GILL, S.R. et al. 1990. Assembly properties of dominant and recessive mutations in the small mouse neurofilament (NF-L) subunit. J. Cell Biol. **111:** 2005–2019.
43. WONG, P.C. et al. 1995. Increasing neurofilament subunit NF-M expression reduces axonal NF-H, inhibits radial growth, and results in neurofilamentous accumulation in motor neurons. J. Cell Biol. **130:** 1413–1422.
44. JOHNSTON, J. et al. 1998. Aggresomes: a cellular response to misfolded proteins. J. Cell Biol. **143:** 1883–1898
45. MARTINDALE, D. et al. 1998. Length of huntingtin and its polyglutamine tract influences localization and frequency of intracellular aggregates. Nature Genet. **18:** 150–154.

46. CLEVELAND, D.W. *et al.* 1996: Mechanisms of selective motor neuron death in transgenic mouse models of motor neuron disease. Neurology **47:** S54–S61
47. VICKERS, J.C. *et al.* 1994. Age-associated and cell-type-specific neurofibrillary pathology in transgenic mice expressing the human midsized neurofilament subunit. J. Neurosci. **14:** 5603–5612.
48. L'ECUYER, T.J. *et al.* 1993. Specific and quantitative immunoprecipitation of tropomyosin and other cytoskeletal proteins by magnetic separation. Biotechniques **14:** 436–441.
49. TU, P.H. *et al.* 1997 Selective degeneration fo Purkinje cells with Lewy body-like inclusions in aged NFHLACZ transgenic mice. J. Neurosci. **17:** 1064–1074.

Bioenergetics in Huntington's Disease

THOMAS GRÜNEWALD AND M. FLINT BEAL[a]

Department of Neurology and Neuroscience, Weill Medical College of Cornell University, New York Presbyterian Hospital, New York, New York 10021, USA

> ABSTRACT: Huntington's disease (HD) is an autosomal dominant inherited neurodegenerative disorder with relentless course and prototypical clinical symptoms. In 1993 HD was associated with an expanded CAG triplet repeat stretch on chromosome 4 in the coding region of its target protein, huntingtin. The length of the resulting polyglutamine extensions correlates with lower age of onset and a higher density of ubiquitine-positive neuronal intranuclear inclusions. Recently it has been proposed that mutant huntingtin induces progressive neuronal cell death by an apoptotic mechanism. There is strong evidence that disturbances in cellular energy homeostasis and oxidative damage contribute to neurodegeneration. This review will summarize and discuss the current concepts that point towards an involvement of free radical–induced oxidative stress, glutamate excitotoxicity and mitochondrial respiratory chain defects in pathogenesis of HD.

INTRODUCTION

More than a century ago, in 1872, George Huntington first described the devastating movement disorder now carrying his name. Huntington's disease (HD) is an autosomal dominant inherited, slowly progressive neurodegenerative disorder with a prevalence of 4 to 8 per 100,000 persons.[1] The onset of symptoms typically occurs between 35 to 40 years of age, though they may appear at any age.[2] The disease is characterized by personality changes as well as motor and cognitive symptoms, including the hallmark feature, chorea.[3,4] The relentless course, leading to death after 10 to 20 years, and the high penetrance in parental generations pose a burden on afflicted families and have challenged generations of physicians and scientists.[5] There is no treatment known so far to postpone disease progression.

The neuropathology of HD shows loss of brain mass (up to 30% of weight), resulting from neuronal cell death, with a direct correlation between brain atrophy and duration and severity of the disease.[6] Initial atrophy is particularly evident in the basal ganglia, with up to 60% loss of mass in the caudate nucleus, putamen, and globus pallidus. The neuronal loss in the striatum is largely confined to spiny striatal GABAergic type II neurons, which project to the globus pallidus and receive glutamatergic signals from cerebral cortex and dopaminergic signals from the substantia nigra; neuronal loss is progressive and associated with astrogliosis.[7]

---

[a]Address for correspondence: M. Flint Beal, M.D., New York Presbyterian Hospital/Weill Medical College of Cornell University, Department of Neurology and Neuroscience, 525 East 68th Street, Room A569, New York, New York 10021. Phone: 212-746-6575; fax: 212-746-8532.
 e-mail: fbeal@mail.med.cornell.edu

With new technology available in molecular medicine, research efforts have been accelerated over the last two decades; they have succeeded in establishing major cornerstones in pathogenesis and created perspectives for treatment.

In 1987 HD was mapped to chromosome 4p16.3, and in 1993 the genetic defect was identified as an unstable CAG trinucleotide repeat within the first exon of the IT-15 gene encoding a protein designated *huntingtin*.[8,9] Huntingtin is a protein widely expressed in the nervous system and other tissues with a still unknown function. It appears to be involved in normal embryological development and particularly in neurogenesis.[10] It is hypothesized that conformational changes in mutant huntingtin will lead to interactions with other proteins, thus altering the participation of huntingtin in vesicular transport, signal transduction, and intracellular trafficking.[11–13]

In healthy subjects the CAG repeat length ranges between 10 and 35; usually an expression of more than 40 CAG repeats leads to phenotypic symptoms of HD. Normal huntingtin protein is found widespread throughout the central nervous system, but the mutated form of huntingtin with enlarged polyglutamine tracts leads to aggregations of N-terminal fragments to form intranuclear inclusions.[14–16] In a recent review Cooper *et al.* suggested that expanded polyglutamine repeats serve as glutamyl donor substrates and lead to a tissue transglutaminase catalyzed crosslinking to form aggregates.[17] Saudou *et al.* have shown that inclusions are not necessary for cell death, while an aberrant decompartmentalization—e.g., translocation of mutant huntingtin into the nucleus—appears to be a more critical pathogenic event, leading to mechanisms of *apoptotic cell death*.[18] The inclusions preferentially might reflect a protective cellular mechanism rather than represent the primary cause of progressive neuronal cell death. Huntingtin was shown to be cleaved during apoptosis by caspase-3, and increasing the length of the polyglutamine tract[19,20] enhances the rate of cleavage. Butterworth *et al.* provided evidence for apoptotic DNA fragmentation in post-mortem striatal tissue of HD patients.[21] Huntingtin is only inconsistently found in striatal projection neurons, a neuronal cell subtype that is supposed to die; surviving striatal cholinergic interneurons and cortical pyramidal neurons express huntingtin abundantly.[22] A similar finding by Sapp *et al.* raises the question as to whether huntingtin preferentially damages the corticostriatal neuronal pathway and via excessive synaptic release of glutamate in its terminal nerve endings leads to *excitotoxic* striatal cell death.[23]

The link between the genetic defect and circumscribed cell death is still missing. It is widely accepted that interactions between defective energy metabolism, free radical–induced oxidative stress, and excitotoxicity contribute to the pathogenesis of HD as well as to other neurodegenerative disorders such as amyotrophic lateral sclerosis, Alzheimer's disease, and Parkinson's disease.[24] This review will briefly summarize current evidence for a bioenergetic failure in HD and highlight the potential role of impaired mitochondrial energy generation (FIG. 1).

EVIDENCE FOR A DEFECTIVE ENERGY METABOLISM IN HD

Brain cells are dependent on a permanent supply of high-energy phosphate compounds to fuel membrane ionic pumps such as Na^+/K^+-ATPases, Na^+/Ca^{++} antiporters or other vital ATP dependent enzymes such as the hexokinase reaction, the initial

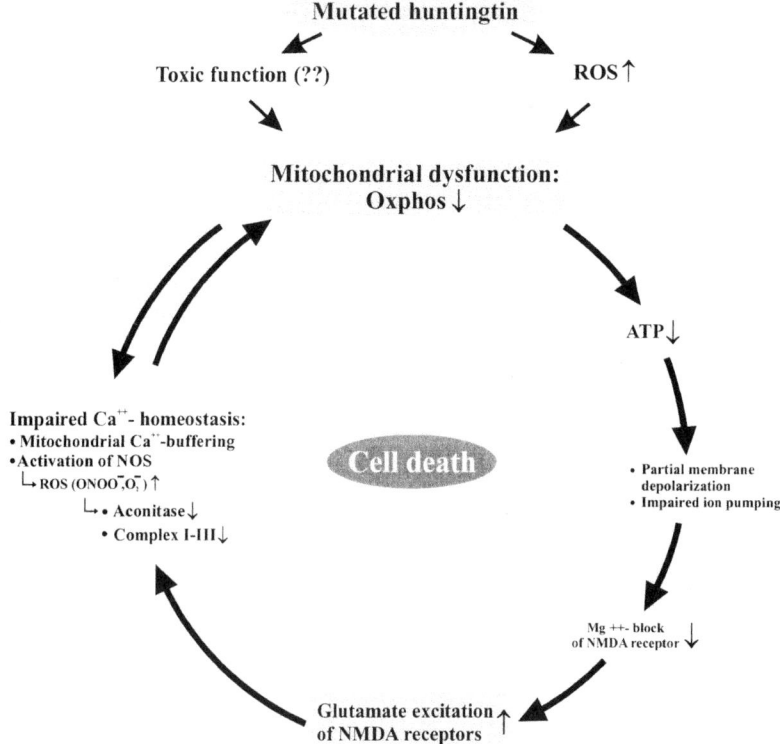

FIGURE 1. Interacting mechanisms contributing to bioenergetic failure in Huntington's disease.

step of glycolysis. It is intriguing to argue that a *defective mitochondrial energy generating system* either primarily or secondarily by means of excitotoxicity leads to progressive neuronal death.[25]

Studies of cerebral glucose metabolism using F-18 *fluorodeoxyglucose positron emission tomography* (FDG-PET) scans provide strong evidence for an impairment of energy metabolism. In HD patients and those at risk of developing this disorder typical patterns of diminished cerebral metabolic rates for glucose metabolism are observed in the caudate nuclei and putamen as well as in frontal, parietal, and striatal regions.[26–28] Basal ganglia metabolism is highly correlated with the functional capacity of individual patients and the degree of their motor dysfunction.[29] Repeated PET scans have proved to be a reliable tool for monitoring the disease progression.[30]

Magnetic resonance spectroscopic (MRS) imaging has emerged as a powerful tool for noninvasively imaging cerebral metabolites *in vivo*. Apart from elucidating the biochemical changes of a disease, it improves diagnostic accuracy and helps to monitor therapy. In conditions when the glycolytic rate exceeds the capacity of oxidative metabolism a lactate peak can be observed in the spectra of the chemical shift.

Jenkins *et al.* demonstrated in occipital cortex an almost threefold elevation of lactate in 31 HD patients as compared to 17 control subjects.[31] Lactate levels in the striatum were also increased and curiously were markedly asymmetric (higher on the left side). Other reports provided findings of elevated lactate in other brain regions.[32] A decrease in PCr/Pi ratio in resting gastrocnemius muscle of HD patients indicates defective energy metabolism in peripheral tissues as well.[33] This observation is consistent with a report of progressive weight loss in HD patients despite increased caloric intake.[34]

In several *biochemical studies* respiratory chain complex activities have been examined in HD brain, platelets, fibroblasts, and muscle tissue. Complex II/III (55% deficiency) and complex IV (25% deficiency) activity was reproducibly impaired in caudate and putamen.[35,36] Tabrizi *et al.* confirmed these results, but could not detect alterations in HD cortex, cerebellum, or fibroblasts.[37] Examinations of complex I (NADH dehydrogenase), however, show inconsistent results. Browne *et al.* did not find changes in complex I or citrate synthase activity. A study of mitochondria of synaptic origin suggested a major contribution of a complex I defect in neurodegenerative disorders.[38] These authors examined the relative contribution of complex I, III, and IV activities to the control of oxidative phosphorylation. Their results suggest that when a threshold of 25% inhibition of complex I is exceeded, energy metabolism is severely impaired. In accordance with that notion complex I activity was reported to be strikingly depressed in platelet mitochondria of patients with HD.[39] Also, defective complex I activity was found in muscle tissue of HD patients.[40] The muscle morphology showed myopathic changes—moth-eaten fibers and enlarged mitochondria with abnormal cristae, changes also to be seen in mitochondrial encephalomyopathies. Sawa *et al.* examined lymphoblastoid cell lines of HD patients and monitored mitochondrial membrane potential after exposure to mitochondrial toxins sodium cyanide (a noncompetitive complex IV inhibitor) and 3-nitropropionic (an irreversible inhibitor of complex II). As compared to normal control subjects the lymphoblasts of HD patients show a CAG repeat–dependent susceptibility to mitochondrial depolarization, which is a critical step of *apoptotic cell death*.[41] This finding provides an interesting model to study mitochondrial dysfunction as a function of increasing CAG repeat length. A prior study in cultured fibroblasts of HD patients and controls investigated ionomycin-induced cytosolic and mitochondrial Ca^{++} fluxes with Fluo-3 and laser scanning confocal microscopy. Ionomycin induced depolarization of the mitochondrial membrane potential, and HD fibroblasts failed to recover fully and rebuild the membrane potential after a second ionomycin stress, whereas control fibroblasts repolarized after the second exposure to ionomycin.[42]

An interaction between the enlarged polyglutamine tracts of mutated huntingtin and other cellular or mitochondrial proteins has been assumed to contribute to the pathogenesis. A variety of huntingtin-binding proteins such as HAP1, HIP1, GAPDH, calmodulin, and an ubiquitin-conjugating enzyme have been identified.[43] Glyceraldehyde-3-phosphate-dehydrogenase (GAPDH) is a huntingtin-binding glycolytic enzyme responsible for the conversion of glyceraldehyde-3-phosphate to 1,3-diphosphoglycerate, which binds to huntingtin. Two independent groups did not find an impairment of GAPDH activity in HD caudate, putamen, cortex, or cerebellum.[36,37] A cellular model with transfected GAPDH and triplet repeat disease genes suggested that GAPDH exerted a toxic function by posttranslational modification by nitric oxide (NO).[44]

ROLE OF OXIDATIVE STRESS AND EXCITOTOXICITY ON PROGRESSIVE CELLULAR ENERGY FAILURE

Putative pathogenic mechanisms in HD such as glutamate excitotoxicity and oxidative stress are tightly linked and targeted to mitochondria. *Defective energy metabolism might result from free radical–induced oxidative damage to mitochondria and other cellular components, and then secondarily contribute to slow excitotoxic death.* Reactive oxygen species (ROS) such a superoxide radicals ($O_2^{\bullet-}$), hydrogen peroxide (H_2O_2), and hydroxyl radicals (OH) are generated during oxidation-reduction reactions, particularly in the mitochondrial electron transport chain. Free radicals can attack virtually all cellular structures such as membrane lipid layers, proteins, and DNA. Radical scavengers such as superoxide dismutase (SOD), gluthathione peroxidase, and catalase provide an efficient defense system. Oxidative damage occurs under conditions when cellular protective mechanisms are exhausted. Due to its location in the mitochondrial matrix, lack of histones, and limited repair mechanisms, mtDNA is a major target of ROS.[45,46] Any disturbances of the sensitive balance between free radical generation and detoxification can result in the formation of DNA strand breaks or formation of DNA adducts, protein carbonylation, and lipid peroxidation. Browne *et al.* found an increased level of 8-hydroxydeoxyguanosine (OH^8dG), a marker of oxidative damage to DNA in HD caudate nuclear DNA (+21%) and in frontal cortex, but the level of cytosolic SOD activity was unchanged.[36] Others found significantly increased levels of oxidized glutathione in the caudate nucleus, but no difference between controls and HD patients was found in the level of reduced GSH.[47]

Animal studies using mitochondrial toxins have highlighted interactions between oxidative stress and excitotoxicity.[48] 3-Nitropropionic acid (3-NP) is an irreversible inhibitor of succinate dehydrogenase. Accidental ingestion in man produces selective basal ganglia lesions and dystonia.[49] Both intrastriatal and systemic administration of 3-nitropropionic acid in rats produced lesions primarily in the caudate and putamen, resembling the neuropathological neurochemical features of HD. Similar observations were made with malonic acid (MA), a reversible inhibitor of the succinate dehydrogenase. Application of both toxins, MA and 3-NP, results in increased levels of oxidative stress markers, such as 3-nitrotyrosine, OH^8dG, and malondialdehyde.[50–52] Otherwise, inhibitors of free radical synthesis such as free radical spin traps attenuate striatal lesions and oxidative stress markers. These and other toxicological approaches using agents such as amino-oxyacetic acid (AOAA) and quinolinic acid in animal models link free radical damage to the hypothesis *of slow or secondary glutamate exitotoxicity.*[53]

Compromise of the components of the electron transport chain leads primarily to reduced ATP synthesis with a concomitant failure of ATP-dependent ion pumps and channels, resulting in depolarization of the cell and inability to restore the resting membrane potential. As a consequence the voltage-gated Mg^{++} block of the N-methyl-D-aspartate (NMDA) channels will be relieved and allow even ambient glutamate levels to further activate NMDA receptors. The spiny neurons, which are the most vulnerable neuronal population in HD, contain abundant NMDA receptors. As Na^+ and Ca^{++} ions enter the intracellular space, Ca^{++}-dependent enzymes such as the neuronal nitric oxide synthase (nNOS) will be activated and increase NO-mediated free radical production. Cell death occurs both by apoptosis and necrosis depending on the severity of the insult.[54] Interestingly, mitochondria harbor two nuclear encoded components, cytochrome c and an "apoptosis inducing factor" (AIF), which are released upon a Ca^{++}-induced opening of the mitochondrial transition pore and play a major role in the apoptotic death cascade.[55,56] Recently Tabrizi *et al.* provided another link between glutamate excitotoxicity and mitochondrial free radical production.[37] They examined the iron-sulfur cluster (4Fe-4S) containing enzyme aconitase, which converts citrate to isocitrate in the TCA cycle. Aconitase activity was markedly decreased in HD brain, most pronounced in caudate (92% reduction) and putamen (73% reduction), but the cerebellar levels were normal. Exitotoxicity is likely to increase the levels of nitric oxide (NO) and thus mediate the production of superoxide and peroxynitrite. Superoxide is a potent oxidant for iron-sulfur cluster–containing enzymes such as aconitase and complexes I and II/III. Liberated Fe^{++} can mediate the Fenton reaction and produce further oxidative damage by generating hydroxyl radicals. Additionally, aconitase is located predominantly inside the mitochondrial matrix; only free radicals generated intramitochondrially will result in pronounced inhibition.

The same authors studied mitochondrial biochemical features in a transgenic mouse model of HD (R6/2). Complex IV activity was decreased in mouse striatum and muscle, with an aconitase reduction of 30% to be observed in striatum.[57] Further evidence for the sensitivity of aconitase as a marker of mitochondrial dysfunction recently was demonstrated in Friedreich's ataxia and in mitochondrial superoxide dismutase mutant mice.[58–62]

THERAPEUTIC CONSIDERATIONS

If energy impairment plays a critical role in cell death in neurodegenerative diseases such as Huntington's disease, a novel therapeutic strategy will be to use compounds to improve mitochondrial metabolism in an attempt to compensate for any disease-related defects. One potential treatment is the electron transport component coenzyme Q_{10} (CoQ_{10}) or ubiquinone. This is a potent antioxidant that protects against glutamate toxicity *in vitro*.[63] We found that it produces dose-dependent protection against striatal lesions produced by the succinate dehydrogenase inhibitor malonate.[64] It also protected against malonate–induced ATP depletions. We have shown that it produces excellent protection against 3-nitropropionic acid neurotoxicity and also protects against depletions of reduced CoQ_9 and reduced CoQ_{10} induced by 3-nitropropionic acid administration.[65] We demonstrated that administration of CoQ_{10} for two months results in increased cortical concentrations in aged rats; and, in particular, it leads to a progressive increase in mitochondrial concentrations of CoQ_{10}. In Huntington's disease patients, oral administration results in significant decreases of occipital cortex lactate concentrations, which were reversed with withdrawal of therapy.[33] This has led to a proposal to test CoQ_{10} and an NMDA receptor antagonist both alone and in combination in the treatment of HD patients. Another novel approach, which we recently examined, is creatine administration to increase phosphocreatine levels and, thereby, buffer against energy depletion. Oral administration of creatine protected against striatal lesions produced by malonate and 3-nitropropionic acid;[66] and we showed that it protects against phosphocreatine and ATP depletions as well as increases in 3-nitrotyrosine concentrations, which, we believe, occur downstream of energy depletion. It also protected against increases in lactate concentrations in the striatum produced by 3-nitropropionic acid as assessed using MRI spectroscopy *in vivo*. It is possible that administration of CoQ_{10} or creatine might exert neuroprotective effects in helping to slow the progression of Huntington's disease.

CONCLUSIONS

There is an increasing body of evidence that suggests that there is energy impairment in Huntington's disease that may play a critical role in its pathogenesis. Energy depletion can lead to cell death by generating free radicals and impairing calcium buffering at a cellular level. This has been associated with both apoptotic and necrotic cell death. If this is the case, then several therapeutic strategies to buffer energy concentrations may prove to be useful in therapy.

REFERENCES

1. HARPER, P.S. 1992. The epidemiology of Huntington's Disease. Hum. Genet. **89:** 365.
2. DECRUYENAERE, M., G. EVERS-KIEBOOMS, A. BOOGAERTS, J.J. CASSIMAN, T. CLOOSTERMANS, K. DEMYTTENAERE, R. DOM, J.P. FRYNS & H. VAN DEN BERGHE. 1995. Predictive for testing Huntington's disease: risk perception, reasons for testing and psychological profile of test applicants. Genet. Couns. **6:** 1.
3. PFIANZ, S., J.A. BESSON, K.P. EBMEIER & S. SIMPSON. 1991. The clinical manifestation of mental disorder in Huntington's disease: a retrospective case record study of disease progression. Acta Psychiatra. Scand. **83:** 53.
4. BAMFORD, K.A., E.D. CAINE, D.K. KIDO, C. COX & I. SHOULSON. 1995. A prospective evaluation of cognitive decline in early Huntington's disease: functional and radiographic correlates. Neurology **45:** 1867.
5. LANSKA, D.J. 1995. George Huntington and hereditary chorea. J. Child Neurol. **10:** 46.
6. VONSATTEL, J.P., R. H. MYERS, T.J. STEVENS, R.J. FERRANTE, E.D. BIRD & E.P.J. RICHARDSON. 1985. Neuropatholigical classification of Huntington's disease. J. Neuropathol. Exp. Neurol. **44:** 559.
7. MYER, R.H., J.P. VONSATTEL, P.A. PASKEVICH, D.K. KIELY, T.J. STEVENS, L.A. CUPPLES, E.P.J. RICHARDSON & E.D. BIRD. 1991. Decease neuronal and increased oligodendroglial densities in Huntington's disease caudate nucleus. J. Neuropathol. Exp. Neurol. **50:** 729.
8. GILLIAM, T.C., M. BUCAN, M.E. MACDONALD, M. ZIMMER, J.L. HAINES, S.V. CHENG, T.M. POHL, R.H. MEYERS, W.L. WHALEY & B.A. ALLITTO. 1987. A DNA segment encoding two genes very tightly linked to Huntington's disease. Science **238:** 950.
9. THE HUNTINGTON'S DISEASE COLLABORATIVE RESEARCH GROUP. 1993. A novel gene containing a trinucleotide repeat that is expended and unstable on Huntington's disease chromosomes. Cell **72:** 971.
10. ZEITLIN, S., J.P. LIU, D.L. CHAPMAN, V.L. PAPAIOANNOU & A. EFSTRATIADIS. 1995. Increased apoptosis and early embryonic lethality in mice nullizygous for the Huntington's disease gene homologue. Nat. Genet. **11:** 155.

11. VELIER, J., M. KIM, C. SCHWARZ, T.W. KIM, E. SAPP, K. CHASE, N. ARONIN & M. DIFIGLIA. 1998. Wild-type and mutant hintingtins function in vesicle trafficking in the secretory and endocytic pathways. Exp. Neurol. **152:** 34.
12. DIFIGLIA, M., E. SAPP, K. CHASE, C. SCHWARZ, A. MELONI, C. YOUNG, E. MARTIN, J.P. VONSATTEL, R. CARRAWAY & S.A. REEVES. 1995. Huntington is a cytoplasmic protein associated with vesicles in human and rat brain neurons. Neuron **14:** 1075.
13. BAO, J., A.H. SHARP, M.V. WAGSTER, M. BECHER, G. SCHILLING, C.A. ROSS, V.L. DAWSON & T.M. DAWSON. 1996. Expansion of polyglutamine repeat in huntingtin leads to abnormal protein interactions involving calmodulin. Proc. Natl. Acad. Sci USA **93:** 5037.
14. BECHER, M.W., J.A. KOTZUK, A.H. SHARP, S.W. DAVIES, G.P BATES, D.L. PRICE & C.A. ROSS. 1998. Intranuclear neuronal inclusions in Huntington's disease and dentatorubral and pallidoluysian atrophy: correlation between the density of inclusions and IT15 CAG triplet repeat length. Neurobiol. Dis. **4:** 387.
15. DAVIES, S.W., M. TURMAINE, B.A. COZENS, M. DIFIGLIA, A.H. SHARP, C.A. ROSS, E. SCHERZINGER, E.E. WANKER, L. MANGIARINI & G.P. BATES. 1997. Formation of neuronal intranuclear inclusions underlies the neurological dysfunction in mice transgenic for the HD mutation. Cell **90:** 537.
16. DIFIGLIA, M., E. SAPP, K.O. CHASE, S.W. DAVIES, G.P. BATES, J.P. VONSATTEL & N. ARONIN. 1997. Aggregation of huntingtin in neuronal intranuclear inclusions and dystrophic neurites in brain. Science **277:** 1990.
17. COOPER, A.J., K.F. SHEU, J.R. BURKE, W.J. STRITTMATTER, V. GENTILE, G. PELUSO & J.P. BLASS. 1999. Pathogenesis of inclusion bodies in (CAG)n/Qn-expansion diseases with special reference to the role of tissue transglutaminase and to selective vulnerability. J. Neurochem. **72:** 889. In press.
18. SAUDOU, F., S. FINKBEINER, D. DEVYS & M.E. GREENBERG. 1998. Huntingtin acts in the nucleus to induce apoptosis but death does not correlate with the formation of intranuclear inclusions. Cell **95:** 55.
19. MARTINDALE D., A. HACKAM, A. WIECZOREK, L. ELLERBY, C. WELLINGTON, K. MCCUTCHEON, R. SINGARAJA, P. KAZEMI-ESFARJANI, R. DEVON, S.U. KIM, D.E. BREDESEN, F. TUFARO & M.R. HAYDEN. 1998. Length of huntingtin and its polyglutamine tract influences localization and frequency of intracellular aggregates. Nat. Genet. **18:** 150.
20. WELLINGTON, C.L., L.M. ELLERBY, A.S. HACKAM, R.L. MARGOLIS, M.A. TRIFIRO, R. SINGARAJA, K. MCCUTCHEON, G.S. SALVESEN, S.S PROPP, M. BROMM, K.J. ROWLAND, T. ZHANG, D. RASPER, S. ROY, N. THORNBERRY, L. PINSKY, A. KAKIZUKA, C.A. ROSS, D.W. NICHOLSON, D.E. BREDESEN & M.R. HAYDEN. 1998. Caspase cleavage of gene products associated with triplet expansion disorders generates truncated fragments containing the polyglutamine tract. J. Biol. Chem. **273:** 9158.
21. BUTTERWORTH, N.J., L. WILLIAMS, J.Y. BULLOCK, D.R. LOVE, R.L. FAULL & M. DRAGUNOW. 1998. Trinucleotide (CAG) repeat length is positively correlated with the degree of DNA fragmentation in Huntington's disease striatum. Neuroscience **87:** 49.
22. FUSCO, F.R., Q. CHEN, W.J. LAMOREAUX, G. FIGUEREDO-CARDENAS, Y. JIAO, J.A. COFFMAN, D.J. SURMEIER, M.G. HONIG, L.R. CARLOCK & A. REINER. 1999. Cellular localization of huntingtin in striatal and cortical neurons in rats: lack of correlation with neuronal vulnerability in Huntington's disease. J. Neurosci. **19:** 1189.
23. SAPP, E., J. PENNEY, A. YOUNG, N. ARONIN, J.P. VONSATTEL & M. DIFIGLIA. 1999. Axonal transport of N-terminal huntingtin suggests early pathology of corticostriatal projections in Huntington disease. J. Neuropathol. Exp. Neurol **58:** 165.
24. BEAL, M.F. 1996. Mitochondria, free radicals, and neurodegeneration. Curr. Opin. Neurobiol. **6:** 661

25. BEAL, M.F. 1992. Does impariment of energy metabolism result in excitotoxic neuronal death in neurodegenerative illness? Ann. Neurol. **31:** 119.
26. ALAVI, A., R. DANN, J. CHAWLUK, J. ALAVI, M. KUSHNER & M. REIVICH. 1986. Positron emission tomography imaging of regional cerebral glucose metabolism. Semin. Nucl. Med. **162:** 2.
27. HAYDEN, M.R., W.R. MARTIN, A.J. STOESSL, C. CLARKE, S. HOLLENBERG, M.J. ADAM, W. AMMANN, R. HARROP, J. ROGERS & T. RUTH. 1986. Positron emission tomograph in the early diagnosis of Huntington's disease. Neurology **36:** 888.
28. GOTO, I., T. TANIWAKI, S. HOSOKAWA, M. OTSUKA, Y. ICHIYA & A. ICHIMIYA. 1993. Positron emission tomographis (PET) studies in dementia. J. Neurol. Sci **114:** 1.
29. YOUNG, A.B., J.B. PENNEY, S. STAROSTA-RUBINSTEIN, D.S. MARKEL, S. BERENT, B. GIORDANI, R. EHRENKAUFER, D. JEWETT & R. HICHWA. 1986. PET scan investigations of Huntington's disease: cerebral metabolic correlates of neurological features and functional decline. Ann. Neurol. **20:** 296.
30. ANDREWS, T.C. & D.J. BROOKS. 1998. Advances in the understanding of early Huntington's disease using the functional imaging techniques of PET and SPET. Mol. Med. Today **4:** 532. In press.
31. JENKINS, B.G., W.J. KOROSHETZ, M.F. BEAL & B.R. ROSEN. 1993. Evidence for impairment of energy metabolism in vivo in Huntington's disease using localized 1H NMR spectroscopy. Neurology **43:** 2689.
32. HARMS, L., H. MEIERKORD, G. TIMM, L. PFEIFFER & A.C. LUDOLPH. 1997. Decreased N-acetyl-aspartate/choline ratio and increased lactate in the frontal lobe of patients with Huntington's disease: a proton magnetic resonance spectroscopy study. J. Neurol. Neurosurg. Psychiatry **62:** 27.
33. KOROSHETZ, W.J., B.G. JENKINS, B.R. ROSEN & M.F. BEAL. 1997. Energy metabolism defects in Huntington's desease and effects of coenzyme Q10. Ann. Neurol. **41:** 160.
34. O'BRIEN, C.F. & C. MILLER. 1999. Extraneural metabolism in early Huntington's Disease. Ann. Neurol. **28:** 300.
35. GU, M., M.T. GASH, V.M. MANN, F. AGID-JAVOY, J.M. COOPER & A.H. SCHAPIRA. 1996. Mitochondrial defect in Huntington's Disease caudate nucleus. Ann. Neurol. **39:** 85.
36. BROWNE, S.E., A.C. BOWLING, U. MACGARVEY, M.J. BAIK, S.C. BERGER, M.M. MUQIT, E.D. BIRD & M.F. BEAL. 1997. Oxidative damage and metabolic dysfunction in Huntington's Disease: selective vulnerability of the basal ganglia. Ann. Neurol. **41:** 646.
37. TABRIZI, S.J., M.W. CLEETER, J. XUEREB, J.W. TAANMAN, J.M. COOPER & A.H. SCHAPIRA. 1999. Biochemical abnormalities and excitotoxicity in Huntington's Disease brain. Ann. Neurol **45:** 25.
38. DAVEY, G.P., S. PEUCHEN & J.B. CLARK. 1998. Energy thresholds in brain mitochondria. Potential involvement in neurodegeneration. J. Biol. Chem. **273:** 12753.
39. PARKER, W.D.J., S.J. BOYSON, A.S. LUDER & J.K. PARKS. 1990. Evidence for a defect in NADH: ubiquinone oxidoreductase (complex I) in Huntington's Disease (see comments). Neurology **40:** 1231.
40. ARENAS, J., Y. CAMPOS, R. RIBACOBA, M.A. MARTIN, J.C. RUBIO, P. ABLANEDO & A. CABELLO. 1998. Complex I defect in muscle from patients with Huntington's Disease. Ann. Neurol. **43:** 397.
41. WIEGAND, C.W., C.A. CALLAHAM, J. LAWLER, L.J. HANLE, R.L. MARGOLIS, A.H. SHARP, C.A. ROSS, S.H. SNYDER & A. SAWA. 1998. Vulnerability of mitochondrial depolarization in lymphoblastoid cells from patients with Huntington's Disease. Soc. Neurosci. **24:** 1454.

42. GUTEKUNST, C.A., T.I. PENG, W.L. WHALEY, B. ROCK, S.M. HERSCH & J.T. GREENAMYRE. 1996. Mitochondrial calcium homeostasis in Huntington's disease fibroblasts. Soc. Neurosci. **22:** 227.
43. WALLING, H. W., J. J.BALDASSARE & T.C. WESTFALL. 1998. Molecular aspects of Huntington's Disease. J. Neurosci. Res. **54:** 301.
44. SAWA, A., L.J. HANLE, L.D. HESTER, J.K. TAKAHASHI, J.K. COOPER, V. COLOMER, S. ENGELENDER, G. SCHILLING, J.D. WOOD, A. KAKIZUKA, R. MARGOLIS, A.H. SHARP, C.A. ROSS & P.J SNYDER. 1999. Glyceraldehyd-3-phosphate dehydrogenase as a positive mediator of cell death; its implications in toxicity of nitric oxide and triplet repeat diseases. Soc. Neurosci. 24.
45. BROWNE, S.E., R.J. FERRANTE & M.F. BEAL. 1999. Oxidative stress in Huntington's Disease. Brain Pathol. **9:** 147. In press.
46. PRZEDBORSKI, S. & V. JACKSON-LEWIS. 1998. Experimental developments in movement disorders: update on proposed free radical mechanisms. Curr. Opin. Neurol. **11:** 335.
47. SIAN, J., D.T. DEXTER, A.J. LEES, S. DANIEL, Y. AGID, F. JAVOY-AGID, P. JENNER & C.D. MARSDEN. 1994. Alterations in glutathione levels in Parkinson's Disease and other neurodegnerative disorders affecting basal ganglia (see comments). Ann. Neurol. **36:** 348.
48. BEAL, M.F. 1994. Neurochemistry and toxin models in Huntington's Disease. Curr. Opin. Neurol. **7:** 542.
49. LUDOLPH, A.C., M. SEELIG, A.G. LUDOLPH, M.I. SABRI & P.C. SPENCER. 1992. ATP deficits and neuronal degeneration induced by 3-nitorpionic acid. Ann. N.Y. Acad. Sci. **648:** 300.
50. BROUILLET, E., P. HANTRAYE, R.J. FERRANTE, R. DOLAN, A. LEROY-WILLIG, N.W. KOWALL & M.F. BEAL. 1995. Chronic mitochondrial energy impairment produces selective striatal degeneration and abnormal choreiform movements in primates. Proc. Natl. Acad. Sci. USA **92:** 7105.
51. SCHULZ, J.B., R.T. MATTHEWS, M.M. MUQIT, S.E. BROWNE & M.F. BEAL. 1995. inhibition of neuronal nitric oxide synthase by 7-nitroindazole protects against MPTP-induced neurotoxicity in mice. J. Neurochem. **64:** 936.
52. GALPERN, W.R., R.T. MATTHEWS, M.F. BEAL & O. ISACSON. 1996. NGF attenuates 3-nitrotyrosine formation in a 3-NP model of Huntington's Disease. Neuroreport **7:** 2639.
53. BEAL, M.F. Huntington's Disease, energy and excitotoxicity. 1994. Neurobiol. Aging **15:** 275.
54. NICOTERA P., M. ANKARCRONA, E. BONFOCO, S. ORRENIUS & S.A. LIPTON. 1997. Neuronal necrosis and apoptosis: two distinct events induced by exposure to glutamate or oxidative stress. Adv. Neurol. **72:** 95-101:95.
55. SUSIN, S.A., H.K. LORENZO, N. ZAMZAMI, I. MARZO, B.E. SNOW, G.M. BROTHERS, J. MANGION, E. JACOTOT, P. CONSTANTINI, M. LOEFFLER, N. LAROCHETTE, D.R. GOODLETT, R. AEBERSOLD, D.P. SIDEROVSKI, J.M. PENNINGER & G. KROEMER. 1999. Molecular characterization of mitochondrial apoptosis-inducing factor (see comments). Nature **397:** 441.
56. SUSIN, S.A., N. ZAMZAMI & G. KROEMER. 1998. Mitochondria as regulators of apoptosis: doubt no more. Biochim. Biophys. Acta **1366:** 151.
57. TABRIZI, S.J., L. MANGIARINI, J. WORKMAN, A. MAHAL, G. BATES, J.M. COOPER & A.H. SCHAPIRA. 1998. Mitochondrial biochemical features of the R6/2 transgenic mouse model of Huntington's disease. Soc. Neurosci. **153**.
58. MELOV, S., P. COSKUM, M. PATEL, R. TUINSTRA, B. COTTRELL, A.S. JUN, T.H. ZASTAWNY, M. DIZDAROGLU, S.I. GOODMAN, T.T. HUANG, H. MIZIORKO, C.J.

ports that in Huntington's disease huntingtin may cause an impairment of glycolysis due to its ability to selectively bind to glyceraldehyde-3-phosphate dehydrogenase (GAPDH).[11–13] For the purposes of illustration, we assume that this fact is true *in vivo*, whether or not it is, and then explore the consequences of this finding using nuclear magnetic resonance (NMR) techniques (similar in concept to mathematical induction).

Since a defect in glycolysis would obviously have profound deleterious consequences for neuronal energy metabolism—and based upon the fact that most ND disorders present with decreased glucose utilization in the affected brain regions[14–20]—a defect in glycolysis is not an unreasonable place to commence our investigation. The postitron emission tomography (PET) studies mentioned above can yield information on the rate at which tissue accumulates glucose. This parameter is usually presumed to be proportional to the rate at which the tissue utilizes glucose. If glucose is *not* being utilized, there could be for many reasons for this, impaired glycolysis being among them. NMR provides a way to selectively dissect this question into a number of the components of the total system for ATP generation. The obvious NMR technique of choice would be ^{13}C NMR spectroscopy. In this technique one starts with an infusion of carbon-13–labeled glucose. Since ^{12}C has a nuclear spin of 0, it is not NMR active. ^{13}C has a spin of 1/2, like protons, and can be readily

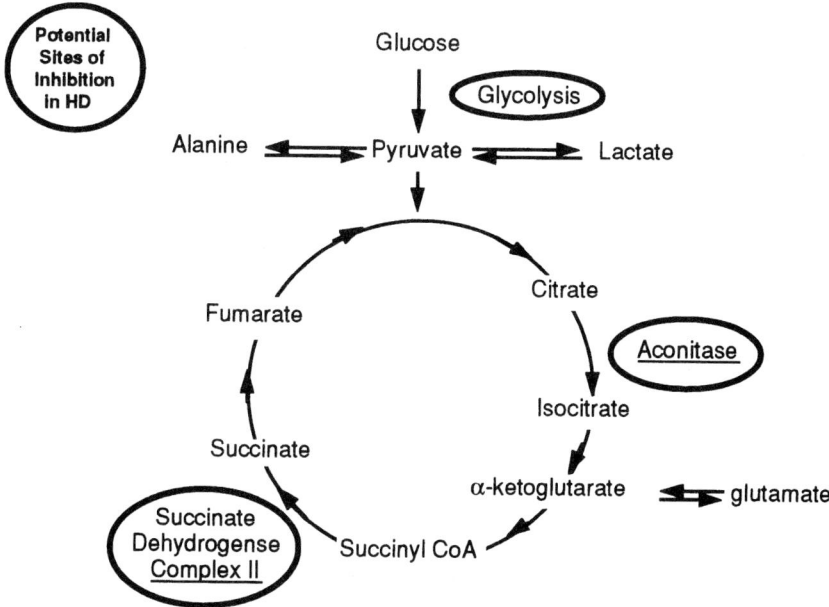

FIGURE 1. Simplified schematic diagram of the TCA cycle and some of the enzymatic defects that have been detected and/or postulated in Huntington's disease. The pathways are also shown with regards to NMR entry points for testing of etiologic hypotheses. For instance, blockade of succinate dehydrogenase may lead to elevations in succinate and lactate. The former due to direct blockade and the latter due to buildup of pyruvate and conversion to lactate.

detected, albeit at considerably lower sensitivity than for ^{1}H NMR. Usually the C1 position of glucose is labeled using ^{13}C; then the metabolism of the glucose molecule is followed as the glucose gets metabolized through the glycolytic pathway, then through the TCA cycle. A schematic diagram of this process is shown in FIGURE 1. On the first pass through the TCA cycle the primary detectable molecule is ^{13}C-labeled glutamate in the C4 position, followed rather quickly by glutamine at the C4 position. Subsequent passes through the TCA cycle produce labeling at the C3 and C2 positions of glutamate and glutamine. One can also detect the absolute amount of ^{13}C-labeled glucose in the brain at the same time. At later time points other metabolites start to appear, such as lactate, gaba, and N-acetylaspartate (NAA).

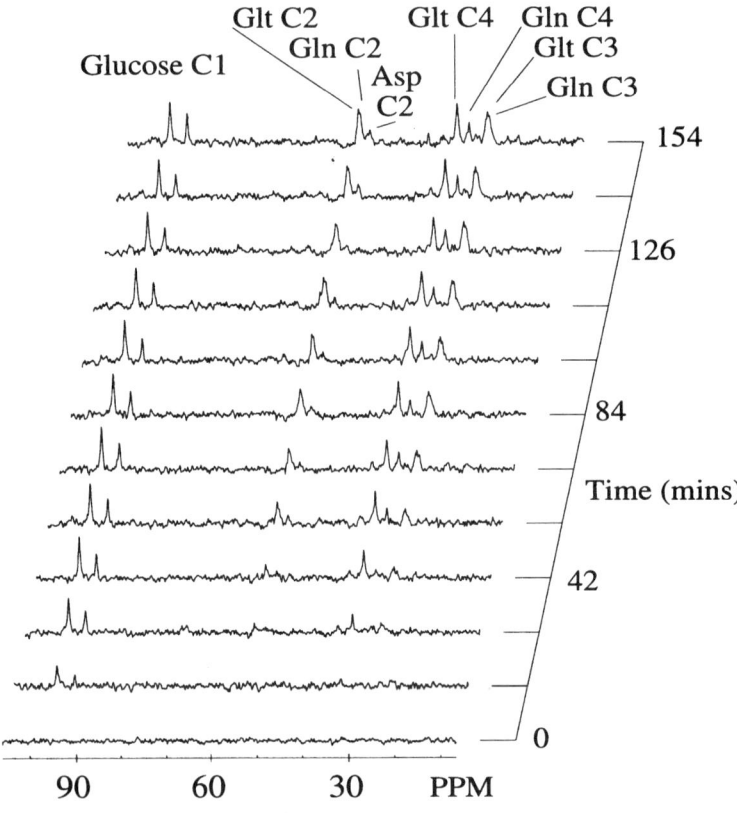

FIGURE 2. Typical dynamic ^{13}C spectra acquired at 4.7T from rat brain frontal cortex after infusion of ^{13}C-labeled glucose. Note that the glucose peaks (α and β anomers) are readily detectable followed by glutamate C4. Subsequent turns through the TCA cycle start to label other glutamate positions (C3 and C2). Glutamine follows a similar pattern, but with a smaller total pool size. Incorporation of these results into a metabolic model[21,23] allows for determination of TCA cycle flux rates. Defects in glycolysis or TCA cycle enzymes will lead to predictable differences in labeling of various intermediates and may allow for comprehensive biochemical evaluation of metabolic defects in various neurodegenerative disorders.

Much work has gone into trying to model the kinetics of the reactions and to determine, for instance, TCA cycle flux from these numbers. Space precludes a full discussion here; but most researchers who study the brain make the assumption that the reaction from oxo-glutarate to glutamate is much faster than the rate at which the substrate comes in (i.e., TCA cycle flux), and therefore the rate of C4 labeling is essentially the rate of TCA cycle flux. While this approach is somewhat controversial, its validity does not affect our qualitative understanding here. As an example of the type of information possible, we show data in FIGURE 2 from a ^{13}C labeling experiment in a wild-type rat. The major metabolites observed along with the corresponding time courses are shown. The rate at which the various TCA cycle intermediates get labeled yields the desired information in the context of a metabolic model.[21–25] Thus, administration of ^{13}C-labeled glucose to a patient with a suspected defect in glycolysis would yield a slower rate for labeling of glutamate compared to a control subject. At this point, such an observation would be indistinguishable from a defect in the TCA cycle. In the case of HD, such an observation would create problems because decreased activity or amounts of complex II or succinate dehydrogenase[26–28] and aconitase[27] have been found in HD striatum (or in transgenic HD mice). Defects in pyruvate dehydrogenase have also been postulated.[9,29,30]

Other experiments would be necessary to more completely specify the metabolic defect. Two candidate molecules would be lactate and alanine. This is so because a block in the TCA cycle (as opposed to glycolysis) should lead to large elevations of lactate and alanine as end points for pyruvate. A block in glycolysis would presumably lead to decreased lactate concentrations. We have performed a series of just such experiments using inhibition of succinate dehydrogenase with 3-nitropropionic acid. Spectra from a rat treated systemically with 3-NP are shown in FIGURE 3. It is clear that increased succinate is visible, as are increased lactate and alanine. The presence of succinate confirms that there is a block of succinate dehydrogenase, and the elevations in lactate and alanine confirm that the pyruvate is being shunted to these two molecules. These observations lead to a number of possibilities. If aconitase were blocked, we should observe elevations in citrate. Likewise, blockade of pyruvate dehydrogenase could also lead to elevations in lactate and alanine. These possibilities could be tested using ^{1}H rather than ^{13}C spectroscopy. If no blockade of the TCA cycle exists, then a defect in glycolysis may be the preferred explanation. For instance, if GAPDH really were impaired in HD *in vivo*, then one might look for elevations in fructose-1,6-bisphosphate. Such a search could be conducted using both proton and phosphorus 31 spectroscopies.

The above lines of reasoning are speculative only, and many holes can be poked in any one of the assumptions. Let us deal with the most likely glaring weakness in such a model. It is very clear that one must work to distinguish between acute and chronic inhibitions of a particular enzyme system. Thus, in the case of the 3-NP data shown in FIGURE 3, we have acutely blocked succinate dehydrogenase. Not only is the blockade acute, but it is likely very nearly complete, leading to a large striatal lesion and necrosis.[8,31,32] In the case of a pathology that evolves much more slowly, it is more than likely that compensatory mechanisms will develop to avert some of the more devastating consequences of the primary defect. It is also true that inhibition of a given enzyme system is likely to be incomplete, and therefore some of the changes postulated above may not be easily visible. This is because molecules such

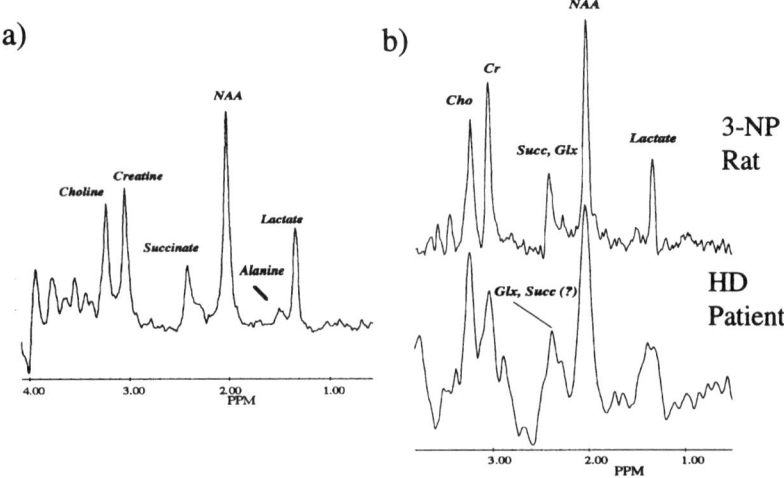

FIGURE 3. a: Typical ^{1}H spectrum from a rat striatum after subacute systemic treatment of the rat with 3-nitropropionic acid compared to a normal control. Note the large buildup of succinate, lactate and alanine in the 3-NP-treated animal. These data imply severe blockade of the TCA cycle and energy impairment. **b:** Comparison of ^{1}H spectra from striata of a 3-NP treated rat (4.7T) and an HD patient (1.5T) with 55 CAG repeats. Note the similar pattern of increased lactate. The succinate peak identifiable in the rat at 4.7T is confounded in the human at 1.5T by the overlap with glutamate and glutamine (glx). Separation of these can be made at higher field strengths (> 4T) and by collecting spectra with multiple TE values. The singlet succinate resonance at 2.4 ppm will not undergo phase modulation as a function of TE. The glutamate and glutamine resonances will undergo phase modulation such that with a PRESS sequence the glutamate and glutamine resonances will be close to zero intensity at a TE of 68 ms.

as lactate, alanine, and succinate are present at such low concentrations in control brain tissue that the changes may be too small to observe using NMR spectroscopy, which is a relatively insensitive technique.

One way to avert this detection sensitivity limit *in vivo* is to run NMR spectroscopy of brain extracts *in vitro*. In this case one can utilize much higher magnetic field strengths, which in turn yield much higher signal-to-noise ratios and optimal detection methodologies (zero TE, optimal rf detection coil). In such cases one can measure at least 12 of the common amino acids with a quantitative precision and accuracy comparable to that of HPLC. Such a comparison is shown in FIGURE 4 for a transgenic HD mouse and a wild-type mouse. It is clear that there are large changes in many of the brain neurochemicals, including decreased NAA and increased glutamine. Such a technique is not in principle different from, say, HPLC detection of brain extracts; and the power of NMR for longitudinal, noninvasive analysis of a given population is lost. This technique is clearly of great use for study of an experimental animal model and ideally would be performed on animals at the end of the *in vivo* study.

Finally, a word about ^{31}P spectroscopy is in order. ^{31}P spectroscopy has the advantage of being able to probe ATP levels as well as the other high-energy phosphate

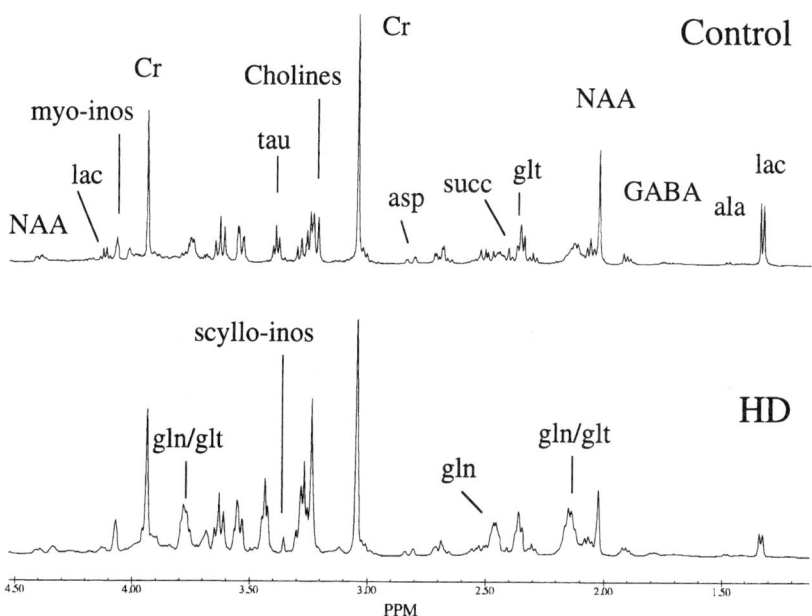

FIGURE 4. High-resolution spectra (500 MHz) from brain extracts of an HD mouse with 141 CAG repeats[140] and a wild-type mouse. Note the large decrease in NAA and increase in glutamine and taurine in the HD mouse. Also note the increased spectral resolution compared to the *in vivo* spectra in FIGURE 3.

store, PCr. Many of the known mitochondrial defects present with decreased PCr levels in the baseline state, which can decrease further upon mild exercise. Unfortunately, ATP turns out to be a poor marker for intracellular energy levels, as ATP concentrations stay very tightly regulated under nearly all but ischemic conditions. Thus, in pathologies where the tissue is not in such immediate danger as it is in ischemia, ATP levels would likely be found unchanged. Evidence of such an assertion is found in HD occipital cortex. In FIGURE 5 we show data from an HD patient and a normal control. It is clear that there is no evidence of a decrease in ATP or phosphocreatine.

Studying Potential Oxidative Stress Using MR Techniques

Oxidative stress is a common entry point for discussion of the etiology and/or progression of a number of neurodegenerative conditions.[33–37] This has been extensively discussed in relation to Parkinson's disease (PD) largely due to the great success of MPTP/MPP+ in replicating the symptoms and dopaminergic neurodegeneration of idiopathic PD.[38] An additional postulated role for oxidative stress in the etiology of PD comes from evidence accumulated over the past 10 years showing increased iron levels in the subtantia nigra with PD (reviewed in Refs. 39, 40). In addition to increased iron levels there are a variety of other abnormalities relevant to iron metabolism such as: increased levels of lactotransferrin in substantia nigra (SN)

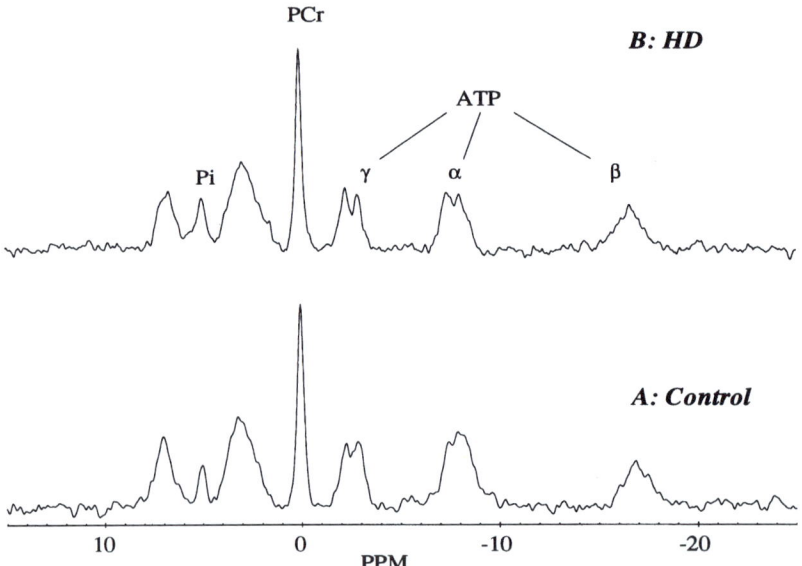

FIGURE 5. ^{31}P spectra (1.5T) taken from occipital cortex in an HD patient and a normal control. Peaks corresponding to ATP, phosphocreatine, inorganic phosphate, and phosphomonoesters and diesters are visible.[117,118] No major differences between the HD and control subject are observed in this brain region implying that any energetic defects seen are less severe than those found in ischemia or in patients with mitochondrial disorders.

and ventral tegmental area (VTA) in PD;[41] decreased numbers of transferrin receptors in SN similar to the level of decrease in nigral neurons in PD;[42] high free iron/ferritin ratios in PD basal ganglia; and decreased glutathione (GSH) levels in SN but not in other structures of the basal ganglia or in other neurodegenerative disorders.[43] The sum total of these results indicates that there is a general consensus that abnormalities in iron levels and iron metabolism exist in PD basal ganglia. What is not clear is whether this is a cause or effect of the disease.

Epidemiologic studies have recently been published showing that occupational exposure to iron, copper, and manganese is associated with an increased risk of PD.[44,45] In addition, injection of iron (III) into the nigra of rats produces a syndrome very similar to that induced by 6-hydroxydopamine lesioning.[46,47] These data are consistent with an etiologic mechanism for iron and possibly other metals. On the other hand, it is well known that high iron loading seen in Hallovorden-Spatz disease produces a syndrome in which dystonia and choreoathetosis are the primary features. Furthermore, high doses of manganese lead to a dystonic syndrome that, while similar to PD, is clearly not PD. Other data that indicate that iron may be a result, rather than a cause, of nigral cell loss are studies reporting increased iron concentrations after MPTP lesioning in monkeys[48] or after 6-hydroxydopamine (6-OHDA) lesioning in rats.[49,50] On the other hand, if free iron is present, regardless of whether it is a cause or an effect, the opportunity exists for catalytic generation of free radicals via the Fenton reaction in the presence of hydrogen peroxide and ferrous iron

(Fe^{2+}). It is not clear exactly what chemical state the increased iron is in in the PD brain. Dexter and colleagues have shown decreased ferritin levels in PD basal ganglia and increased free iron; similar changes were not noted in patients with either progressive supranuclear palsy (PSP) or multiple system atrophy (MSA).[51–53] Since the iron in ferritin is relatively inert biochemically, the implication is that in PD basal ganglia there could be enough free iron present to catalyze generation of reactive oxygen species (ROS). Even if the increased iron in PD brains is not a direct etiologic agent, increased levels of free iron could lead to a cycle of further neuronal degeneration via catalysis of ROS. The dopamine neurons appear to be particularly vulnerable to the effects of iron-catalyzed ROS. The reasons for this are multifarious and can include[39] high activity of iron-dependent tyrosine hydroxylase and high rates of autooxidation of dopamine (with subsequent generation of hydroxyl radical), as well as iron-melanin interactions that can change the redox state of the iron, making it a more potent ROS catalytic agent. Investigation of the role iron may play in both the etiology and/or progression of PD is an important step in developing rational therapies for PD. For instance, the iron chelator desferoxamine can protect against 6-OHDA–induced lesions of the nigrostriatal dopamine neurons.[54]

Several investigators have demonstrated MRI signal changes in the basal ganglia in PD that are consistent with increased iron levels.[55–59] These changes are especially pronounced in the substantia nigra compared to the putamen, where the results have been more variable (some investigators show an increase, some a decrease). It has long been known that the globus pallidus, the substantia nigra, and the red nuclei present a dark appearance on T2-weighted MRI scans. This is known to correlate with the relative iron concentrations in these structures.[60–62] Some groups have claimed that the field dependence of the T2 relaxation rates (1/T2 = R2) in these brain regions are reflective of the relaxation behavior of ferritin rather than paramagnetic iron (i.e., free iron) or heme-iron. In the former case the R2 scales linearly with field strength, while in the latter case the dependence is quadratic. A linear dependence of R2 with field strength has been found in basal ganglia *in vivo*.[62–64] The implication of these observations is that the predominant cause of the increased R2 values seen in the brain is the iron in ferritin, though this contention has not been proved unequivocally. The reason ferritin produces a linear field strength dependence of R2 is not known. The conversion of R2 relaxation rates into iron concentrations is thus rather subtle. In addition, it is well known that in the presence of magnetic field inhomogenetites there is a dramatic dependence of the T2 measurements on the interecho spacing due to diffusion of water during the echo times, allowing sampling of a variable distribution of magnetic fields. The implication of this is that the R2 measurements are quite sensitive to the choice of parameters in the detection sequence as well as static magnetic field homogeneities. For this reason many investigators choose to break up the observed relaxation rate into two components:

$$R2^* = R2' + R2,$$

where R2* is the observed relaxation rate (1/T2*); R2' (1/T2') is the static field line-broadening term, which is dependent on magnetic field disturbances caused, for instance, by iron; and R2 (1/T2) is the residual dipolar contribution to the observed relaxation time. Ordidge *et al.*[58] concluded that R2' lent greater precision to differences both between differing brain regions and between PD and control subjects

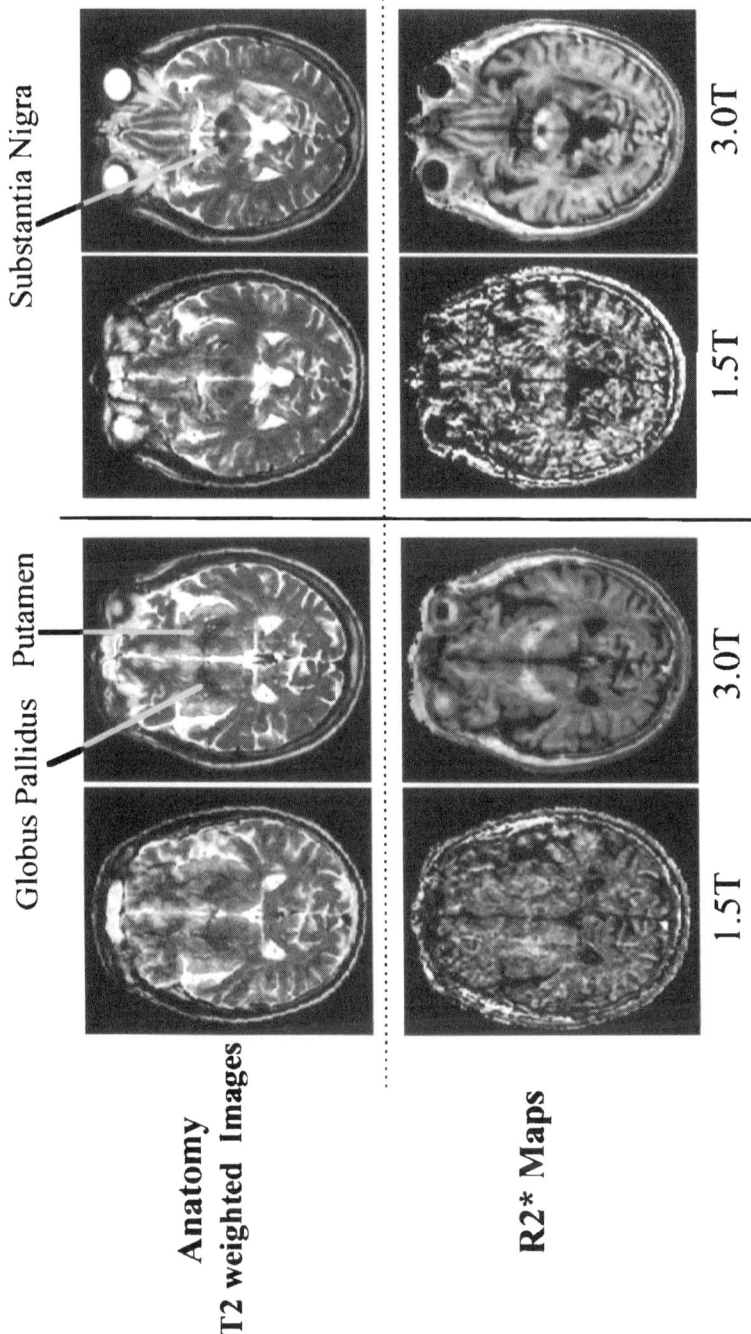

FIGURE 6. Maps of R2* as a function of field strength in the same PD patient measured on the same day. Note the much greater conspicuousness of the basal ganglia at the higher field strength owing to the effects of paramagnetic iron. The effect of field strength on the measured R2* value was supralinear.

than did R2*; and both R2* and R2' were significantly better than just R2. An alternative approach is to measure the interecho time dependence of the relaxation rates, and this has been claimed to correlate well with the relative brain iron concentrations in normal controls.[59,65] As shown in FIGURE 6, one can detect large increases in the R2* relaxation times in putamen, globus pallidus, and substantia nigra in the PD patient shown. The contrast-to-noise ratio increases dramatically as one goes from 1.5T to 3.0T. The increases measured correlated to an increase in R2* of much greater than 2, suggesting the possibility that the iron is not all chelated in the form of, for instance, ferritin, but may actually be free paramagnetic iron.

A link between ^{1}H MRS and measures of iron comes from studies of macrophage infiltration after various types of brain injury. Reactive microglia/macrophages are potent sources of iron generation, and studies indicate that a large part of the increased iron in PD can be found in these cells.[39,66] However, iron and ferritin reactivity in microglia is a relatively nonspecific finding in a large number of cerebral pathologies, and it has been speculated that "... all destructive processes of gray matter release iron ...."[67] Macrophages and reactive glia themselves are potent sources of cellular oxidative stress[68] and can generate superoxide and nitric oxide and bring about degradation of ferritin.[39,66,68] Thus, if iron is increased (by whatever mechanism) in PD brain, it has the potential to add to the progressive destruction of the nigrostriatal tract that is the hallmark of this disease. The fact that the reactive glia

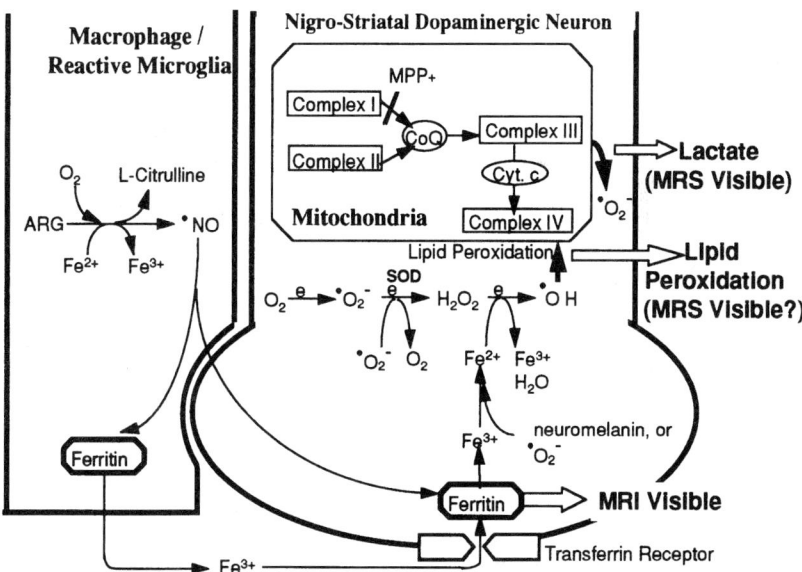

FIGURE 7. *Some* of the metabolic pathways and multiple interactive effects of both iron catalyzed free radical generation and inhibition of mitochondrial respiration by binding of MPP+ to complex I. Note that the inhibition of mitochondrial respiration can lead to free radicals as well as lactate production. Iron-based catalysis of free radicals mediated by reactive glia can lead to inhibition of mitochondrial respiration via lipid peroxidation of the mitochondrial membrane. Potential MRI and MRS visible markers are shown.

incorporate lipid-laden phagocytes means that there is a good chance of detecting them using short–echo time spectroscopy as shown below and in our studies of MPTP-treated monkeys.[69] The correlation of the iron measurement with the macromolecular peaks may lead to an *in vivo*, noninvasive measure of gliotic processes in PD. If these then correlate with the elevated lactate seen, we may have a powerful measure of events related to oxidative stress in the PD brain. A schematic diagram of the biochemical basis of this approach is shown in FIGURE 7.

NATURAL HISTORY AND THERAPY

The cascade of events that results from, for instance, a defect in energy metabolism will often lead to a predictable sequence the natural history of which may provide insight into both etiological mechanisms and potential therapeutic entry points. There are a number of MR parameters that provide unique windows into these processes, and clearly the nonnvasive and safe nature of the MR experiment lends itself well to longitudinal studies. Use of MR techniques can be applied to either follow single subjects longitudinally or to follow select cohorts at various stages in the disease. The progressive nature of most neurodegenerative illnesses makes study of single time points much less useful than is often recognized. We have recently shown that chronic, low-dose MPTP lesioning in nonhuman primates leads to a progressive decline in dopaminergic binding potential measured using PET that continues long after MPTP treatment stops.[69] These changes were accompanied by changes in NAA levels in the striatum that continued to decline after MPTP treatment. In addition lactate and macromolecular resonances, characteristic of macrophage infiltration,[70] increased, reaching a peak several months after cessation of MPTP treatment, and then returned back to baseline values. The changes in behavior, NAA levels, and dopamine binding were permanent.[69] From this study it is clear that a single time point would be inadequate to precisely describe the pathological changes. The same thing is likely to hold true in human illnesses as well. Many studies have found decreased NAA concentrations in frontal, temporal, and even occipital cortex in Alzheimer's disease,[71,72] and more recently have shown the value of NAA as a longitudinal marker of progression.[73] NAA would provide a valuable marker for following in any potential AD therapy.

Another molecule of potential for following in therapy is lactate. Elevated lactate concentrations have been detected in a number of pathologies but, perhaps not surprisingly, are often quite large in mitochondrial encephalopathies. These disorders are progressive degenerative disorders that affect multiple organs. Among the most common of these disorders is mitochondrial myopathy, encephalopathy, lactic acidosis, and stroke-like episodes (MELAS syndrome). A hallmark of MELAS is elevated intracerebral lactate because of the dysfunctional oxidative metabolism. Monitoring lactate levels has potential as a therapeutic outcome measure. De Stefano *et al.* studied 11 patients with various mitochondrial disorders in a short-term clinical treatment trial.[74] The patients received sodium dichloroacetate in a placebo-controlled, double-blinded crossover study. Dichloroacetate can cause stimulation of pyruvate dehydrogenase activity and subsequent decreases in lactate. Seven of these patients showed a significant decrease of the Lac/Cr ratio after one week of treat-

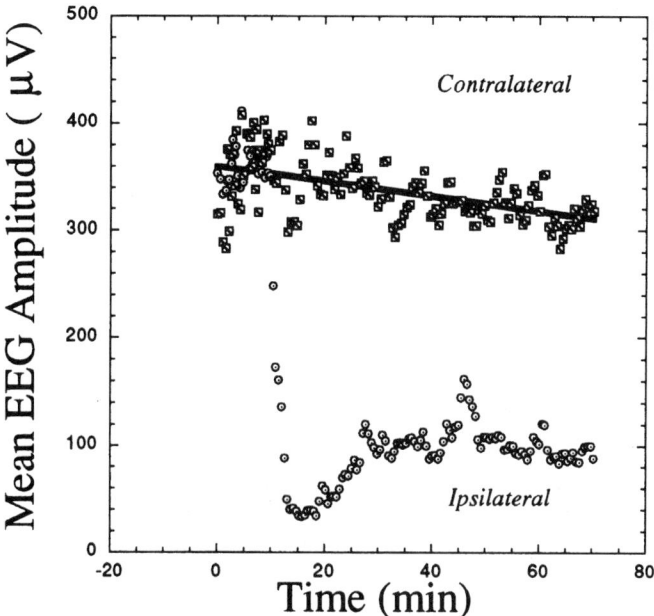

FIGURE 8. Mean EEG amplitude in the striatum after unilateral injection of 50 µmols of iodoacetate at time zero. Note the precipitous drop in EEG amplitude at the injected site after about 10 minutes. Such a drop is characteristic of anoxic depolarization.

ment. Their clinical disability remained unchanged, likely because a large fraction of the neuronal damage was not due to lactic acidosis (i.e., the lactate was a symptom, not a cause, of the encephalopathy).

As an example of the multiple parameters that can be followed to assess the natural history of a neurodegenerative condition, data is presented here for the natural history of an acute, focal energy impairment in an animal model induced by intrastriatal injection of iodoacetate, an inhibitor of GAPDH.[141] In this model an initial block in glycolysis leads to a decrease in ATP levels (in the absence of increased lactate). This leads to anoxic depolarization, which can be measured using depth electrodes in the striatum (as shown in FIG. 8). The drop in EEG power coincides with a decrease in the apparent diffusion coefficient, similar to what occurs in spreading depresssion and focal ischemia.[75,76] The decrease in the diffusion coefficient correlates quite strongly with increases in intracellular water content (cell swelling). Then, commencing approximately 3 hours later, there is a precipitous opening of the blood-brain barrier, as measured by injection of a contrast agent that enhances T1 relaxation rates. This period is followed by NAA loss that is apparent at the lesion site, as well as a drop in regional cerebral blood volume reflective of neuronal death. These data are shown in FIGURE 9. At much later time points, lactate starts to increase over baseline, likely a reflection of the "penumbral" tissue in which glycolysis is not completely impaired. Clearly, all these MR measurements yield a comprehensive package for evaluation of the natural history of an acute lesion. We

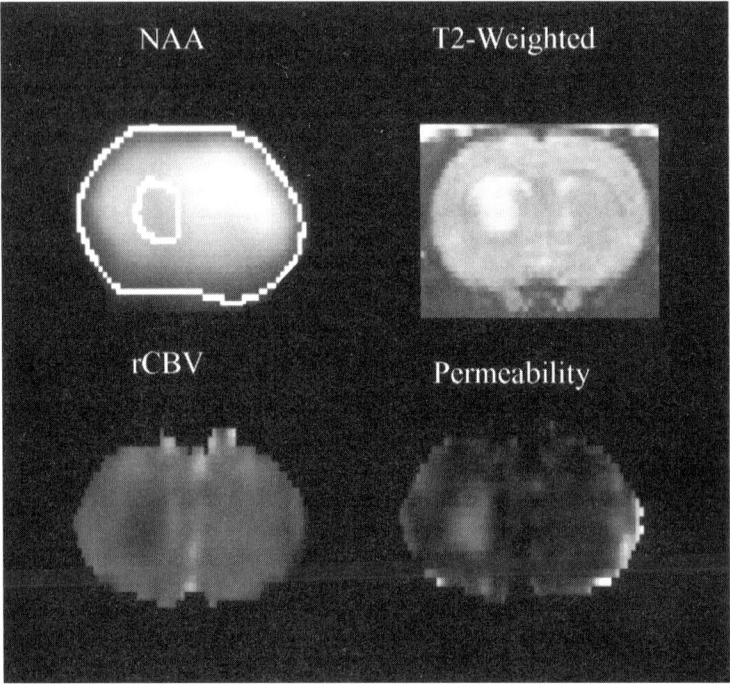

FIGURE 9. Maps of various parameters measured in a single animal 3–7 hours after injection of 50 μmols of iodoacetate. At the top left is a map of NAA showing an approximately 40% decrease compared to the contralateral striatum; a map of rCBV showing a drop of about 80% compared to the contralateral side; and a map of permeability changes at the lesion site showing a relative permeability surface area product (compared to skeletal muscle) of about 1.5 ml/min/100 g tissue. For anatomic reference a T2 weighted MRI image of the animal is also shown.

have also made similar, but less comprehensive, measurements in a primate model using quinolinic acid lesioning in the striatum. In these animals there is a large drop in NAA (as well as CBV) four months after lesioning. There is also a precipitous increase in macromolecular peaks at this time point (see FIG. 10), reminiscent of what we have found in the MPTP lesion model.[69] These peaks seem to correlate with macrophage infiltration into the lesion site,[70] and may represent neuronal membrane breakdown, as well as lipids from the macrophages (see FIG. 7). In these same animals at one year after lesioning the macromolecular peaks had disappeared back into the baseline, while the NAA levels remained depressed (FIG. 10).

These spectropscopic changes may provide useful therapeutic markers in Parkinson's disease. One promising therapy is transplantation with fetal dopamine cells. Due to the fact that fetal cells have little or no NAA whereas adult cells possess very large concentrations, MRS has been used to follow neural cell grafting in both primate models of PD[77] and, more recently, in PD patients.[78] These results have led to the conclusion that MRS may be useful for following the process of cell maturation

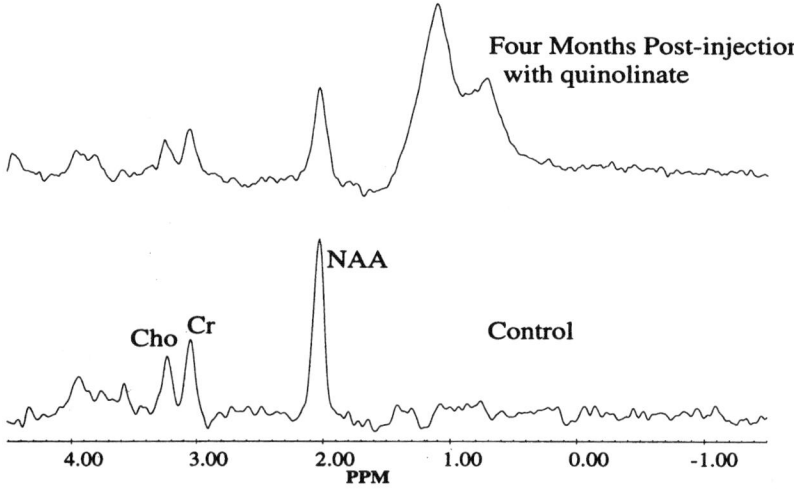

FIGURE 10. Proton spectra acquired from a 0.7-cc voxel in the striatum of a control monkey and a monkey lesioned with quinolinic acid (PRESS TR/TER 2000/272 ms). There is a decrease in NAA in the lesioned monkey (NAA/Cr = 1.68 ± 0.27 [$n = 4$] vs. 2.38 ± 0.11 in controls [$n = 10$], $p < 0.001$). In addition there is a large increase in the lipid/lactate/macromolecular region between 1–1.5 ppm. Similar peaks have been seen in MPTP lesioned monkeys[69] and also in strokes.[70,134] These peaks likely represent macrophage/microglia activation. At later time points they return to baseline noise.

(and increases in NAA) during growth of the graft. Failed grafts may also provide unique spectral signatures characterized by inflammatory processes and macrophage infiltration. These signatures include lactate and macromolecular resonances. Even better MR methods for following graft survival are discussed below.

These data once again indicate the utility of MR for potentially following specific processes. We have found a progressive decline in NAA in striatum (but not occipital cortex) in HD and an increase in lactate as a function of duration of symptoms in HD.[79,80] We further found that there was a dramatic correlation with CAG repeat length when normalized to age (to control for the age-dependent onset of HD).[80] We have also found that a similar progressive decline in striatal volume occurs with both CAG repeat length and duration of symptoms in HD. These MR parameters may able to accurately and objectively stage the neuronal decay that accompanies HD progression. We have evidence that a smaller decline may also occur in normal adults as a function of age in the striatum (but not cortex). Shown in FIGURE 11 is a plot of the change in NAA levels in the striatum as a function of age when normalized to creatine (creatine is often unchanged in neurodegenerative conditions and thus serves as a good reference in the voxel of interest). These data raise the provocative question of whether the NAA loss indicates neuronal loss, or just a decline in neuronal health. Unfortunately, the role of NAA in the brain is unclear (see discussion of this in Ref. 81), so the question is unanswerable at present. A similar decline in caudate volume (but not putaminal volume) also accompanies the NAA loss as

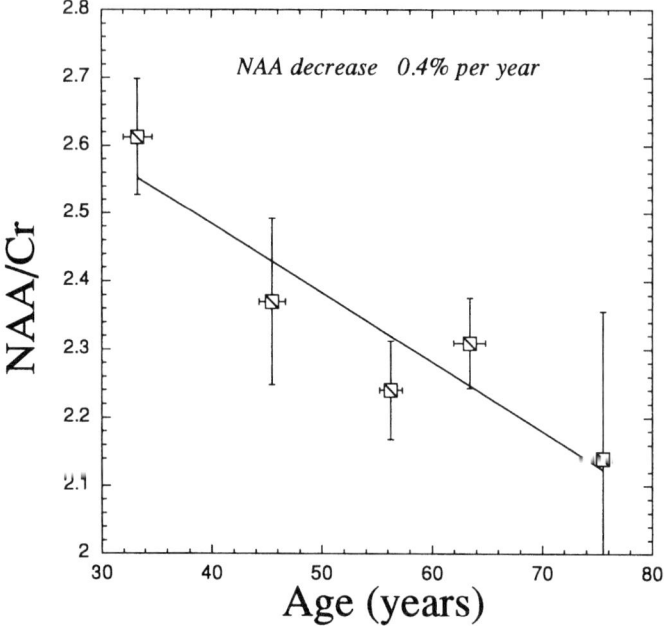

FIGURE 11. Plot of the decrease in NAA in normal striata as a function of age from 31 neurologically normal controls. Note that although the NAA decrease is quite significant, the standard deviations are also large at the later ages.

determined by others[82] and confirmed in our own studies. Similar age-related decreases in total numbers of striatal dopamine transporters have been found in SPECT or PET studies.[83,84]

In addition to brain volumetric morphometry, a new technique utilizes the anisotropic diffusional mobility of water molecules in various brain structures to map out white matter fiber tract orientations.[85–88] Water in a myelinated axon will diffuse more rapidly along its long axis compared to its short axis, and this can be used to generate orientation-dependent contrast sensitive to this phenomenon. In cases where ATP levels may be impaired or white matter structure is destroyed, membrane depolarization can lead to increases in the intracellular water content, leading to changes in the diffusivity of the water.[75] This tool will likely prove to find great utility in mapping changes in areas axonally connected to a specific brain region that undergoes degeneration. Preliminary studies using this technique have shown exquisite detail relevant to the degeneration of the corticospinal tracts in ALS[88,89] and of focal lesions in stroke.[88] Again, this marker could be expected to mirror the continuing degenerative process, at not only the areas directly affected by the pathology, but also at areas that have axonal projections to and from the primary sites of pathology.

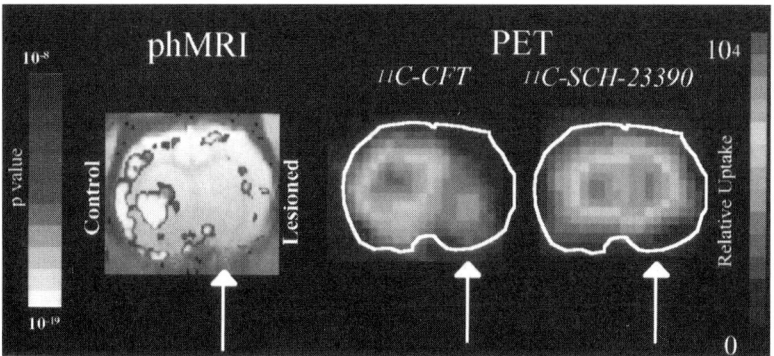

FIGURE 12. Pharmacologic MRI and PET images of the same rat unilaterally lesioned ith 6-hydroxydopamine in the substantia nigra. The image on the left is an image through he striatal region of statistically significant changes in CBV after stimulation with 3 mg/kg .p. amphetamine administration. The CBV is only stimulated with amphetamine on the inact unlesioned side, indicative of the depletion of dopamine on the lesioned side. On the ight are two PET images in the same animal showing loss of dopamine transporters on the esioned side using ^{11}C-CFT. Postsynaptic D1 receptors are still intact as measured by binding of ^{11}C-SCH23390, a selective D1 antagonist.

Another useful technique for determining the progression of, say, dopaminergic cell loss in PD is functional MRI in the presence of a pharmacological challenge. We have demonstrated the utility of this in an animal model of PD using unilateral lesioning with 6-hydroxydopamine. In this case the loss of a hemodynamic response on the ipsilateral side correlates with loss of PET binding of ^{11}C-labeled dopamine transporter binding.[90] Such a phenomenon is shown in FIGURE 12, where the hemodynamic response to a pharmacologic challenge using CFT (2β-carbomethoxy-3β-[4-fluoropheny] tropane, WIN 36,528), a potent blocker of the dopamine transporter, is shown in a control and unilaterally lesioned 6-hydroxydopamine animal. It is clear that the response is lost on the ipsilateral side. In addition, one can also follow potential therapies—such as neural transplantation of fetal dopamine cells where regeneration of dopaminergic function can easily be followed—with much higher sensitivity than MRS.[91] This technique would certainly be applicable to patients with PD, in which case one could follow loss of the remaining dopamine fibers after diagnosis and therapy with fetal cell transplantation. Similar experiments could be proposed for the acetylcholine system in AD and the glutamatergic system in ALS.

Functional challenges, as opposed to pharmacologic challenges, may also be of great benefit in a number of ND disorders. A large body of literature has accumulated relevant to PD using PET scanning (see reviews in Refs. 18, 92, 93) of both pharmacologic and functional studies. The functional studies reveal recruitment of lateral premotor areas as well as altered response of supplementary motor cortex. An example of a similar functional study using fMRI instead of PET is shown in FIGURE 13, where a normal control and a PD patient are compared for performance of a motor task. It is quite clear that the PD patient shows recruitment of additional cortical areas compared to the control. Such recruitment of additional cortical areas has been found often in stroke patients for both motor[94,95] and language tasks,[96] especially in

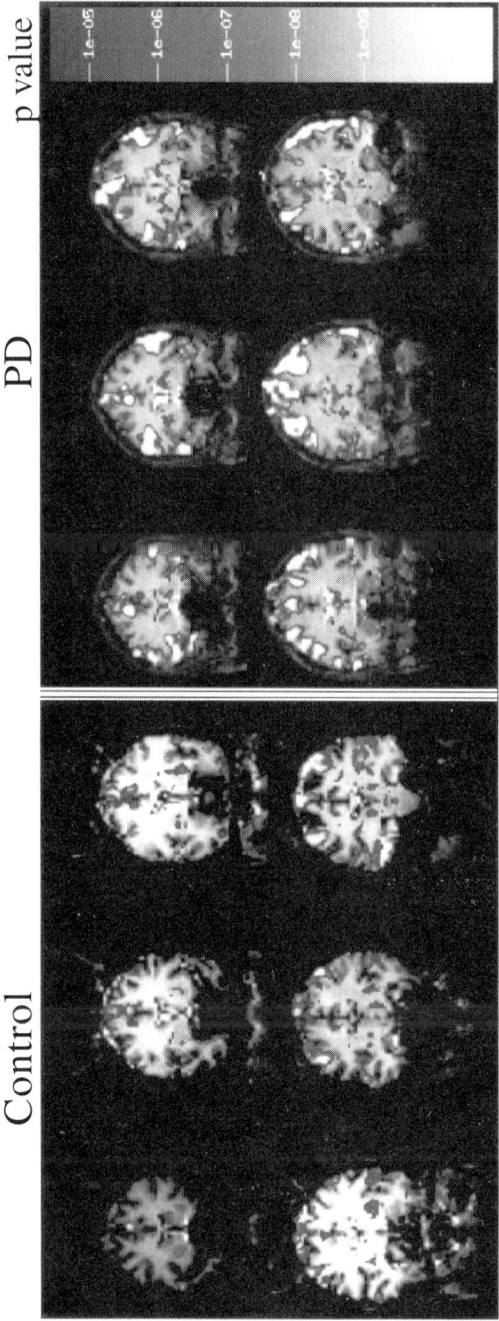

FIGURE 13. Two statistical maps of significant changes in BOLD fMRI of a motor task in a normal control (*left*) or a PD patient (*right*). The motor task was a bilateral sequential finger tapping task. Note the significantly increased cortical activation in the PD patient.

contralateral regions of homologous cortex. Longitudinal analysis of the progression of functional changes in neurodegenerative patients may provide quite a detailed picture of the consequences of the specific pathology. For instance, in a disease such as ALS one might expect recruitment of additional cortical regions in the early stages of illness to compensate for motor neuron loss. Ultimately, the pattern might be expected to reflect the paralysis of the subjects, and little activation would be observed. This may provide an additional therapeutic marker to follow as well.

Additional hemodynamic experiments for following the basal metabolic state in a given ND illness may also provide great utility in following the natural history of progression of the pathology. These types of scans could be considered analogs of PET FDG scans. A number of techniques capable of measuring regional cerebral blood flow (CBF) or blood volume (CBV) using MR have been developed.[97–102] These techniques could be used much as have prior PET studies of reduced glucose utilization rates in ND. This may be of particular relevance to AD as well as vascular dementias. A number of recent studies show that blood flow and volume may be compromised in AD, and that this may aid in following progression of the disease.[103–106] Such a hypotheis remains to be tested in other ND diseases.

HYPOTHETICAL INTEGRATED MR NEUROEXAM

In the sections above we described a series of different MR exams to address a myriad of questions, which may or may not be of relevance to one another. In order to pull together these disparate pieces we will demonstrate the possibility of integrating components relating to anatomy, physiology, and metabolism into one scanning session. Obviously, one cannot possibly hope to perform all the exams discussed above in one scan session, nor would such a thing be useful even if it were possible. One must tailor the choice of techniques to be those most relevant to the disorder or questions one wishes to examine. Let us suppose that we wish to determine longitudinal progression of a disease such as ALS (amyotrophic lateral sclerosis). A study could, in principle, be performed in under two hours, using the following protocol:

(i) Scout images, ≈3 min;
(ii) Rapid volumetric homogenization of the magnetic field (shimming) over the entire brain volume,[107] ≈2 min;
(iii) 3D Morphometric images (inversion recovery weighted T1 images such as GE SPGR or Siemen's MPRAGE), ≈10 min;
(iv) 2D Spectroscopic imaging of motor cortex and brain stem (medulla) using two sequential PRESS selected slices and volume, ≈30 min;
(v) Diffusion tensor mapping collecting six separate directions from motor cortex and brain stem, ≈30 min
(vi) fMRI of hand flexion (right, left, and both), ≈20 min.

Naturally, the protocol has to be debugged sufficiently to be of value, but we have recently implemented and successfully run patients using such a protocol. To follow the longitudinal progression one can imagine a vector set up from the parameters derived from the MR exam. Such a vector could be determined on a voxel-by-voxel basis in various brain regions depending upon the disease. Let us discuss a vector from hand-motor cortex composed of the following parameters: NAA levels (either in motor cortex and medulla, or over many regions determined from spectroscopic imaging experiments); glutamate or glx levels; myoinositol levels; choline levels; fractional anisotropy (a measure of the degree of anisotropy of water diffusion tensor eigenvalues compared to the magnitude of the diffusion values determined from diffusion tensor imaging); volume of cortex and volume of cortex activated during an fMRI task such as simple hand flexion. NAA levels have been shown to be a marker for the regional pattern of neuronal dysfunction in ALS,[108] as a marker for riluzole therapy in ALS,[109,110] and aberrant glutamate levels are implicated in the etiology of ALS.[111–113] Reductions in fractional anisotropy have been found in corticospinal tracts in ALS.[114] We have found increased motor cortical activation in asymmetric ALS patients with greater activated volumes on the hemisphere contralateral to the most affected side. Prior PET studies have indicated similar findings of enhanced CBF during task activation in contralateral primary sensorimotor cortex and in the adjacent contralateral ventral premotor and parietal association cortices as well as decreased CBF at rest in sensorimotor cortex.[115] The rationale for all these markers is thus apparent.

This vector, then, forms a multidimensional space whose surfaces can be followed and quantified over time. Multivariate discriminant analysis[116] can be employed to test for which parameters are most sensitive to the longitudinal progression of the ALS, and which are the best markers for potential therapeutic interventions. These parameters can then be compared to clinical markers, if desired; but they form an independent, objective exam on their own that may provide either earlier or more reliable markers than current clinical indices. Such an hypothesis remains to be tested.

Similar exams can be designed for other neurodegenerative disorders. For instance, in PD one may wish to substitute iron mapping for the fiber tract anisotropy measures in the ALS exam, and obviously change the locations of the MRS slices to include nigral and striatal regions. We firmly believe that in the future, integrated MR exams will provide enough added value to evaluation of the neurodegenerative disorders that they may one day become routine.

REFERENCES

1. SIESJO, B.K. 1992. Pathophysiology and treatment of focal cerebral ischemia. Part II: Mechanisms of damage and treatment. J. Neurosurg. **77**: 337–354.
2. HOSSMANN, K.A. 1994. Glutamate-mediated injury in focal cerebral ischemia: the excitotoxin hypothesis revised. Brain Pathol. **4**: 23–36.
3. FISHER, M. 1995. Potentially effective therapies for acute ischemic stroke. Eur. Neurol. **35**: 3–7.
4. FISHER, M. 1997. Characterizing the target of acute stroke therapy. Stroke **28**: 866–872.
5. KOGURE, T., & K. KOGURE. 1997. Molecular and biochemical events within the brain subjected to cerebral ischemia (targets for therapeutical intervention). Clin. Neurosci. **4**: 179–183.
6. ENDRES, M., S. NAMURA, M. SHIMIZU-SASAMATA, C. WAEBER, L. ZHANG, T. GOMEZ-ISLA et al. 1998. Attenuation of delayed neuronal death after mild focal ischemia in mice by inhibition of the caspase family. J. Cereb. Blood Flow Metab. **18**: 238–247.
7. KELLER, J.N., M.S. KINDY, F.W. HOLTSBERG, D.K. ST CLAIR, H.C. YEN, A. GERMEYER et al. 1998. Mitochondrial manganese superoxide dismutase prevents neural apoptosis and reduces ischemic brain injury: suppression of peroxynitrite production, lipid peroxidation, and mitochondrial dysfunction. J. Neurosci. **18**: 687–697.

8. JENKINS, B., E. BROUILLET, Y. CHEN, E. STOREY, J. SCHULZ, P. KIRSCHNER et al. 1996. Non-Invasive neurochemical analysis of focal excitotoxic lesions in models of neurodegenerative illness using spectroscopic imaging. J. Cereb. Blood Flow Metab. **16:** 450–461.
9. BLASS, J.P., R.K. SHEU & J.M. CEDARBAUM. 1988. Energy metabolism in disorders of the nervous system. Rev. Neurol. **144:** 543–563.
10. BEAL, M.F. 1995. Aging, energy, and oxidative stress in neurodegenerative diseases. Ann. Neurol. **38:** 357–366.
11. BURKE, J.R., J.J. ENGHILD, M.E. MARTIN, Y.S. JOU, R.M. MYERS, A.D. ROSES et al. 1996. Huntingtin and DRPLA proteins selectively interact with the enzyme GAPDH [see comments]. Nat. Med. **2:** 347–350.
12. KISH, S.J., I. LOPES-CENDES, M. GUTTMAN, Y. FURUKAWA, M. PANDOLFO, G.A. ROULEAU et al. 1998. Brain glyceraldehyde-3-phosphate dehydrogenase activity in human trinucleotide repeat disorders. Arch. Neurol. **55:** 1299–1304.
13. COOPER, A.J., K.F. SHEU, J.R. BURKE, W.J. STRITTMATTER & J.P. BLASS. 1998. Glyceraldehyde 3-phosphate dehydrogenase abnormality in metabolically stressed Huntington disease fibroblasts. Dev. Neurosci. **20:** 462–468.
14. BERENT, S., B. GIORDANI, S. LEHTINEN, D. MARKEL, J.B. PENNEY, H.A. BUCHTEL et al. 1988. Positron emission tomographic scan investigations of Huntington's disease: cerebral metabolic correlates of cognitive function. Ann. Neurol. **23:** 541–546.
15. RAPOPORT, S.I., B. HORWITZ, C.L. GRADY, J.V. HAXBY, C. DECARLI & M.B. SCHAPIRO. 1991. Abnormal brain glucose metabolism in Alzheimer's disease, as measured by position emission tomography. Adv. Exp. Med. Biol. **291:** 231–248.
16. RAPOPORT, S.I., K. HATANPAA, D.R. BRADY & K. CHANDRASEKARAN. 1996. Brain energy metabolism, cognitive function and down-regulated oxidative phosphorylation in Alzheimer disease. Neurodegeneration **5:** 473–476.
17. GOTO, I., T. TANIWAKI, S. HOSOKAWA, M. OTSUKA, Y. ICHIYA & A. ICHIMIYA. 1993. Positron emission tomographic (PET) studies in dementia. J. Neurol. Sci. **114:** 1–6.
18. BROOKS, D.J. 1998. Positron emission tomography studies in movement disorders. Neurosurg. Clin. N. Am. **9:** 263–282.
19. ANTONINI, A., K.L. LEENDERS, R. SPIEGEL, D. MEIER, P. VONTOBEL, M. WEIGELL-WEBER et al. 1996. Striatal glucose metabolism and dopamine D2 receptor binding in asymptomatic gene carriers and patients with Huntington's disease. Brain **119:** 2085–2095.
20. IBANEZ, V., P. PIETRINI, G.E. ALEXANDER, M.L. FUREY, D. TEICHBERG, J.C. RAJAPAKSE et al. 1998. Regional glucose metabolic abnormalities are not the result of atrophy in Alzheimer's disease. Neurology **50:** 1585–1593.
21. MASON, G.F., D.L. ROTHMAN, K.L. BEHAR & R.G. SHULMAN. 1992. NMR determination of the TCA cycle rate and alpha-ketoglutarate/glutamate exchange rate in rat brain. J. Cereb. Blood Flow Metab. **12:** 434–447.
22. GRUETTER, R., E.R. SEAQUIST, S. KIM & K. UGURBIL. 1998. Localized in vivo ^{13}C-NMR of glutamate metabolism in the human brain: initial results at 4 tesla. Dev. Neurosci. **20:** 380–388.
23. MASON, G.F., R. GRUETTER, D.L. ROTHMAN, K.L. BEHAR, R.G. SHULMAN & E.J. NOVOTNY. 1995. Simultaneous determination of the rates of the TCA cycle, glucose utilization, alpha-ketoglutarate/glutamate exchange, and glutamine synthesis in human brain by NMR. J .Cereb. Blood Flow Metab. **15:** 12–25.
24. SIBSON, N.R., A. DHANKHAR, G.F. MASON, K.L. BEHAR, D.L. ROTHMAN & R.G. SHULMAN. 1997. In vivo ^{13}C NMR measurements of cerebral glutamine synthesis as evidence for glutamate-glutamine cycling. Proc. Natl. Acad. Sci USA **94:** 2699–2704.
25. VAN ZIJL, P.C., D. DAVIS, S.M. ELEFF, C.T. MOONEN, R.J. PARKER & J.M. STRONG. 1997. Determination of cerebral glucose transport and metabolic kinetics by dynamic MR spectroscopy. Am. J. Physiol. **273:** E1216–E1227.

26. BROWNE, S.E., A.C. BOWLING, U. MACGARVEY, M.J. BAIK, S.C. BERGER, M.M. MUQIT *et al.* 1997. Oxidative damage and metabolic dysfunction in Huntington's disease: selective vulnerability of the basal ganglia. Ann. Neurol. **41:** 646–653.
27. TABRIZI, S.J., M.W. CLEETER, J. XUEREB, J.W. TAANMAN, J.M. COOPER & A.H. SCHAPIRA. 1999. Biochemical abnormalities and excitotoxicity in Huntington's disease brain. Ann. Neurol. **45:** 25–32.
28. GU, M., M.T. GASH, V.M. MANN, F. JAVOY-AGID, J.M. COOPER & A.H. SCHAPIRA. 1996. Mitochondrial defect in Huntington's disease caudate nucleus. Ann. Neurol. **39:** 385–389.
29. SORBI, S., E.D. BIRD & J.P. BLASS. 1983. Decreased pyruvate dehydrogenase complex activity in Huntington and Alzheimer brain. Ann. Neurol. **13:** 72–78.
30. YATES, C.M., J. BUTTERWORTH, M.C. TENNANT & A. GORDON. 1990. Enzyme activities in relation to pH and lactate in postmortem brain in Alzheimer-type and other dementias. J. Neurochem. **55:** 1624–1630.
31. BROUILLET, E., B.G. JENKINS, B.T. HYMAN, R.J. FERRANTE, N.W. KOWALL, R. SRIVASTAVA *et al.* 1993. Age-dependent vulnerability of the striatum to the mitochondrial toxin 3-nitropropionic acid. J. Neurochem. **60:** 356–359.
32. BEAL, M.F., E. BROUILLET, B.G. JENKINS, R.J. FERRANTE, N.W. KOWALL, J.M. MILLER *et al.* 1993. Neurochemical and histologic characterization of striatal excitotoxic lesions produced by the mitochondrial toxin 3-nitropropionic acid. J. Neurosci. **13:** 4181–4192.
33. GORMAN, A.M., A. MCGOWAN, C. O'NEILL & T. COTTER. 1996. Oxidative stress and apoptosis in neurodegeneration. J. Neurol. Sci. **139** (Suppl.): 45–52.
34. SCHAPIRA, A.H. 1996. Oxidative stress and mitochondrial dysfunction in neurodegeneration. Curr. Opin. Neurol. **9:** 260–264.
35. JENNER, P. & C.W. OLANOW. 1998. Understanding cell death in Parkinson's disease. Ann. Neurol. **44:** S72–S84.
36. BEAL, M.F. 1998. Mitochondrial dysfunction in neurodegenerative diseases. Biochim. Biophys. Acta **1366:** 211–223.
37. BUSCIGLIO, J., J.K. ANDERSEN, H.M. SCHIPPER, G.M. GILAD, R. MCCARTY, F. MARZATICO *et al.* 1998. Stress, aging, and neurodegenerative disorders. Molecular mechanisms. Ann. N.Y. Acad. Sci. **851:** 429–443.
38. NICKLAS, W.J., M. SAPORITO, A. BASMA, H.M. GELLER & R.E. HEIKKILA. 1992. Mitochondrial mechanisms of neurotoxicity. Ann. N.Y. Acad. Sci. **648:** 28–36.
39. GERLACH, M., D. BEN-SHACHAR, P. RIEDERER & M.B. YOUDIM. 1994. Altered brain metabolism of iron as a cause of neurodegenerative diseases? J. Neurochem. **63:** 793–807.
40. YOUDIM, M.B., D. BEN-SHACHAR & P. RIEDERER. 1993. The possible role of iron in the etiopathology of Parkinson's disease [published erratum appears in Mov. Disord. 1993 Apr; 8(2): 255]. Mov. Disord. **8:** 1–12.
41. LEVEUGLE, B., B.A. FAUCHEUX, C. BOURAS, N. NILLESSE, G. SPIK, E.C. HIRSCH *et al.* 1996. Cellular distribution of the iron-binding protein lactotransferrin in the mesencephalon of Parkinson's disease cases. Acta Neuropathol. **91:** 566–572.
42. FAUCHEUX, B.A., M.T. HERRERO, J. VILLARES, R. LEVY, F. JAVOY-AGID, J.A. OBESO *et al.* 1995. Autoradiographic localization and density of [^{125}I]ferrotransferrin binding sites in the basal ganglia of control subjects, patients with Parkinson's disease and MPTP-lesioned monkeys. Brain Res. **691:** 115–124.
43. JENNER, P. & C.W. OLANOW. 1996. Oxidative stress and the pathogenesis of Parkinson's disease. Neurology **47:** S161–S170.
44. GORELL, J.M., C.C. JOHNSON, B.A. RYBICKI, E.L. PETERSON, G.X. KORTSHA, G.G. BROWN *et al.* 1996. Occupational exposures to metals as risk factors for Parkinson's disease. Neurology **48:** 650–658.

45. GORELL, J.M., C.C. JOHNSON, B.A. RYBICKI, E.L. PETERSON, G.X. KORTSHA, G.G. BROWN et al. 1999. Occupational exposure to manganese, copper, lead, iron, mercury and zinc and the risk of Parkinson's disease. Neurotoxicology **20:** 239–247.
46. BEN-SHACHAR, D. & M.B. YOUDIM. 1991. Intranigral iron injection induces behavioral and biochemical "parkinsonism" in rats. J. Neurochem. **57:** 2133–2135.
47. SENGSTOCK, G.J., C.W. OLANOW, A.J. DUNN & G.W. ARENDASH. 1992. Iron induces degeneration of nigrostriatal neurons. Brain Res. Bull. **28:** 645–649.
48. MOCHIZUKI, H., H. IMAI, K. ENDO, K. YOKOMIZO, Y. MURATA, N. HATTORI et al. 1994. Iron accumulation in the substantia nigra of 1-methyl-4-phenyl-1,2,3,6-tetrahydropyridine (MPTP)–induced hemiparkinsonian monkeys. Neurosci. Lett. **168:** 251–253.
49. HALL, S., J.N. RUTLEDGE & T. SCHALLERT. 1992. MRI, brain iron and experimental Parkinson's disease. J. Neurol. Sci. **113:** 198–208.
50. HE, Y., P.S. THONG, T. LEE, S.K. LEONG, C.Y. SHI, P.T. WONG et al. 1996. Increased iron in the substantia nigra of 6-OHDA induced parkinsonian rats: a nuclear microscopy study. Brain Res . **735:** 149–153.
51. DEXTER, D.T., A. CARAYON, M. VIDAILHET, M. RUBERG, F. AGID, Y. AGID et al. 1990. Decreased ferritin levels in brain in Parkinson's disease. J. Neurochem. **55:** 16–20.
52. DEXTER, D.T., A. CARAYON, F. JAVOY-AGID, Y. AGID, F.R. WELLS, S.E. DANIEL et al. 1991. Alterations in the levels of iron, ferritin and other trace metals in Parkinson's disease and other neurodegenerative diseases affecting the basal ganglia. Brain **114:** 1953–1975.
53. DEXTER, D.T., J. SIAN, P. JENNER & C.D. MARSDEN. 1993. Implications of alterations in trace element levels in brain in Parkinson's disease and other neurological disorders affecting the basal ganglia. Adv. Neurol. **60:** 273–281.
54. BEN-SHACHAR, D., G. ESHEL, J.P. FINBERG & M.B. YOUDIM. 1991. The iron chelator desferrioxamine (Desferal) retards 6-hydroxydopamine–induced degeneration of nigrostriatal dopamine neurons. J. Neurochem. **56:** 1441–1444.
55. DRAYER, B.P., W. OLANOW, P. BURGER, G.A. JOHNSON, R. HERFKENS & S. RIEDERER. 1986. Parkinson plus syndrome: diagnosis using high field MR imaging of brain iron. Radiology **159:** 493–498.
56. DUGUID, J.R., R. DE LA PAZ & J. DEGROOT. 1986. Magnetic resonance imaging of the midbrain in Parkinson's disease. Ann. Neurol. **20:** 744–747.
57. ANTONINI, A., K.L. LEENDERS, D. MEIER, W.H. OERTEL, P. BOESIGER & M. ANLIKER. 1993. T2 relaxation time in patients with Parkinson's disease. Neurology **43:** 697–700.
58. ORDIDGE, R.J., J.M. GORELL, J.C. DENIAU, R.A. KNIGHT & J.A. HELPERN. 1994. Assessment of relative brain iron concentrations using T2-weighted and T2*-weighted MRI at 3 Tesla. Magn. Reson. Med. **32:** 335–341.
59. YE, F.Q., P.S. ALLEN & W.R. MARTIN. 1996. Basal ganglia iron content in Parkinson's disease measured with magnetic resonance. Mov. Disord. **11:** 243–249.
60. CHEN, J.C., P.A. HARDY, M. CLAUBERG, J.G. JOSHI, J. PARRAVANO, J.H. DECK et al. 1989. T2 values in the human brain: comparison with quantitative assays of iron and ferritin. Radiology **173:** 521–526.
61. CHEN, J.C., P.A. HARDY, W. KUCHARCZYK, M. CLAUBERG, J.G. JOSHI, A. VOURLAS et al. 1993. MR of human postmortem brain tissue: correlative study between T2 and assays of iron and ferritin in Parkinson and Huntington disease. Am. J. Neuroradiol. **14:** 275–281.
62. SCHENCK, J.F. 1995. Imaging of brain iron by magnetic resonance: T2 relaxation at different field strengths. J. Neurol. Sci. **134** (Suppl.)**:** 10–18.
63. BIZZI, A., R.A. BROOKS, A. BRUNETTI, J.M. HILL, J.R. ALGER, R.S. MILETICH et al. 1990. Role of iron and ferritin in MR imaging of the brain: a study in primates at different field strengths. Radiology **177:** 59–65.

64. VYMAZAL, J., R.A. BROOKS, N. PATRONAS, M. HAJEK, J.W. BULTE & G. DI CHIRO. 1995. Magnetic resonance imaging of brain iron in health and disease. J. Neurol. Sci. **134** (Suppl.): 9–26.
65. YE, F.Q., W.R. MARTIN & P.S. ALLEN. 1996. Estimation of brain iron in vivo by means of the interecho time dependence of image contrast. Magn. Reson. Med. **36:** 153–158.
66. YOUDIM, M.B., D. BEN-SHACHAR, G. ESHEL, J.P. FINBERG & P. RIEDERER. 1993. The neurotoxicity of iron and nitric oxide. Relevance to the etiology of Parkinson's disease. Adv. Neurol. **60:** 259–266.
67. KOEPPEN, A.H. 1995. The history of iron in the brain. J. Neurol. Sci. **134** (Suppl.):1–9.
68. COLTON, C.A. & D.L. GILBERT. 1987. Production of superoxide anions by a CNS macrophage, the microglia. FEBS Lett. **223:** 284–288.
69. BROWNELL, A.L., B.G. JENKINS, D.R. ELMALEH, T.W. DEACON, R.D. SPEALMAN & O. ISACSON. 1998. Combined PET/MRS brain studies show dynamic and long-term physiological changes in a primate model of Parkinson disease. Nat. Med. **4:** 1308–1312.
70. PETROFF, O.A., G.D. GRAHAM, A.M. BLAMIRE, M. AL-RAYESS, D.L. ROTHMAN, P.B. FAYAD *et al.* 1992. Spectroscopic imaging of stroke in humans: histopathology correlates of spectral changes. Neurology **42:** 1349–1354.
71. SHONK, T.K., R.A. MOATS, P. GIFFORD, T. MICHAELIS, J.C. MANDIGO, J. IZUMI *et al.* 1995. Probable Alzheimer disease: diagnosis with proton MR spectroscopy [see comments]. Radiology **195:** 65–72.
72. SCHUFF, N., D.L. AMEND, D.J. MEYERHOFF, J.L. TANABE, D. NORMAN, G. FEIN *et al.* 1998. Alzheimer disease: quantitative H-1 MR spectroscopic imaging of frontoparietal brain. Radiology **207:** 91–102.
73. ROSE, S.E., G.I. DE ZUBICARAY, D. WANG, G.J. GALLOWAY, J.B. CHALK, S.C. EAGLE *et al.* 1999. A 1H MRS study of probable Alzheimer's disease and normal aging: implications for longitudinal monitoring of dementia progression. Magn. Reson. Imaging **17:** 291–299.
74. DE STEFANO, N., P.M. MATTHEWS, B. FORD, A. GENGE, G. KARPATI & D.L. ARNOLD. 1995. Short-term dichloroacetate treatment improves indices of cerebral metabolism in patients with mitochondrial disorders. Neurology **45:** 1193–1198.
75. MOSELEY, M.E., J. MINTOROVITCH, Y. COHEN, H.S. ASGARI, N. DERUGIN, D. NORMAN *et al.* 1990. Early detection of ischemic injury: comparison of spectroscopy, diffusion-, T2-, and magnetic susceptibility–weighted MRI in cats. Acta Neurochir. Suppl. **51:** 207–209.
76. VAN DER TOORN, A., E. SYKOVA, R.M. DIJKHUIZEN, I. VORISEK, L. VARGOVA, E. SKOBISOVA *et al.* 1996. Dynamic changes in water ADC, energy metabolism, extracellular space volume, and tortuosity in neonatal rat brain during global ischemia. Magn. Reson. Med. **36:** 52–60.
77. JENKINS, B.G., L. BURNS, P. PAKZABAN, Y.I. CHEN, T. DEACON, B.R. ROSEN *et al.* 1994. Spectroscopic studies of neurochemical changes in a primate model of neurodegeneration and neural transplantation. *In* Proceedings of the Society of Magnetic Resonance Second Annual Meeting, Vol. 3: 1423. Society of Magnetic Resonance. San Francisco.
78. ROSS, B.D., T.Q. HOANG, S. BLUML, D. DUBOWITZ, O.V. KOPYOV, D.B. JACQUES *et al.* 1999. In vivo magnetic resonance spectroscopy of human fetal neural transplants. NMR Biomed . **12:** 221–236.
79. JENKINS, BG, W.J. KOROSHETZ, M.F. BEAL & B.R. ROSEN. 1993. Evidence for impairment of energy metabolism in vivo in Huntington's disease using localized 1H NMR spectroscopy. Neurology **43:** 2689–2695.
80. JENKINS, B.G., H.D. ROSAS, Y.C. CHEN, T. MAKABE, R. MYERS, M. MACDONALD *et al.* 1998. 1H NMR spectroscopy studies of Huntington's disease: correlations with CAG repeat numbers. Neurology **50:** 1357–1365.

81. JENKINS, B., Y. CHEN & B. ROSEN, Eds. 1997. Investigating the Neurochemistry and Etiology of Neurodegenerative Disorders Using Magnetic Resonance Spectroscopy. New York. Wiley-Liss.
82. MURPHY, D.G., C. DECARLI, M.B. SCHAPIRO, S.I. RAPOPORT & B. HORWITZ. 1992. Age-related differences in volumes of subcortical nuclei, brain matter, and cerebrospinal fluid in healthy men as measured with magnetic resonance imaging [published erratum appears in Arch. Neurol. 1994, Jan; **51**(1): 60]. Arch. Neurol. **49:** 839–845.
83. VAN DYCK, C.H., J.P. SEIBYL, R.T. MALISON, M. LARUELLE, E. WALLACE, S.S. ZOGHBI et al. 1995. Age-related decline in striatal dopamine transporter binding with iodine-123-beta-CITSPECT [see comments]. J. Nucl. Med. **36:** 1175–1181.
84. VOLKOW, N.D., Y.S. DING, J.S. FOWLER, G.J. WANG, J. LOGAN, S.J. GATLEY et al. 1996. Dopamine transporters decrease with age. J. Nucl. Med. **37:** 554–559.
85. BASSER, P.J., J. MATTIELLO & D. LEBIHAN D. 1994. MR diffusion tensor spectroscopy and imaging. Biophys. J. **66:** 259–267.
86. BASSER, P.J. & C. PIERPAOLI. 1996. Microstructural and physiological features of tissues elucidated by quantitative-diffusion-tensor MRI. J. Magn. Reson. B **111:** 209–219.
87. PIERPAOLI, C., P. JEZZARD, P.J. BASSER, A. BARNETT & G. DI CHIRO. 1996. Diffusion tensor MR imaging of the human brain. Radiology **201:** 637–648.
88. MAKRIS, N., A.J. WORTH, A.G. SORENSEN, G.M. PAPADIMITRIOU, O. WU, T.G. REESE et al. 1997. Morphometry of in vivo human white matter association pathways with diffusion-weighted magnetic resonance imaging. Ann. Neurol. **42:** 951–962.
89. ISHIHARA, T., K. HIRATA, J. KUBO, K. YAMAZAKI & T. SATO. 1998. [Clinical application of multi-shot diffusion EPI in neurological disease]. Rinsho Shinkeigaku **38:** 453–456.
90. CHEN, Y.I., W.R. GALPERN, A.L. BROWNELL, R.T. MATTHEWS, M. BOGDANOV, O. ISACSON et al. 1997. Detection of dopaminergic neurotransmitter activity using pharmacologic MRI: correlation with PET, microdialysis, and behavioral data. Mag. Reson. Med. **38:** 389–398.
91. CHEN, Y.I., A.L. BROWNELL, W. GALPERN, O. ISACSON, M. BOGDANOV, M.F. BEAL et al. 1999. Detection of dopaminergic cell loss and neural transplantation using pharmacologic MRI, PET and behavioral assessment. NeuroReport **10:** 2881–2886.
92. BROOKS, D.J. 1997. Motor disturbance and brain functional imaging in Parkinson's disease. Eur. Neurol. **38:** 26–32.
93. BROOKS, D.J. 1998. The early diagnosis of Parkinson's disease. Ann. Neurol. **44:** S10–S18.
94. ROSSINI, P.M., C. CALTAGIRONE, A. CASTRIOTA-SCANDERBEG, P. CICINELLI, C. DEL GRATTA, M. DEMARTIN et al. 1998. Hand motor cortical area reorganization in stroke: a study with fMRI, MEG and TCS maps. NeuroReport **9:** 2141–2146.
95. CRAMER, S.C., G. NELLES, R.R. BENSON, J.D. KAPLAN, R.A. PARKER, K.K. KWONG et al. 1997. A functional MRI study of subjects recovered from hemiparetic stroke. Stroke **28:** 2518–2527.
96. THULBORN, K.R., P.A. CARPENTER & M.A. JUST. 1999. Plasticity of language-related brain function during recovery from stroke. Stroke **30:** 749–754.
97. VILLRINGER, A., B.R. ROSEN, J.W. BELLIVEAU, J.L. ACKERMAN, R.B. LAUFFER, R.B. BUXTON et al. 1988. Dynamic imaging with lanthanide chelates in normal brain: contrast due to magnetic susceptibility effects. Magn. Reson. Med. **6:** 164–174.
98. BELLIVEAU, J.W., D.N. KENNEDY, JR., R.C. MCKINSTRY, B.R. BUCHBINDER, R.M. WEISSKOFF, M.S. COHEN et al. 1991. Functional mapping of the human visual cortex by magnetic resonance imaging. Science **254:** 716–719.
99. DETRE, J.A., J.S. LEIGH, D.S. WILLIAMS & A.P. KORETSKY. 1992. Perfusion imaging. Magn. Reson. Med. **23:** 37–45.
100. KWONG, K.K., J.W. BELLIVEAU, D.A. CHESLER, I.E. GOLDBERG, R.M. WEISSKOFF, B.P. PONCELET et al. 1992. Dynamic magnetic resonance imaging of human brain activity during primary sensory stimulation. Proc. Natl. Acad. Sci. USA **89:** 5675–5679.
101. OGAWA, S., D.W. TANK, R. MENON, J.M. ELLERMANN, S.G. KIM, H. MERKLE et al. 1992. Intrinsic signal changes accompanying sensory stimulation: functional brain mapping with magnetic resonance imaging. Proc. Natl. Acad. Sci. USA **89:** 5951–5955.

102. MANDEVILLE, J.B., J.J. MAROTA, B.E. KOSOFSKY, J.R. KELTNER, R. WEISSLEDER, B.R. ROSEN *et al.* 1998. Dynamic functional imaging of relative cerebral blood volume during rat forepaw stimulation. Magn. Reson. Med. **39:** 615–624.
103. BARCLAY, L., A. ZEMCOV, J.P. BLASS & F. MCDOWELL. 1984. Rates of decrease of cerebral blood flow in progressive dementias. Neurology **34:** 1555–1560.
104. HARRIS, G.J., R.F. LEWIS, A. SATLIN, C.D. ENGLISH, T.M. SCOTT, D.A. YURGELUN-TODD *et al.* 1996. Dynamic susceptibility contrast MRI of regional cerebral blood volume in Alzheimer's disease. Am. J. Psychiatry **153:** 721–724.
105. MEYER, J.S., T. SHIRAI & H. AKIYAMA. 1996. Neuroimaging for differentiating vascular from Alzheimer's dementias. Cerebrovasc. Brain Metab. Rev. **8:** 1–10.
106. MAAS, L.C., G.J. HARRIS, A. SATLIN, C.D. ENGLISH, R.F. LEWIS & P.F. RENSHAW. 1997. Regional cerebral blood volume measured by dynamic susceptibility contrast MR imaging in Alzheimer's disease: a principal components analysis. J. Magn. Reson. Imaging **7:** 215–219.
107. REESE, T.G., T.L. DAVIS & R.M. WEISSKOFF. 1995. Automated shimming at 1.5 T using echo-planar image frequency maps. J. Magn. Reson. Imaging **5:** 739–745.
108. ROONEY, W.D., R.G. MILLER, D. GELINAS, N. SCHUFF, A.A. MAUDSLEY & M.W. WEINER. 1998. Decreased N-acetylaspartate in motor cortex and corticospinal tract in ALS. Neurology **50:** 1800–1805.
109. KALRA, S., N.R. CASHMAN, A. GENGE & D.L. ARNOLD. 1998. Recovery of N-acetylaspartate in corticomotor neurons of patients with ALS after riluzole therapy. Neuroreport **9:** 1757–761.
110. KALRA, S., D.L. ARNOLD & N.R. CASHMAN. 1999. Biological markers in the diagnosis and treatment of ALS. J. Neurol. Sci. **165** (Suppl. 1): S27–S32.
111. PIORO, E.P., A.W. MAJORS, H. MITSUMOTO, D.R. NELSON & T.C. NG. 1999. 1H-MRS evidence of neurodegeneration and excess glutamate + glutamine in ALS medulla. Neurology **53:** 71–79.
112. BRUIJN, L.I., M.W. BECHER, M.K. LEE, K.L. ANDERSON, N.A. JENKINS, N.G. COPELAND *et al.* 1997. ALS-linked SOD1 mutant G85R mediates damage to astrocytes and promotes rapidly progressive disease with SOD1-containing inclusions. Neuron **18:** 327–338.
113. BRISTOL, L.A. & J.D. ROTHSTEIN. 1996. Glutamate transporter gene expression in amyotrophic lateral sclerosis motor cortex. Ann. Neurol. **39:** 676–679.
114. ELLIS, C.M., A. SIMMONS, D.K. JONES, J. BLAND, J.M. DAWSON, M.A. HORSFIELD *et al.* 1999. Diffusion tensor MRI assesses corticospinal tract damage in ALS. Neurology **53:** 1051–1058.
115. KEW, J.J., P.N. LEIGH, E.D. PLAYFORD, R.E. PASSINGHAM, L.H. GOLDSTEIN, R.S. FRACKOWIAK *et al.* 1993. Cortical function in amyotrophic lateral sclerosis. A positron emission tomography study. Brain **116:** 655–680.
116. DECARLI, C., D.G. MURPHY, A.R. MCINTOSH, D. TEICHBERG, M.B. SCHAPIRO & B. HORWITZ. 1995. Discriminant analysis of MRI measures as a method to determine the presence of dementia of the Alzheimer type. Psychiatry Res. **57:** 119–130.
117. RADDA, G.K. 1992. Control, bioenergetics, and adaptation in health and disease: noninvasive biochemistry from nuclear magnetic resonance. FASEB J. **6:** 3032–3038.
118. BARBIROLI, B., P. MONTAGNA, P. MARTINELLI, R. LODI, S. IOTTI, P. CORTELLI *et al.* 1993. Defective brain energy metabolism shown by in vivo ^{31}P MR spectroscopy in 28 patients with mitochondrial cytopathies. J. Cereb. Blood Flow Metab. **13:** 469–474.
119. BARBIROLI, B., S. IOTTI & R. LODI. 1998. Aspects of human bioenergetics as studied in vivo by magnetic resonance spectroscopy. Biochimie **80:** 847–853.
120. MATHEWS, P.M., F. ANDERMANN, K. SILVER, G. KARPATI & D.L. ARNOLD. 1993. Proton MR spectroscopic characterization of differences in regional brain metabolic abnormalities in mitochondrial encephalomyopathies. Neurology **43:** 2484–2490.

121. DAGER, S.R., W.L. STRAUSS, K.I. MARRO, T.L. RICHARDS, G.D. METZGER & A.A. ARTRU. 1995. Proton magnetic resonance spectroscopy investigation of hyperventilation in subjects with panic disorder and comparison subjects. Am. J. Psychiatry **152:** 666–672.
122. GYNGELL, M.L., T. MICHAELIS, D. HORSTERMANN, H. BRUHN, W. HANICKE, K.D. MERBOLDT et al. 1991. Cerebral glucose is detectable by localized proton NMR spectroscopy in normal rat brain in vivo. Magn. Reson. Med. **19:** 489–495.
123. SIBSON, N.R., J. SHEN, G.F. MASON, D.L. ROTHMAN, K.L. BEHAR & R.G. SHULMAN. 1998. Functional energy metabolism: in vivo ^{13}C-NMR spectroscopy evidence for coupling of cerebral glucose consumption and glutamatergic neuronal activity. Dev. Neurosci. **20:** 321–330.
124. GRUETTER, R., E.J. NOVOTNY, S.D. BOULWARE, D.L. ROTHMAN & R.G. SHULMAN. 1996. ^{1}H NMR studies of glucose transport in the human brain. J. Cereb. Blood Flow Metab. **16:** 427–438.
125. SONNEWALD, U., L.R. WHITE, E. ODEGARD, N. WESTERGAARD, I.J. BAKKEN, J. AASLY et al. 1996. MRS study of glutamate metabolism in cultured neurons/glia. Neurochem. Res. **21:** 987–993.
126. ROTHMAN, D.L., N.R. SIBSON, F. HYDER, J. SHEN, K.L. BEHAR & R.G. SHULMAN. 1999. In vivo nuclear magnetic resonance spectroscopy studies of the relationship between the glutamate-glutamine neurotransmitter cycle and functional neuroenergetics [In Process Citation]. Philos. Trans. R. Soc. Lond. B Biol. Sci. **354:** 1165–1177.
127. SHEN, J., K.F. PETERSEN, K.L. BEHAR, P. BROWN, T.W. NIXON, G.F. MASON et al. 1999. Determination of the rate of the glutamate/glutamine cycle in the human brain by in vivo ^{13}C NMR. Proc. Natl. Acad. Sci. USA **96:** 8235–8240.
128. YUDKOFF, M., I. NISSIM, Y. DAIKHIN, Z.P. LIN, D. NELSON, D. PLEASURE et al. 1993. Brain glutamate metabolism: neuronal-astroglial relationships. Dev. Neurosci. **15:** 343–530.
129. BATES, T.E., M. STRANGWARD, J. KEELAN, G.P. DAVEY, P.M. MUNRO & J.B. CLARK. 1996. Inhibition of N-acetylaspartate production: implications for 1H MRS studies in vivo. NeuroReport **7:** 1397–1400.
130. THURSTON, J.H., W.R. SHERMAN, R.E. HAUHART & R.F. KLOEPPER. 1989. myo-inositol: a newly identified nonnitrogenous osmoregulatory molecule in mammalian brain. Pediatr. Res. **26:** 482–485.
131. BRAND, A., C. RICHTER-LANDSBERG & D. LEIBFRITZ. 1993. Multinuclear NMR studies on the energy metabolism of glial and neuronal cells. Dev. Neurosci. **15:** 289–298.
132. PETTEGREW, J.W., W.E. KLUNK, K. PANCHALINGAM, R.J. MCCLURE & J.A. STANLEY. 1997. Magnetic resonance spectroscopic changes in Alzheimer's disease. Ann. N. Y. Acad. Sci. **826:** 282–306.
133. KLUNK, W.E., K. PANCHALINGAM, R.J. MCCLURE, J.A. STANLEY & J.W. PETTEGREW. 1998. Metabolic alterations in postmortem Alzheimer's disease brain are exaggerated by Apo-E4. Neurobiol. Aging **19:** 511–515.
134. HWANG, J.H., G.D. GRAHAM, K.L. BEHAR, J.R. ALGER, J.W. PRICHARD & D.L. ROTHMAN. 1996. Short echo time proton magnetic resonance spectroscopic imaging of macromolecule and metabolite signal intensities in the human brain. Magn. Reson. Med. **35:** 633–639.
135. KERMODE, A.G., A.J. THOMPSON, P. TOFTS, D.G. MACMANUS, B.E. KENDALL, D.P. KINGSLEY et al. 1990. Breakdown of the blood-brain barrier precedes symptoms and other MRI signs of new lesions in multiple sclerosis. Pathogenetic and clinical implications. Brain **113:** 1477–1489.
136. TOFTS, P.S. & A.G. KERMODE. 1991. Measurement of the blood-brain barrier permeability and leakage space using dynamic MR imaging. 1. Fundamental concepts. Magn. Reson. Med. **17:** 357–367.

137. DAVIS, T.L., K.K. KWONG, R.M. WEISSKOFF & B.R. ROSEN. 1998. Calibrated functional MRI: mapping the dynamics of oxidative metabolism. Proc. Natl. Acad. Sci. USA **95:** 1834–1839.
138. WILLIAMS, D.S., J.A. DETRE, J.S. LEIGH & A.P. KORETSKY. 1992. Magnetic resonance imaging of perfusion using spin inversion of arterial water [published erratum appears in Proc. Natl. Acad. Sci. USA, 1992, May 1; **89**(9)**:** 4220]. Proc. Natl. Acad. Sci. USA **89:** 212–216.
139. DETRE, J.A., D.C. ALSOP, L.R. VIVES, L. MACCOTTA, J.W. TEENER & E.C. RAPS. 1998. Noninvasive MRI evaluation of cerebral blood flow in cerebrovascular disease. Neurology **50:** 633–641.
140. MANGIARINI, L., K. SATHASIVAM, M. SELLER, B. COZENS, A. HARPER, C. HETHERINGTON *et al.* 1996. Exon 1 of the HD gene with an expanded CAG repeat is sufficient to cause a progressive neurological phenotype in transgenic mice. Cell **87:** 493–506.
141. MATTHEWS, R.T., R.J. FERRANTE, B.G. JENKINS, S.E. BROWNE, K. GOETZ, S. BERGER, Y.I. CHEN & M.F. BEAL. 1997. Iodoacetate produces striatal excitotoxic lesions. J. Neurochem. **69:** 285–289.

Apoptosis and Necrosis in Cerebrovascular Disease

B. JOY SNIDER,[a] FRANK J. GOTTRON, AND DENNIS W. CHOI

Center for the Study of Nervous System Injury and the Department of Neurology, Washington University School of Medicine, St. Louis, Missouri 63110, USA

ABSTRACT: **Neuronal death following ischemic insults has been thought to reflect necrosis. However, recent evidence from several labs suggests that programmed cell death, leading to apoptosis, might additionally contribute to this death. We have used both** *in vitro* **and** *in vivo* **models to study the role of apoptosis in ischemic cell death. Some features of apoptosis (TUNEL staining, internucleosomal DNA fragmentation, sensitivity to cycloheximide) were observed following transient focal ischemia in rats. Brief transient focal ischemia was followed by delayed infarction more than 3 days later; this delayed infarction was sensitive to cycloheximide. A cycloheximide-sensitive component of neuronal cell death was also observed in cultured murine neocortical neurons deprived of oxygen-glucose in the presence of glutamate receptor antagonists. This presumed ischemic apoptosis was attenuated by caspase inhibitors, or by homozygous deletion of the** *bax* **gene. Neurons may undergo both apoptosis and necrosis after ischemic insults, and thus it may be therapeutically desirable to block both processes.**

INTRODUCTION

In recent years, concepts of cell death have evolved, and the idea has emerged that cells can die via two different pathways: a more fulminant necrotic process or a more tightly regulated process called programmed cell death or apoptosis. These types of cell death can, in their purest forms, be distinguished by morphologic criteria, as well as by their sensitivity to pharmacologic and genetic manipulations (see TABLES 1 and 2). The main features of necrosis are early cell and organelle swelling with eventual loss of cytoplasmic membrane integrity and cellular disintegration. Late in necrosis, chromatin may disappear entirely. Because necrotic cells lose membrane integrity and spill their contents into the extracellular space, inflammation often accompanies necrosis.[1] In contrast, apoptosis is characterized by organelle preservation, chromatin and cytoplasmic condensation, and convolution of the nuclear and plasma membranes.[1] Apoptosis was originally described in lymphoid cells[2,3] but appears to be the primary process by which normal cell death occurs during development, or in the maintenance of cell number in tissues with rapid cell turnover such as intestinal epithelium or skin (reviewed in References 4–8).

---

[a]Address for correspondence: B. Joy Snider, Department of Neurology, Campus Box 8111, 660 S. Euclid, St. Louis, Missouri 63110. Phone: 314-747-2107; fax: 314-362-9462.
e-mail: sniderj@neuro.wustl.edu

TABLE 1. Typical morphologic features of necrosis and apoptosis

Necrosis	Apoptosis
Early	
Margination of chromatin	chromatin aggregates in large masses convoluted nuclear outline
Pyknosis (uniformly compacted chromatin)	
Dilatation of the ER	cytoplasmic condensation
Dispersal of ribosomes	blunt protuberances appear on the cell surface
	loss of cell-cell attachment
	some cytoplasmic vacuoles may appear
	organelles crowded but maintain integrity
Later	
Swelling of mitochondrial matrix	nucleus breaks up into discrete fragments
Rupture of nuclear, organelle, plasma membrane	cell surface protuberances seal off to produce apoptotic bodies
Karyorrhexis, small discrete masses of marginated chromatin	
Karyolysis - nucleus loses basophilia leaving nuclear ghost	
Cytoplasmic boundaries indistinct	
Flocculent, granular matrix densities in mitochondria	
Ribosomes, lysosomes dissolve	

Although necrosis is thought to be the predominant form of death triggered by severe cellular injury or extreme environmental perturbation,[1] cell death with many features of apoptosis can occur after cellular injury. It seems likely that in at least some pathological conditions these forms of cell death might coexist in the same tissues, and that elements of both processes might even occur in the same cells. Indeed, this concurrence may account for observed difficulties in arriving at definitive identification of necrosis vs. apoptosis after pathological insults. No single individual morphological, physiological, or biochemical feature appears to be diagnostic of either necrosis or apoptosis, and mixed patterns of features have been described (e.g., see Refs. 9–14).

Whether necrosis or apoptosis predominates and eventually accounts for a given cell's demise may depend on both intrinsic (e.g., cell type or developmental state of the cell[15]) and extrinsic factors (e.g, the nature and severity of the insult, pharmacological interventions[16,17]). If necrosis and apoptosis occur in parallel, blocking one form of cell death may cause a cell to die by the alternative pathway (e.g., see Ref. 18). Further complicating the task of distinguishing necrosis from apoptosis, these death cascades may share some common mediators (FIG. 1). For example, a prominent contributor to ischemic neuronal necrosis is likely excitotoxicity, triggered by glutamate receptor overactivation and leading to intracellular calcium overload[19–22] and the downstream activation of deleterious events including the activation of catabolic enzymes and overproduction of free radicals (reviewed in Refs. 23–25). In cell cultures, even low levels of glutamate receptor overactivation preferentially trigger neuronal necrosis,[26–28] although neuronal apoptosis may also occur.[29] Neuronal

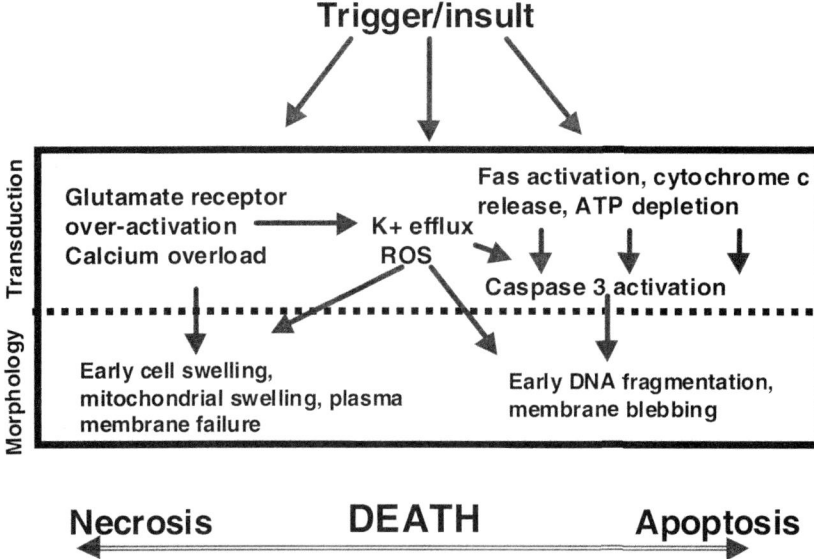

FIGURE 1. Cell death pathways may function in parallel in injured neurons. ROS = reactive oxygen species. See text for discussion.

propensity to undergo apoptosis may be enhanced by glutamate receptor–mediated K^+ efflux[30,31] and free radical production.[32–34]

GLOBAL CEREBRAL ISCHEMIA AND SELECTIVE NEURONAL DEATH

Ischemic insults in the intact brain can result in either the selective death of certain highly vulnerable neurons or tissue infarction involving both neurons and astrocytes (see below). Selective neuronal death is typically induced by transient global ischemia, an interruption of the vascular supply of the entire brain such as occurs during cardiac arrest or severe hypotension. Under these conditions certain neurons— for example, the large pyramidal neurons in the CA1 region of the hippocampus—die, while other nearby neurons and astrocytes remain viable.[35]

A significant body of evidence has accumulated to support a role for programmed cell death in the selective neuronal death that occurs in hippocampal neurons after global ischemia. For example, DNA fragmentation,[36] sensitivity to caspase inhibition,[37,38] and activation of apoptotic cell death pathways occur in vulnerable neurons after global ischemia, in particular involving caspases, a family of proteases that mediate most if not all forms of apoptosis (reviewed in Ref. 39). Caspase 9 is released from mitochondria,[40] and caspase 3 is activated.[38] On the other hand, electron microscopic evidence of neuronal necrosis after global ischemia has also been reported.[41]

TABLE 2. Manipulations capable of reducing central neuronal apoptosis

Manipulation	References
Pharmacological	
Neurotrophins	(75–77)
Protein synthesis inhibitors	(53,54)
Caspase inhibitors	(58,59,61,60)
Free radical scavengers	(78–80)
Elevating extracellular K^+	(30,31)
Raising intracellular Ca^{2+}	(81)
Inhibiting *fas* pathways	(50)
Genetic	
Deletion of proapoptotic genes	(82)
Overexpression of antiapoptotic genes (e.g., *bcl-2*)	(65,67,66)

APOPTOSIS IN *IN VIVO* MODELS OF FOCAL ISCHEMIA

Unlike the selective neuronal death observed in the hippocampus and other particularly vulnerable brain regions after global ischemia, cerebral infarction is a fulminant injury that results in death of both neurons and nonneuronal cells in contiguous areas of brain, usually representing a single vascular territory. This type of injury is most commonly produced experimentally by temporarily or permanently occluding the middle cerebral artery. The well-described pathologic features of infarction—the "ischemic cell process"—include early cell shrinkage and swelling of mitochondria followed by dispersal of the Nissl substance, shrinkage of the nucleus, and the formation of cytoplasmic projections. Finally, a shrunken, pyknotic nucleus without surrounding cytoplasm is the last remnant of the dead neuron.[35] The early mitochondrial swelling and compromise of the cytoplasmic membrane with preservation of nuclear membranes are classical features of necrotic cell death,[42–44] but more recently evidence supporting a role for apoptosis in focal ischemia has surfaced. Even in injuries where frank infarction occurs in the middle of the affected vascular territory, or "core" of the infarction, delayed neuronal death with many features of apoptosis has been observed in the areas surrounding the ischemic core, referred to as the *penumbra*.[17,45] These features include evidence of DNA fragmentation,[46–49] activation of the Fas death-inducing ligand/receptor system,[50] release of cytochrome *c* from injured mitochondria,[51] and caspase 3 activation.[52]

Evidence that apoptotic mechanisms are involved in neuronal death after ischemia is provided by studies demonstrating reduction in neuronal death, and infarct volume, using antiapoptotic maneuvers (summarized in TABLE 2). For example, treatment with the protein synthesis inhibitor cycloheximide reduces infarct volume in rats subjected to transient middle cerebral artery occlusion (MCAO).[53,54] We have subsequently demonstrated that this treatment can be delayed for up to six hours after the onset of ischemia; this suggests that antiapoptotic strategies could be effective even when given long after the onset of ischemia (Du & Choi, manuscript in preparation). These results must be interpreted cautiously, as protein synthesis inhibition could have other effects, such as suppression of prostaglandin synthesis, enhancement of free radical scavengers, and enhanced cerebral vasodilatation after ischemia.[55–57]

Furthermore, ischemic brain damage has been reduced by pharmacological or genetic manipulations that affect molecules involved in the apoptosis cascade, such as caspases or members of the *bcl-2* family. Caspase inhibition reduces infarction volume in transient focal ischemia.[58–60] Interestingly, even when the caspase inhibitors were applied 9 hours after the onset of the ischemia, infarction was still attenuated,[61] raising the hope that such a long treatment window may be possible in human stroke. In addition to their antiapoptotic effects, caspase inhibitors reduce the edema and inflammation that occur following cerebral ischemia. Mice that lack the gene for caspase-1, or interleukin-1b converting enzyme (ICE), have less cerebral edema within 4 hours after middle cerebral artery occlusion and have smaller infarctions 24 hours later.[60] Blockade of the action of interleukin-1b, a proinflammatory cytokine, also reduces infarct volume in rodent models of focal ischemia.[62,63]

An expanding family of proteins with homology to the protooncogene *bcl-2* has been identified in recent years; some members of this family (e.g., *bcl-2*) promote apoptosis, and some (e.g., *bax*) inhibit it (for review see Ref. 64). Transgenic overexpression of *bcl-2* reduces injury in mouse models of both focal and global ischemia,[65,66] as does infection with a recombinant herpes virus directing expression of *bcl-2*.[67] Recent experiments in our laboratory have demonstrated that the deletion of *bax* also reduces infarct volume after focal ischemia (Gottron & Choi, submitted).

Consistent with the idea that both excitotoxic necrosis and apoptosis occur after brain ischemia, combining antiexcitotoxic and antiapoptotic strategies can have additive therapeutic effects. We demonstrated that combining cycloheximide and MK-801 pretreatment reduced the infarct volume after transient MCAO more than either treatment alone.[54] Of particular import for therapeutic applications, caspase inhibition increases the efficacy of NMDA antagonists and prolongs their therapeutic window in mice subjected to focal ischemia.[68,69]

APOPTOSIS IN *IN VITRO* MODELS OF ISCHEMIA

We have used the *in vitro* model of combined oxygen glucose deprivation in cultured murine neocortical neurons to test the hypothesis that apoptotic pathways mediate neuronal death after ischemic insults. When cultures containing neurons and astrocytes are deprived of oxygen and glucose for 30 to 60 minutes, neurons become swollen, the plasma membrane breaks down, and necrosis ensues.[18] This neuronal death is attenuated by blockade of ionotropic glutamate receptors.[70] When cultures are exposed to prolonged (90–120-min) deprivation of oxygen and glucose in the presence of glutamate receptor blockade, a more slowly evolving neuronal death is observed that has features of apoptosis including shrinkage of the neuronal soma, DNA fragmentation, and sensitivity to cycloheximide.[18] The caspase inhibitor Z-Val-Ala-Asp-fluormethylketone (Z-VAD.FMK) also attenuates this death, while the cathepsin B protease inhibitor Z-Phe-Ala fluormethylketone does not.[71]

To determine which caspases were activated in this model of ischemic apoptosis we measured caspase activity in extracts from dying neurons. Each extract was

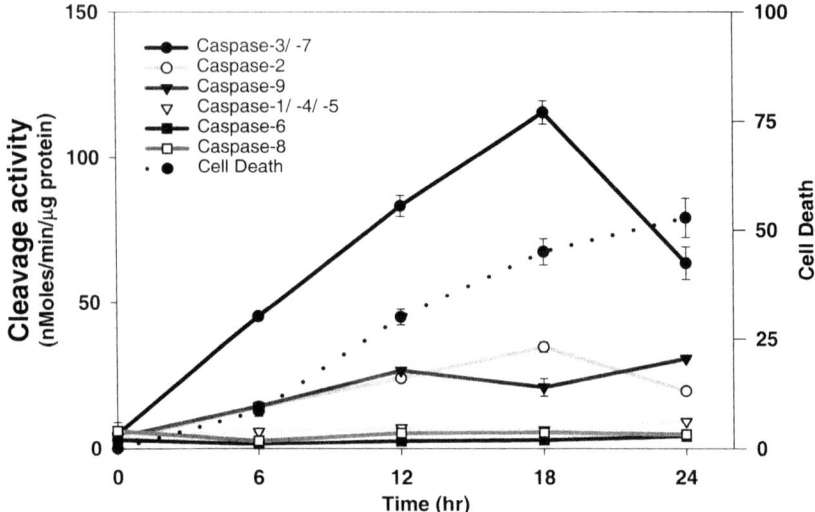

FIGURE 2. Activation of caspases in an *in vitro* model of ischemic apoptosis. Murine neocortical cultures were deprived of oxygen and glucose[74] for 110 minutes in the presence of 10 μM MK-801 and 100 μM 6-cyano-7-nitroquinoxaline-2,3-dione (CNQX). At the indicated time after the termination of oxygen-glucose deprivation, cell lysates were prepared. Lysates were assayed for caspase activity using a battery of fluorometric substrates that are preferentially cleaved by certain caspases. Caspase-1, -4, and -5 activity was detected using WEHD-AFC; caspase-9 by LEHD-AFC; caspase-6 by VEHD-AFC; caspase-8 by LETD-AFC; caspase-2 by DEHD-AFC; and caspase-3, and -7 by DEVD-AFC.[72] Cell death was analyzed by propidium iodide staining. Data is representative of five independent experiments.

tested against a series of substrates that are preferred by various members of the caspase family.[72] During the course of oxygen-glucose deprivation–induced apoptosis there were small increases in caspase-2–like and caspase-9–like activity. There was a large increase in caspase-3–like activity that peaked 18 hours after the end of the oxygen-glucose deprivation, at about the same time as the majority of cells were dying, as evidenced by propidium iodide staining (FIG. 2). There was little or no cleavage of substrates preferred by caspase-1, -4, -5, -6, or -8.

Supporting the idea that ischemia-induced necrosis and apoptosis may share some mechanisms, enhanced expression of stress proteins attenuated both components of cortical neuronal death after oxygen-glucose deprivation *in vitro*.[73]

CONCLUSIONS

Necrosis and apoptosis likely both contribute to cell death after cerebral ischemia. As new treatments for cerebral ischemia are devised, their ability to attenuate both types of cell death should be determined. Therapies targeting shared mechanisms might attenuate both necrosis and apoptosis, or separate therapies might be administered at different points in space or time. For example, glutamate antagonists might be used immediately after ischemia to attenuate excitotoxic necrosis, while antiapoptotic strategies might be employed later on to block delayed neuronal death.

REFERENCES

1. KERR, J.F.R. & B.V. HARMON. 1991. Definition and incidence of apoptosis: an historical perspective. *In* Apoptosis: the Molecular Basis Of Cell Death. L. D. Tomei & F. O. Cope, Eds.: 5–29. Cold Spring Harbor Laboratory Press. Plainview, NY.
2. KERR, J.F., A.H. WYLLIE & A.R. CURRIE. 1972. Apoptosis: a basic biological phenomenon with wide-ranging implications in tissue kinetics. Br. J. Cancer **26:** 239–257.
3. WYLLIE, A.H., J.F.R. KERR & A.R. CURRIE. 1990. Cell death: the significance of apoptosis. Int. Rev. Cytol. **68:** 251–306.
4. WALKER, N.I., B.V. HARMON, G.C. GOBE *et al.* 1988. Patterns of cell death. Methods Achiev. Exp. Pathol. **13:** 18–54.
5. JOHNSON, E.M., JR. & T.L. DECKWERTH. 1993. Molecular mechanisms of developmental neuronal death. Ann. Rev. Neurosci. **16:** 31–46.
6. HENGARTNER, M.O. & H.R. HORVITZ. 1994. Programmed cell death in Caenorhabditis elegans. Curr Opin Genet Dev **4:** 581–586.
7. JACOBSON, M.D., M. WEIL & M.C. RAFF. 1997. Programmed cell death in animal development. Cell **88:** 347–354.
8. VAUX, D.L. & S.J. KORSMEYER. 1999. Cell death in development. Cell **96:** 245–254.
9. CHARRIAUT-MARLANGUE, C. & Y. BEN-ARI. 1995. A cautionary note on the use of the TUNEL stain to determine apoptosis. Neuroreport **7:** 61–64.
10. PORTERA-CAILLIAU, C., J.C. HEDREEN, D.L. PRICE *et al.* 1995. Evidence for apoptotic cell death in Huntington disease and excitotoxic animal models. J. Neurosci **15:** 3775–3787.
11. PETITO, C.K., J. TORRES-MUNOZ, B. ROBERTS *et al.* 1997. DNA fragmentation follows delayed neuronal death in CA1 neurons exposed to transient global ischemia in the rat. J. Cereb. Blood Flow Metab. **17:** 967–976.
12. PORTERA-CAILLIAU, C., D.L. PRICE & L.J. MARTIN. 1997. Excitotoxic neuronal death in the immature brain is an apoptosis–necrosis morphological continuum. J. Comp. Neurol. **378:** 70–87.
13. MEHMET, H., X. YUE, J. PENRICE *et al.* 1998. Relation of impaired energy metabolism to apoptosis and necrosis following transient cerebral hypoxia-ischaemia. Cell Death Differ. **5:** 321–329.
14. FUKUDA, T., H. WANG, H. NAKANISHI *et al.* 1999. Novel non-apoptotic morphological changes in neurons of the mouse hippocampus following transient hypoxic-ischemia. Neurosci. Res. **33:** 49–55.
15. YUE, X., H. MEHMET, J. PENRICE *et al.* 1997. Apoptosis and necrosis in the newborn piglet brain following transient cerebral hypoxia-ischaemia. Neuropathol. Appl. Neurobiol. **23:** 16–25.
16. BONFOCO, E., D. KRAINC, M. ANKARCRONA *et al.* 1995. Apoptosis and necrosis: two distinct events induced, respectively, by mild and intense insults with N-methyl-D-aspartate or nitric oxide/superoxide in cortical cell cultures. Proc. Natl. Acad. Sci. USA **92:** 7162–7166.
17. LI, Y., C. POWERS, N. JIANG ET AL. 1998. Intact, injured, necrotic and apoptotic cells after focal cerebral ischemia in the rat. J. Neurol. Sci. **156:** 119–132.
18. GWAG, B.J., D. LOBNER, J.Y. KOH *et al.* 1995. Blockade of glutamate receptors unmasks neuronal apoptosis after oxygen-glucose deprivation *in vitro*. Neuroscience **68:** 615–619.

19. MELDRUM, B., M. EVANS, T. GRIFFITHS et al. 1985. Ischaemic brain damage: the role of excitatory activity and of calcium entry. Br. J. Anaesth. **57:** 44–46.
20. ROTHMAN, S.M. & J.W. OLNEY. 1986. Glutamate and the pathophysiology of hypoxic-ischemic brain damage. Ann. Neurol. **19:** 105–111.
21. CHOI, D.W. 1988. Calcium-mediated neurotoxicity: relationship to specific channel types and role in ischemic damage. Trends Neurosci. **11:** 465–469.
22. SIESJO, B. & F. BENGTSSON. 1989. Calcium fluxes, calcium antagonists, and calcium-related pathology in brain ischemia, hypoglycemia, and spreading depression: a unifying hypothesis. J. Cereb. Blood Flow Metab. **9:** 127–140.
23. CHOI, D.W. 1988. Glutamate neurotoxicity and diseases of the nervous system. Neuron **1:** 623–634.
24. HALL, E.D. 1997. Brain attack. Acute therapeutic interventions. Free radical scavengers and antioxidants. Neurosurg. Clin. N. Am. **8:** 195–206.
25. SAMDANI, A.F., T.M. DAWSON & V.L. DAWSON. 1997. Nitric oxide synthase in models of focal ischemia. Stroke **28:** 1283–1288.
26. IKEDA, J., S. TERAKAWA, S. MUROTA et al. 1996. Nuclear disintegration as a leading step of glutamate excitotoxicity in brain neurons. J. Neurosci. Res. **43:** 613–622.
27. GWAG, B.J., J.-Y. KOH, J.A. DEMARO et al. 1997. Slowly-triggered excitotoxicity occurs by necrosis in cortical cultures. Neuroscience **77:** 393–401.
28. MACMANUS, J.P., I. RASQUINHA, M.A. BLACK et al. 1997. Glutamate-treated rat cortical neuronal cultures die in a way different from the classical apoptosis induced by staurosporine. Exp. Cell Res. **233:** 310–320.
29. BONFOCO, E., D. KRAINC, M. ANKARCRONA et al. 1995. Apoptosis and necrosis—two distinct events induced, respectively, by mild and intense insults with N-methyl-D-aspartate or nitric oxide superoxide in cortical cell cultures. Proc. Natl. Acad. Sci. USA **92:** 7162–7166.
30. BORTNER, C.D., F.M. HUGHES, JR. & J.A. CIDLOWSKI. 1997. A primary role for K^+ and Na^+ efflux in the activation of apoptosis. J. Biol. Chem. **272:** 32436–32442.
31. YU, S.P., C.H. YEH, S.L. SENSI et al. 1997. Mediation of neuronal apoptosis by enhancement of outward potassium current. Science **278:** 114–117.
32. RATAN, R.R., T.H. MURPHY & J.M. BARABAN. 1994. Oxidative stress induces apoptosis in embryonic cortical neurons. J. Neurochem. **62:** 376–379.
33. WHITTEMORE, E.R., D.T. LOO & C.W. COTMAN. 1994. Exposure to hydrogen peroxide induces cell death via apoptosis in cultured rat cortical neurons. Neuroreport **5:** 1485–1488.
34. GREENLUND, L.J., T.L. DECKWERTH & E.M. JOHNSON, JR. 1995. Superoxide dismutase delays neuronal apoptosis: a role for reactive oxygen species in programmed neuronal death. Neuron **14:** 303–315.
35. ADAMS, J.H. & L.W. DUCHEN. 1992. Greenfield's Neuropathology: 1557. Oxford University Press. New York.
36. MACMANUS, J.P., A.M. BUCHAN, I.E. HILL et al. 1993. Global ischemia can cause DNA fragmentation indicative of apoptosis in rat brain. Neurosci. Lett. **164:** 89–92.
37. GILLARDON, F., B. BOTTIGER, B. SCHMITZ et al. 1997. Activation of CPP-32 protease in hippocampal neurons following ischemia and epilepsy. Brain Res. Mol. Brain Res. **50:** 16–22.
38. CHEN, J., T. NAGAYAMA, K. JIN et al. 1998. Induction of caspase-3-like protease may mediate delayed neuronal death in the hippocampus after transient cerebral ischemia. J. Neurosci. **18:** 4914–4928.
39. NUNEZ, G., M.A. BENEDICT, Y. HU et al. 1998. Caspases: the proteases of the apoptotic pathway. Oncogene **17:** 3237–3245.
40. KRAJEWSKI, S., M. KRAJEWSKA, L.M. ELLERBY et al. 1999. Release of caspase-9 from mitochondria during neuronal apoptosis and cerebral ischemia. Proc. Natl. Acad. Sci. USA **96:** 5752–5757.

41. COLBOURNE, F., G.R. SUTHERLAND & R.N. AUER. 1999. Electron micoscopic evidence against apoptosis as the mechanism of neuronal death in global ischemia. J. Neurosci. **19:** 4200–4210.
42. DEGIROLAMI, U., R.M. CROWELL & F.W. MARCOUX. 1984. Selective necrosis and total necrosis in focal cerebral ischemia. Neuropathologic observations on experimental middle cerebral artery occlusion in the macaque monkey. J. Neuropathol. Exp. Neurol. **43:** 57–71.
43. NAKANO, S., K. KOGURE & H. FUJIKURA. 1990. Ischemia-induced slowly progressive neuronal damage in the rat brain. Neuroscience **38:** 115–124.
44. GARCIA, J.H., K.F. LIU & J.K. RELTON. 1995. Interleukin-1 receptor antagonist decreases the number of necrotic neurons in rats with middle cerebral artery occlusion. Am. J. Pathol. **147:** 1477–1486.
45. DERESKI, M.O., M. CHOPP, R.A. KNIGHT et al. 1993. The heterogeneous temporal evolution of focal ischemic neuronal damage in the rat. Acta Neuropathol.**85:** 327–333.
46. TOMINAGA, T., S. KURE, K. NARISAWA et al. 1993. Endonuclease activation following focal ischemic injury in the rat brain. Brain Res. **608:** 21–26.
47. LI Y., M. CHOPP, N. JIANG et al. 1995. Induction of DNA fragmentation after 10 to 120 minutes of focal cerebral ischemia in rats. Stroke **26:** 1252–1257; discussion 7–8.
48. LINNIK, M.D., J.A. MILLER, J. SPRINKLE-CAVALLO et al. 1995. Apoptotic DNA fragmentation in the rat cerebral cortex induced by permanent middle cerebral artery occlusion. Brain Res. Mol. Brain Res. **32:** 116–124.
49. DU, C., R. HU, C.A. CSERNANSKY et al. 1996. Very delayed infarction after mild focal cerebral ischemia: a role for apoptosis? J. Cereb. Blood Flow Metab. **16:** 195–201.
50. MARTIN-VILLALBA, A., I. HERR, I. JEREMIAS et al. 1999. CD95 ligand (Fas-L/APO-1L) and tumor necrosis factor–related apoptosis-inducing ligand mediate ischemia-induced apoptosis in neurons. J. Neurosc.i **19:** 3809–3817. In press.
51. FUJIMURA, M., Y. MORITA-FUJIMURA, K. MURAKAMI et al. 1998. Cytosolic redistribution of cytochrome c after transient focal cerebral ischemia in rats. J. Cereb. Blood Flow Metab. **18:** 1239–1247.
52. NAMURA, S., J. ZHU, K. FINK et al. 1998. Activation and cleavage of caspase-3 in apoptosis induced by experimental cerebral ischemia. J. Neurosci. **18:** 3659–3668.
53. LINNIK. M.D., R.H. ZOBRIST & M.D. HATFIELD. 1993. Evidence supporting a role for programmed cell death in focal cerebral ischemia in rats. Stroke **24:** 2002–2008.
54. DU, C., R. HU, C.A. CSERNANSKY et al. 1996. Additive neuroprotective effects of dextrorphan and cycloheximide in rats subjected to transient focal cerebral ischemia. Brain Res. **718:** 233–236.
55. RATAN, R.R., T.H. MURPHY & J.M. BARABAN. 1994. Macromolecular synthesis inhibitors prevent oxidative stress-induced apoptosis in embryonic cortical neurons by shunting cysteine from protein synthesis to glutathione. J. Neurosci. **14:** 4385–4392.
56. NAM, M.J., C. THORE & D. BUSIJA. 1995. Rapid induction of prostaglandin synthesis in piglet astroglial cells by interleukin 1 alpha. Brain Res. Bull. **36:** 215–218.
57. VELTKAMP, R., F. DOMOKI, F. BARI et al. 1999. Inhibitors of protein synthesis preserve the N-methyl-D-aspartate–induced cerebral arteriolar dilation after ischemia in piglets. Stroke **30:** 148–152.
58. LODDICK, S.A., A. MACKENZIE & N.J. ROTHWELL. 1996. An ICE inhibitor, z-VAD-DCB attenuates ischaemic brain damage in the rat. Neuroreport **7:** 1465–1468.
59. HARA, H., R.M. FRIEDLANDER, V. GAGLIARDINI et al. 1997. Inhibition of interleukin 1 beta converting enzyme family proteases reduces ischemic and excitotoxic neuronal damage. Proc. Natl. Acad. Sci. USA **94:** 2007–2012.

60. SCHIELKE, G.P., G.Y. YANG, B.D. SHIVERS *et al.* 1998. Reduced ischemic brain injury in interleukin-1 beta converting enzyme– deficient mice. J. Cereb. Blood Flow Metab. **18:** 180–185.
61. FINK, K., J. ZHU, S. NAMURA *et al.* 1998. Prolonged therapeutic window for ischemic brain damage caused by delayed caspase activation. J. Cereb. Blood Flow Metab. **18:** 1071–1076.
62. RELTON, J.K. & N.J. ROTHWELL. 1992. Interleukin-1 receptor antagonist inhibits ischaemic and excitotoxic neuronal damage in the rat. Brain Res. Bull. **29:** 243–246.
63. YAMASAKI, Y., T. SUZUKI, H. YAMAYA *et al.* 1992. Possible involvement of interleukin-1 in ischemic brain edema formation. Neurosci. Lett. **142:** 45–47.
64. YANG, E. & S.J. KORSMEYER. 1996. Molecular thanatopsis: a discourse on the BCL2 family and cell death. Blood **88:** 386–401.
65. MARTINOU, J.C., M. DUBOIS-DAUPHIN, J.K. STAPLE *et al.* 1994. Overexpression of BCL-2 in transgenic mice protects neurons from naturally occurring cell death and experimental ischemia. Neuron **13:** 1017–1030.
66. KITAGAWA, K., M. MATSUMOTO, Y. TSUJIMOTO *et al.* 1998. Amelioration of hippocampal neuronal damage after global ischemia by neuronal overexpression of BCL-2 in transgenic mice. Stroke **29:** 2616–2621.
67. LINNIK, M.D., P. ZAHOS, M.D. GESCHWIND *et al.* 1995. Expression of bcl-2 from a defective herpes simplex virus–1 vector limits neuronal death in focal cerebral ischemia. Stroke **26:** 1670–1674; discussion 5.
68. MA, J., M. ENDRES & M.A. MOSKOWITZ. 1998. Synergistic effects of caspase inhibitors and MK-801 in brain injury after transient focal cerebral ischaemia in mice. Br. J. Pharmacol. **124:** 756–762.
69. SCHULZ, J.B., M. WELLER, R.T. MATTHEWS *et al.* 1998. Extended therapeutic window for caspase inhibition and synergy with MK-801 in the treatment of cerebral histotoxic hypoxia. Cell Death Differ. **5:** 847–857.
70. GOLDBERG, M.P., J.H. WEISS, P.C. PHAM *et al.* 1987. N-methyl-D-aspartate receptors mediate hypoxic neuronal injury in cortical culture. J. Pharmacol. Exp.Ther. **243:** 784–791.
71. GOTTRON, F.J., H.S. YING & D.W. CHOI. 1997. Caspase inhibition selectively reduces the apoptotic component of oxygen-glucose deprivation–induced cortical neuronal cell death. Mol. Cell. Neurosci. **9:** 159–169.
72. THORNBERRY, N.A., T.A. RANO, E.P. PETERSON *et al.* 1997. A combinatorial approach defines specificities of members of the caspase family and granzyme B. Functional relationships established for key mediators of apoptosis. J. Biol. Chem. **272:** 17907–17911.
73. SNIDER, B.J., D. LOBNER, K. YAMADA *et al.* 1998. Conditioning heat stress reduces excitotoxic and apoptotic components of oxygen-glucose deprivation–induced neuronal death in vitro. J. Neurochem. **77:** 120–129.
74. GOLDBERG, M.P. & D.W. CHOI. 1993. Combined oxygen and glucose deprivation in cortical cell culture: calcium-dependent and calcium-independent mechanisms of neuronal injury. J. Neurosci. **13:** 3510–3524.
75. BECK, T., D. LINDHOLM, E. CASTREN *et al.* 1994. Brain-derived neurotrophic factor protects against ischemic cell damage in rat hippocampus. J. Cereb. Blood Flow Metab. **14:** 689–692.
76. CHAN, K.M., D.T. LAM, K. PONG *et al.* 1996. Neurotrophin-4/5 treatment reduces infarct size in rats with middle cerebral artery occlusion. Neurochem. Res. **21:** 763–767.
77. PRINGLE, A.K., L.E. SUNDSTROM, G.J. WILDE *et al.* 1996. Brain-derived neurotrophic factor, but not neurotrophin-3, prevents ischaemia-induced neuronal cell death in organotypic rat hippocampal slice cultures. Neurosci. Lett. **211:** 203–206.

78. TAGAMI, M., K. YAMAGATA, K. IKEDA *et al.* 1998. Vitamin E prevents apoptosis in cortical neurons during hypoxia and oxygen reperfusion. Lab. Invest. **78:** 1415–1429.
79. RAVATI, A., V. JUNKER, M. KOUKLEI *et al.* 1999. Enalapril and moexipril protect from free radical–induced neuronal damage in vitro and reduce ischemic brain injury in mice and rats. Eur. J. Pharmacol. **373:** 21–33.
80. TAGAMI, M., K. IKEDA, K. YAMAGATA *et al.* 1999. Vitamin E prevents apoptosis in hippocampal neurons caused by cerebral ischemia and reperfusion in stroke-prone spontaneously hypertensive rats. Lab. Invest. **79:** 609–615.
81. JOHNSON, E.M., JR., T. KOIKE & J. FRANKLIN. 1992. A "calcium set-point hypothesis" of neuronal dependence on neurotrophic factor. Exp. Neurol. **115:** 163–166.
82. DECKWERTH, T.L., J.L. ELLIOTT, C.M. KNUDSON *et al.* 1996. BAX is required for neuronal death after trophic factor deprivation and during development. Neuron **17:** 401–411.

Effects of Ebselen, a Glutathione Peroxidase Mimic, in Several Models of Mitochondrial Dysfunction

ALAIN BOIREAU, PIERRE DUBEDAT, FRANÇOISE BORDIER,
MADELEINE COIMBRA, MIREILLE MEUNIER, ASSUNTA IMPERATO,
AND SALIHA MOUSSAOUI

Rhône-Poulenc Rorer SA, CRVA, 13 Quai Jules Guesde, 94403 Vitry-sur-Seine, France

It has been hypothesized that mitochondrial-derived pathological free radicals are implicated in various neurodegenerative diseases. Ebselen, a seleno-organic antioxidant with glutathione peroxidase-like action,[1] has been shown to be a potent neuroprotective agent both in humans (in stroke[2] and subarachnoid hemorrhage[3]) and in animal models of neurodegenerative diseases.[4,5] Mitochondria use oxygen at a high rate and are at risk of injury from reactive oxygen species, and peroxidation of membrane lipids may represent a key component of mitochondrial dysfunction.[6] It is also well accepted that the mitochondrial permeability transition (MPT) constitutes a critical event in apoptosis and necrosis.[7] Ca^{2+}, Pi, and oxidants promote onset of the MPT. The aim of the present work was to explore further the mechanism of action of ebselen by using several models of mitochondrial dysfunction in rat liver preparations. First, Fe^{2+}/ascorbate-induced lipid peroxidation was assessed by thiobarbituric reactive substances. Second, Ca^{2+}/Pi-induced mitochondrial swelling due to the opening of the megachannel was estimated by measuring change in optical density. Third, Fe^{2+}/citrate-induced mitochondrial damage was estimated by measuring the decrease in absorbance. The effects of cyclosporine A (CsA, a potent antagonist of the MPT) and butylated hydroxytoluene (BHT, a well-known antioxidant inhibitor of lipid peroxidation, are also reported.

Rat liver mitochondria were prepared from fasted animals by a standard differential centrifugation procedure[8] including three washes in Tris-HCl 0.15 M, pH7.4. Freshly prepared mitochondria were incubated in a medium containing Tris-HCl 0.15 M, in the presence of KH_2PO_4 1 mM at pH 7.4. Drugs were added to the medium. After 3.5 min, oxidative stress was induced by addition of 25 µM $FeSO_4$ followed two minutes later by 0.2 mM ascorbate. Exposure was for 90 min at 37°C; then peroxidation was stopped by lowering the temperature to 0°C. Thiobarbituric acid (final concentration: 3 mM), trichloracetic acid (15%), and HCl 0.25 M were added to each sample in the presence of 0.1% BHT to avoid artifactual malondialdehyde (MDA) formation, and the samples were boiled for 30 min. Then the samples were cooled and centrifuged (20,000 g for 10 min), and the absorbance was measured spectrophotometrically at 532 nm. The standard curve was obtained with MDA as standard. Lipid peroxidation was expressed as nmol MDA per mg of protein. When appropriate, the concentration of compound inhibiting 50% of lipid peroxidation (IC_{50}) was calculated by computer-assisted iterative nonlinear regression analysis, for which the ENZFITTER software package was used.

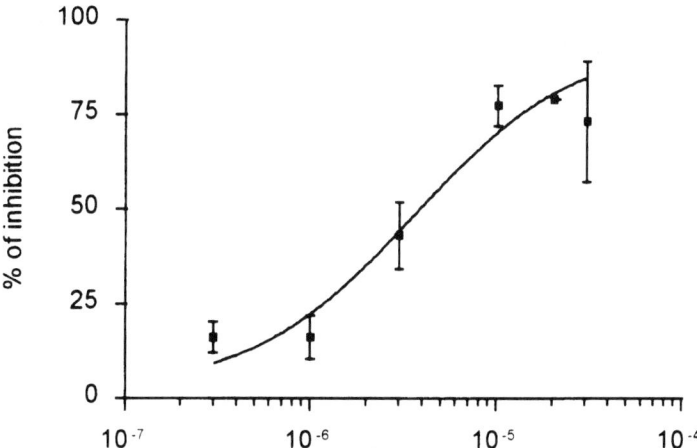

FIGURE 1. Dose-response curve for the inhibitory effect of ebselen on lipid peroxidation induced by Fe^{2+}/ascorbate. Results are data from 12 different experiments.

For the studies of mitochondrial swelling induced by Ca^{2+}/Pi as well as damage induced by Fe^{2+}/citrate, all incubations were conducted at 25°C with about 0.5 mg of mitochondrial protein/ml. The medium contained 200 mM sucrose, 10 mM HEPES, and 1.25 µM rotenone. Swelling was induced by a combination of Ca^{2+} (100 µM) and Pi (1 mM) in the presence of 5 mM succinate. Mitochondrial damage was induced by a combination of Fe^{2+} (25 µM) and citrate (500 µM) in the absence of succinate. In both cases, apparent absorbance was measured at 540 nm in a Hitachi U-2010 spectrophotometer. Quantification of the magnitude of swelling was calculated as the absorbance at the time corresponding to the second-order derivative of the Fe^{2+}/citrate function and expressed as a percentage of baseline absorbance. When appropriate, the IC_{50} value was calculated as indicated above.

Mitochondria exposed to 25 µM Fe^{2+}/0.2 mM ascorbate showed an extent of lipid peroxidation of 31 ± 1 nmol MDA/mg protein (control levels in the absence of the two oxidants: 0.2 ± 0.02 nmol MDA/mg protein). Ebselen potently inhibited iron-induced lipid peroxidation (IC_{50} = 3.5 ± 1.9 µM)(FIG. 1). Lipid peroxidation was also potently inhibited by BHT (IC_{50} = 2.55 ± 0.30 µM). Up to 10 µM, CsA was inactive against lipid peroxidation. CsA, by blocking the megachannel, inhibited the swelling induced by a combination of Ca^{2+} and Pi in the nanomolar range (IC_{50} = 56 ± 6 nM). Ebselen, up to 10 µM and BHT at 100 µM did not antagonize the swelling of mitochondria. As illustrated by FIGURE 2, ebselen protected dose-dependently mitochondria from Fe^{2+}/citrate-induced damage (IC_{50} = 1.4 ± 0.1 µM), but CsA was found to be inactive in this model. BHT also antagonized the effect of Fe^{2+}/citrate (IC_{50} = 0.24 ± 0.08 µM).

This study shows that ebselen protects mitochondria from lipid peroxidation induced by Fe^{2+}/ascorbate. In the present study, ebselen protected mitochondria when membrane damage was induced by Fe^{2+}/citrate, a combination under which a permeabilization is associated with lipid peroxidation.[9] Ebselen does not antagonize

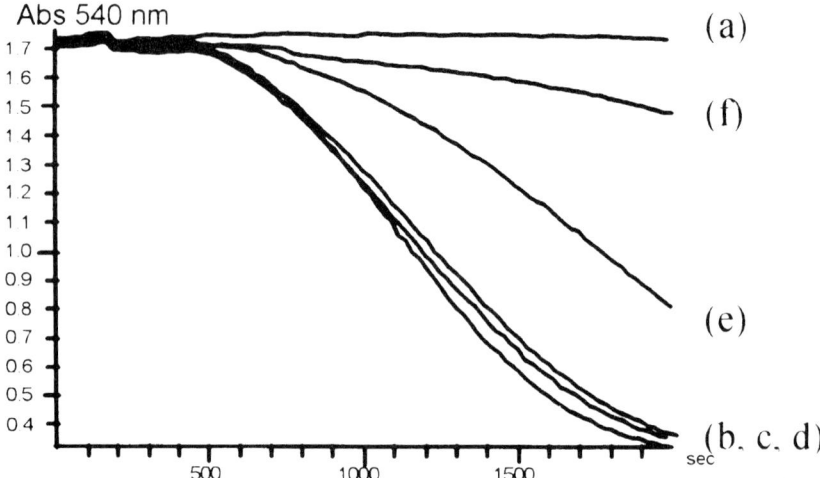

FIGURE 2. Effect of ebselen on Fe^{2+}/citrate-induced mitochondrial damage. Changes in absorbance were measured at 540 nm either in the absence (**b**) or in the presence of various concentrations of ebselen (**c**: 0.1 µM, **d**: 0.3 µM, **e**: 1 µM, or **f**: 3 µM); in (**a**) no Fe^{2+}/citrate.

swelling induced by Ca^{2+} in combination with 1 mM Pi, an experimental situation that favors MPT opening and reveals the potent activity of CsA. As BHT is also inactive, this suggests that an oxidative stress is not involved in Ca^{2+}/Pi-induced swelling.

In conclusion, our findings demonstrate that ebselen protects mitochondria from damage through its direct antioxidant properties and not through a blockade of the megachannel. This suggests that the neuroprotective effect of ebselen previously demonstrated in humans and in animals may be due, at least in part, to its antioxidant properties at the mitochondrial level.

REFERENCES

1. SCHEWE, T. 1995. Molecular actions of ebselen, an antiinflammatory antioxidant. Gen. Pharmacol. **26:** 1153–1169.
2. YAMAGUCHI, T., K. SANO, K. TAKAKURA, I. SAITO, Y. SHINOHARA, T. ASANO & H. YASUHARA. 1998. Ebselen in acute ischemic stroke. A placebo-controlled, double-blind clinical trial. Stroke **29:** 12–17
3. SAITO, I., T. ASANO, K. SANO, K. TAKAKURA, H. ABE, T. YOSHIMOTO, H. KIKUCHI, T. OHTA & S. ISHIBASHI. 1998. Neuroprotective effect of an antioxidant, ebselen, in patients with delayed neurological deficits after aneurysmal subarachnoid hemorrhage. Neurosurgery **42:** 269–278.
4. DAWSON, D.A., H. MASAYASU, D.I. GRAHAM & I.M. MACRAE. 1995. The neuroprotective efficacy of ebselen (a glutathione peroxidase mimic) on brain damage induced by transient focal cerebral ischaemia in the rat. Neurosci. Lett. **185:** 65–69.
5. TAKASAGO, T., E.E. PETERS, D.I. GRAHAM, H. MASAYASU & I.M. MACRAE. 1997. Neuroprotective efficacy of ebselen, an anti-oxidant with anti-inflammatory actions, in a

rodent model of permanent middle cerebral artery occlusion. Br. J. Pharmacol. **122:** 1251–1256.
6. BINDOLI, A. 1988. Lipid peroxidation in mitochondria. Free Radical Biol. Med. **5:** 247–261.
7. ZAMZAMI, N., T. HIRSCH, B. DALLAPORTA, P.X. PETIT & G. KROEMER. 1997. Mitochondrial implication in accidental and programmed cell death: apoptosis and necrosis. J. Bioenerg. Biomemb. **29:** 185–193.
8. PUSHPENDRAN, C.K., M. SUBRAMANIAN, T.P. DEVASAGAYAM & B.B. SINGH. 1998. Study on lipid peroxidation potential in different tissues induced by ascorbate-Fe^{2+}: possible factors involved in their differential susceptibility. Mol. Cell. Biochem. **178:** 197–202.

EDTA-Induced Monovalent Fluxes through the Ca^{2+} Uniporter in Brain Mitochondria

NICKOLAY BRUSTOVETSKY AND JANET M. DUBINSKY[a]

Departments of Physiology and Neuroscience, University of Minnesota, Minneapolis, Minnesota 55455, USA

The low ion permeability of the inner mitochondrial membrane is a crucial prerequisite for normal mitochondrial functioning. Considering the high, negative mitochondrial membrane potential ($\Delta\psi$), induction of a permeability pathway for protons and/or monovalent cations might have disastrous consequences, inhibiting oxidative phosphorylation and altering mitochondrial Ca^{2+} buffering capacity. A rapid increase of cytosolic Ca^{2+}, accompanying neurodegenerative disorders, may cause mitochondrial dysfunction due to various permeability mechanisms involving activation of Ca^{2+} cycling and/or induction of the mitochondrial permeability transition (mPT) pore. Induction of the mPT is a candidate intermediate step in both necrotic and apoptotic cell death pathways.[1] Recently, we reported Ca^{2+}-dependent induction of a novel low-conductance proton-permeable pathway linked to the Ca^{2+} uniporter.[2,3] Under normal conditions with low cytosolic calcium the Ca^{2+} uniporter acts as a Ca^{2+}-selective *proton-impermeable* channel. Under some circumstances the Ca^{2+} uniporter becomes permeable for monovalent cations as well. A rapid increase in extramitochondrial Ca^{2+} caused sustained mitochondrial depolarization, probably due to an induction of proton permeability.[2,3] Repolarization by ruthenium red (RR), a potent inhibitor of the Ca^{2+} uniporter, linked this pathway to the uniporter.[2,3] An inverse correlation between Ca^{2+} fluxes and the sustained depolarization ruled out a substantial contribution of Ca^{2+} cycling to the sustained depolarization.[2,3] Mg^{2+} also significantly influences operation of the Ca^{2+} uniporter. Bernardi *et al.*[4] reported an electrogenic Na^+ influx into liver mitochondria that could be induced by removal of external Mg^{2+}. Because this Na^+ influx was sensitive to inhibition by RR and La^{3+}, Kapus *et al.*[5] suggested that the Mg^{2+}-dependent Na^+ channel and the Ca^{2+} uniporter could be identical. In the present work we investigated whether Mg^{2+}-regulated, monovalent cation permeability could be detected in brain mitochondria, similarly linked to the Ca^{2+} uniporter. Mitochondrial membrane permeability was evaluated by monitoring membrane potential with tetraphenylphosphonium (TPP^+) and a TPP^+-sensitive electrode in the mitochondrial suspension. Sequestration of external Mg^{2+} by 0.5 mM EDTA in NaCl medium partially depolarized mitochondria, suggesting induction of a proton or Na^+ permeability (FIG. 1A). RR prevented the EDTA-induced depolarization and partially repolarized EDTA-treated mitochondria. Pretreatment with the divalent cation ionophore A23187 depleted intramitochondrial Mg^{2+} by providing a pathway for matrix Mg^{2+} efflux. In the presence of

---

[a]Address for correspondence: Janet M. Dubinsky, Ph.D. University of Minnesota, Department of Neuroscience, 6-145 Jackson Hall, 321 Church Street SE, Minneapolis, Minnesota 55455. Phone: 612-635-8447; fax: 612-625-5009.
   e-mail: dubin001@tc.umn.edu

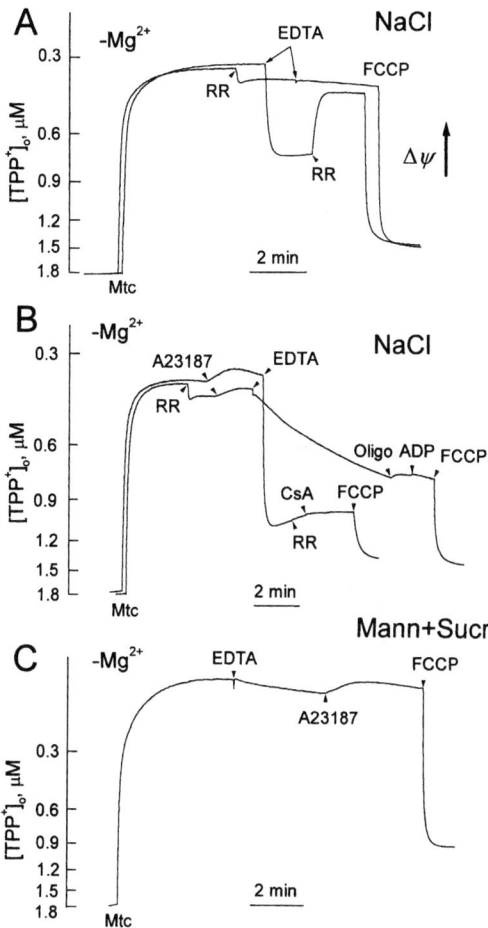

FIGURE 1. Monovalent cation fluxes through the Ca^{2+} uniporter. (**A**) EDTA induced depolarizations in isolated brain mitochondria in NaCl medium that was antagonized by RR. (**B**) A23187 exacerbated the EDTA-induced depolarization, diminishing the ability of RR to prevent depolarization or restore Dy. (**C**) In sucrose-mannitol medium, EDTA failed to depolarize CNS mitochondria. Experiments were performed in medium containing either 100 mM NaCl (**A,B**) or 215 mM mannitol, 50 mM sucrose (**C**), and 3 mM Pi, 3 mM succinate, 10 mM HEPES/Tris, pH 7.4, at 30°C under continuous stirring. $\Delta\psi$ was followed by monitoring the distribution of TPP^+ between the external medium and the mitochondrial matrix with a TPP^+-sensitive electrode. Additions: 0.5 mM EDTA, 1 μM ruthenium red (RR), 1 μM A23187, 1 μM cyclosporin A (CsA), 1 μM oligomycin (Oligo), 100 μM ADP, 1 μM FCCP.

A23187, EDTA sequestered both external and internal Mg^{2+}, inducing a stronger depolarization (FIG. 1B). In these conditions RR slowed the EDTA-induced depolarization but neither completely prevented depolarization nor repolarized mitochondria. Cyclosporin A, an inhibitor of the mPT pore, or a combination of oligomycin plus ADP was also unable to restore potential. A similar depolarization, but with 1.25 mM EDTA, was observed in a medium where 100 mM KCl replaced the NaCl. In contrast, in mannitol-sucrose medium EDTA alone or in combination with A23187 did not depolarize mitochondria (FIG. 1C). In this nonionic medium, protons apparently could not flow through the EDTA-activated pathway. The EDTA-induced depolarizations were attributable to an increased Na^+ (or K^+) permeability. The effects of RR indicated involvement of the Ca^{2+} uniporter in the increased permeability. Removal of Mg^{2+} from brain mitochondria transformed the Ca^{2+} uniporter into a channel for monovalent cations, but not for protons, comparable to liver mitochondria.[5] In contrast to liver mitochondria, EDTA alone induced moderate depolarization of CNS mitochondria in KCl medium. The responses of brain mitochondria differed from those of liver mitochondria by the lack of RR-induced repolarization in KCl medium after treatment with A23187 and EDTA.[5] These results suggest that the structure of the Ca^{2+} uniporter may be flexible and perhaps more vulnerable in brain than in liver. Mg^{2+}, or Mg^{2+} plus RR, appeared to be required to maintain the uniporter in a closed state. In the absence of Mg^{2+}, monovalent cations could traverse the inner mitochondrial membrane. One might expect that Ca^{2+} could also interact with the uniporter at this divalent-sensitive site to alter uniporter selectivity. Following rapid elevation of cytosolic Ca^{2+}, proton leak through the uniporter was observed under conditions restricting electron flow through the respiratory chain.[2,3] Since Ca^{2+} and Mg^{2+} often interact at the same site, under normal conditions, the balance of these divalent cations may regulate the selectivity of the Ca^{2+} uniporter. Since Ca^{2+} rises are more common than Mg^{2+} depletions, the proton permeability may be the physiologically relevant variant of uniporter function. Ca^{2+}-induced proton leak through the Ca^{2+} uniporter may depolarize mitochondria, inhibit oxidative phosphorylation, slow Ca^{2+} sequestration by mitochondria, decrease production of reactive oxygen species, and diminish the probability of mPT pore induction.

ACKNOWLEDGMENTS

This work was supported by NIA AG10034 to J.M.D. and by AHA Fellowship (Minnesota Affiliate) 9804691X to N.B.

REFERENCES

1. SCHINDER, A.F. et al. 1996. Mitochondrial dysfunction is a primary event in glutamate neurotoxicity. J. Neurosci. **16:** 6125–6133.
2. BRUSTOVETSKY, N. & J.M. DUBINSKY. 1998. Does the Ca^{2+} uniporter form the mitochondrial permeability transition pore? Biophys. J. **74:** A384.
3. BRUSTOVETSKY, N. & J.M. Dubinsky. 1999. Role of the Ca^{2+} uniporter in the Ca^{2+}-induced proton permeability of brain mitochondria. J. Neurochem. **72** (Suppl.): S39.
4. BERNARDI, P. et al. 1990. A gated pathway for electrophoretic Na^+ fluxes in rat liver mitochondria: regulation by surface Mg^{2+}. Eur. J. Biochem. **188:** 91–97.
5. KAPUS, A. et al. 1990. Ruthenium red inhibits mitochondrial Na^+ and K^+ uniports induced by magnesium removal. J. Biol. Chem. **265:** 18063–18066.

Signaling Events in NMDA Receptor–Induced Apoptosis in Cerebrocortical Cultures

SAMANTHA L. BUDD[a] AND STUART A. LIPTON

*Center for Neuroscience and Aging, The Burnham Institute,
10901 North Torrey Pines Road, La Jolla, California 92073, USA*

Increasing evidence suggests that the selective loss of neurons in neurodegenerative and acute neurologic disorders is mediated by apoptosis.[1–3] The overstimulation of glutamate receptors can stimulate the death of neurons by either necrosis or apoptosis, depending on the severity of the insult.[4] Activation of the *N*-methyl-D-aspartate (NMDA) glutamate receptor in particular, results in a prolonged increase in the intracellular free calcium concentration $[Ca^{2+}]_i$. The increase in $[Ca^{2+}]_i$ leads to the activation of many potentially toxic Ca^{2+}-dependent events. Little is known about the relative importance of each of these events in stimulating the death of the neurons. Mitochondria, which have potentially deleterious calcium loading following NMDA receptor activation,[5] are also recognized as having an important function in signaling to apoptosis. Mitochondria can activate caspases by releasing proapoptotic molecules such as cytochrome c[6,7] and the apoptosis-inducing factor (AIF).[8] The mechanism by which signaling molecules are released from the mitochondria is currently unknown. One possibility includes release following opening of the mitochondrial permeability transition pore (mPTP).[9] While not all of the molecular components of the mPTP have been defined, the adenine nucleotide translocator (ANT), which resides at the inner mitochondrial membrane, is implicated as a member of this complex. Two compounds, bongkrekic acid (BA) and cyclosporin A (CsA) (through its interaction with a matrix cyclophilin), both inhibit mPT. BA, which is produced by *P. cocovenans*, is a specific inhibitor of the ANT and is toxic when this organism contaminates a food product. Also, BA has been shown to prevent apoptosis in a variety of cell types.[8,9]

A unifying element in the induction of mPT seems to be the presence of high Ca^{2+}. The present study has investigated whether NMDA receptor–induced apoptosis, known to involve high $[Ca^{2+}]$, is mediated by mitochondrial signaling. The model is compared with staurosporine-induced apoptosis. The use of these two systems has facillitated the elucidation of separate pathways leading to neuronal cell death.

RESULTS AND DISCUSSION

Neurons in mixed cerebrocortical cultures die by apoptosis when incubated for 20 minutes with 300 μM NMDA plus 5 μM glycine in the absence of extracellular Mg^{2+} (FIG. 1a). At 18 hours 20% of cells have condensed nuclei, a hallmark of apo-

---

[a]Corresponding author. Phone: 858-646-3100; fax: 858-713-6273.
e-mail: Sbudd@burnham-inst.org

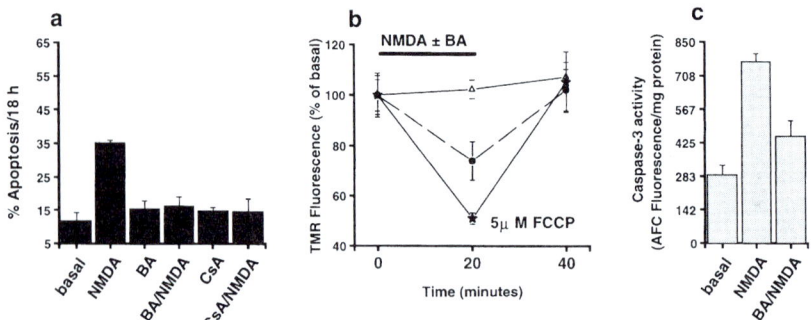

FIGURE 1. (a) Effect of BA on NMDA–induced apoptosis. Cultures were exposed to 300 μM NMDA/5 μM glycine ± 1 μM bongkrekic acid or 10 μM (20 min). At 18 hours the number of cells with condensed nuclei was assessed by staining with propidium iodide. (b) Determination of $\Delta\psi_m$ (mitochondrial TMR fluorescence). NMDA receptor activation results in a partial depolarization of $\Delta\psi_m$ (●), which is prevented by inclusion of BA (△). FCCP (★) further decreases TMR fluorescence intensity. (c) Effect of BA on NMDA receptor–induced activation of caspase-3 in cerebrocortical cell lysates. NMDA induces activation of caspase-3 at 6 hours ($p < 0.05$). No increase in caspase-3 activity is detected following NMDA + BA.

ptosis (FIG. 1a). Inhibition of the mitochondrial ANT with 1 μM BA for 10 minutes prior to and during exposure to NMDA/glycine reduces the extent of apoptosis (FIG. 1a, mean number of apoptotic nuclei 4.81% above control). CsA at 10 μM is also able to decrease the extent of NMDA receptor–induced apoptosis (FIG. 1a). CsA is known to affect multiple sites, and its protection cannot be simply attributed to its inhibition of the mPT. One consequence of NMDA receptor activation in neurons is a large influx of Ca^{2+}, resulting in an increase of $[Ca^{2+}]_i$ to at least 1 μM in these cultures. It is known that mitochondrial Ca^{2+} accumulation occurs above an external free $[Ca^{2+}]$ of 0.3–0.5 μM, and may interfere with $\Delta\psi_m$.[10] For this reason, an estimation of $\Delta\psi_m$ was made by assessing changes in the fluorescence intensity of tetramethylrhodamine (TMR) accumulated by individual mitochondria within cerebrocortical neurons. At low, nonquenching concentrations of TMR, depolarizations of $\Delta\psi_m$ are accompanied by both a decrease in TMR fluorescence and a loss of punctate localization. NMDA receptor activation produces a 26% decrease in TMR fluorescence intensity (Fig. 1b); the depolarization of $\Delta\psi_m$ induced by NMDA receptor activation is not complete, because the protonophore FCCP can further decrease TMR fluorescence intensity and fully dissipate its punctate localization (FIG. 1b). Following removal of NMDA or FCCP, TMR fluorescence returns to basal levels within 30 seconds, indicating a rapid repolarization of $\Delta\psi_m$. Incubation with BA completely abrogates the NMDA-induced decrease in $\Delta\psi_m$ (FIG. 1b). However, BA does not prevent an FCCP-induced $\Delta\psi_m$ depolarization (not shown).

Morphological criteria such as nuclear condensation mark relatively late events in the cell death pathway. The activation of cysteine proteases (caspase enzymes) occurs prior to these changes. Caspase-3 is recognized as an effector caspase, and its activation represents an important commitment to the subsequent death of cells.

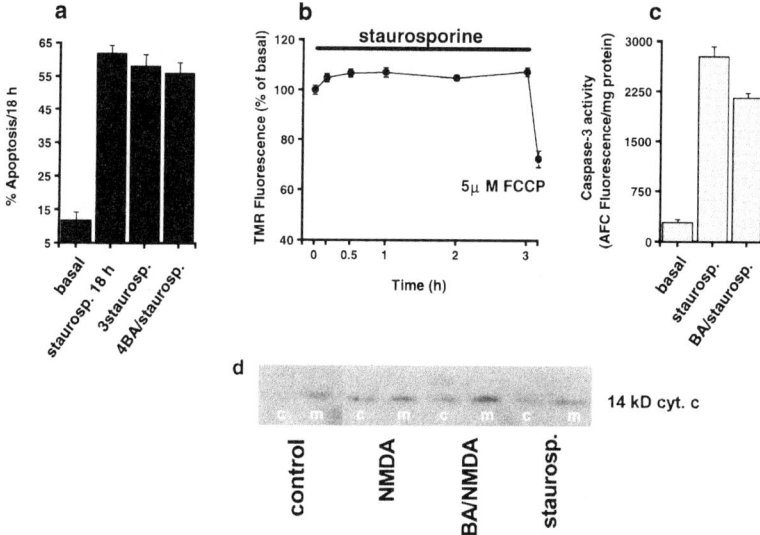

FIGURE 2. (a) Effect of BA on staurosporine-induced apoptosis. Cultures were exposed to 0.5 µM staurosporine ± 1 µM bongkrekic acid (60 min). At 18 hours the number of cells with condensed nuclei was assessed by staining with propidium iodide. BA does not prevent staurosporine-induced apoptosis. (b) Determination of $\Delta\psi_m$. Staurosporine does not depolarize mitochondria for at least 3 hours, at which time $\Delta\psi_m$ is still sensitive to FCCP. (c) Effect of BA on staurosporine-induced activation of caspase-3 in cerebrocortical cell lysates. Staurosoporine induces a large increase in caspase-3 activation that is unaffected by the presence of 1 µM BA. (d) Cytochrome c is detected in cytoplasmic fractions (c) at 6 hours after NMDA or staurosporine. BA does not prevent NMDA-induced release of cytochrome c.

Caspase-3 proteolytic activity in cerebrocortical culture lysates is detected 6 hours after removal of NMDA (FIG. 1c), whereas cultures treated with BA during NMDA receptor activation do not show a significant increase in caspase-3 activity (FIG. 1c). The release of proapoptotic factors such as cytochrome c from mitochondria is implicated in the stimulation of apoptosis by a variety of inducing agents. In the present study, cytochrome c is found in cytoplasmic fractions 6 hours after NMDA, though most remains in the mitochondrial or membrane-associated fraction (FIG. 2d). BA does not prevent the movement of cytochrome c into the cytoplasm (FIG. 2d), which indicates that release of cytochrome c in the present preparation is not mediated by the mPTP. The presence of cytochrome c in the cytoplasmic compartment is not sufficient for the activation of caspase-3 and induction of apoptosis.

The mycotoxin staurosporine induces apoptosis in many different cell types. While its precise modus operandi is unknown, it is often used as a positive standard. When added to cerebrocortical cultures, 0.5 µM staurosporine stimulates the death of up to 50% of cells (FIG. 2a). This greater effect than NMDA presumably reflects the ability of staurosporine to induce apoptosis in all cell types in these cultures (neuronal content approximately 20–30%). Inclusion of 1 µM BA during a 1-hour exposure to 0.5 µM staurosporine, does not reduce the extent of apoptosis (FIG. 2b). Staurosporine has no effect on $\Delta\psi_m$ for up to 3 hours (FIG. 2b); at the same time caspase-3 activation is greatly enhanced (FIG. 2c). After 3 hours' incubation with staurosporine, the mitochondria are still able to respond to depolarization by FCCP (FIG. 2b). This indicates that mitochondrial $\Delta\psi_m$ depolarization is not required for the induction of apoptosis. Incubation with staurosporine stimulates a greater increase in caspase-3 activity than that seen in NMDA-treated cultures (FIG. 2c). BA has little effect on staurosporine-induced activation of caspase-3 (FIG. 2c). Staurosporine is also associated with a partial release of cytochrome c into the cytoplasm (FIG. 2d).

The present results show that NMDA receptor–induced apoptosis is associated with $\Delta\psi_m$ depolarization, caspase-3 activation, and cytochrome c release. BA protection against NMDA receptor–induced apoptosis is related to its ability to prevent $\Delta\psi_m$ depolarization, but this is not correlated with prevention of release of cytochrome c. The lack of BA effect upon staurosporine-induced apoptosis, coupled with the observation that staurosporine does not induce $\Delta\psi_m$ depolarization, indicates that an alternative, or nonmitochondrial, pathway exists to induce apoptosis, and mitochondrial $\Delta\psi_m$ depolarization per se is not required for cytochrome c release or caspase-3 activation.

REFERENCES

1. PORTERA-CAILLIAU, C. *et al.* 1995. Evidence for apoptotic cell death in Huntington disease and excitotoxic animal models. J. Neurosci. **15:** 3775–3787.
2. BARINAGA, M. 1998. Is apoptosis key in Alzheimer's disease? Science **271:** 1303–1304.
3. BURKE, R.E. & N.G. KHOLODILOV. 1998. Programmed cell death: does it play a role in Parkinson's disease? Ann. Neurol. **44:** S126–S133.
4. BONFOCO, E. *et al.* 1995. Apoptosis and necrosis: two distinct events induced respectively by mild and intense insults with NMDA or nitric oxide/superoxide in cortical cell cultures. Proc. Natl. Acad. Sci. USA **92:** 7162–7166.
5. BUDD, S.L. & D.G. NICHOLLS. 1996. Mitochondria, calcium and acute glutamate excitotoxicity in cultured cerebellar granule cells. J. Neurochem. **66:** 403–411.
6. KLUCK, R.M. *et al.* 1997. The release of cytochrome c from mitochondria: a primary site for Bcl-2 regulation of apoptosis. Science **275:** 1132–1136.
7. YANG, J. *et al.* 1997. Prevention of apoptosis by Bcl-2: release of cytochrome c from mitochondria blocked. Science **275:** 1129–1132.
8. MARCHETTI, P. *et al.* 1996. Mitochondrial permeability transition is a central coordinating event of apoptosis. J. Exp. Med. **184:** 1155–1160.
9. MARCHETTI, P. *et al.* 1996. Mitochondrial permeability transition triggers lymphocyte apoptosis. J. Immunol. **157:** 4830–4836.
10. ÅKERMAN, K. 1978. Changes in membrane potential during calcium ion influx and efflux across the mitochondrial membrane. Biochim. Biophys. Acta **502:** 359–366.

In Vitro and *in Vivo* Protein Oxidation Induced by Alzheimer's Disease Amyloid β-Peptide (1-42)

D. ALLAN BUTTERFIELD,[a,c] SERVET M. YATIN,[a] AND CHRISTOPHER D. LINK[b]

[a]*Department of Chemistry, Center of Membrane Sciences, and Sanders-Brown Center on Aging, University of Kentucky, Lexington, Kentucky 40506, USA*

[b]*Institute for Behavioral Genetics, University of Colorado, Boulder, Colorado 80309, USA*

Amyloid β-peptide (Aβ) is thought by many researchers to be central to the pathogenesis of Alzheimer's disease (AD) (reviewed in Ref. 1). In addition, oxidative stress, manifested by protein oxidation and lipid peroxidation, is apparent in AD brain.[2,3] Our laboratory developed a comprehensive hypothesis for neurotoxicity in AD brain that unites these two observations and provides a testable framework for much of the AD literature. We proposed an Aβ-associated free radical oxidative stress model for neuronal death in AD brain[2] (FIG. 1). In AD brain, the predominant forms of Aβ are Aβ(1-40) and Aβ(1-42). Consistent with our model and in ways completely inhibited by free radical scavengers (antioxidants), Aβ leads to lipid peroxidation[4,5] and protein oxidation[6–8] in various brain membrane systems; generates reactive oxygen species (ROS);[7,8] inhibits hippocampal neuronal and cortical synaptosomal membrane ion-motive ATPases, including Na^+/K^+-ATPase and Ca^{2+}-ATPase; blocks glutamate uptake and inhibits the activity of glutamine synthetase (both of the latter Aβ-induced alterations have the effect of increasing excitotoxic glutamate levels); causes intracellular Ca^{2+} levels to increase dramatically;[8] and leads to neurotoxicity in hippocampal neuronal or astrocytic cultures (reviewed in Ref. 2).

A prediction of the Aβ-associated free radical oxidative stress model for neurotoxicity in AD brain is that Aβ(1-42), the predominant form of Aβ found in AD, will induce protein oxidation. A key marker of protein oxidation is protein carbonyl content.[9] Previous studies showed increased antioxidant-inhibited protein oxidation in hippocampal neuronal cultures induced by Aβ(1-40)[8] and Aβ(25-35).[6,7] In the current study, we provide evidence for Aβ(1-42)–induced ROS generation *in vitro* and protein oxidation *in vitro* and *in vivo*. In agreement with our model (FIG. 1), 10 μM Aβ(1-42) added to cultured hippocampal neurons led to ROS formation that was inhibited by vitamin E (Fig. 2A) and induced significantly greater protein oxidation than in controls (Fig. 2B). In addition to the *in vitro* studies, *in vivo* studies were carried out. We reported earlier that AD brain regions rich in Aβ-containing senile plaques had significantly increased protein oxidation but Aβ-poor cerebellum did

---

[c]Address for correspondence: Professor D. Allan Butterfield, Department of Chemistry, Center of Membrane Sciences, and Sanders-Brown Center on Aging, University of Kentucky, Lexington, Kentucky 40506-0055. Phone: 606-257-3184; fax: 606-257-5876.

e-mail: dabcns@pop.uky.edu

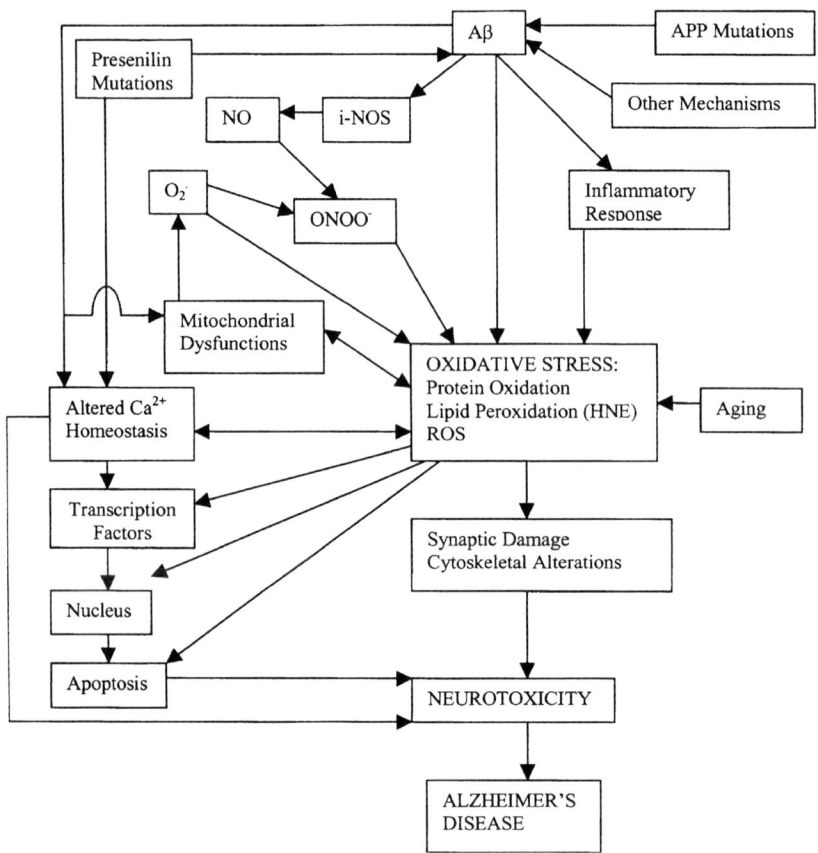

FIGURE 1. Flow diagram of our comprehensive model for Aβ-associated free radical oxidative stress–induced neurotoxicity in Alzheimer's disease brain. See Ref. 2 for a review and greater details.

not.[10] If our model is correct, then one may predict that transgenic animals overexpressing Aβ(1-42) should show increased protein oxidation *in vivo*. *Caenorhabditis elegans (C. elegans)* transgenic animals expressing full-length Aβ(1-42) were produced,[11] and protein oxidation was determined. In agreement with predictions of our model and with our earlier studies in AD brain,[10] Aβ(1-42)–expressing animals had significantly increased protein oxidation *in vivo* (FIG. 2C). To gain some insight into potential molecular mechanisms by which Aβ(1-42) led to protein oxidation *in vivo*, methionine was mutated to cys in this *in vivo* model of Aβ(1-42) expression. Consistent with previous *in vitro* studies of methionine substitution in Aβ(25-35) and Aβ(1-40) (2,13), no *in vivo* protein oxidation was found.

These findings are consistent with the Aβ-associated free radical oxidative stress model of neurotoxicity in AD brain[2] (FIG. 1). Other sequelae of Aβ(1-42)–induced *in vitro* and *in vivo* oxidative stress and their inhibition by antioxidants are currently

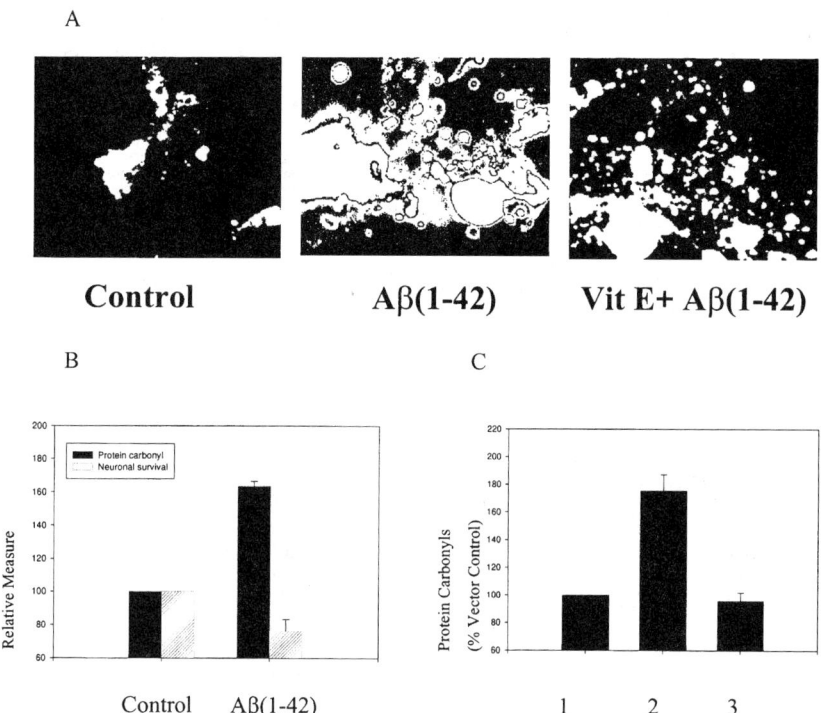

FIGURE 2. A. Reactive oxygen species production in cultured hippocampal neurons to which Aβ(1-42) had been added. ROS are assessed by fluorescence of 2,7-dicholorofluorescein, formed by reaction of peroxyl radicals or hydrogen peroxide to the DCF dye employed. **B.** Protein carbonyls *(dark bars)*, a measure of protein oxidation, and cell survival *(lighter bars)* of hippocampal neurons to which Aβ(1-42) had been added. Percent increased protein carbonyls in Aβ(1-42)–treated neurons over that of controls; mean ± SEM: 163 ± 2%, $p < 0.01$, $n = 3$. Percent cell survival was decreased significantly in Aβ(1-42)–treated neurons (76.3% of control cells, $p < 0.01$, $n = 3$). **C.** *In vivo* protein oxidation was found in *C. elegans* transgenic animals expressing full-length Aβ(1-42). (1) Protein carbonyls in vector control animals were assigned a value of 100%, $n = 5$. (2) Percent increased protein carbonyls over that of vector control; mean ± SEM: 176 ± 3%, $p < 0.001$, $n = 5$). (3) Protein carbonyls in transgenic animals in which methionine residue 35 in Aβ(1-42) was mutated to cysteine were equal to those of vector controls—e.g., no increase in protein oxidation was found.

under investigation. These current and ongoing studies may provide additional insight into AD pathogenesis and therapeutic strategies.

ACKNOWLEDGMENTS

This work was supported in part by NIH grants AG-051191 and AG-10836.

REFERENCES

1. SELKOE, D. 1994. Alzheimer's disease: a central role of amyloid. J. Neuropathol. Exp. Neurol **53:** 438–447.
2. BUTTERFIELD, D.A. 1997. β-amyloid-associated free radical oxidative stress and neurotoxicity: implications for Alzheimer's disease. Chem. Res. Toxicol. **10:** 495–506.
3. MARKESBERY. W.R. 1997. Oxidative stress hypothesis in Alzheimer disease. Free Radical Biol. Med. **23:**134–147.
4. BUTTERFIELD, D.A., K. HENSLEY, M. HARRIS, M. MATTSON & J. M. CARNEY. 1994. β-Amyloid peptide free radical fragments initiate synaptosomal lipoperoxidation in a sequence-specific fashion: implications to Alzheimer's disease. Biochem. Biophys. Res. Commun. **200:** 710–715.
5. KOPPAL, T., R. SUBRAMANIAM, J. DRAKE, M.R. PRASAD & D.A. BUTTERFIELD. 1998. Vitamin E protects against Alzheimer's amyloid peptide (25-35)–induced changes in neocortical synaptosomal membrane lipid structure and composition. Brain Res. **786:** 270–273.
6. SUBRAMANIAM, R., T. KOPPAL, M. GREEN, S. YATIN, B. JORDAN, J. DRAKE & D.A. BUTTERFIELD. 1998. The free radical antioxidant vitamin E protects cortical synaptosomal membranes from amyloid beta-peptide (25-35) toxicity but not from hydroxynonenal toxicity: relevance to the free radical hypothesis of Alzheimer's disease. Neurochem. Res. **23:** 1403–1410.
7. YATIN, S.M., M. AKSENOV & D.A. BUTTERFIELD. 1999. The antioxidant vitamin E modulates amyloid β-peptide–induced creatine kinase activity inhibition and increased protein oxidation: implications for the free radical hypothesis of Alzheimer's disease. Neurochem. Res. **24:** 427–435.
8. HARRIS, M., K. HENSLEY, D.A. BUTTERFIELD, R.A. LEEDLE & J.M. CARNEY. 1995. Direct evidence of oxidative injury produced by the Alzheimer's amyloid beta-peptide (1-40) in cultured hippocampal neurons. Exp. Neurol. **131:** 193–202.
9. BUTTERFIELD, D.A. & E.R. STADTMAN. 1997. Protein oxidation processes in aging brain. Adv. Cell Aging Gerontol. **2:** 161–191.
10. HENSLEY, K., N. HALL, R. SUBRAMANIAM, P. COLE, M. HARRIS, M. AKSENOV, M. AKSENOVA, S.P. GABBITA, J.F. WU, J.M. CARNEY, M. LOVELL, W.R. MARKESBERY & D.A. BUTTERFIELD. 1995. Brain regional correspondence between Alzheimer's disease histopathology and biomarkers of protein oxidation, J. Neurochem. **65:** 2146–2156
11. FAY, D., A. FLUET, C. JOHNSON & C. LINK. 1998. In vivo aggregation of beta-amyloid peptide variants. J. Neurochem. **71:** 1616–1625.
12. VARADARAJAN, S., S. YATIN, J. KANSKI, F. JAHANSHAKI & D.A. BUTTERFIELD. 1999. Methionine residue 35 is important in amyloid b-peptide–associated free radical oxidative stress. Submitted for publication.

Depolarization of *in Situ* Mitochondria by Hydrogen Peroxide in Nerve Terminals

CHRISTOS CHINOPOULOS AND VERA ADAM-VIZI[a]

Department of Medical Biochemistry, Semmelweis University of Medicine, P.O. Box 262, Budapest, H-1444, Hungary

INTRODUCTION

A large number of data presented recently in the literature has suggested that mitochondria play a critical role in the fate of cells exposed to oxidative stress.[1] It is well known that free radicals are generated in mitochondria in the respiratory chain[2,3] and mitochondria could be also one of the targets by which reactive oxygen species deteriorate cellular functions.[1,4,5] In the present work the effect of oxidative stress was studied on the membrane potential of *in situ* mitochondria ($\Delta\Psi m$) in isolated nerve terminals. Oxidative stress was induced by hydrogen peroxide, which has been reported to increase $[Ca^{2+}]_i$[6] and impair the energy status[7] in this preparation. We used the fluorescent probe JC-1 by which $\Delta\Psi m$ in cells can be monitored.[8]

MATERIALS AND METHODS

Preparation of Synaptosomes. Isolated nerve terminals (synaptosomes) were prepared from brain cortex of guinea pigs as detailed previously.[9] For the experiments synaptosomes were incubated in a medium containing NaCl 140 mM, KCl 3 mM, $MgCl_2$ 2 mM, $CaCl_2$ 2 mM, PIPES (pH 7.38) 10 mM, and glucose 10 mM. All the experiments were carried out at 37°C.

Determination of Mitochondrial Membrane Potential. Synaptosomes were suspended in Ca^{2+}-free standard medium (4 mg/ml) and loaded with JC-1 (30 mM) for 15 min at 37°C. After sedimentation synaptosomes were resuspended in standard medium (8 mg/ml), and fluorescence intensity was determined in a PTI Deltascan fluorescence spectrophotometer at 535-nm emission wavelength (with excitation at 490 nm) corresponding to the fluorescence peak of the monomer.

ATP and ADP Measurement. ATP and ADP were determined in synaptosomes (1 mg/ml) incubated at 37°C in the standard medium by the luciferin/luciferase method.[10]

Materials. Standard laboratory chemicals were obtained from Sigma. JC-1 was obtained from Molecular Probes.

Statistics. Results are expressed as mean ± SEM values. Statistical analysis used unpaired Student's *t* test. Differences were considered significant when $p < 0.05$.

---

[a]Corresponding author. Phone: 361-266 2773; fax: 361-267 0031.
e-mail: av@puskin.sote.hu

ABBREVIATIONS: JC-1,5,5',6,6'-tetrachloro-1,1,3,3'-tetraethylbenzimidazolyl-carbocyanin iodide; FCCP, carbonyl cyanide-p-trifluoromethoxyphenyl-hydrazon.

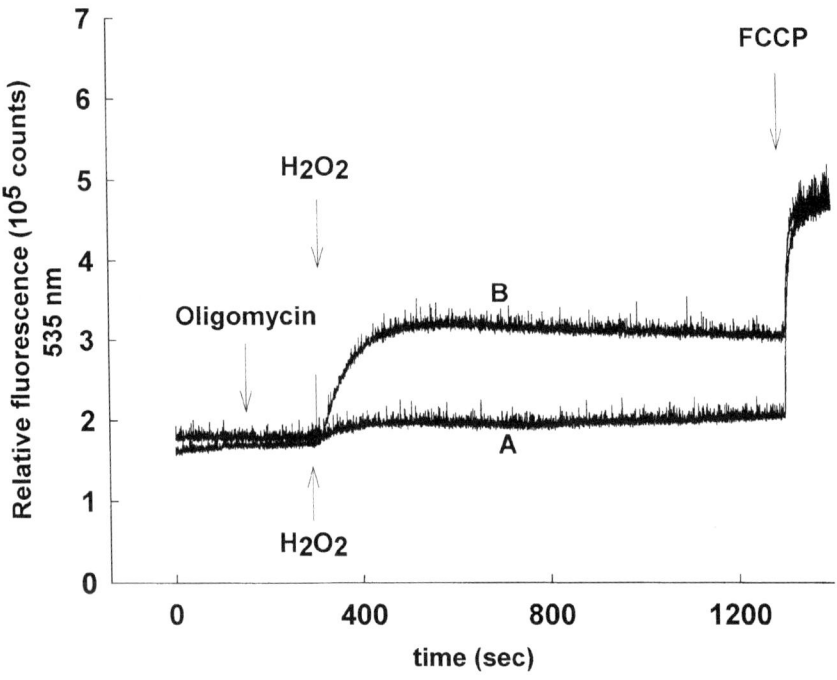

FIGURE 1. Fluorescence change of JC-1 at 535 nm due to H_2O_2. The oxidant was given at 300 s without oligomycin (**A**) or 200 s after application of 10 μM oligomycin (**B**). Fluorescence was monitored for 20 minutes. At the end of the experiments 1 μM FCCP was given to collapse $\Delta\Psi m$. *Traces* are representatives of four independent experiments.

RESULTS AND DISCUSSION

Given our recent report that H_2O_2 produce a nonspecific change in the J-aggregate fluorescence at 595 nm, which is unrelated to $\Delta\Psi m$,[11] in the present work only fluorescence at 535 nm was considered as an indication of $\Delta\Psi m$. FIGURE 1 shows that the fluorescence of JC-1 at 535 nm was only very slightly changed upon exposure of nerve terminals to 0.5 mM H_2O_2. Application of the oxidant in a lower (0.1-mM) concentration gave a similar result (not shown). Fluorescence signal due to 1 μM FCCP indicates a deenergized state of mitochondria, in which $\Delta\Psi m$ is collapsed. The result that $\Delta\Psi m$ is maintained in nerve terminals acutely exposed to H_2O_2-induced oxidative stress is consistent with the report that components of the respiratory chain are resistant to H_2O_2,[12] but is in contrast with a study in which the ratio fluorescence of JC-1 was monitored in the presence of the oxidant.[13] When we applied oligomycin (10 μM) to block the F_0-F_1-ATPase, no change in $\Delta\Psi m$ was observed (FIG. 1). The addition of H_2O_2 (0.5 mM) after oligomycin induced a large increase in the monomer fluorescence indicating a fall in $\Delta\Psi m$. Proton flux through the F_0-F_1-ATPase has been shown to contribute to the maintenance of $\Delta\Psi m$ when the complex I in the respiratory chain is inhibited by rotenone.[11,14] It could be sug-

TABLE 1. Decrease in the ATP level and ATP/ADP ratio in the presence of H_2O_2

	ATP nmol/mg protein	[ATP]/[ADP] ratio
Control	3.54 ± 0.02	6.83 ± 0.27
+ H_2O_2 50 mM	3.06 ± 0.05	5.38 ± 0.27
+ H_2O_2 100 mM	2.80 ± 0.10	4.29 ± 0.08
+ H_2O_2 500 mM	1.42 ± 0.10	2.53 ± 0.10

NOTE: Synaptosomes were incubated in the presence of different concentrations of H_2O_2 for 5 minutes; ATP and ADP levels were then determined.

gested from the results shown in FIGURE 1 that in the presence of the oxidant the F_0-F_1–ATPase serves partly to maintain $\Delta\Psi m$, by working in reverse and pumping protons out of the mitochondrial matrix at the expense of ATP hydrolysis.

We have also measured the ATP content and the [ATP]/[ADP] ratio in nerve terminals exposed to H_2O_2. TABLE 1 demonstrates that after incubation for 5 minutes with the oxidant, ATP level and ATP/ADP ratio were significantly lowered. This effect was slow; the maximal effect was achieved after incubation for 10 minutes in the presence of H_2O_2.[7] These data are consistent with those on $\Delta\Psi m$ and indicate that during acute exposure to H_2O_2 maintenance of $\Delta\Psi m$ is limited and requires the reverse function of the F_0-F_1–ATPase. The underlying mechanism by which H_2O_2 impairs the ability of mitochondria to maintain $\Delta\Psi m$ is very likely the inhibition of α-ketoglutarate dehydrogenase and a consecutive decrease in the steady state NADH level.[11] Given the crucial role of α-ketoglutarate dehydrogenase in providing NADH for the respiratory chain and in determining the rate of respiratory chain,[15,16] inhibition of this enzyme by H_2O_2 could limit the function of the respiratory chain, which in this case can maintain $\Delta\Psi m$ only with the contribution of proton pump by the F_0-F_1–ATPase. The decrease in ATP level is very likely a reflection of the impaired respiratory capacity and ATP hydrolysis by the F_0-F_1–ATPase, but a possible inhibition of glycolysis induced by the oxidant should also be considered.[17]

REFERENCES

1. BEAL, M.F. 1996. Mitochondria, free radicals and neurodegeneration. Curr. Opin. Neurobiol. **6:** 661–666.
2. BOVERIS, A., N. OSHINO & B. CHANCE. 1972. The cellular production of hydrogen peroxide. Biochem. J. **128:** 617–630.
3. LOSCHEN, G., A. AZZI, C. RICHTER & L. FLOHE. 1974. Superoxide radicals as precursors of mitochondrial hydrogen peroxide. FEBS Lett. **42:** 68–72.
4. BINDOLI, A. 1988. Lipid peroxidation in mitochondria. Free Radical Biol. Med. **5:** 247–261.
5. COYLE, J.T. & P. PUTTFARCHEN. 1993. Oxidative stress, glutamate, and neurodegenerative disorders. Science **262:** 689–694.
6. TRETTER, L. & V. ADAM-VIZI. 1996. Early events in free radical–mediated damage of isolated nerve terminals: effect of peroxides on membrane potential and intracellular Na^+ and Ca^{2+} concentrations. J. Neurochem. **66:** 2057–2066.
7. TRETTER L., C. CHINOPOULOS & V. ADAM-VIZI. 1997. Enhanced depolarization-evoked calcium signal and reduced ATP/ADP ratio are unrelated events induced by oxidative stress in synaptosomes. J. Neurochem. **69:** 2529–2537.

8. REERS, M., T.W. SMITH & L.B. CHEN. 1991. J-Aggregate formation of a carbocyanine as a quantitative fluorescent indicator of membrane potential. Biochemistry **30:** 4480–4486.
9. ADAM-VIZI, V. & E. LIGETI. 1986. Calcium uptake of synaptosomes as a function of membrane potential under different depolarizing conditions. J. Physiol.(Lond.) **372:** 363–377.
10. KAUPPINEN, R. A. & D.G. NICHOLLS. 1986. Failure to maintain glycolysis in anoxic nerve terminals. J. Neurochem. **47:** 1864–1869.
11. CHINOPOULOS, C. et al. 1999. Depolarization of in situ mitochondria due to hydrogen peroxide–induced oxidative stress in nerve terminals: inhibition of α-ketoglutarate dehydrogenase. J. Neurochem. **73:** 220–228.
12. ZHANG, Y., O. MARCILLAT, C. GIULIVI, L. ERNSTER & K.J.A. DAVIES. 1990. The oxidative inactivation of mitochondrial electron transport chain components and ATPase. J. Biol. Chem. **265:** 16330–16336.
13. SCANLON, J.M. & I.J. REYNOLDS. 1998. Effect of oxidants and glutamate receptor activation on mitochondrial membrane potential in rat forebrain neurons. J. Neurochem. **71:** 2392–2400.
14. SCOTT, I.D. & D.G. NICHOLLS. 1980. Energy transduction in intact synaptosomes. Influence of plasma-membrane depolarization on the respiration and membrane potential of internal mitochondria determined in situ. Biochem. J. **186:** 21–33.
15. COONEY, G.J., H. TAEGTMEYER & E.A. NEWSHOLME. 1981. Tricarboxylic acid cycle flux and enzyme activities in the isolated working rat heart. Biochem. J. **200:** 701–703.
16. MORENO-SANCHEZ, R., B.A. HOGUE & R.A. HANSFORD. 1990. Influence of NAD-linked dehydrogenase activity on flux through oxidative phosphorylation. Biochem. J. **268:** 421–428.
17. LITTLE, C. & P.J. O'BRIEN. 1969. Mechanism of peroxide inactivation of the sulphydryl enzyme glyceraldehyde-3-phosphate dehydrogenase. Eur. J. Biochem. **10:** 533–538.

Monoamine Oxidase Inhibits Mitochondrial Respiration

GERALD COHEN[a] AND NATASA KESLER

Department of Neurology and Center for Neurobiology, Mount Sinai School Medicine, New York, New York 10029, USA

INTRODUCTION

Parkinson's disease (PD) is characterized by a prominent defect in the mitochondrial respiratory chain—namely, a 35% deficiency of mitochondrial complex I in the substantia nigra.[1,2] The substantia nigra pars compacta is a pigmented portion of the brain that contains the melanized dopamine (DA) cell bodies that degenerate in PD. The same mitochondrial lesion is evoked by exposure to the neurotoxin 1-methyl-4-phenyl-2,3,5,6-tetrahydropyridine (MPTP), which induces the selective destruction of dopaminergic pars compacta neurons in human subjects and primates.[3,4] It is widely believed that the complex I deficiency is translated into a deficiency in mitochondrial respiration, leading to decreased energy (ATP) production. In turn, diminished respiration contributes to, or acts as a driving force for, apoptotic and necrotic cell death,[5] leading to a progressive loss of DA neurons.

We describe experiments with isolated rat brain mitochondria in which a defect in mitochondrial respiration is evoked by the action of monoamine oxidase (MAO). MAO is a marker for the outer mitochondrial membrane. Its enzymatic role is to remove DA from the cell cytoplasm. Under normal circumstances, released neurotransmitter (DA) is taken back into the presynaptic nerve terminal by means of the DA transporter. A major fraction is subsequently stored in vesicles, but MAO metabolizes a significant amount, generating hydrogen peroxide (H_2O_2) in the process, so that the presynaptic turnover of DA is associated with the intraneuronal formation of H_2O_2, a potential toxin. The amount of H_2O_2 generated increases as the metabolic turnover of DA increases.

In PD, a loss of 80% of DA neurons is required to evoke impaired motor function, such as tremor and rigidity. The surviving DA neurons are activated in a compensatory response, and, therefore, they accelerate their turnover of DA.[6] It follows that these DA neurons are exposed to increased steady state levels of H_2O_2 within the Parkinson brain.

As a direct consequence of DA turnover (Eq. 1), glutathione disulfide (GSSG) is produced from glutathione (GSH) via the enzymatic activity of GSH peroxidase (Eq. 2). The reaction of GSSG with protein thiols (PrSH) results in the formation of GSH-protein mixed-disulfides (PrSSG; Eq. 3). The latter reaction occurs spontaneously and is also catalyzed by tissue thiol transferases and thioredoxins.[7,8]

---

[a]Address for correspondence: Dr. Gerald Cohen, Department of Neurology (Box 1137), Mount Sinai School of Medicine, One Gustave L. Levy Place, New York, New York 10029. Phone: 212-241-7312.
  e-mail: gerald.cohen@mssm.edu

$$DA + O_2 \rightarrow H_2O_2 + NH_3 + 3,4\text{-dihydroxyphenylacetaldehyde} \qquad (1)$$

$$H_2O_2 + 2GSH \rightarrow GSSG + 2H_2O \qquad (2)$$

$$GSSG + PrSH \rightarrow PrSSG + GSH \qquad (3)$$

MATERIALS AND METHODS

Rat brain mitochondria were isolated by a modification[9] of the method of Clark and Nicklas[10] and suspended in cold buffer, which consisted of 5 mM Mops, containing 0.225 M mannitol, 0.075 M sucrose, and 1.0 mM EGTA, adjusted to pH 7.4 with KOH. Incubations with and without tyramine (500 µM) and MAO inhibitors (2 µM pargyline plus 2 µM clorgyline) were conducted after dilution of an aliquot of the mitochondrial preparation to 0.5 mg or 1.0 mg mitochondrial protein/ml in the respiration buffer (pH 7.2), which consisted of 5 mM Hepes, 125 mM sucrose, 50 mM KCl, 2 mM KH_2PO_4, 1 mM $MgCl_2$,[11] and 0.5 mg BSA/ml at 27°C. Incubations were carried out at 27°C with gentle shaking. Samples were processed individually. Incubation was followed by the immediate assessment of respiration, with rotation among the experimental groups. Samples not incubated with MAO inhibitors received additions of the inhibitors after the incubation was complete, just prior to the measurement of respiration. Each experiment consisted of 10–12 samples (3–4 samples per group).

Respiration was measured with an oxygen electrode immediately after the experimental incubation period. The mitochondria were loaded into the well of a miniature chamber system (0.6-ml capacity), equipped with a magnetic stirrer and maintained at 27°C. Oxygen consumption was measured sequentially with a biological oxygen monitor after the addition of 5 mM pyruvate and 5 mM malate (state 4 respiration), followed by 0.4 mM ADP (state 3 respiration), and, last, 10 µM FCCP (state 5 or uncoupled respiration). The respiratory control ratios (state 3/state 4) of freshly isolated mitochondria were in the range 5.5–7.0; state 3 respiration was in the range 72–103 ng–atoms oxygen/min/mg protein. Respiratory activity of the stock mitochondrial preparation, which was held on ice, was well maintained and did not change over the course of 3–4 hours.

PrSSG was measured with a modification of the method of Akerboom and Sies.[12] The liberated GSH was measured on the plate reader with a modification of the enzymatic recycling method of Tietze.[13] Data for both respiration and PrSSG were expressed per mg of mitochondrial protein.

RESULTS AND DISCUSSION

A schematic representation of the experiments is presented in FIGURE 1. Tyramine was used as a representative substrate for MAO. Tyramine, like DA, is a substrate for both the A and B isoforms of MAO. Tyramine was used to avoid formation of quinoidal products via autoxidation, such as occurs with DA; tyramine does not autoxidize and does not form quinones. The concentration of tyramine

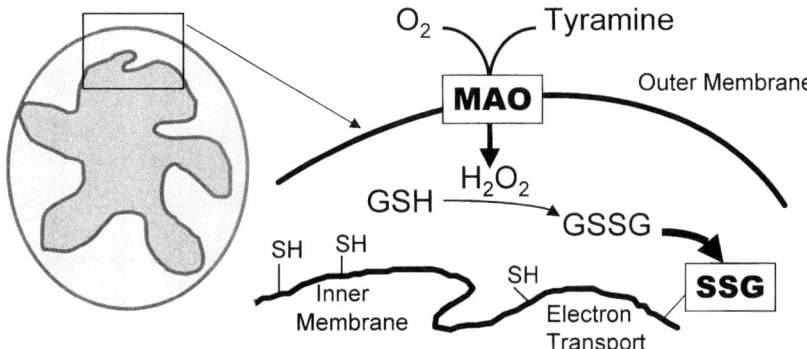

FIGURE 1. Schematic representation of the reaction mechanism for forming PrSSG at the inner membrane and inhibiting mitochondrial respiration. Tyramine at the outer mitochondrial membrane produces H_2O_2, which is removed by glutathione peroxidase in the intermembrane space, forming GSSG. GSSG, in turn, reacts with protein thiols of the inner membrane, inhibiting thiol-dependent enzymes, such as complex I.

(500 µM) closely approximates the level of monoamine neurotransmitter in the cytosol.[9] FIGURE 1 shows that MAO activity at the outer mitochondrial membrane oxidatively deaminates tyramine in the presence of oxygen. In this process, oxygen is reduced to H_2O_2. The H_2O_2 is detoxified by GSH peroxidase, forming GSSG. A portion of the GSSG reacts with protein thiols, forming protein mixed-disulfides (PrSSG). It is assumed that the GSSG can diffuse across the intermembrane space to react with protein thiols of the inner mitochondrial membrane, which is the site for respiratory enzymes. Formation of PrSSG at the inner membrane can result in the inhibition of SH-dependent enzymes, such as complex I[14] and succinate dehydrogenase.[15] Therefore, MAO activity at the outer membrane can affect respiratory activity at the inner membrane.

In a previous study, mitochondrial electron flow associated with the electron transport chain was measured. MAO suppressed both pyruvate-supported activity, which is initiated at complex I, and succinate-supported activity, initiated at complex II.[9] Complex I was more severely affected, in keeping with observations of a relatively specific decrease in complex I in the PD brain.

In the current experiments, we studied respiration directly with an oxygen electrode. FIGURE 2 shows that exposure of mitochondria to 500 µM tyramine for 15 min at 27°C resulted in suppression of ADP-stimulated respiration (state 3), as well as uncoupled respiration (state 5). The inclusion of a mixture of clorgyline (2 µM, MAO-A inhibitor) and pargyline (2 µM, MAO-B inhibitor) in the medium completely protected mitochondria from damage by MAO. Suppression of respiration is not due to the physical presence of tyramine per se, but to a product of the MAO reaction. MAO inhibitors by themselves did not enhance respiratory activity (not shown).

In other experiments, formation of PrSSG (Eq. 3) was measured. After the brief incubation of mitochondria for 15 min at 27°C with 500 µM tyramine or 500 µM dopamine, PrSSG levels were elevated by more than 30-fold, from 20 ± 17 to 647 ±

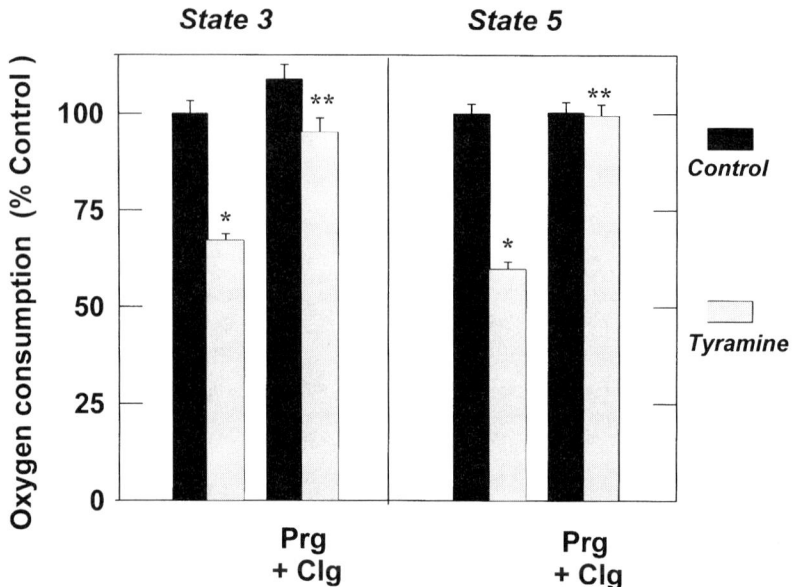

FIGURE 2. Suppression of state 3 and state 5 respiration after exposure of mitochondria to 500 μM tyramine for 15 minutes at 27°C. Where indicated, a mixture of 2 μM pargyline (MAO-B inhibitor) and 2 μM clorgyline (MAO-A inhibitor) was present. MAO inhibitors were also added to the remaining samples just prior to the measurement of mitochondrial respiration. Results are the mean ± SEM for $n = 6$ (state 3, two experiments) or $n = 11$ (state 5, 4 experiments). *$p < 0.01$ versus control and **$p < 0.01$ versus tyramine alone (Tukey-Kramer multiple comparison test).

9 (DA) and 839 ± 48 (tyramine) pmoles/mg mitochondrial protein (mean ± SEM, $p < 0.001$). These observations are in keeping with a mechanism of mitochondrial damage that is dependent upon the formation of PrSSG by essential protein thiol groups of respiratory chain enzymes. Because pyruvate was used as the mitochondrial substrate in these experiments, respiration was initiated at the level of complex I (pyruvate dehydrogenase and NADH-coenzyme Q reductase).

These experiments illustrate that damage to complex I–linked respiration by MAO is associated with a sharp rise in PrSSG. Diminished respiration of nonsynaptic rat brain mitochondria, such as those used in the current experiments, requires greater than 72% inhibition of complex I.[16] Turnover of monoamine substrates by MAO also induces damage to mitochondrial DNA.[17]

We conclude that DA turnover possesses the potential to adversely affect mitochondrial respiration at the level of complex I. Increased turnover, such as that naturally associated with the surviving DA neurons in PD,[6] places these neurons at increased risk. The risk will be further amplified during treatment with L-dopa, which elevates cerebral DA levels and contributes to a "wasteful" turnover of a portion of the DA (i.e., turnover not necessarily associated with synaptic neurotransmission). Damage to complex I in the substantia nigra has been observed after long-range exposure to L-dopa.[18] In PD, the increased turnover of DA by surviving DA neurons may contribute to the complex I deficiency seen at autopsy, and this effect may be exacerbated for patients undergoing treatment with L-dopa.

Recent studies have described gene defects[19,20] in a small fraction of Parkinson patients. It is not clear how the proteins encoded by these genes interact with other biologic factors to induce a parkinsonian state. One possibility is interplay with the oxidant stress associated with the turnover of DA. In this paradigm, DA becomes an endogenous stressor (or toxin) that explains why the neurodegenerative lesion is localized to neurons that utilize DA as neurotransmitter.

ACKNOWLEDGMENTS

This study was supported by a grant (DAMD17-98-1-8624) from the U.S. Army Medical Research and Materiel Command (USAMRMC) and by a grant from the Parkinson's Disease Foundation. Support by USAMRMC does not constitute endorsement by the U.S. Government or the U.S. Army.

REFERENCES

1. SCHAPIRA, A.H.V., J.M. COOPER, D. DEXTER, J.B. CLARK, P. JENNER & C.D. MARSDEN. 1990. Mitochondrial complex I deficiency in Parkinson's disease. J. Neurochem. **54:** 823–827.
2. SCHAPIRA, A.H. 1998. Mitochondrial dysfunction in neurodegenerative disorders. Biochim. Biophys. Acta **1366:** 225–233.
3. BURNS, R.S., C.C. CHIUEH, S.P. MARKEY, M.P. EBERT, D.M. JACOBOWITZ & I.J. KOPIN. 1983. A primate model of parkinsonism: selective destruction of dopaminergic neurons in the pars compacta of the substantia nigra by N-methyl-4-phenyl-1,2,3,6-tetrahydropyridine, Proc. Natl. Acad. Sci. USA **80:** 4546–4550.
4. LANGSTON, J.W., P. BALLARD, J.W. TETRUD & I. IRWIN. 1983. Chronic parkinsonism in humans due to a product of meperidine-analog synthesis, Science **219:** 979–980.
5. ZAMZAMI, N., T. HIRSCH, B. DALLAPORTA, P.X. PETIT & G. KROEMER. 1997. Mitochondrial implication in accidental and programmed cell death: apoptosis and necrosis. J. Bioenerg. Biomembr. **29:** 185–193.
6. HORNYKIEWICZ, O. & S.J. KISH. 1996. Biochemical pathophysiology of Parkinson's disease. Adv. Neurol. **45:** 19–34.
7. REED, D.J. 1990. Glutathione: toxicological implications. Annu. Rev. Pharmacol. Toxicol **30:** 603–631.
8. HOLMGREN, A. 1985. Thioredoxin. Annu. Rev. Biochem. **54:** 237–271.
9. COHEN, G., R. FAROOQUI & N. KESLER. 1997. Parkinson disease: a new link between monoamine oxidase and mitochondrial electron flow. Proc. Natl. Acad. Sci. USA **94:** 4890–4894.
10. CLARK, J.B. & W.J. NICKLAS. 1970. The metabolism of rat brain mitochondria. Preparation and characterization. J. Biol. Chem. **245:** 4724–4731.
11. MOREADITH, R.W. & G. FISKUM. 1984. Isolation of mitochondria from ascites tumor cells permeabilized with digitonin. Anal. Biochem. **137:** 360–367.
12. AKERBOOM, T.P.M. & H. SIES. 1981. Assay of glutathione, glutathione disulfide, and glutathione mixed disulfides in biological samples. Methods Enzymol. **77:** 373–382.
13. TIETZE, F. 1969. Enzymic method for the quantitative determination of nanogram amounts of total and oxidized glutathione: applications to mammalian blood and other tissues. Anal. Biochem. **27:** 502–522.

14. GUTMAN, M., H. MERSMANN, J. LUTHY & T.P. SINGER. 1970. Action of sulfhydryl inhibitors on different forms of the respiratory chain–linked reduced nicotinamide-adenine dinucleotide dehydrogenase. Biochemistry **9:** 2678–2687.
15. BENARD, O. & K.A. BALASUBRAMANIAN. 1995. Effect of oxidized glutathione on intestinal mitochondria and brush border membrane. Int. J. Biochem. Cell Biol. **27:** 589–595
16. DAVEY, G.P., S. PEUCHEN & J.B. CLARK. 1998. Energy thresholds in brain mitochondria. Potential involvement in neurodegeneration. J. Biol. Chem. **273:** 12753–12757.
17. HAUPTMANN, N., J. GRIMSBY, J.C. SHIH & E. CADENAS. 1996. The metabolism of tyramine by monoamine oxidase A/B causes oxidative damage to mitochondrial DNA. Arch. Biochem. Biophys. **335:** 295–304.
18. PRZEDBORSKI, S., V. JACKSON-LEWIS, U. MUTHANE, H. JIANG, M. FERREIRA, A.B. NAINI & S. FAHN. 1993. Chronic levodopa administration alters cerebral mitochondrial respiratory chain activity. Ann. Neurol. **34:** 715–23.
19. POLYMEROPOULOS, M.H., C. LAVEDAN, E. LEROY, S.E. IDE, A. DEHEJIA, A. DUTRA, B. PIKE, H. ROOT, J. RUBENSTEIN, R. BOYER, E.S. STENROOS, S. CHANDRASEKHARAPPA, A. ATHANASSIADOU, T. PAPAPETROPOULOS, W.G. JOHNSON, A.M. LAZZARINI, R.C. DUVOISIN, G. DI IORIO, L.I. GOLBE & R.I. NUSSBAUM. 1997. Mutation in the alpha-synuclein gene identified in families with Parkinson's disease. Science **276:** 2045–2047.
20. KITADA, T., S. ASAKAWA, N. HATTORI, H. MATSUMINE, Y. YAMAMURA, S. MINOSHIMA, M. YOKOCHI, Y. MIZUNO & N. SHIMIZU. 1998. Mutations in the parkin gene cause autosomal recessive juvenile parkinsonism. Nature **392:** 605–608.

Enhanced Acetate and Glucose Utilization during Graded Photic Stimulation

Neuronal-Glial Interactions *in Vivo*

GERALD A. DIENEL,[a] KENIAN LIU, DAVID POPP, AND NANCY F. CRUZ

Department of Neurology, University of Arkansas for Medical Sciences, Little Rock, Arkansas 72205, USA

Physiological activity alters neuronal signaling, generates the demand for energy in working brain cells, and activates metabolic interactions between neurons and glia. Local glial responses to changes in neuronal activity are poorly understood, in part, because functional activities of specific cell types are difficult to assay in brain of conscious subjects *in vivo*. Because local changes in metabolic rate are usually closely correlated with altered neural function, compounds known to be metabolized mainly in glia might be useful to assess functional responses of glia to neuronal stimulation in brain *in vivo*.

Acetate is preferentially metabolized by glia, and radiolabeled acetate has been used in many laboratories as a "glial reporter molecule"—i.e., a tool to examine the specialized metabolic activities and functions of one major class of brain cells *in vitro* and *in vivo*. [^{14}C]Acetate is rapidly (i.e., within minutes) incorporated into amino acids derived from the tricarboxylic acid cycle in brain *in vivo*, producing labeled glutamine with a higher specific activity than that of glutamate, its obligatory precursor; this finding was explained by a "small" (glial) pool of glutamate which turns over rapidly and is segregated from the "large" (neuronal) glutamate pool[1] and by cellular localization of glutamine synthetase mainly in glial cells.[2] Selective uptake and metabolism of acetate by cultured astrocytes compared to synaptosomes and cultured neurons has been confirmed both *in vitro* and *in vivo*,[3–6] and is due to preferential transport of acetate into astrocytes.[7] In the present study, [^{14}C]acetate was used to investigate functional neuronal-glial interactions in mature brain of the conscious rat *in vivo* by assessing the metabolic responses of glia to altered neuronal signaling in the visual system.

Because direct retinal input to the dorsal region of the superior colliculus of the rat is derived mainly from the contralateral eye, left-right comparisons of [^{14}C]acetate uptake can be made in each animal during unilateral visual stimulation. On the day of the experiment, normal fed male rats were anesthetized with halothane, and the right eye was removed. Then catheters were inserted in a femoral vein and artery, xylocaine ointment was applied to all wounds, and 3 hours allowed for recovery. To examine responses of glia in the superior colliculus to a noninvasive stimulus of retinal neurons, local metabolic activity in brain of normoxic conscious rats was as-

---

[a]Address for correspondence: Gerald A. Dienel, Ph.D., Department of Neurology, Slot 500, University of Arkansas for Medical Sciences, 4301 W. Markham St., Little Rock, AR 72205. Phone: 501-603-1167; fax: 501-686-8689.
 e-mail: dienelgeralda@exchange.uams.edu

TABLE 1. Influence of photic stimulation on acetate uptake and glucose utilization in the superior colliculus

Metabolic Tracer	Photic Stimulation Rate (flashes/sec)	Number of Rats	Ratio of Metabolic Activity (Right/Left Superior Colliculus)	
			Dorsal Region of Superior Colliculus	Ventral Region of Superior Colliculus
[^{14}C]Acetate	0	4	1.09 ± 0.01^a	1.03 ± 0.05
	8	6	1.18 ± 0.02^c	1.02 ± 0.01
	16	6	1.20 ± 0.03^c	1.02 ± 0.02
[^{14}C]Deoxyglucose	0	4	1.51 ± 0.08^b	1.04 ± 0.04
	8	3	2.21 ± 0.39^a	1.10 ± 0.13
	16	4	2.50 ± 0.23^c	1.11 ± 0.05^a

NOTE: To remove retinal input to the left dorsal superior colliculus the right eye was removed under anesthesia. Three h later metabolic assays were carried out either in the dark or during on-off photic stimulation of the left eye to activate metabolism in the right dorsal superior colliculus. The ventral region of the superior colliculus does not receive direct input from the retina.

Values are means ± S.D. $^a p < 0.05$; $^b p < 0.01$; $^c p < 0.001$, paired t test.

sayed in the dark or during on-off photic stimulation with a Grass PS22 photic stimulator; photic flashing was initiated just before an intravenous pulse of [^{14}C]acetate or [^{14}C] deoxyglucose (125 µCi/kg) and maintained for the 5- or 30-min experimental period, respectively. Local ^{14}C levels in brain were assayed by quantitative autoradiography.

Glucose utilization (CMR$_{glc}$) and [^{14}C]acetate uptake in the innervated (i.e., right) dorsal superior colliculus exceeded that in the contralateral acutely deafferented structure when assayed in the dark, and the ratio of metabolic activity in the right compared to left dorsal superior colliculus increased with photic stimulation. When experiments were carried out under resting conditions in the dark, both CMR$_{glc}$ and acetate utilization fell in the deafferented tissue; CMR$_{glc}$ was reduced by a greater proportion than acetate uptake (TABLE 1). During on-off photic stimulation at 8 or 16 flashes/sec, acetate uptake increased by about 20%, and CMR$_{glc}$ rose 2.2–2.5-fold (TABLE 1). The ventral region of the superior colliculus, which does not receive direct input from the retina, was unresponsive to changes in visual stimuli (TABLE 1).

Acetate is metabolized via the glial tricarboxylic acid (TCA) cycle, and local flux of substrate into this metabolic pathway rises when retinal neuronal signaling increases in response to sensory stimulation, whereas it falls during acute visual deprivation. Local changes in acetate utilization were in the same direction but smaller in magnitude than those of stimulus-induced consumption of glucose, which is a fuel for all brain cells. Altered flux of [^{14}C]acetate into the glial TCA cycle during functional activation can reflect variations in the demand for oxidative and biosynthetic processes in working glia; ATP is produced by both processes, and the energy yield is higher when oxidation is complete.

To summarize, altered sensory input changes neuronal signaling and influences the flow of substrate into the glial tricarboxylic acid cycle in brain *in vivo*. The local energy or biosynthetic demands of working glia are governed, in part, by stimulus intensity and neuronal activity.

REFERENCES

1. BERL, S., D.D. CLARK & D. SCHNEIDER, EDS. 1975. Metabolic Compartmentation and Neurotransmission: Relation to Brain Structure and Function. Plenum Press. New York.
2. MARTINEZ-HERNANDEZ, A., K. BELL & M. NORENBERG. 1977. Glutamine synthetase: glial localization in brain. Science **195:** 1356–1358.
3. HASSEL, B. & F. FONNUM. 1992. Selective inhibition of glial cell metabolism in vivo by fluorocitrate. Brain Res. **576:** 120–124.
4. HASSEL, B., U. SONNEWALD & F. FONNUM. 1995. Glial-neuronal interactions as studied by cerebral metabolism of [2-^{13}C]acetate and [1-^{13}C]glucose: an ex vivo ^{13}C NMR spectroscopic study. J. Neurochem. **64:** 2773–2782.
5. SONNEWALD, U., N. WESTERGAARD, A. SCHOUSBOE, J. SVENDSEN, G. UNSGÅRD & S.B. PETERSEN. 1993. Direct demonstration by [^{13}C]NMR spectroscopy that glutamine from astrocytes is a precursor for GABA synthesis in neurons. Neurochem. Int. **22:** 19–29.
6. ZEEVALK, G.D., N. DAVIS, A.G. HYNDMAN & W.J. NICKLAS. 1998. Origins of the extracellular glutamate released during total metabolic blockade in the immature retina. J. Neurochem. **71:** 2373–2381.
7. WANIEWSKI, R.A. & D.L. MARTIN. 1998. Preferential utilization of acetate by astrocytes is attributable to transport. J. Neurosci. **18:** 5225–5233.

Calcium Overload Triggers Rod Photoreceptor Apoptotic Cell Death in Chemical-Induced and Inherited Retinal Degenerations

DONALD A. FOX,[a–c] ANN T. POBLENZ,[a] AND LIHUA HE[a]

[a]*Department of Biology and Biochemistry and* [b]*College of Optometry, University of Houston, Houston, Texas 77204-6052, USA*

Impairment and loss of vision are major human health problems. The majority of these cases have a retinal origin.[1] For example, selective rod photoreceptor cell apoptosis occurs in humans and animals with different forms of inherited retinal degenerations, cancer-associated retinopathy (CAR), lead exposure during development and adulthood, mild hypoxic-ischemia injury, or light-induced damage (TABLE 1).[1–9] Alterations in cGMP phosphodiesterase (PDE) metabolism underlie many types of apoptotic retinal degenerations. In some humans with autosomal recessive retinitis pigmentosa there are mutations in the genes encoding the α- and β-subunits of rod cGMP PDE while in lead poisoning there is a competitive inhibition of the cGMP PDE enzyme.[8,10–13] *In vitro* exposure of isolated retinas to 3-isobutyl-1-methylxanthine (IBMX), a competitive inhibitor of cGMP PDE, results in selective rod degeneration.[11,14] The loss or inhibition of rod cGMP PDE activity results in elevated [cGMP][8,10,15] that is localized almost exclusively to rods.[16,17] An increase in rod [cGMP] results in more rod cGMP-gated nonselective cation channels being opened and a sustained elevation of the rod and retinal intracellular [Ca^{2+}].[18–21]

Numerous studies report that a sustained elevation of intracellular [Ca^{2+}] results in apoptotic cell death.[22,23] During the effector phase of apoptosis the mitochondrial permeability transition pore (PTP) is irreversibly opened by sustained increases in matrix Ca^{2+} leading to mitochondrial depolarization, release of cytochrome *c*, activation of caspases, chromatin cleavage and apoptotic nuclear morphology.[24] The open conformation of the mitochondrial PTP is blocked by cyclosporine A (CsA).[24,25] Accumulating evidence suggests that sustained elevations of rod intracellular [Ca^{2+}], following the loss or inhibition of rod cGMP PDE activity as well as in other types of retinal degenerations, may trigger apoptosis. Rod and retinal intracellular [Ca^{2+}] are elevated in isolated retinas exposed to IBMX or Pb^{2+} and in lead-exposed rats.[19–21,26] Ca^{2+} overload blockers protect against rod apoptosis induced by constant light and mild hypoxia-ischemia.[4,27] In CAR an increase in rod intracellular [Ca^{2+}] appears to be the apoptotic triggering event[9] as there is an increase in an autoantibody against recoverin: a cytoplasmic Ca^{2+}-binding protein primarily localized to the photoreceptors and bipolars cells.[28]

Our goals were to determine the retinal [Ca^{2+}] and examine high molecular weight (HMW) DNA fragmentation in developing retinal degeneration (rd) mice and

---

[c]Address for correspondence: Donald A. Fox, Ph.D., College of Optometry, University of Houston, 4901 Calhoun Blvd. Houston, TX 77204-6052. Phone: 713-743-1964; fax: 713-743-2053.
  e-mail: dafox@uh.edu

TABLE 1. Comparison of biochemical, cellular and molecular events involved in apoptotic cell death pathway in the retinas of rd mice, lead-exposed rats, rats exposed to recoverin antibody and isolated rat retinas exposed to elevated Ca^{2+a}

Measures	rd Mice	Lead Rats	Recoverin MAb	In vitro Ca^{2+}
Rod cell loss	+	+	+	+
Cone cell loss	−	−	−	−
Bipolar cell loss	−	+	+	−
Ganglion cell loss	−	−	−	−
Apoptotic nuclei present	+	+	+	+
HMW DNA fragmentation	+	+	ND	+
Oligonucleosomal fragmentation	+	ND	+	ND
Decreased retinal cGMP PDE activity	+	+	ND	+
Increased retinal [Ca^{2+}]	+	+	ND	+
Mitochondrial transition pore opening	ND	ND	ND	+
Caspase activation	ND	ND	ND	+

[a] All data from papers cited in text or are results presented in this paper. +: present; −: not present; ND: not determined.

to elucidate the mechanisms underlying Ca^{2+}-induced rod apoptosis in rat retinas to understand how this might occur *in vivo*.

The retinal [Ca^{2+}] of developing rd mice and age-matched control mice were determined as described.[21] At 3 days of age (PN3), the retinal elemental [Ca^{2+}] was similar in rd and control mice. In contrast, the retinal [Ca^{2+}] at PN5, PN7, PN10, PN15 and PN17 increased 18%, 40%, 122%, 189% and 67% in rd mice, respectively, whereas at PN20 it was decreased 42%. Retinas from PN7-17 rd mice exhibited increases in 50–600 kilobase pair (kbp) HMW DNA fragments as detected by field inversion gel electrophoresis (FIGE).[29] This pattern of sustained elevation in retinal [Ca^{2+}] and increased HMW DNA fragmentation in rd mice paralleled the observed increases in retinal [cGMP], TUNEL positive rod cells and DNA internucleosomal fragmentation.[6,7,16] The decrease in retinal [Ca^{2+}] at PN20 most likely reflects the almost complete loss of rod photoreceptor cells by this age.[15]

Lead exposure during development or adulthood produces apoptotic cell death in rat rods and rod bipolar cells.[3,8] The molecular mechanism causing this selective cell loss has been postulated to be partially triggered by Ca^{2+} overload.[8] The current studies utilized whole isolated adult rat retinas incubated for 15 minutes in a physiological buffer containing IBMX (to increase the number of open cGMP channels) with or without Ca^{2+},[22] to examine this hypothesis and to determine the apoptotic signaling cascade. Retinas incubated in Ca^{2+}:

(1) had elevated levels of intracellular [Ca^{2+}] that were localized exclusively to the photoreceptors (assessed with atomic absorption spectrometry[21] and confocal microscopy and Fluo-3),[30]
(2) exhibited rod selective apoptosis as evidenced by electron microscopy,[8] DNA chromatin condensation and fragmentation (assessed with acridine orange/ethidium bromide staining and FIGE), and

(3) had increased CPP32-like caspase activity (assessed with DEVD-pNA by Clontech Assay Kit).

CsA, DEVD-fmk [carbobenzoxy-Asp-Glu-Val-Asp-fluorotrimethylketone], and diltiazem (a rod cGMP blocker[18]) completely blocked these effects. There was no evidence of reactive oxygen species.

In summary, our results suggest that Ca^{2+} entered the rod through the cGMP channel and then induced the opening of the mitochondrial PTP by binding to the internal metal regulatory site of the mitochondrial PTP[31] and switched it from its normal low- to its high-conductance open conformational state.[25] Following the opening of the PTP cytochrome c is released, caspases are activated, and DNA chromatin is fragmented and condensed. Thus, Ca^{2+} overload induced by alterations in cGMP PDE metabolism, or other mechanisms, may underlie the selective apoptotic rod cell death observed in rd mice, humans with cGMP PDE mutations, lead-exposed animals, humans and animals with CAR, and possibly lead-exposed children and workers. The therapeutic drugs CsA and diltiazem may be useful cytoprotective agents for these retinal degenerations.

REFERENCES

1. NATIONAL EYE ADVISORY COUNCIL REPORT. 1998. A National Plan: 1999–2003.
2. THIRKILL, C.E., A.M. ROTH & J.L. KELTNER. 1987. Cancer-associated retinopathy. Arch. Ophthalmol. **105:** 372–375.
3. FOX, D.A. & L.W.F. CHU. 1988. Rods are selectively altered by lead: II. Ultrastructure and quantitative histology. Exp. Eye Res. **46:** 613–625.
4. EDWARD, D.P., T.T. LAM, S. SHAHINFAR, J. LI & M.O. TSO. 1991. Amelioration of light-induced retinal degeneration by a calcium overload blocker: flunarizine. Arch. Ophthalmol. **109:** 554–562.
5. BUCHI, E.R. 1992. Cell death in rat retina after pressure-induced ischaemia-reperfusion insult: electron microscopic study. II. Outer nuclear layer. Jpn. J. Ophthalmol. **36:** 62–68.
6. CHANG, G.Q., Y. HAO & F. WONG. 1993. Apoptosis: final common pathway of photoreceptor death in rd, rds, and rhodopsin mutant mice. Neuron **11:** 595–605.
7. PORTERA-CAILLIAU, C., C.H. SUNG, J. NATHANS & R. ADLER. 1994. Apoptotic photoreceptor cell death in mouse models of retinitis pigmentosa. Proc. Natl. Acad. Sci. USA **91:** 974–978.
8. FOX, D.A., M.L. CAMPBELL & Y.S. BLOCKER. 1997. Functional alterations and apoptotic cell death in the retina following developmental or adult lead exposure. Neurotoxicology **18:** 645–665.
9. ADAMUS, G., M. MACHNICKI, H. ELERDING, B. SUGDEN, Y.S. BLOCKER & D.A. FOX. 1998. Antibodies to recoverin induce apoptosis of photoreceptor and bipolar cells in vivo. J. Autoimmun. **11:** 523–533.
10. FOX, D.A., L.M. KATZ & D.B. FARBER. 1991. Low-level developmental lead exposure decreases the sensitivity, amplitude and temporal resolution of rods. Neurotoxicology **12:** 641–654.
11. SRIVASTAVA, D., D.A. FOX & R.L. HURWITZ. 1995. Effects of magnesium on cGMP hydrolysis by the bovine retinal rod cGMP phosphodiesterase. Biochem. J. **308:** 653–658.
12. SRIVASTAVA, D., R.L. HURWITZ & D.A. FOX. 1995. Lead- and calcium-mediated inhibition of bovine rod cGMP phosphodiesterase: interactions with magnesium. Toxicol. Appl. Pharmacol. **134:** 43–52.

13. VAN SOEST, S., A. WESTERVELD, P.T. DE JONG, E.M. BLEEKER-WAGEMAKERS & A.A. BERGEN. 1999. Retinitis pigmentosa: defined from a molecular point of view. Surv. Ophthalmol. **43:** 321–334.
14. LOLLEY, R.N., D.B. FARBER, M.E. RAYBORN & J.G. HOLLYFIELD. 1977. Cyclic GMP accumulation causes degeneration of photoreceptor cells: simulation of an inherited disease. Science **196:** 664–666.
15. FARBER, D.B. & T.A. SHUSTER. 1986. Cyclic nucleotides in retinal function and degeneration. *In* The Retina, Part 1. R. Adler & D.B. Farber, Eds. : 239–296. Academic Press. New York.
16. FARBER, D.B. & R.N. LOLLEY. 1974. Cyclic guanosine monophosphate: elevation in degenerating photoreceptor cells of the C3H mouse retina. Science **186:** 449–451.
17. GOTZES, S., J. DE VENTE & F. MULLER. 1998. Nitric oxide modulates cGMP levels in neurons of the inner and outer retina in opposite ways. Vis. Neurosci. **15:** 945–955.
18. YAU, K.W. & D.A. BAYLOR. 1989. Cyclic GMP-activated conductance of retinal photoreceptor cells. Annu. Rev. Neurosci. **12:** 289–327.
19. RATTO, G.M., R. PAYNE, W.G. OWEN & R.Y. TSIEN. 1988. The concentration of cytosolic free calcium in vertebrate rod outer segments measured with fura-2. J. Neurosci. **8:** 3240–3246.
20. MCCARTHY, S.T., J.P. YOUNGER & W.G. OWEN. 1994. Free calcium concentrations in bullfrog rods determined in the presence of multiple forms of Fura-2. Biophys. J. **67:** 2076–2089.
21. MEDRANO, C.J. & D.A. FOX. 1994. Substrate-dependent effects of calcium on rat retinal mitochondrial respiration: physiological and toxicological studies. Toxicol. Appl. Pharmacol. **125:** 309–321.
22. TRUMP, B.F. & I.K. BEREZESKY. 1996. The role of altered $[Ca^{2+}]_i$ in apoptosis, oncosis and necrosis. Biochim. Biophys. Acta **1313:** 173–178.
23. NICOTERA, P. & S. ORRENIUS. 1998. The role of calcium in apoptosis. Cell Calcium **23:** 173–180.
24. GREEN, D. & G. KROEMER. 1998. The central executioners of apoptosis: caspases or mitochondria? Trends Cell. Biol. **8:** 267–271.
25. ICHAS, F. & J.P. MAZAT. 1998. From calcium signaling to cell death: two conformations for the mitochondrial permeability transition pore. Switching from low- to high-conductance state. Biochim. Biophys. Acta **1366:** 33–50.
26. FOX, D.A. & L.M. KATZ. 1992. Developmental lead exposure selectively alters the scotopic ERG component of dark and light adaptation and increases rod calcium content. Vision Res. **32:** 249–255.
27. CROSSON, C.E., J.A. WILLIS & D.E. POTTER. 1990. Effect of the calcium antagonist, nifedipine, on ischemic retinal dysfunction. J. Ocul. Pharmacol. **6:** 293–299.
28. MILAM, A.H., D.M. DACEY & A.M. DIZHOOR. 1993. Recoverin immunoreactivity in mammalian cone bipolar cells. Vis. Neurosci. **10:** 1–12.
29. BROWN, D.G., X.M. SUN & G.M. COHEN. 1993. Dexamethasone-induced apoptosis involves cleavage of DNA to large fragments prior to internucleosomal fragmentation. J. Biol. Chem. **268:** 3037–3039.
30. HUANG, B. & D.A. REDBURN 1996. GABA-induced increases in $[Ca^{2+}]_i$ in retinal neurons of postnatal rabbits. Vis. Neurosci. **13:** 441–447.
31. SZABO, I., P. BERNARDI & M. ZORATTI. 1992. Modulation of the mitochondrial megachannel by divalent cations and protons. J. Biol. Chem. **267:** 2940–2946.

Molecular Mechanisms of Free Radical Production and Protective Efficacies of Antioxidants in *in Vitro* Ischemia-Reperfusion

MARINA V. FRANTSEVA,[a] PETER L. CARLEN, AND
JOSE L. PEREZ VELAZQUEZ

Playfair Neuroscience Unit, Toronto Western Hospital, 399 Bathurst St., Toronto, Ontario M5T 2S8, Canada

Compelling evidence implicates free radicals (FRs) as major contributors to ischemic and excitotoxic tissue injury in the CNS.[1] The biochemical mechanisms leading to FR production during ischemic brain injury remain unclear owing to methodological difficulties in detecting short-lived FRs in *in vivo* experiments, and a lack of neuronal circuitry required for FR generation in most of the *in vitro* model systems. We investigated the mechanisms of FR overproduction in rat CA1 pyramidal neurons of organotypic hippocampal slices during ischemia-reperfusion injury (IRI). IRI was initiated by superfusion of cultured hippocampal slices with glucose-free deoxygenated artificial cerebrospinal fluid (ACSF) for 8 minutes, as described in detail in reference 2. Ischemia-induced free radical generation (measured as changes in the fluorescence emission of dihydrorhodamine123) temporally correlated with intracellular calcium elevation, as measured by injection of fluo-3 in individual pyramidal cells employing patch electrodes. Both FR generation and intracellular calcium accumulation were markedly diminished in the presence of glutamate receptor blockers AP-5 and CNQX, implicating glutamate-mediated $[Ca^{2+}]_i$ rises as a direct cause of FR overproduction resulting from IRI. FR generation was greatly decreased by the mitochondrial complex I blocker, rotenone, indicating that mitochondria are the principal source of ischemic FR production. Measurements of mitochondrial calcium with the mitochondrial calcium probe rhod-2, revealed that peaks of FR production during and after the anoxic episode always followed the step-like increases of mitochondrial calcium, suggesting that mitochondrial calcium overload is linked to ischemic mitochondrial FR generation.

Biochemical cascades initiated by oxidative stress and excitotoxic intracellular calcium rises are thought to converge on mitochondrial dysfunction (mitochondrial permeability transition[3]). The mitochondrial permeability transition (MPT) is a gradual loss of mitochondrial potential accompanied by the swelling of the organelles, prevented by cyclosporine A (CsA).[3] A gradual, CsA-sensitive decrease of the mitochondrial potential, indicative of the MPT, has been previously observed in our ischemia model.[4] Because the MPT resulting from oxidative stress has been shown to cause secondary FR generation in hepatocytes,[5] we sought to investigate the possible contribution of IRI-induced mitochondrial dysfunction to FR genera-

---

[a]Address for correspondence: Marina V. Frantseva, Playfair Neuroscience Unit, McL 12-413, Toronto Western Hospital,399 Bathurst Street, Toronto, Ontario M5T 2S8, Canada. Phone: 416-603-5040; fax: 416-603-5745.

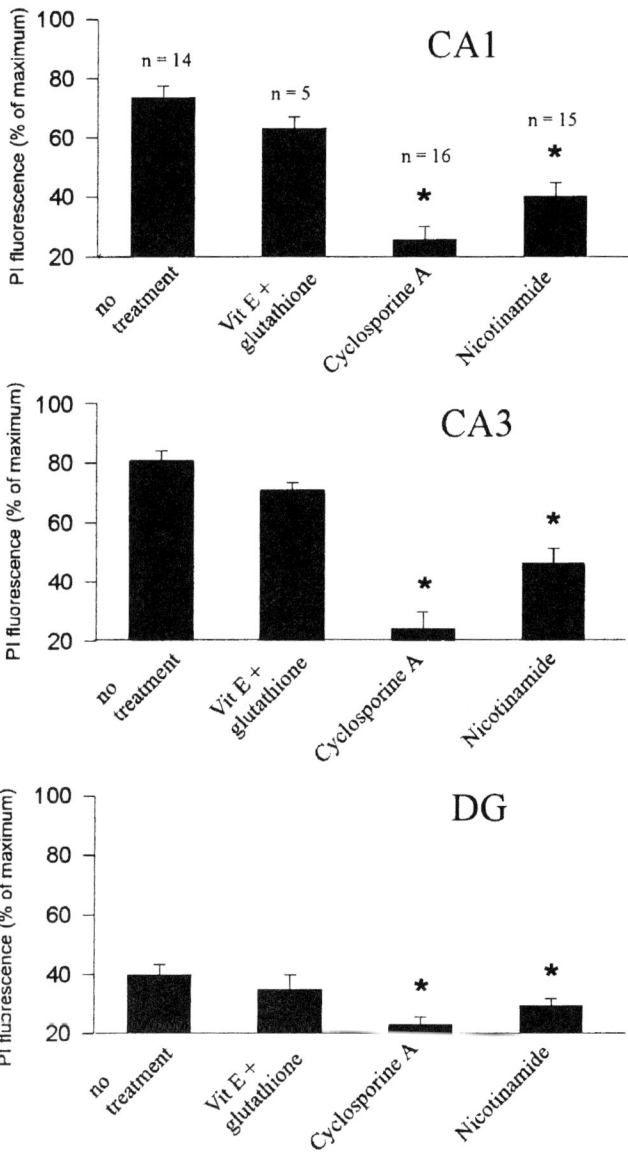

FIGURE 1. Inhibitors of pyridine nucleotide hydrolysis, but not free radical scavengers, attenuate neuronal injury in an *in vitro* model of ischemia-reperfusion. Organotypic hippocampal slice cultures were used after 14 days *in vitro*. The ischemic insult was initiated by submerging the cultured slices in glucose-free balanced salt solution, and was terminated by transferring the slices back to normal culture medium after 1 hour. Cell death was assessed in the three main areas of the slice (CA1, CA3, and DG) by the fluorescence of the viability marker propidium iodide (PI) over a 48-hour period. Cell death at 48 hours was

tion. Ischemic injury was initiated in the slices pretreated with 5 µM CsA in the continuous presence of the drug. CsA completely arrested both ischemic FR generation and mitochondrial calcium overload, while having no effect on intracellular calcium levels, suggesting that, in addition to arresting the MPT, CsA has an additional site of action involved in blocking FR generation. Available evidence suggests that CsA prevents mitochondrial calcium disregulation via inhibition of pyridine nucleotides (NAD^+) hydrolysis.[6] The NAD^+ hydrolysis inhibitor nicotinamide[7] (25 mM), reproduced CsA effects on FR generation, mitochondrial calcium accumulation and cytoplasmic calcium increases.

Electrophysiological whole-cell recordings from CA1 neurons revealed membrane potential depolarization and decreased input resistance during IRI insult. Glutamate receptor blockade or CsA prevented neuronal depolarization and input resistance decrease during the ischemic episode. To further elucidate possible protective role of these agents, we used an *in vitro* ischemia model, developed in our laboratory and described in reference 8 (see also FIG. 1). Cell death was assessed with propidium iodide as described[8] over a 48-hour period in the presence or absence of nicotinamide (25 mM), CsA (5 µM), or antioxidants (vitamin E, 50 µM and glutathione, 1 mM). There was no significant reduction of cell death with the FR scavengers, whereas both CsA and nicotinamide dramatically attenuated IRI-induced neuronal loss (FIG. 1).

These observations indicate that IRI-induced FR generation occurs because of mitochondrial calcium overload, caused by glutamate-mediated $[Ca^{2+}]_i$ rises. Nicotinamide and CsA, agents known to prevent NAD^+ hydrolysis and mitochondrial dysfunction, prevented mitochondrial calcium accumulation, FR generation, and significantly attenuated neuronal loss resulting from the ischemic insult.

REFERENCES

1. DUBINSKY, J.M., B.S. KRISTAL & M. ELIZONDO-FOURNIER. 1995. An obligate role for oxygen in the early stages of glutamate-induced, delayed neuronal death. J. Neurosci. **15** (11): 7071–7078.
2. PEREZ VELAZQUEZ, J.L., M.V. FRANTSEVA & P.L. CARLEN. 1997. In vitro ischemia promotes glutamate-mediated free radical generation and intracellular calcium accumulation in pyramidal neurons of cultured hippocampal slices. J. Neurosci. **17** (23): 9085–9094.
3. ZORATTI, M. & I. SZABO. 1995. The mitochondrial permeability transition. Biochem. Biophys. Acta **1241**: 139–176.
4. PEREZ VELAZQUEZ, J.L. *et al.* Porin required for ischemia-induced mitochondrial dysfunction and neuronal damage. Neuroscience. Submitted.
5. NIEMINEN, A.L. *et al.* 1997. Mitochondrial permeability transition in hepatocytes induced by t-BuOOH: NAD(P)H and reactive oxygen species. Am. J. Physiol. **272** (4 Pt .1): C1286–1294.

---

quantified as per cent of maximal PI fluorescence, obtained by incubation of the slices at 4°C in the presence of PI for 24 hours.[8] Images were acquired by an MRC-600 Bio-Rad confocal microscope and were stored and analyzed using the Bio-Rad software. Cyclosporine A (5 µM), nicotinamide (25 mM), and antioxidants (glutathione, 1 mM and vitamin E, 50 µM) were applied 6 hours prior to the insult and maintained throughout the experiment. Asterisks denote statistical significance ($p < 0.01$) between treated and untreated groups. *Bars* represent means ± standard errors.

6. RICHTER, C., M. THEUS & J. SCHLEGEL. 1990. Cyclosporine A inhibits mitochondrial pyridine nucleotide hydrolysis and calcium release. Biochem. Pharmacol. **40** (4): 779–782.
7. LÖTSCHER, H.R. *et al.* 1980. Hydroperoxide-induced loss of pyridine nucleotides and release of calcium from rat liver mitochondria. J. Biol. Chem. **255** (19): 9325–9330.
8. FRANTSEVA, M.V., P.L. CARLEN & H. EL-BEHEIRY. 1999. A static-perfusion method to induce hypoxic damage in organotypic hippocampal cultures. J. Neurosci. Methods **89:** 25–31.

A Disturbance in the Neuronal Insulin Receptor Signal Transduction in Sporadic Alzheimer's Disease

L. FRÖLICH,[a,d] D. BLUM-DEGEN,[b] P. RIEDERER,[b] AND S. HOYER[c]

[a]*Department of Psychiatry I, University of Frankfurt/Main,*
60528 Frankfurt am Main, Germany

[b]*Department of Psychiatry, University of Würzburg, 97080 Würzburg, Germany*

[c]*Department of Pathochemistry and General Neurochemistry, University of Heidelberg,*
69120 Heidelberg, Germany

INTRODUCTION

In sporadic dementia of Alzheimer type (SDAT), the hypothesis has been forwarded that defects in the regulation of glucose metabolism, i.e., due to changes in CNS insulin receptor function, might be an early contributing event to the onset of SDAT.[1–3] Brain insulin regulates enzymes of cerebral glucose metabolism via specific high-affinity insulin receptors.[4,5] Furthermore, insulin also binds to insulin-like growth factor I receptors and via these receptors possibly exerts trophic effects on neuronal cells and interacts with cholinergic neurotransmission.[6]

We have investigated, whether the binding properties of insulin receptors in the brain in SDAT change in relation to middle-aged and aged controls and whether this relates to changes in the activity of tyrosine receptor kinase, a common output system to insulin receptors.

MATERIAL AND METHODS

Brains from persons with Alzheimer's disease diagnosed clinically and histopathologically and control brains without a history of neurological or psychiatric diseases were investigated. The sample consisted of 17 patients with SDAT (mean age: 79.8 ± 2.0 years, postmortem delay: 24.7 ± 4.1 hours, brain lactate content: 20.7 ± 1.5 mg/dl) and 21 controls (mean age: 64.5 ± 5.1 year, postmortem delay: 25.4 ± 4.0 hours, brain lactate content: 21.7 ± 1.4 mg/dl). A middle-aged ($n = 8$) and an aged control group ($n = 13$) was selected form these controls. Brain tissue was obtained using an anatomical atlas and followed a standard procedure.[7] For ELISA the tissue was homogenized at 4°C in a medium containing 10 mM MOPS (pH 7.6), 120 mM NaCl, 1 mM EDTA, 0.1 mM benzethonium chloride, 1 mM benzamidine and 0.1%

---

[d]Address for correspondence: Priv.-Doz. Dr. Lutz Frölich, Department of Psychiatry, University of Frankfurt/Main Medical Center, Heinrich-Hoffmannstr. 10, D-60528 Frankfurt am Main, Germany; Phone: +49-69-6301-7094; fax: + 49-69-6301-5189.
e-mail: froelich@em.uni-frankfurt.de

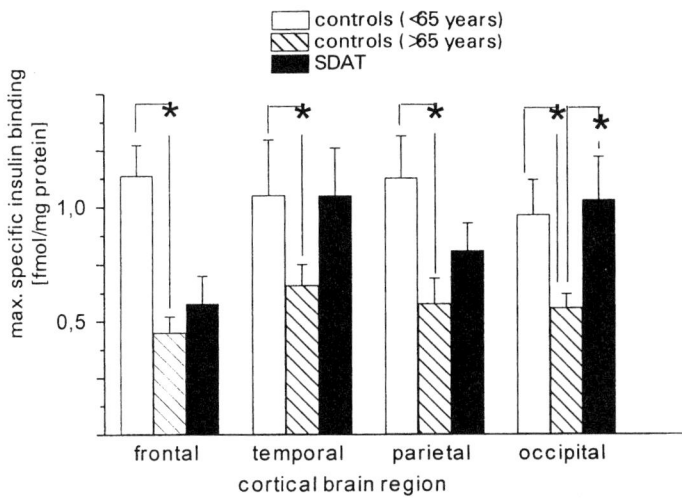

FIGURE 1. Density (B_{max}) of insulin receptor-binding in post-mortem human brain cortex in SDAT compared to middle-aged and age-matched controls. Values are means ± SEM, given as fmol/mg protein for receptor density; sample size is $n = 13$ for aged controls (> 65 years), $n = 17$ for SDAT. Significant differences of SDAT group vs. control values or of middle-aged vs. aged control group ($p < 0.05$ by Mann-Whitney U-test) are marked by asterisk.

trasylol. For radioligand binding assays, the tissue was homogenized in MOPS/sucrose buffer and washed twice. Radioligand binding assays were performed as "cold" saturation assays with ^{125}J-insulin or ^{125}J-IGF-I, respectively. 10^{-10} M tracer were displaced by increasing amounts of cold ligand, unspecific binding was determined by 10^{-7} M cold ligand. Bound and free ligand were separated by centrifugation, binding parameters were calculated by EBDA and LIGAND software. Tyrosine kinase activity was determined in brain homogenates with a modified and adapted ELISA system following the methods and using antibodies according to Rijksen and Staal.[8]

RESULTS

In comparison to the middle-aged control group, region-specific changes of insulin receptors in SDAT became evident. A slight decrease in receptor densities was found in frontal and parietal cortices only, whereas no changes in temporal and occipital cortices could be observed. With respect to insulin receptors in SDAT compared to age-matched controls, we showed for the first time that the number of insulin receptors was increased in all regions of brain cortex, with a significant increase in the occipital cortex. Tyrosine kinase activity in SDAT was reduced in all cortical regions studied, compared to both control groups. This loss of activity was significant in temporal and occipital cortex ($p < 0.05$) in comparison to the middle-aged, and significant for occipital cortex only ($p < 0.05$) in comparison to the aged control group.

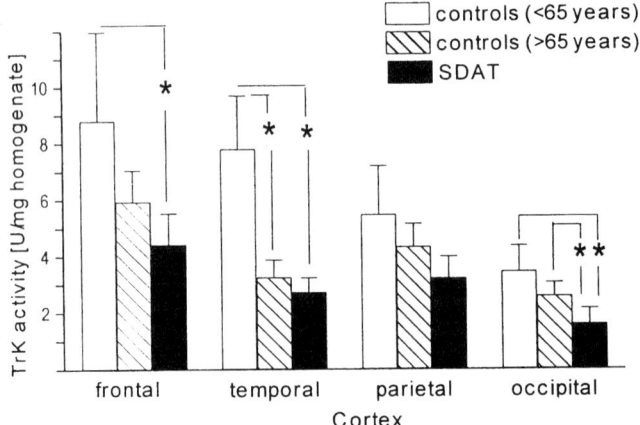

FIGURE 2. Tyrosine kinase (TrK) activity in various postmortem human brain areas in SDAT compared to middle-aged and age-matched controls. Values are means ± SEM, given as U/mg homogenate for TrK activity; sample size is $n = 13$ for aged controls (> 65 years), $n = 17$ for SDAT. Significant differences of SDAT group vs. control values or of middle-aged vs. aged control group ($p < 0.05$ by Mann-Whitney U-test) are marked by an *asterisk*.

DISCUSSION

Cell culture and animal experiments show that insulin in the brain has potent effects on neuronal glucose metabolism and cell differentiation.[9,10] In our brain samples, the number of insulin receptors decreased with advancing age. During human ontogenesis, brain insulin receptors have also been shown to decrease.[11] In this study, we compared brain samples with histopathologically comfirmed Alzheimer's disease with an age-matched control group without neuropsychiatric disorders. Close matching of both groups with respect to several important factors in addition to age, i.e. sex, postmortem delay, storage time of the brain tissue, and lactate content of the brain as measure of the agonal period excluded any major artifacts (data not shown). With respect to insulin receptors in SDAT compared to age-matched controls, we showed for the first time that the number of insulin receptors was increased in all regions of brain cortex, with a significant increase in the occipital cortex. It is suggested that this finding is due to an upregulation of the insulin receptor system in SDAT on top of a general loss of insulin receptors with normal aging, possibly because of an impaired signal transduction system in SDAT. This is supported by the reduced tyrosine kinase activity, the major effector system for insulin and IGF-1 receptors. Other experiments with postmortem human brain have provided additional evidence for a biological role of insulin in the brain in Alzheimer's disease,[12] as well as an involvement of insulin-degrading enzyme in amyloid plaques.[13] Our data are compatible with the hypothesis that in sporadic Alzheimer's disease brain, the known disturbances in glucose metabolism are, at least in part, due to a reduced activity of the insulin receptor–dependent signal transduction, most likely at the level of one of the kinase/phosphatases involved. This abnormality may be of pivotal significance for the early pathogenesis of SDAT.

SUMMARY

Disturbances of glucose and energy metabolism are hypothesized as pathogenetic factors in sporadic dementia of Alzheimer type (SDAT). Insulin and is receptors play an important role in the regulation of brain glucose metabolism and neuronal growth. In postmortem brain cortex in SDAT, the densities of brain insulin receptors were decreased compared to adult controls, but were increased in relation to aged controls. Tyrosine kinase activity, a signal transduction mechanism common to insulin and IGF-1 receptors, was reduced in SDAT in comparison to middle-aged and age-matched control groups. The data are consistent with a neurotrophic role of insulin in the human brain and an upregulation of insulin receptors in SDAT brain as a compensatory mechanism, possibly due to impaired signal transduction mechanism.

REFERENCES

1. HOYER, S. 1998. Risk factors for Alzheimer's disease during aging. Impacts of glucose/energy metabolism. J. Neural Transm. Suppl. **54:** 187–194.
2. VANHANEN, M & H. SOININEN. 1998. Glucose intolerance, cognitive impairment and Alzheimer's disease. Curr. Opin. Neurol. **11:** 673–677.
3. WICKELGREN, I. 1998. Tracking insulin to the mind. Science **280:** 517–519.
4. DE PABLO, F. & E. DE LA ROSA. 1995. The developing CNS: a scenario for the action of proinsulin, insulin and insulin-like growth factors. Trends Neurosci. **18:** 143–150.
5. WOZNIAK, M., et al. 1993. The cellular and physiological actions of insulin in the central nervous system. Neurochem. Int. **22:** 1–10.
6. QUIRION, R., et al. 1991. Growth factors and lymphokines: modulators of cholinergic neuronal activity. Can. J. Neurol. Sci. **18:** 390–393.
7. GSELL, W., et al. 1993. How to run a brain bank. A report from the Austro-German brain bank. J. Neural Transm. (Gen. Sect.) Suppl. **39:** 31–70.
8. RIJKSEN, G. & G.E.J. STAAL. 1992. Methodologies for the detection and nonradioactive assay of protein-tyrosine kinase activities in human brain and gliomas. Neuroprotocols **1:** 201–206.
9. HOYER, S. et al. 1993. Stimulation of glycolytic key enzymes in cerebral cortex by insulin. NeuroReport **4:** 991–993.
10. KNUSEL, B., et al. 1990. Selective and nonselective stimulation of central cholinergic and dopaminergic development in vitro by nerve growth factor, basic fibroblast growth factor, epidermal growth factor, insulin and the insulin-like growth factors I and II. J. Neurosci. **10:** 558–570.
11. POTAU, N., et al. 1991. Ontogenesis of insulin receptors in human cerebral cortex. J Endocrinol. Invest. **14:** 53–58.
12. FRÖLICH, L. et al. 1998. Insulin and insulin receptors in the brain in aging and in sporadic Alzheimer's disease. J. Neural Transm. **105:** 423–438.
13. BERNSTEIN, H.G., et al. 1999. Insulin-degrading enzyme in the Alzheimer's disease brain: prominent localization in neurons and senile plaques. Neurosci. Lett. **263:** 161–164.

Modulation of Presenilin-1 Processing by Nitric Oxide during Apoptosis Induced by Serum Withdrawal and Glucose Deprivation

LAURA GASPARINI,[a] ROBERTA GHIDONI,[a,e] ANTONELLA C. ALBERICI,[a]
LUISA BENUSSI,[a] DANIELE MORATTO,[a] MARCO TRABUCCHI,[b]
JOHN H. GROWDON,[c] ROGER M. NITSCH,[d] AND GIULIANO BINETTI[a]

[a]*Neurobiology Lab, IRCCS "Centro S. Giovanni di Dio-Fatebenefratelli," Brescia, Italy*

[b]*Department of Experimental Medicine and Biochemical Sciences, University of Rome Tor Vergata, Rome, Italy*

[c]*Department of Neurology, MGH Harvard Medical School, Boston, Massachusetts, USA*

[d]*Center for Molecular Neurobiology, University of Hamburg, Hamburg, Germany*

Mutations in presenilin genes account for a major portion of early onset familial Alzheimer's disease (AD). The biological functions of presenilins are still unknown, but increasing evidence points out a pivotal role in apoptosis. It was demonstrated that presenilins overexpression increases susceptibility to apoptosis. Presenilins have an alternative cleavage by caspase3 during apoptosis.[1] Oxidative stress, and, particularly, nitric oxide (NO) are involved in regulating the molecular mechanisms leading to apoptosis and neurodegeneration. Markers of NO-derived nitrogen reactive species formation were detected in AD brain.[2–4] NO is a very reactive substance that could be both neurotoxic[5,6] or neuroprotective[7,8] depending on the redox state of the cell. In this study we set out to examine the effect of nitric oxide on apoptosis induced by serum withdrawal (SW) and/or glucose deprivation (GD, serum and glucose deprivation + 50 mM 2-deoxy-D-glucose) for 24 hours in confluent H4 human neuroglioma cells and how apoptosis and NO may influence presenilin 1 (PS1) metabolism.

Sodium nitroprusside (SNP, 1mM) and SNAP (500 µM) were used as NO-donors. Some experiments were performed coincubating SNP or SNAP with 1 mM Carboxy-PTIO (C-PTIO), a specific NO-scavenger. PS1 full length protein and 21 kD fragment were detected by western blot analysis using respectively a C-terminal (4318) and an anti-loop (9672) polyclonal antibodies.

NO-donors treatment enhanced cell death by apoptosis under both SW or GD as shown by the increased cytochrome *c* release from mitochondria (FIG. 1) and TUNEL assay (not shown), and this was associated with an increased immunoreactivity for nitrotyrosine (not shown) demonstrating the formation of peroxynitrite. C-PTIO reduced cytochrome *c* release from mitochondria under SW but not under GD. Incubation with NO-donors for 24 hours was able to increase PS1 21 kD intracellular fragment production during apoptosis induced by SW or GD. A redistribution of full

---

[e]*Address correspondence to: Dr. R.Ghidoni, IRCCS "San Giovanni di Dio-FBF," Via Pilastroni, 4-25125 Brescia, Italy. Phone: +39-030-3501710; fax: +39 -030 -3501366.
e-mail: rghidoni@oh-fbf.it*

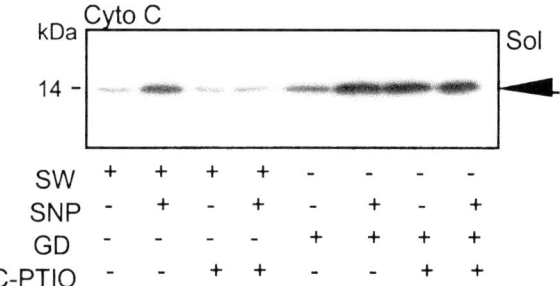

FIGURE 1. Apoptosis induced by SW and GD and neurotoxic effect of 1 mM SNP. Cytochrome c release from mitochondria was increased by treatment with SNP 1 mM under both SW and GD for 24 h. 1 mM C-PTIO reduced cytochrome c release from mitochondria under SW but not under GD.

length PS1 from the particulate fraction to a cytoskeletal fraction was observed following NO-donors treatment (FIG. 2). The full-length PS1 redistribution and 21-kD PS1 fragment formation were blocked by a specific NO-scavenger (C-PTIO) under SW, but not under GD conditions (FIG. 2), suggesting the formation of different kind of nitrogen reactive species in these apoptotic conditions. These data suggest a triggering role of NO in SW conditions but not in apoptosis induced by GD where another kind of NO-derived reactive species may be formed. Full length PS1 level was not altered (not shown) in total lysates from H4 cells during apotosis induced by SW or GD in absence and presence of NO-donors, suggesting a recompartimentalization of the protein rather than an effect on its expression. Bcl-2 seems to have no protective effect in the apoptotic conditions used in this study. No differences in cytochrome c release from mitochondria, PS1 cleavage and subcellular distribution were observed both under SW and GD in absence and presence of SNP for 24 hours between H4 cells overexpressing Bcl-2 and wild type cells (not shown). Since it has been demonstrated that NO-donors can induce cytoskeletal breakdown[9] during apoptosis and presenilin 1 can bind with proteins directly (tau protein) or indirectly (β-catenin, filamin, ABP 280) involved in cytoskeleton formation and stabilization,[10–13] according to the redistribution of PS1 protein in the cytoskeletal fraction after NO-donors treatment, it seems interesting to analyze the cytoskeletal status as well as PS1 localization by immunocytochemical analysis in the experimental conditions adopted in this study. Following NO-donors incubation under both SW and GD, immunocytochemical analysis of α-tubulin, F-actin, and tau protein (monoclonal antibody anti-tau 7.51 was a kind gift of Dr. C. Harrington) demonstrated a wide cytoskeletal breakdown while PS1 immunoreactivity was more concentrated in a nuclear and perinuclear area of H4 cells. These results show that NO and/or its related nitrogen reactive species enhanced neurotoxicity during apoptosis induced by SW and GD and modulate PS1 processing and subcellular distribution in H4 neuroglioma cells, but the significance of PS1 metabolism alterations during apoptosis and its relationship with cytoskeletal breakdown has not been established.

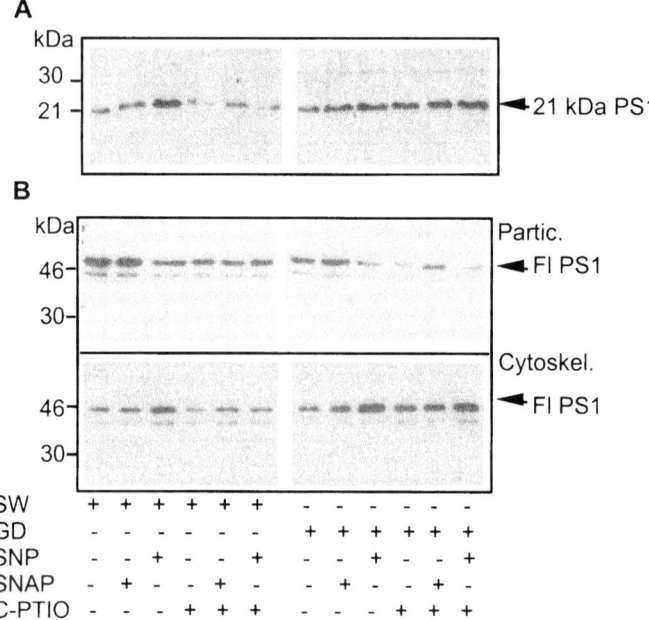

FIGURE 2. PS1 21-kD fragment and full length protein immunoreactivity in H4 cells treated for 24 h with NO-donors under SW and GD. **A:** PS1 21-kD C-terminal fragment was detected using 9672 anti-loop polyclonal antibody. Treatment with NO-donors (1 mM SNP; 500 μM SNAP) increased PS1 21 kD fragment in the membrane fraction of H4 cells under both SW and GD. 1 mM C-PTIO had a rescue effect during apoptosis induced by SW, but not under GD. **B:** Full-length PS protein redistribution from membrane to cytoskeletal fraction in H_4 cells treated with NO-donors under SW and GD. Full-length PS1 was analyzed using 4318 polyclonal antibody directed against the C-terminal domain of the protein. PS1 immunoreactivity decreased in the membrane fraction (Partic) and increased in the cytoskeletal fraction (Cytoskel) following NO-donors treatment under both SW and GD for 24 h. 1 mM C-PTIO had a rescue effect during apoptosis induced by SW, but not under GD.

ACKNOWLEDGMENTS

We are grateful to Dr. C. Harrington, University of Aberdeen, Aberdeen, Scotland for the kind gift of 7.51 antibody.

REFERENCES

1. MATTSON, M.P. *et al.* 1998. Presenilins, the endoplasmic reticulum, and neuronal apoptosis in Alzheimer's disease. J. Neurochem. **70:** 1–14.
2. SMITH, M.A. *et al.* 1997. Widespread peroxynitrite–mediated damage in Alzheimer's disease. J. Neurosci. **17:** 2653–2657.
3. SU, J.H. *et al.* 1997. Neuronal DNA damage precedes tangle formation and is associated with up-regulation of nitrotyrosine in Alzheimer's disease brain. Brain Res. **774:** 193–199.

4. HENSLEY, K. *et al.* 1998. Electrochemical analysis of protein nitrotyrosine and dityrosine in the Alzheimer brain indicates region-specific accumulation. J. Neurosci. **18:** 8126–8132.
5. DAWSON, V.L. *et al.* 1993. Mechanisms of nitric oxide-mediated neurotoxicity in primary brain cultures. J. Neurosci. **16:** 2651–2661.
6. ZHANG, J. *et al.* 1994. Nitric oxide activation of poly-(ADP-ribose) synthetase in neurotoxicity. Science **263:** 687–689.
7. LIPTON, S.A. *et al.* 1993. A redox-based mechanism for the neuroprotective and neurodestructive effects of nitric oxide and related nitroso-compounds. Nature **364:** 626–632.
8. WINK, D.A. *et al.* 1995. Nitric oxide protects against cytotoxic effects of reactive oxygen species. Ann. N.Y. Acad. Sci. **738:** 265–278.
9. BONFOCO, E. *et al.* 1996. Cytoskeletal breakdown and apoptosis elicited by NO donors in cerebellar granule cells require NMDA receptor activation. J. Neurochem. **67:** 2484–2493.
10. MURAYAMA, M. *et al.* 1998. Direct association of presenilin-1 with β-catenin. FEBS Lett. **433:** 73–77.
11. TAKASHIMA, A. 1998. Presenilin 1 associates with glycogen synthase kinase-3β and its substrate tau. Proc. Natl. Acad. Sci. USA **95:** 9637–9641.
12. YU, G. *et al.* 1998. The presenilin 1 protein is a component of a high molecular weight intracellular complex that contains β-catenin. J. Biol. Chem. **273:** 16470–16475.
13. ZHANG, W. *et al.* 1998. Interaction of presenilins with the filamin family of actin-binding proteins. J. Neurosci. **18:** 914–922.

Metabolic and Glutamatergic Disturbances in the Huntington's Disease Transgenic Mouse

DONALD S. HIGGINS,[a] KARI R. HOYT, CORINNE BAIC, JESSICA VENSEL, AND MATTHEW SULKA

Departments of Neurology, and Cell Biology, Neurobiology and Anatomy
The Ohio State University, Columbus, Ohio 43210, USA

Huntington's disease (HD) is an autosomal dominant neurodegenerative disorder characterized by involuntary movements, mood disturbance and dementia. Excitotoxicity appears to play an essential role in striatal projection neuron degeneration. Accumulating evidence suggests augmentation of neuronal injury by metabolic dysfunction.[1] Inefficient ATP synthesis will compromise the resting membrane potential and induce aberrant glutamate receptor activation and excitotoxic neuronal injury.

Recently, the trinucleotide repeat expansion accounting for autosomal dominant inheritance has been inserted into the laboratory mouse.[2] This transgenic strain (R6/2) has provided a valuable tool with which to investigate the pathogenesis of HD. We have examined, using enzyme histochemistry and receptor autoradiography, the integrity of oxidative metabolism, intermediary metabolism and ionotropic glutamate receptors in brain of this strain and wildtype controls ($n = 4$/group).

Transgenic animals were significantly smaller than nontransgenic animals at 8 weeks of age (24.6 g vs. 30.1 g, $p = 0.008$). Metabolic and receptor alterations were evident in all brain regions examined (TABLE 1). A generalized reduction (−42.7–58.8%) in lactate dehydrogenase (LDH) activity was observed in transgenic animals. Disturbances of lactate metabolism have been reported in HD.[3] Reduced LDH activity may represent the enzymatic correlate of cerebral lactic acidosis. [^{3}H]Dihydrorotenone binding to complex I of the electron transport chain was increased in striatum (+14.5%, $p = 0.04$) and molecular layer of the cerebellum (+10.8%, $p = 0.008$) of the HD transgenic mice. Decreased complex I activity has been reported in HD platelets.[4] Increased complex I protein may reflect compensation for impaired function. Abnormalities of complex II/III and complex IV have been previously described in HD[5,6] and recently in the R6/2 line.[7] While no change was observed in the current experiments, the optimized histochemical methods used to visualize cytochrome oxidase (complex IV) and succinate dehydrogenase (complex II) may have obscured subtle, yet physiologically relevant, changes in these enzymes.

Phosphate-activated glutaminase (PAG) plays a crucial role in transmitter glutamate metabolism. Increased striatal PAG activity (+30.0%, $p = 0.004$) in the transgenic mice may reflect an increase in synaptic glutamate. In accord with this

---

[a]Address for correspondence: Donald S. Higgins, Jr., MD, The Ohio State University, 188 Medical Research Facility, 420 West 12th Ave., Columbus, OH 43210. Phone: 614-688-4048; fax: 614-688-8755.
e-mail: higgins.84@osu.edu

TABLE 1. Histochemical (relative optical density) and autoradiographic (pmol/mg protein) examination of oxidative metabolism, intermediary metabolism and glutamate receptors in the HD transgenic mouse (R6/2)

	Metabolism		Glutamate Metabolism and Receptors	
	Transgenic	Wildtype	Transgenic	Wildtype
	Lactate Dehydrogenase		PO_4 Activated Glutaminase	
Striatum	0.153 ± 0.011^a	0.260 ± 0.029	0.338 ± 0.011^b	0.260 ± 0.014
Dentate Gyrus	0.124 ± 0.012^b	0.240 ± 0.022	0.324 ± 0.027	0.307 ± 0.014
Molecular Layer	0.071 ± 0.006^a	0.166 ± 0.030	0.250 ± 0.018	0.209 ± 0.014
	Isocitrate Dehydrogenase		Glutamate dehydrogenase	
Striatum	0.871 ± 0.088	0.703 ± 0.032	0.121 ± 0.003	0.114 ± 0.004
Dentate Gyrus	0.995 ± 0.105	0.877 ± 0.035	0.138 ± 0.007	0.144 ± 0.003
Molecular Layer	0.760 ± 0.093	0.630 ± 0.009	0.071 ± 0.004	0.076 ± 0.004
	[^{3}H]Dihydrorotenone		Nitric Oxide Synthase	
Striatum	9.47 ± 0.29^a	8.27 ± 0.35	0.154 ± 0.011	0.153 ± 0.008
Dentate Gyrus	11.90 ± 0.82	10.93 ± 0.38	0.129 ± 0.015^a	0.183 ± 0.015
Molecular Layer	10.90 ± 0.19^a	9.84 ± 0.19	0.125 ± 0.006	0.123 ± 0.020
	Succinate Dehydrogenase		[^{3}H]AMPA	
Striatum	0.366 ± 0.013	0.372 ± 0.036	2.19 ± 0.19^a	2.72 ± 0.06
Dentate Gyrus	0.524 ± 0.027	0.484 ± 0.023	3.04 ± 0.43	4.41 ± 0.55
Molecular Layer	0.411 ± 0.039	0.478 ± 0.035	0.53 ± 0.09	1.07 ± 0.26
	Cytochrome Oxidase		[^{3}H]Kainate	
Striatum	0.154 ± 0.009	0.182 ± 0.019	7.14 ± 0.22^b	10.06 ± 0.28
Dentate Gyrus	0.182 ± 0.005	0.190 ± 0.016	8.81 ± 0.37	8.92 ± 0.23
Molecular Layer	0.174 ± 0.016	0.240 ± 0.027	3.22 ± 0.28^b	5.50 ± 0.29
	Na, K,-ATPase		[^{3}H]MK-801	
Striatum	0.117 ± 0.011	0.132 ± 0.04	7.00 ± 0.29^a	9.45 ± 0.54
Dentate Gyrus	0.169 ± 0.011	0.162 ± 0.035	19.51 ± 0.37^b	25.27 ± 0.82
Molecular Layer	0.199 ± 0.016	0.189 ± 0.038		

NOTE: Values represent Mean ± SEM ($n = 4$/group). $^a p < 0.05$, $^b p < 0.005$.

metabolic disturbance, glutamate receptor binding was altered in the transgenic strain. Consistent with the recent report of Cha et al.,[8] [^{3}H]AMPA (-19.5%, $p = 0.042$) and [^{3}H]kainate (-29.0%, $p = 0.0002$) binding were decreased in transgenic striatum. NMDA receptor binding ([^{3}H]MK801 binding) was reduced in striatum (-25.9%, $p = 0.007$) in contrast to the results of Cha et al. This disparity likely arises from methodological differences ([^{3}H]MK801 vs. NMDA-sensitive [^{3}H]glutamate autoradiography).

As neurodegeneration has not been identified in this strain at 2 months of age, the present findings are unlikely to result from cell loss. While extrastriatal changes were also observed, conspicuous metabolic and glutamate receptor alterations in the striatum reinforces the relevance of "weak excitotoxicity" to the pathogenesis of HD. Recent demonstration of glucose intolerance in this transgenic strain complicates interpretation of these results.[9] Whether the observed changes in oxidative metabolism, intermediary metabolism and glutamate metabolism are pathophysiologically relevant, or simply reflect chronic hyperglycemia, remains to be determined.

ACKNOWLEDGMENTS

Research presented in this paper was supported by the Vaughan Family Huntington's Disease Research Fund.

REFERENCES

1. BEAL, M.F. 1995. Aging, energy and oxidative stress in neurodegenerative diseases. Ann. Neurol. **38**(3): 357–366.
2. BATES, G.P. *et al.* 1997. Transgenic models of Huntington's disease. Human Molec. Gen. **6**(10): 1633–1637.
3. JENKINS, B.G. *et al.* 1993. Evidence for impairment of energy metabolism in vivo in Huntington's disease using localized 1H NMR spectroscopy. Neurology **43**: 2689–2695.
4. PARKER, W.D. *et al.* 1990. Evidence for a defect in NADH:ubiquinone oxidoreductase (complex I) in Huntington's disease. Neurology **40**: 1231–1234.
5. BRENNAN, W.A., E.D. BIRD & J.R. APRILLE. 1985. Regional mitochondrial respiratory activity in Huntington's disease brain. J. Neurochem. **44**: 1948–1950.
6. TABRIZI, S.J. *et al.* 1999. Biochemical abnormalities and excitotoxicity in Huntington's disease brain. Ann. Neurol. **45**: 25–32.
7. TABRIZI, S.J. *et al.* 1998. Mitochondrial biochemical features of the R6/2 transgenic mouse model of Huntington's disease. Movement Disorders **13** (Suppl. 2): 153.
8. CHA, J.-H.J. *et al.* 1998. Altered brain neurotransmitter receptors in transgenic mice expressing a portion of an abnormal human Huntington disease gene. Proc. Natl. Acad. Sci. USA **95**: 6480–6485.
9. HURLBERT, M.S. *et al.* 1999. Mice transgenic for an expanded CAG repeat in the Huntington's disease gene develop diabetes. Diabetes **48**: 649–651.

Inhibition of the Neuronal Insulin Receptor Causes Alzheimer-like Disturbances in Oxidative/Energy Brain Metabolism and in Behavior in Adult Rats

SIEGFRIED HOYER AND HEINRICH LANNERT

Department of Pathochemistry and General Neurochemistry, University of Heidelberg, D-69120 Heidelberg, Germany

INTRODUCTION

Sporadic Alzheimer's disease (SAD) is as yet of unknown etiology. Its pathogenesis, however, is multifactorial. Based on *in vivo* and post mortem findings in SAD patients the working hypothesis was forwarded that SAD is the brain type of diabetes mellitus II.[1,2] To prove this hypothesis, an animal model was established in which the function of the neuronal insulin receptor was inhibited by the intracerebroventricularly (icv) injected diabetogenic substance streptozotocin (STZ). In short-term studies (3 weeks), after a single icv STZ, abnormalities in oxidative/energy metabolism, phospholipid composition of membranes, cholinergic and catecholaminergic functions and in learning and memory were found (for review see Ref. 3) To induce a long-term and progressive deterioration in behavior and in cerebral oxidative/energy metabolism such as is characteristic for SAD, a triplicate icv STZ challenge was performed. Learning, memory and cognition were investigated over a period of 2 and 3 months, and energy metabolism was studied at the end of these periods. We were interested to investigate whether or not estradiol and Ginkgo biloba extract (EGb761) exert beneficial effects on this neuronal damage. Estradiol was found to affect the CNS (for review see Ref. 4) and to regulate glycolysis in the brain,[5] like insulin. EGb 761 demonstrated its therapeutic usefulness for the treatment of dementia and other cerebral insufficiences (for review see Ref. 6) which was clearly demonstrated in a long-term study in SAD.[7]

RESULTS AND CONCLUSIONS

Mental training performed over 6 days enhances and maintains learning, memory and cognition capacities. Triplicate icv STZ induced disturbances in the neuronal insulin signal transduction cascade and caused long-term and progressive abnormalities in learning, memory, and cognition abilities accompanied by permanent and ongoing reductions in neuronal oxidative/energy metabolism. These disturbances resembled those ones found in SAD.[8]

Triplicate icv STZ-induced deficits in learning, memory, and cognitive behavior in adult male rats were improved by estradiol which also normalized the STZ-induced deficits in energy-rich phosphates. Estradiol may be assumed to improve de-

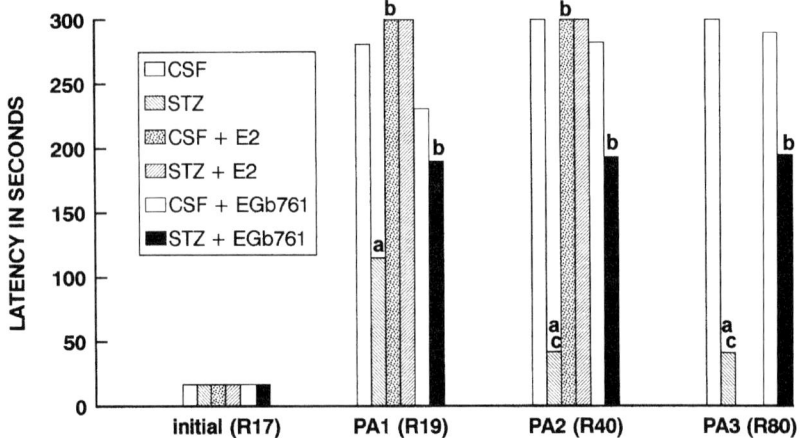

FIGURE 1. Passive avoidance behavior of 1-year-old rats in different experimental groups: after triplicate icv STZ damage and subsequent treatment with 17β-estradiol (E2) s.c. (200 μg/rat/day) over 40 days, and subsequent EGb761 (50 mg/day) over 80 days. STZ/CSF was administered for the first time after habituation (7 days) and training (7 days) at retest days 1, 3, and 20 before the initial passive avoidance behavior at retest day 17 (R17). The step through latency in all six groups studied ($n = 10$/group) was low. Foot shock was applied at retest day 18. Retention tests were conducted at retest days R19, R40, and R80. **a:** $p \leq 0.05$ vs CSF; **b:** $p \leq 0.05$ between STZ + E2 and STZ and STZ + EGb761; **c:** $p \leq 0.05$ between PA1 and PA2/PA3. Experiments were conducted during the dark period in a reversed 12:12 hour light-dark cycle (lights on at 19:00). CSF: control; STZ: triplicate icv streptozotocin damage; E2: 17β-estradiol; EGb761: Ginkgo biloba extract 761.

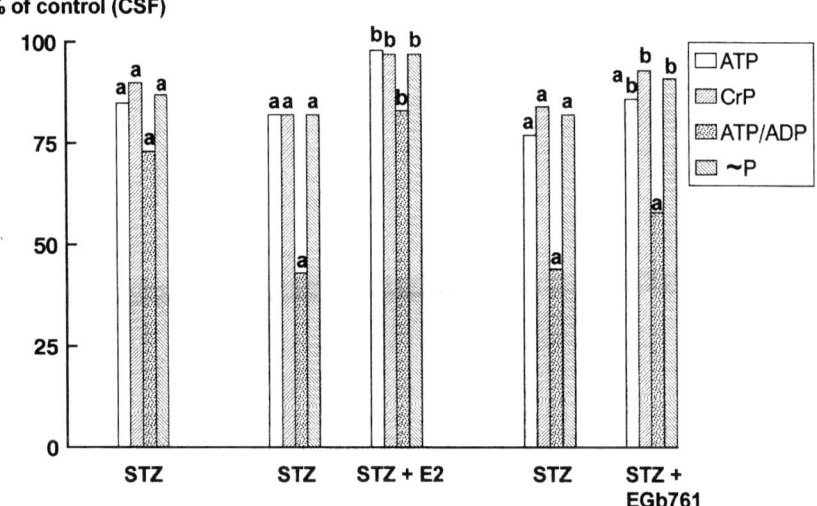

FIGURE 2. Percent changes of energy rich phosphates in parietotemporal cerebral cortex in different experimental groups of 1-year-old rats: after triplicate icv STZ damage and subsequent treatment with 17β-estradiol (E2) s.c. (200 μg/rat/day) over 40 days, and subse-

mentia symptoms in general, and also in males in particular, and may be assumed to become a substance in the treatment strategy of SAD.

Triplicate icv STZ-induced deficits in memory, learning and cognition behavior were partially compensated by EGb761, which also improved the disturbances in cerebral energy metabolism to almost normal values. This antidementive drug has a rational basis in the therapeutic strategy for SAD because of its capability to reduce the progress of the disease.

REFERENCES

1. FRÖLICH, L., D. BLUM-DEGEN, H.G. BERNSTEIN, S. ENGELSBERGER, J. HUMRICH, S. LAUFER, D. MUSCHNER, A. THALHEIMER, A. TÜRK, S. HOYER, H. BECKMANN & P. RIEDERER. 1998. Insulin and insulin receptors in the brain in aging and in sporadic Alzheimer's disease. J. Neural Transm. **105:** 423–438.
2. HOYER, S. 1998. Is sporadic Alzheimer disease the brain type of non-insulin dependent diabetes mellitus? A challenging hypothesis. J. Neural Transm. **105:** 415–422.
3. HOYER, S., D. MÜLLER & K. PLASCHKE. 1994. Desensitization of brain insulin receptor. Effect on glucose/energy and related metabolism. J. Neural Transm. (Suppl.) **44:** 259–268.
4. LANNERT, H., P. WIRTZ, V. SCHUHMANN & R. GALMBACHER. 1998. Effects of estradiol (-17β) on learning, memory and cerebral energy metabolism in male rats after intracerebroventricular administration of streptozotocin. J. Neural Transm. **105:** 1045–1063.
5. KOSTANYAN, A. & K. NAZARYAN. 1992. Rat brain glycolysis regulation by estradiol-17ß. Biochim. Biophys. Acta **1133:** 301–306.
6. DE FEUDIS, F.V. 1998. Ginkgo Biloba Extract (EGb761). From Chemistry to Clinic. Ullstein Medical Verlagsgesellschaft. Wiesbaden, Germany.
7. LE BARS, P.L., M.M. KATZ, N. BERMAN, T.M. ITIL, A.M. FREEDMAN & A.F. SCHATZBERG. 1997. A placebo-controlled, double blind, randomized trial of an extract of Ginkgo biloba for dementia. J. Am. Med. Assoc. **278:** 1327–1332.
8. LANNERT, H. & S. HOYER. 1998. Intracerebroventricular administration of streptozotocin causes long-term diminutions in learning and memory abilities and in cerebral energy metabolism in adult rats. Behav. Neurosci. **112:** 1199–1208.

---

quent EGb761 (50 mg/day) over 80 days. The animals were killed by *in situ* freezing of the brains with liquid nitrogen in arterial normotension, normoxemia and normocapnia, and in normothermia. Energy-rich phosphates were determined by high-performance liquid chromatography. ADP, adenosinediphosphate; ATP, adenosinetriphosphate; CrP, phosphocreatine; ~P, sum of available phosphate.

Temporary Axonal Conduction Block and Axonal Loss in Inflammatory Neurological Disease

A Potential Role for Nitric Oxide?

R. KAPOOR, M. DAVIES, AND K. J. SMITH[a]

Neuroinflammation Research Group, Department of Clinical Neurosciences, Guy's, King's & St. Thomas' School of Medicine–Guy's Campus, King's College, London, SE1 9RT, United Kingdom

INTRODUCTION

Several neurological disorders, including multiple sclerosis (MS), are associated with prominent inflammation within the nervous system. In MS the inflammation has been linked with both the transient expression of symptoms in the relapsing-remitting phase of the disease,[1,2] and, by inference, with the gradual accumulation of permanent deficits in progressive MS. The transient expression of symptoms as severe as blindness, paralysis and numbness has been attributed to temporary periods of conduction block in the relevant pathways, while the permanent deficit has been attributed largely to axonal loss.[3] This loss is most pronounced in lesions with prominent inflammation.[4,5] The mechanisms linking inflammation to reversible conduction block, and to axonal loss, are not known. However, NO is known to be produced in raised concentrations within MS lesions (reviewed in Ref. 6), and we have now found that NO may play a role in each process, via different mechanisms.

METHODS

Rats (Lewis, Wistar and Sprague-Dawley, male or female, 250–350g) were anesthetized (1.5–2% halothane in an $O_2:N_2O$ (30:70) mixture), and the caudal spinal cord and spinal roots exposed beneath a mineral oil pool maintained at $37 \pm 0.1°C$. Three dorsal roots were raised separately onto different pairs of stimulating and recording electrodes arranged rostrally and caudally on each root, and compound action potentials were obtained every 2 or 4 minutes in response to twice supramaximal electrical stimulation. The spinal roots were left in continuity to maintain their vascular supply, but a portion of the conduction pathway was exposed to the NO donors DETA NONOate and spermine NONOate: the NO concentration was measured using a NO-selective probe (NO ISO mark 2, WPI). In some experiments the exposure to NO was combined with the conduction of sustained trains of impulses at physiological frequencies. The roots were exposed to NO for a 2–3 hour

---

[a]Corresponding author. Phone: +44 (0)20 7848 6121; fax: +44 (0)20 7848 6123.
e-mail: kenneth.smith@kcl.ac.uk

KAPOOR *et al.*: INFLAMMATORY NEUROLOGICAL DISEASE AND NO 305

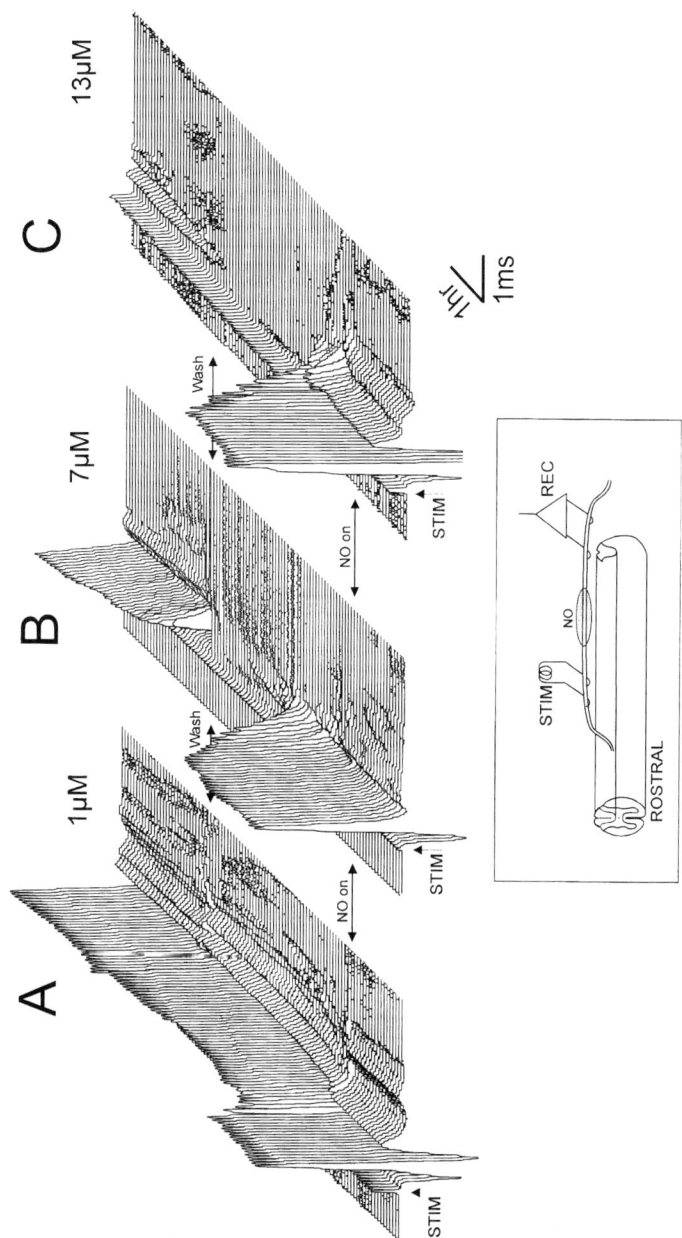

FIGURE 1. *Caption on following page.*

period following an initial, control period of 1–2 hours. Following NO exposure, the treated region of the roots was washed in tissue culture medium, and then maintained in the same medium for at least 2 hours to assess the degree of recovery.

RESULTS

Exposure of spinal roots to NO at concentrations of between 1 and 7 µM resulted in a dose-dependent block of conduction which was fully reversed upon withdrawal of the NO (FIG. 1). At 7 µM concentration, block was induced in many axons within 30 minutes, and reversed in some axons within 4 minutes of its withdrawal. In other experiments (not shown) it was found that axons which were experimentally demyelinated prior to NO exposure were exceptionally vulnerable to this blocking effect.[7]

Exposure of axons to higher concentrations of NO (13 µM) also caused conduction block, but the block persisted when the NO was removed. Hence, in most axons, exposure to these higher levels of NO converted the block from transient, to persistent (FIG. 1). Interestingly, the block was also rendered persistent even at low concentrations of NO, if the NO exposure occurred in conjunction with a sustained train of impulses at frequencies within the physiological range (e.g., 50 or 100 Hz) (FIG. 2). This block persisted throughout the period of observation (up to 12 hours) following removal of the NO.

DISCUSSION

We have shown that relatively brief exposure of myelinated axons to low micromolar concentrations of NO can lead to a reversible block of axonal impulse conduction. Although the concentration of NO within active MS lesions is not known,

---

FIGURE 1. Data obtained simultaneously from 3 dorsal roots exposed in parallel to different concentrations of NO in a terminally anesthetized, normal rat. The inset shows the recording arrangement. The roots were left in continuity to maintain their vascular supply, and to ensure that they were in as physiological a state as possible. The roots were raised individually on pairs of stimulating and recording electrodes within a mineral oil recording pool maintained at $37 \pm 0.1°C$. Each root passed through its own 7-mm bath in which it was exposed to a different concentration of the same NO donor (DETA NONOate). Stimuli ("STIM") at twice supramaximal intensity were applied at 3 Hz and records were obtained of the evoked compound action potentials. Each plot shows a series of compound action potentials obtained at 4-minute intervals by computer averaging. The records are plotted with 3-dimensional perspective, in the order in which they were obtained, with the earliest records displayed at the front. Each plot represents about 7 hours of recording time. After a control period of approximately 1 hour to ensure that the preparation was stable, the medium in the bath was changed to one releasing NO. After 3 hours exposure, the NO was replaced with tissue culture medium for the remainder of the experiment. The plots show that exposure of spinal roots to NO concentrations of 1 µM (**A**) and 7 µM (**B**) results in conduction block, in some (**A**) or all (**B**) axons, which reverses when NO is withdrawn. However, exposure to a higher concentration of NO (13 µM) causes conduction block which persists when the NO is removed (**C**).

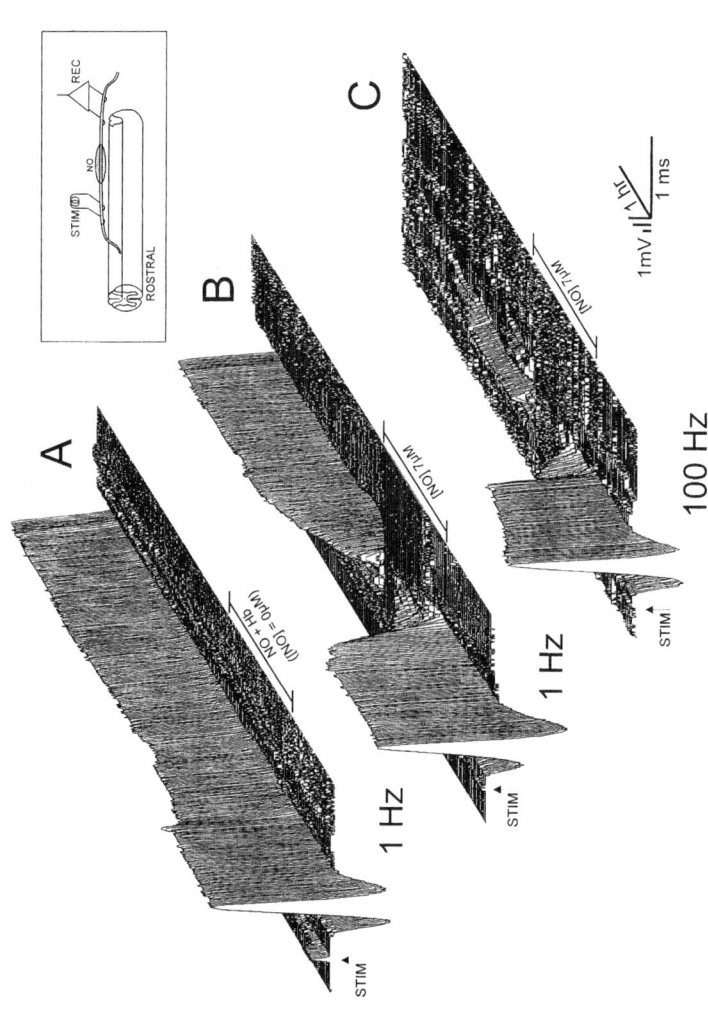

FIGURE 2. Data obtained simultaneously from 3 dorsal roots exposed to the same concentration of NO donor, using the experimental arrangement indicated. The records were obtained every 2 minutes. In the control data (**A**), hemoglobin was included with the donor (DETA NONOate) to scavenge the NO released. The other plots (**B,C**) show that 7 μM NO causes conduction block in all axons during a 2-h exposure period. However, the block is reversible in all, or nearly all, axons at 1 Hz stimulation (**B**), but it is rendered persistent (for > 5 h) when NO exposure occurs in conjunction with sustained impulse activity (100 Hz) (**C**).

observations in other systems suggest that it is likely to be in the low micromolar range, as examined in these experiments. A reversible block of conduction induced by NO may contribute to the transient disability which is seen in relapses of MS.

We have also shown that exposure to higher concentrations of NO (13 µM) can render the block persistent. Perhaps more relevant to human disease, we have shown that even low concentrations of NO cause persistent conduction block, if the axons are concomitantly firing at physiological frequencies. These observations may help to explain how permanent disability can occur in some patients with MS, since it seems likely that the persistent conduction block is due to the degeneration of the axons at the site of NO exposure.

The mechanisms by which NO exerts the effects on axons described in this study remain unclear. Apart from its inhibitory effects on mitochondrial respiration,[8] NO is also known to affect membrane currents,[9] and to act indirectly through the damaging effects of its derivatives, including peroxynitrite and its breakdown products.[6] Although other inflammatory mediators may also have pathogenic effects on axonal function, the present findings raise the possibility that the inhibition of NO production could offer a novel approach to the therapy of MS.

REFERENCES

1. MOREAU, T., A. COLES, M. WING et al. 1996. Transient increase in symptoms associated with cytokine release in patients with multiple sclerosis. Brain **119:** 225–237.
2. YOUL, B.D., G. TURANO, D.H. MILLER et al. 1991. The pathophysiology of acute optic neuritis. An association of gadolinium leakage with clinical and electrophysiological deficits. Brain **114:** 2437–2450.
3. LOSSEFF, N.A., S.L. WEBB, J.I. O'RIORDAN et al. 1996. Spinal cord atrophy and disability in multiple sclerosis. A new reproducible and sensitive MRI method with potential to monitor disease progression. Brain **119:** 701–708.
4. FERGUSON, B., M.K. MATYSZAK, M.M. ESIRI et al. 1997. Axonal damage in acute multiple sclerosis lesions. Brain **120:** 393–399.
5. TRAPP, B.D., J. PETERSON, R.M. RANSOHOFF et al. 1998. Axonal transection in the lesions of multiple sclerosis. N. Engl. J. Med. **338:** 278–285.
6. SMITH, K.J., R. KAPOOR & P.A. FELTS. 1999. Demyelination: the role of reactive oxygen and nitrogen species. Brain Pathol. **9:** 69–92.
7. REDFORD, E.J., S.M. HALL, K.J. SMITH. 1995. Vascular changes and demyelination induced by the intraneural injection of tumour necrosis factor. Brain **118:** 869–878.
8. BOLANOS, J.P., A. ALMEIDA, V. STEWART et al. 1997. Nitric oxide-mediated mitochondrial damage in the brain: mechanisms and implications for neurodegenerative diseases. J. Neurochem. **68:** 2227–2240.
9. LI, Z., M.W. CHAPLEAU, J.N. BATES et al. 1998. Nitric oxide as an autocrine regulator of sodium currents in baroreceptor neurons. Neuron **20:** 1039–1049.

Maturational Changes in Rabbit Brain Phosphocreatine and Creatine Kinase

T. KEKELIDZE,[a,d] I. KHAIT,[a] A. TOGLIATTI,[a] J. BENZYCRY,[a] R. MULKERN,[b] AND D. HOLTZMAN[a,c]

[a]*Department of Radiology, Brigham and Women's Hospital and Harvard Medical School, Boston, Massachusetts 02115, USA*

[b]*Department of Radiology, Children's Hospital and Harvard Medical School, Boston, Massachusetts 02115, USA*

[c]*Department of Neurology, Massachusetts General Hospital and Harvard Medical School, Boston, Massachusetts 02114, USA*

INTRODUCTION

Brain ATP metabolism shows regional differences, with gray matter (GM) showing high and variable rates compared to more consistent lower rates in white matter (WM).[1] The phosphocreatine (PCr)/creatine kinase (CK)/creatine (Cr) systems in rats and piglets show similar regional differences, with low PCr/nucleoside triphosphate (NTP) ratio, high Cr concentration, and high mitochondrial CK (Mi-CK) in GM compared to WM.[2,3] In the postnatal rat, brain slice and homogenate respiratory rates and rates of aerobic glycolysis increase between about 5–25 days of age parallel to an *in vivo* increase in PCr/NTP ratio measured by surface coil localized ^{31}P NMR spectroscopy.[2,4,5] In order to understand the physiology of the CK system in development of ATP metabolism in GM and WM, regional development of the PCr/CK/Cr system has been studied *in vivo* and *in vitro* in the rabbit, which shows marked postnatal metabolic development and has a brain large enough to use regional ^{31}P NMR techniques.

METHODS

New Zealand white rabbit pups aged 5, 10, 15, 20, and 30 days were used for all studies. Cerebral hemisphere PCr/NTP ratios were obtained from fully relaxed surface coil localized ^{31}P NMR spectra. Ratios in predominantly GM and WM slices were obtained with the one-dimensional chemical shift ^{31}P NMR imaging (1D-CSI) technique.[3] All NMR studies used a 4.7 T horizontal bore spectrometer (General Electric Medical Systems, Milwaukee, WI). CK isoenzymes and creatine transporter protein (CREAT) were analyzed using the Western blot procedure employing anti–brain-specific CK (BB-CK, rabbit anti-human) (Cortex Biochem, CA), anti–Mi-CK

---

[d]Corresponding author: Téa Kekelidze, Ph.D., Department of Radiology, Brigham and Women's Hospital, LMRC, 221 Longwood Avenue, Boston, Massachusetts 02115. Phone: 617-732-5623; fax: 617- 278-0610.
e-mail: tkeke@bwh.harvard.edu

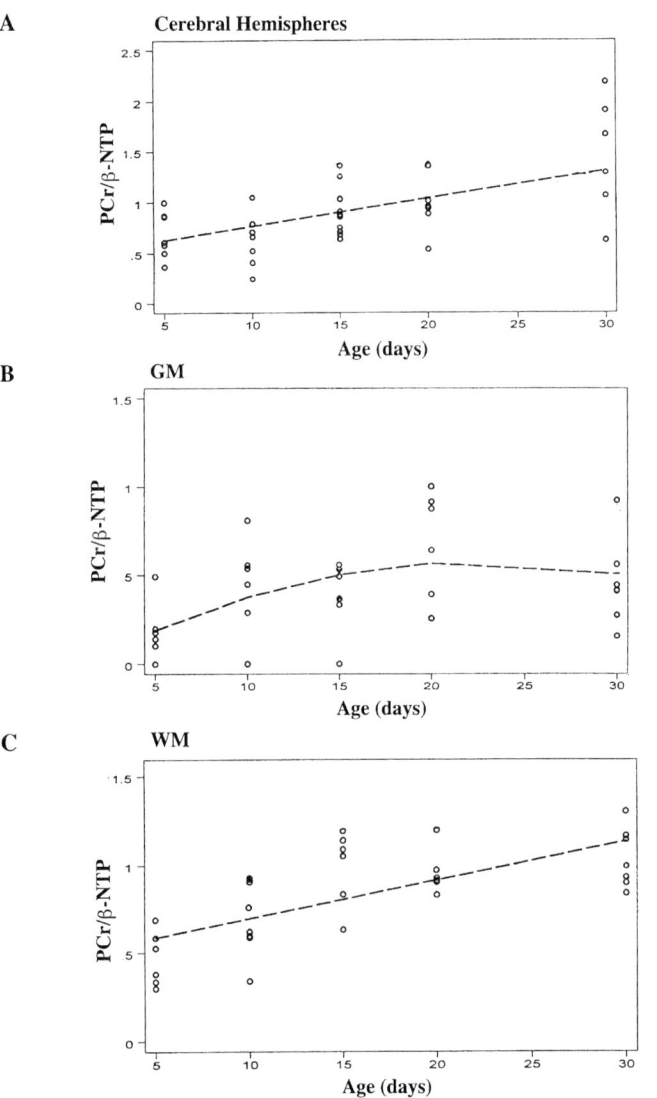

FIGURE 1. Brain PCr/NTP ratios in 5–30-day-old rabbits. Surface coil localized ^{31}P NMR spectra (fully relaxed, 10-s recycle time) were used for measuring the ratio in cerebral hemispheres (**A**). One-dimensional chemical shift imaging (1D-CSI) was used to acquire the partially saturated spectra (2-s recycle time) for predominantly GM (**B**) and for predominantly WM (**C**) brain slices. Six animals in each age group were tested. The best fit for the values in brain and in each region is shown. Maturational increase in PCr/NTP was analyzed using analysis of covariance, and was significant in brain hemispheres (twofold increase, $p < 0.0005$), GM ($p < 0.001$), and WM (twofold increase, $p < .0005$) regions over the time period studied. PCr/NTP ratio was significantly higher in WM compared to GM ($p < 0.0005$) at all ages.

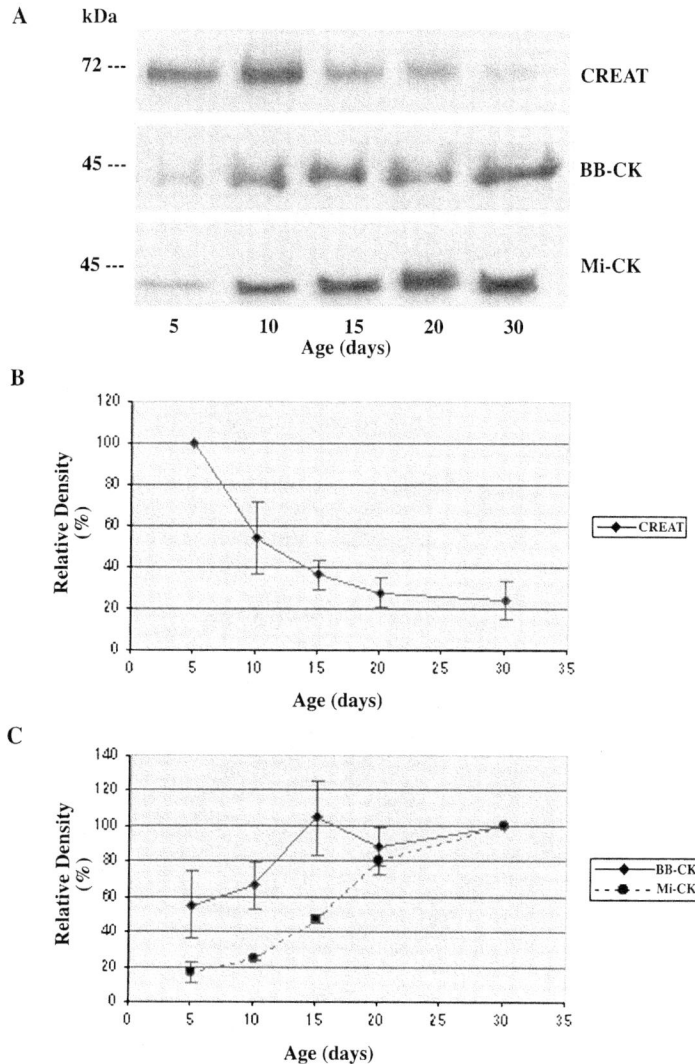

FIGURE 2. Maturation of brain CK isoenzymes and creatine transporter in predominantly GM of 5–30-day-old rabbits. Homogenates from brain GM regions were resolved by SDS–polyacrylamide gel electrophoresis (25 μg of homogenate protein/lane) and transferred to polyvinylidene fluoride membrane (Biorad, CA). Anti–BB-CK, –Mi-CK, and –CREAT polyclonal antibodies were used for Western blot procedure, and immunoreactivity was visualized using enhanced chemiluminescence (Amersham, IL) (**A**). Bands were semiquantified using NIH densitometry software, and densities were expressed as percentage of the mean value at 5 days (**B**, creatine transporter) or 30 days (C, BB-CK, and Mi-CK). Each data point is a mean of 4 animals ± SEM at each age. All developmental changes over the age period studied were significant ($p < 0.0002$ for BB-CK, $p < 0.00007$ for Mi-CK, and $p < 0.0007$ for CREAT).

(rabbit anti-chick, from Dr. T.Wallimann), and anti-CREAT fusion protein polyclonal antibodies.[6]

RESULTS AND DISCUSSION

Cerebral hemisphere PCr/NTP ratios from surface coil localized spectra increased linearly between 5 and 30 days of age (FIG. 1A). This increase probably represents increased PCr since in the rat brain NTP concentration does not change during this developmental period.[7]

Results of the 1D-CSI experiment show different time courses for PCr/NTP increases in GM and WM (FIG. 1B,C). In GM the PCr signal was not seen at 5 days of age. This ratio reached adult level by 15 days. In WM, PCr was present at 5 days and the PCr/NTP ratio steadily increased between 5 and 30 days. This increase in WM PCr/NTP closely paralleled the nonlocalized NMR signal from the cerebral hemispheres. At all ages the PCr/NTP ratio was higher in WM than in GM, as previously shown in the piglet.[3]

Developmental changes in BB-CK, Mi-CK, and CREAT proteins in tissue that is predominantly GM are shown in FIGURE 2. The CK isoforms were both present at 5 days of age, but increased with different time courses. On a semiquantitative basis, cytosolic BB-CK reached adult levels by 15 days. The Mi-CK increased between 5 and 30 days, reaching higher adult level compared to BB-CK. This greater relative increase of Mi-CK is similar to that seen in postnatal rat.[2] CREAT protein was present at high levels at 5 days of age before a 60% decrease between 5 and 15 days, followed by an additional slower decrease (20%) between 15 and 30 days.

In conclusion, maturational changes in the cerebral hemisphere PCr/CK/Cr system, measured by surface coil localized ^{31}P NMR spectroscopy, are the same as seen in the rat pup but occur over a slightly longer postnatal age period.[4,8] For the first time, brain regional developmental changes in PCr/NTP have been measured *in vivo*. The PCr signal first appears in the predominantly WM slices, in which it increases with the same time course seen in nonlocalized spectra. In marked contrast, in GM the PCr appears later but increases more rapidly to adult levels by 15 days. The similarity of maturational time courses suggests that the flattened configuration of the rabbit brain causes signal predominantly from WM to be received by the surface coil. The PCr/NTP increase in predominantly GM slices is temporally coincident with the decrease in CREAT in GM, suggesting that the transporter may contribute to a regional increase in Cr and, secondarily, PCr. Interestingly, the PCr/NTP concentration ratio increases in GM in parallel with BB-CK between 5–15 days. The large brain and prolonged postnatal metabolic development makes the rabbit appropriate for *in vivo* developmental study of brain energy metabolism.

ACKNOWLEDGMENTS

This work was supported by a National Institutes of Health Research grant to D.H. (NS 26371) and a Fogarty International Fellowship to T.K. (NS 10176).

REFERENCES

1. CLARK, D. & L. SOKOLOFF. 1999. Circulation and energy metabolism. *In* Basic Neurochemistry, 5th edit. J. Siegal, B. Arganoff, R. Albers, S. Fisher & M. Uhler, Eds.: 637–669. Lippincott-Raven. Philadelphia.
2. HOLTZMAN, D. *et al.* 1993. Functional maturation of creatine kinase in rat brain. Dev. Neurosci. **15:** 261–270
3. TSUJI, M. *et al.* 1996. Relative phosphocreatine and nucleoside triphosphate concentrations in cerebral gray and white matter measured in vivo by ^{31}P nuclear magnetic resonance. Brain Res. **707:** 146–154.
4. TOFTS, P. & S. WRAY. 1985. Changes in brain phosphorous metabolites during the postnatal development of the rat. J. Physiol. **359:** 417–429.
5. HIMWICH, H. 1970. Historical review. *In* Developmental Neurobiology. W. Himwich, Ed.: 22–44. Thomas. Springfield, IL.
6. KEKELIDZE, T. *et al.* 1997. Analysis of creatine transporter expression in rat brain and cell line using specific polyclonal antisera. Soc. Neurosci. Abstr. **23**(586): 15.
7. LOLLEY, R. *et al.* 1961. The high energy phosphates in developing brain. J. Neurochem. **7:** 289–297.
8. HOLTZMAN, D. *et al.* 1991. Maturational increase in mouse brain creatine kinase reaction rates shown by phosphorus magnetic resonance. Dev. Brain Res. **58:** 181–188.

Neurotoxicity and Oxidative Damage of Beta Amyloid 1-42 versus Beta Amyloid 1-40 in the Mouse Cerebral Cortex

AUTUMN M. KLEIN, NEIL W. KOWALL, AND ROBERT J. FERRANTE[a]

Departments of Neurology, Pathology, and Psychiatry, Boston University School of Medicine, Boston, Massachusetts 02118, USA, and
Geriatric Research Education Clinical Center, Veterans Affairs Medical Center, Bedford, Massachusetts 01730, USA

Senile plaques (SP), a neuropathological hallmark of Alzheimer's disease (AD), are characterized by extracellular accumulations of beta amyloid (Aβ). SP predominantly contain Aβ42 with a small amount of associated Aβ40. We determined the neurotoxic properties of Aβ42 as compared to Aβ40 by injections into the frontal cortex of three month old C57BL/6 mice. Aβ42 was associated with a significantly larger area of glial fibrillary acidic protein (GFAP) immunoreactivity and a greater density of reactive astrocytes than Aβ40. Immunohistochemical staining for markers of oxidative damage against 3-nitrotyrosine (3-NT) and 8-hydroxydeoxyguanosine (8-OHDG) were significantly more intense around the Aβ42 injection compared to the Aβ40 injection sites. These findings are consistent with previous *in vitro* studies and suggest that Aβ42 is more neurotoxic and may generate more free radical damage than Aβ40.

INTRODUCTION

Beta Amyloid Peptide

Beta amyloid (Aβ), a 38-42 amino acid peptide derived from the transmembrane amyloid precursor protein (APP), accumulates extracellularly in the brains of patients with Alzheimer's disease (AD) to form SP. Aβ is neurotoxic both *in vitro*[1] and *in vivo*,[2] and its neurotoxicity is dependent upon the aggregation state of the peptide.[3]

Aβ and Free Radicals

Aβ's toxicity *in vitro* is mediated through oxidative mechanisms[3] and can be ameliorated with antioxidant agents.[4–7] Free radical production promotes the aggregation and deposition of Aβ.[8] Aβ may generate free radicals via direct and indirect pathways. Mass spectroscopy and electron paramagnetic resonance (EPR) demon-

---

[a]Address for correspondence: Dr. Robert J. Ferrante, GRECC Unit 182B, Building 18, Bedford VA Medical Center, 200 Springs Road, Bedford, MA 01730. Phone: 781-687-2908; fax: 781-687-3515.
e-mail: rjferr@bu.edu

strated that Aβ25-35 in a metal-independent aqueous solution fragments and generates free radicals. Aβ also has direct effects on enzymes sensitive to oxidative stress, such as glutamine synthetase and creatine kinase, which may damage energy-dependent cellular processes.[9] Impairment of energy production causes dysfunction of the electron transport chain and an accumulation of free radicals, establishing a vicious cycle of energy impairment and free radical generation. Finally, Aβ may activate cells that produce free radicals. Aβ causes endothelial cells to produce an excess of superoxide radicals[10] and also stimulates microglia to produce nitric oxide (NO$^\bullet$).[11] Numerous studies point to oxidative damage to lipids, proteins, and DNA in the AD brain.

METHODS

Preparation of Aβ40 and Aβ42

Aβ40 and Aβ42 were supplied by Quality Controlled Biochemicals, Inc. (QCB, Hopkinton, MA). Aβ40 was dissolved in sterile distilled water to a concentration of 6 mg/mL and then diluted with PBS without calcium to a concentration of 1 mg/mL. Aβ42 was dissolved in sterile distilled water to a concentration of 1 mg/mL. Each solution of Aβ was then incubated for one week at 37°C to ensure aggregation. Aggregation of the peptide is crucial to Aβ's neurotoxicity and has been associated with free radical generation.

ANIMAL CHARACTERISTICS AND SURGICAL PROCEDURES

Ten three-month-old male C57BL/6 mice (Charles River, Portage, MI) were used. Mice were deeply anesthetized with isoflurane and transferred to a Kopf stereotactic apparatus and maintained at a surgical plane of anesthesia with an inhalation mask. Injections were made 1 mm lateral to midline and 1 mm anterior to Bregma in each hemisphere through a burr hole in the cranium, 1 mm into the cortex. Into one hemisphere, 1 µL of a 1-mg/mL solution of Aβ42 was infused over a two-minute period. Into the contralateral hemisphere, 1 µL of a 1-mg/mL solution of Aβ40 was injected in a similar fashion. Animals were allowed to survive for one week, were deeply anesthetized, and were perfused transcardially with phosphate-buffered saline followed by 2% paraformaldehyde-lysine-periodate. The brains were removed, postfixed, and cryoprotected in increasing concentrations of glycerol.

Tissue Preparation and Antibodies

Brains were cut at 35 µm on a freezing microtome and stained immunocytochemically for GFAP, 3-NT, 8-OHDG, and heme-oxygenase-1 (HO-1). Anti-GFAP (1:100, Boehringer-Mannheim, Indianapolis, IN) is a mouse monoclonal antibody that specifically immunoprecipitates glial fibrillary acidic protein. GFAP labels reactive astrocytes, a well-established marker for brain lesions. Anti–3-NT (1:200, Upstate Biotechnology, Lake Placid, NY) is a mouse monoclonal antibody that specifically recognizes nitrated albumin. Anti–8-OHDG (1:1500, Pharmigen, San Di-

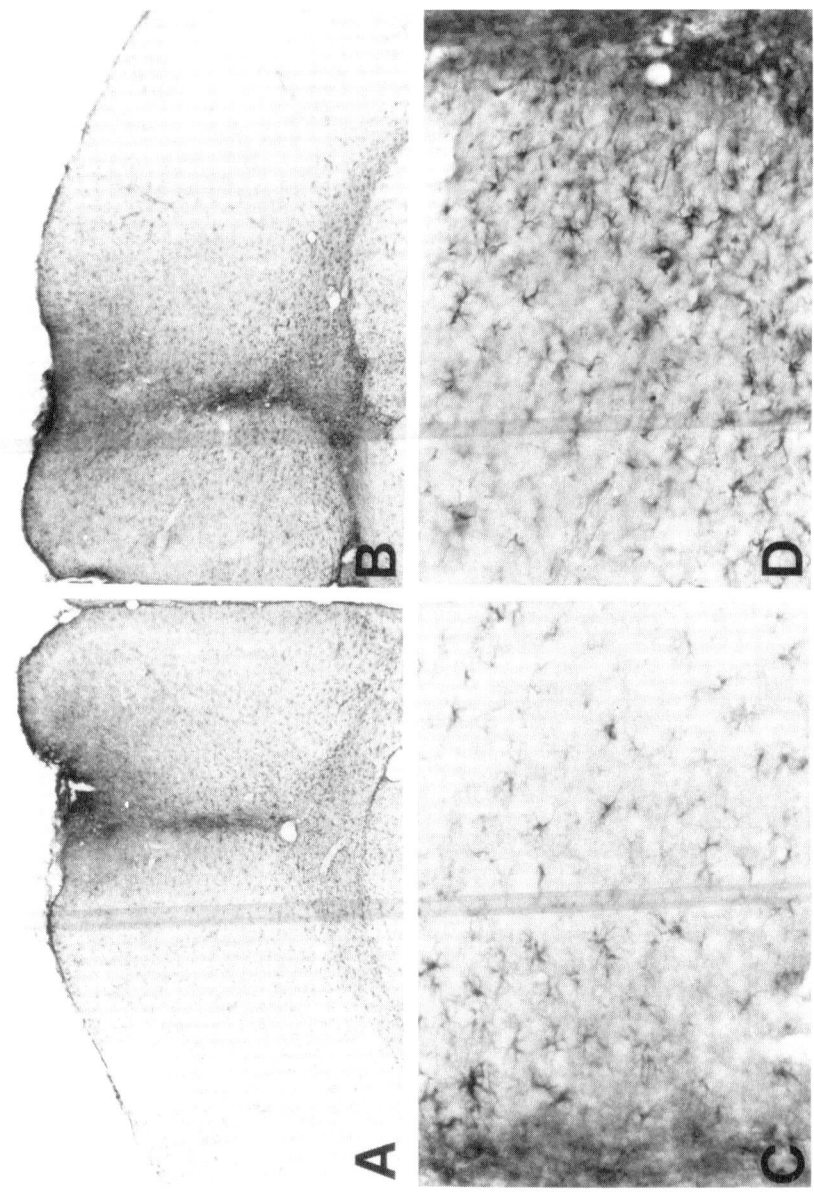

ego, CA) is a mouse monoclonal antibody that is highly specific for detecting hydroxyl radical damage to DNA bases. Anti–HO-1 (1:1000, Stress Gen, Victoria, British Columbia) is a polyclonal rabbit antibody that detects the microsomal enzyme HO-1, which is induced by oxidative stress.

Quantitation

Areal measurements from GFAP-stained sections were made using the NIH Image 1.62 software. Astroglial immunoreactivity, defined by thickened, tortuous processes, greatly enlarged cell somas, and intense GFAP staining, was used as criteria to determine the area of the lesion and to count the number of reactive glia. The maximal lesion area was determined by two independent observers (A.M.K. and R.J.F.) and was scanned into an NIH Image library. For each animal, the area of maximal reactivity for each oxidative marker was also identified. Intensity of immunoreactivity for each oxidative marker was graded by two blinded observers (A.M.K. and R.J.F.). Immunostaining was graded according to the intensity of the lesioned area, using a semiquantitative scale of 0 (absent), 1 (trace), 2 (weak), 3 (moderate), and 4 (intense).

Glial quantitation was determined from an area of maximal glial reactivity adjacent to the needle tract. The number of reactive astrocytes was counted within a 250-mm^2 area on a manual counter.

Statistics

Statistical analysis was performed using the PC-based program Statview 512+ (Abacus Software, Calabasas, CA). Paired two-tailed t-tests were used to determine significance between two groups.

RESULTS

Qualitative microscopic evaluation of both lesions demonstrated cellular disruption, neuronal death, and astrogliosis. Cellular disruption on the Aβ42 side extended into white matter and underlying striatum, even though the needle tract was within the expected parameters. Aβ42 showed a significantly greater area of glial activation as compared to Aβ40 (Aβ42: 1.431 + 0.356 mm^2; Aβ40: 0.938 + 0.200 mm^2, $p <$ 0.0005). There was a significantly greater density of reactive astroglia on the side of

---

FIGURE 1. Glial fibrillary acidic protein (GFAP) immunoreactivity of cortical injection sites of aggregated Aβ42 (**B** and **D**) and a contralateral injection of aggregated Aβ40 (**A** and **C**) in the cerebral cortex of a C57BL/6 mouse. Injections of aggregated Aβ42 caused increased astrogliosis as compared to an injection of aggregated Aβ40. This figure demonstrates the point of maximal GFAP reactivity surrounding each injection site. The area of GFAP immunoreactivity associated with an injection of Aβ42 (**B**) was significantly larger than the area of GFAP immunoreactivity associated with a contralateral injection of Aβ40 (**A**). A high-power view of GFAP immunoreactivity of cortical injection sites demonstrates that reactive astroglial density, as evidenced by greatly enlarged cell somas and thickened, tortuous processes, is greater on the side of the Aβ42 injection (**D**) as compared to the Aβ40 injection (**C**).

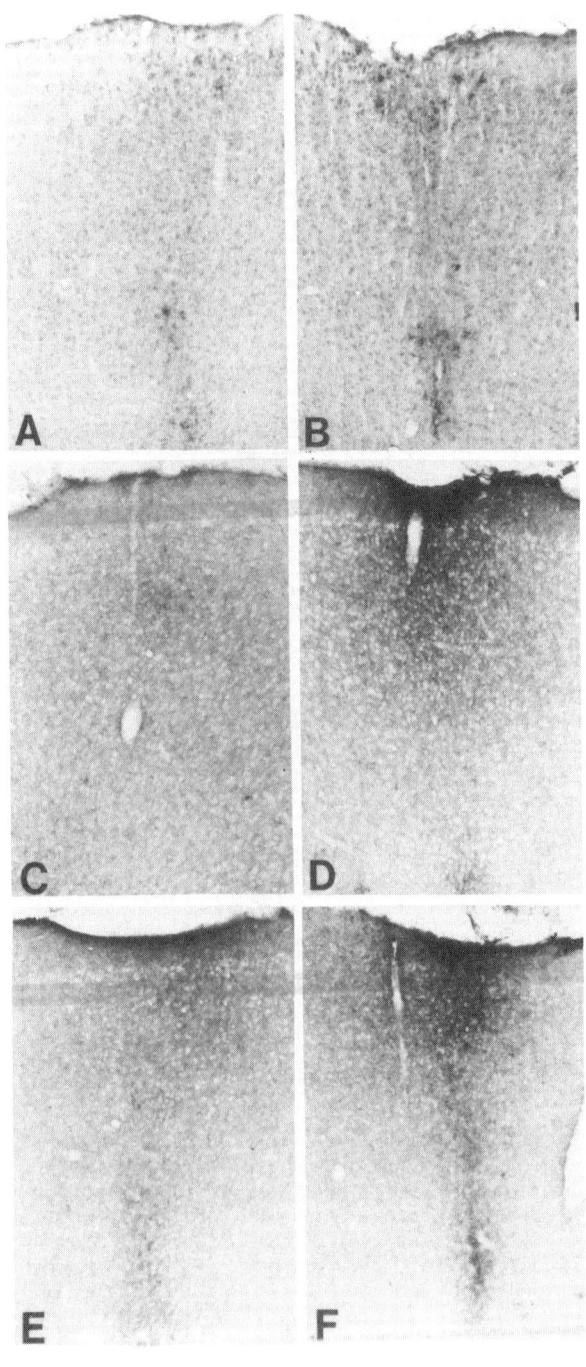

the Aβ42 lesion as compared to the side of the Aβ40 injection (Aβ42: 176 + 22.7/ 250 mm^2; Aβ40: 129.6 + 9.9/250 mm^2, $p < 0.002$).

Significantly greater intensity of 3-NT immunoreactivity was associated with the Aβ42 than the Aβ40 injection sites (Aβ42: 2.6 + 0.5 mm^2; Aβ40: 1.8 + 0.8 mm^2, $p < 0.005$). The distribution of 3-NT label was located primarily in the neuropil with some cellular labeling. Immunostaining for 8-OHDG was also found to be significantly more intense around the Aβ42 lesion than the Aβ40 lesion (Aβ42: 2.6 + 0.7 mm^2; Aβ40: 2.1 + 0.8 mm^2, $p < 0.018$). Immunoreactivity for 8-OHDG was mostly within the cytoplasm of neurons with minimal labeling of the neuropil. In contrast to 3-NT and 8-OHDG, there was no significant difference in the intensity of HO-1 immunoreactivity between the Aβ42 and the Aβ40 injection sites (Aβ42: 1.5 + 0.5 mm^2; Aβ40: 1.3 + 0.4 mm^2, $p < 0.363$). HO-1 immunostaining was found to be intensely neuronal with very little immunostaining of the neuropil.

DISCUSSION

The sequence and length of Aβ is crucial to the aggregation and neurotoxicity of Aβ. Aggregation of Aβ42 was faster than Aβ1-39 regardless of concentration, temperature, pH, or solvent.[12] Supersaturated solutions of Aβ26-42 and Aβ36-43 are unstable, and amyloid formation is instantaneous; supersaturated solutions of Aβ26-40 and Aβ1-39 do not form amyloid for 12 hours and 2 days, respectively.[13] Aβ's toxicity *in vitro* depends on its aggregation state.[3] Aβ fragments containing Aβ25-35 were toxic to neurons only after aggregation, not when freshly solubilized and applied to cultures.[14] Aβ25-35 itself was found to aggregate and be neurotoxic immediately upon dissolution.[14,15]

Aβ42 versus Aβ40 and Free Radicals

Previous studies have directly compared the free radical generating abilities of the two peptides. The toxic effects of both peptides were blocked with antioxidants.[16,17] This suggests that both peptides mediate toxicity through oxidative mechanisms. If the degree of neurotoxicity is directly related to the free radical gen-

---

FIGURE 2. Immunostaining for oxidative markers heme-oxygenase-1 (HO-1), 8-hydroxydeoxyguanosine (8-OHDG), and 3-nitrotyrosine (3-NT) in the cerebral cortex of a C57BL/6 mouse injected with aggregated Aβ42 (**B, D,** and **F**) and a contralateral injection of aggregated Aβ40 (**A, C,** and **E**). Immunostaining with HO-1, an enzyme induced with oxidative stress, is intensely cellular but is also apparent in the neuropil. This panel demonstrates that there is no significant difference in HO-1 immunoreactivity between the Aβ42 injection (**B**) and the Aβ40 injection (**A**). Antibodies against 8-OHDG, a marker of oxidative damage to DNA bases, stain the cytoplasm of neurons as well as the neuropil. This figure demonstrates more intense 8-OHDG immunoreactivity on the side associated with the Aβ42 injection (**D**) than on the side associated with the Aβ40 injection (**C**). Antibodies raised against 3-NT, a marker of oxidative damage to proteins, primarily stain the neuropil with occasional cellular labeling. This figure demonstrates more intense 3-NT immunoreactivity on the side associated with the Aβ42 injection (**F**) than on the side associated with the Aβ40 injection (**E**).

erating capability of the peptide, then one would expect Aβ42 to produce more free radicals than Aβ40. The present study presents evidence to support this.

REFERENCES

1. YANKNER, B., L. DUFFY *et al.* 1990. Neurotrophic and neurotoxic effects of amyloid β protein: reversal by tachykinin neuropeptides. Science **250:** 279–282.
2. KOWALL, N.W., M.F. BEAL *et al.* 1991. An in vivo model for the neurodegenerative effects of beta amyloid and protection by substance P. Proc. Natl. Acad. Sci. USA **88:** 7247–7251.
3. PIKE, C.J., A.J. WALENCEWICZ *et al.* 1991. In vitro aging of b-amyloid protein causes peptide aggregation and neurotoxicity. Brain Res. **563:** 311–314.
4. BEHL, C., J. DAVIS *et al.* 1992. Vitamin E protects nerve cells from amyloid beta protein toxicity. Biochem. Biophys. Res. Commun. **186**(2): 944–950.
5. GOODMAN, Y. & M.P. MATTSON. 1994. Staurosporine and K-252 compounds protect hippocampal neurons against amyloid beta-peptide toxicity and oxidative injury. Brain Res. **650**(1): 170–174.
6. MANELLI, A.M. & P.S. PUTTFARCKEN. 1995. β-Amyloid–induced toxicity in rat hippocampal cells: in vitro evidence for the involvement of free radicals. Brain Res. Bull. **38**(6): 569–576.
7. PUTTFARCKEN, P.S., A.M. MANELLI *et al.* 1996. Inhibition of age-induced beta-amyloid neurotoxicity in rat hippocampal cells. Exp. Neurol. **138**(1): 73–81.
8. DYRKS, T., E. DYRKS *et al.* 1992. Amyloidogenicity of beta A4 and beta A4-bearing amyloid protein precursor fragments by metal-catalyzed oxidation. J. Biol. Chem. **267**(25): 18210–18217.
9. HENSLEY, K., J.M. CARNEY *et al.* 1994. A model for beta-amyloid aggregation and neurotoxicity based on free radical generation by the peptide: relevance to Alzheimer disease. Proc. Natl. Acad. Sci. USA **91**(8): 3270–3274.
10. THOMAS, T., G. THOMAS *et al.* 1996. Beta-amyloid–mediated vasoactivity and vascular endothelial damage [see comments]. Nature **380:** 168–171.
11. MCDONALD, D.R., K. BRUNDEN *et al.* 1997. Amyloid fibrils activate tyrosine kinase-dependent signalling and superoxide production in microglia. J. Neurosci. **17**(7): 2284–2294.
12. BARROW, C.J. & M.G. ZAGORSKI. 1991. Solution structures of beta peptide and its constituent fragments: relation to amyloid deposition. Science **253**(5016): 179–182.
13. JARRETT, J. & P. LANSBURY. 1993. The carboxy terminus of the β amyloid protein is critical for the seeding of amyloid formation: implications for the pathogenesis of Alzheimer's disease. Biochemistry **32:** 4693–4697.
14. PIKE, C.J., D. BURDICK *et al.* 1993. Neurodegeneration induced by β-amyloid peptides in vitro: the role of peptide assembly state. J. Neurosci. **13:** 1676–1687.
15. SHEARMAN, M.S., C.I. RAGAN *et al.* 1994. Inhibition of PC12 cell redox activity is a specific, early indicator of the mechanism of beta-amyloid–mediated cell death. Proc. Natl. Acad. Sci. USA **91**(4): 1470–1474.
16. CRAWFORD, F., Z. SUO *et al.* 1997. The vasoactivity of A beta peptides. Ann. N.Y. Acad. Sci. **826:** 35–46.
17. SUO, Z., C. FANG *et al.* 1997. Superoxide free radical and intracellular calcium mediate A beta(1-42) induced endothelial toxicity. Brain Res. **762**(1-2): 144–152.

Ganglioside GD3, the Mitochondrial Permeability Transition, and Apoptosis

BRUCE S. KRISTAL[a] AND ABRAHAM M. BROWN

Dementia Research Service, Burke Medical Research Institute, 785 Mamaroneck Avenue, White Plains, New York 10605, USA, and Department of Biochemistry, Cornell University Medical College, 1300 York Avenue, New York, New York 10021, USA

De Maria and colleagues have recently shown that ganglioside GD3 appears to be *both necessary and sufficient* for CD95- and ceramide-mediated apoptosis.[1,2] The biochemical events between GD3 synthesis and cell death remain unknown. Clues to a potential mechanism include observations that GD3 is transported from the Golgi to the mitochondria,[3] and that GD3-mediated apoptosis includes loss of mitochondrial membrane potential ($\Delta\Psi^2$), a classic hallmark of the mitochondrial permeability transition (PT).[4,5]

The PT involves the opening of a pore in the inner mitochondrial membrane that allows free diffusion of solutes with mass under 1500 daltons.[4,5] PT induction abolishes oxidative phosphorylation, leads to loss of the mitochondrial proton gradient, and allows efflux of mitochondrially sequestered calcium into the cytoplasm.[4,5] These consequences on energy metabolism and calcium homeostasis, as well as the biochemical identity of the first PT inducers studied (high calcium, oxidants, inorganic phosphate), led to a recognition of the probable role of PT in ischemia-reperfusion injury.[6] The possible involvement of PT in apoptosis has been recognized more recently.[7] The PT has been proposed to propagate the apoptotic cascade by triggering mitochondrial release of AIF (apoptosis inducing factor, which stimulates nuclear fragmentation) and cytochrome *c*, which activates the downstream caspases that effect the end stage of apoptosis.[8]

We have begun to test the hypothesis that GD3 can directly mediate PT induction and/or release of cytochrome *c*.

METHODS

PT induction by Ca^{+2} was studied in liver mitochondria isolated from 4–6-month-old male Fischer 344 rats as previously described.[9] Livers were rapidly homogenized in ice-cold buffer [250 mM mannitol, 75 mM sucrose, 100 µM K-EDTA, 10 mM K-HEPES (pH 7.4)] supplemented with 500 µM K-EGTA (pH 7.4). Homogenates were centrifuged at 1000 *g* for 10 min. Supernatants were removed and centrifuged at 10,000 *g* for 15 min. Pellets were washed three times in the above buffer supplemented with 0.5% fatty acid–free bovine serum albumin (Sigma A-6003). The

---

[a]Address for correspondence: Dr. Bruce Kristal, Dementia Research Service, Burke Medical Research Institute, 785 Mamaroneck Avenue, White Plains, New York 10605. Phone: 914-597-2333; fax: 914-597-2757.

e-mail: bkristal@burke.org

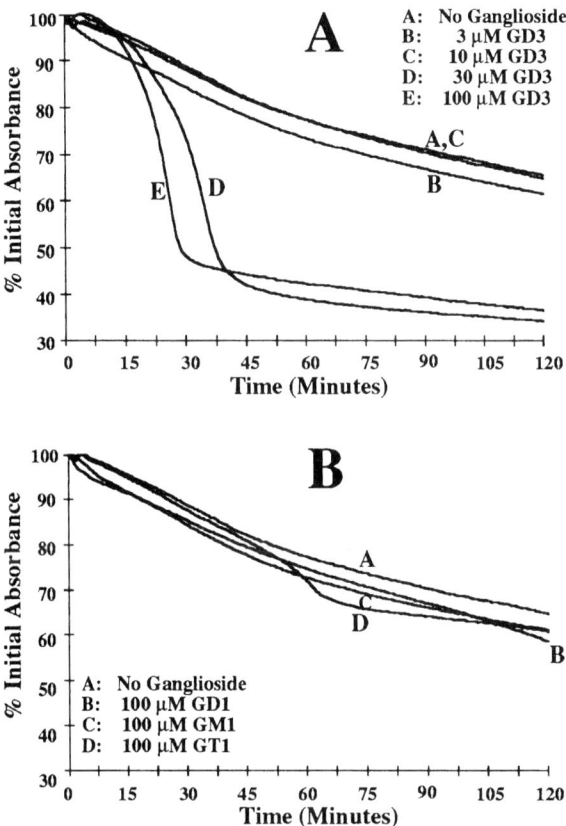

FIGURE 1. GD3 promotes PT induction. Representative traces showing the effects of GD3 (**A**) or GD1a, GM1, and GT1b (**B**) on PT induction. The efficacy of 30 μM GD3 varied from rat to rat; the greatest response observed for this dose is shown.

first wash was also supplemented with 500 μM EGTA. The final mitochondrial pellet was resuspended in the same buffer.

PT induction was assessed spectrophotometrically as previously described[9] by suspending ~0.2 mg mitochondria at room temperature in 200 μl of 215 mM mannitol, 71 mM sucrose, 5 mM K-HEPES (pH 7.4), and 10 mM sodium succinate. Changes in absorbance at 540 nm (A_{540}) were followed for two hours.

Western blotting was done using minor modifications of established procedures.

RESULTS AND DISCUSSION

Concentrations of GD3 ≥ 30 μM can accelerate PT induction (FIG. 1A). The ability of gangliosides GD1a, GM1, and GT1b to induce PT was examined as controls

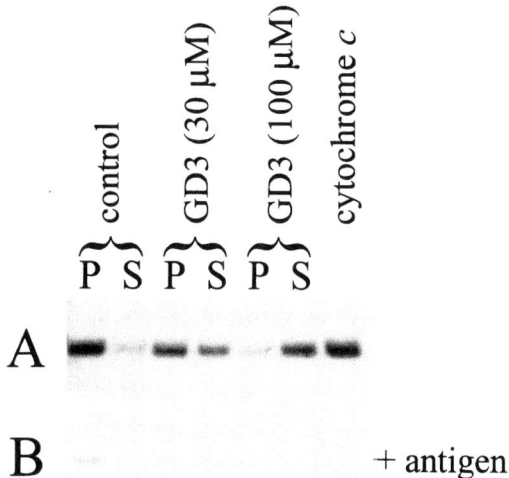

FIGURE 2. GD3, but not C_2-ceramide, releases cytochrome c. Mitochondrial pellets (0.1–0.2 mg protein, *P lanes* in figure) and their respective supernatants (*S lanes* in figure) were analyzed by Western blotting using an antibody that recognizes a denatured epitope of cytochrome c (Zymed, **panel A**). Antibody specificity is demonstrated by loss of immunostaining following preincubation of the primary antibody with 50-fold molar excess of denatured, purified rat cytochrome c (Sigma, **panel B**). *Lane pairs* are as labeled; *final lane* is purified cytochrome c.

for nonspecific effects of gangliosides, such as bulk effects on lipid organization or on ion permeability. These other gangliosides do not induce apoptosis in the HuT78 system.[1] Unlike GD3, 100-μM concentrations of gangliosides GD1a, GT1b, and GM1 had only minimal effects (FIG. 1B). The structural specificity of the sialogangliosides in PT induction is reminiscent of that seen with hydroxyalkenals, where 4-hydroxyhexenal induces the PT at femtomolar levels, whereas 4-hydroxynonenal requires micromolar concentrations.[9] GD3-mediated effects on absorbance were inhibited by well-characterized inhibitors of the PT—cyclosporin A (CsA), $MgCl_2$, ADP, and trifluoperazine[4,5] (not shown). These data confirm that the effects observed were caused by induction of the PT.

Further support for the hypothesis that GD3 can directly enlist mitochondria into the apoptotic cascade is that GD3-mediated PT induction was associated with release of mitochondrial cytochrome c (FIG. 2). In contrast, cytochrome c was not released from mitochondria in control (untreated) samples.

These results are consistent with a model in which the formation of GD3 is followed by transport of GD3 to mitochondria where this ganglioside stimulates PT induction and/or cytochrome c release. The proposed activity of GD3 would provide a direct mechanistic link between the early and late stages of CD95-mediated and ceramide-induced apoptosis.

REFERENCES

1. DE MARIA, R., M.R. RIPPO, E.H. SCHUCHMAN & R. TESTI. 1998. Acidic sphingomyelinase (ASM) is necessary for *Fas*-induced GD3 ganglioside accumulation and efficient apoptosis of lymphoid cells. J. Exp. Med. **187:** 897–902.
2. DE MARIA, R. *et al.* 1997. Requirement for GD3 ganglioside in CD95- and ceramide-induced apoptosis. Science **277:** 1652–1655.
3. MATYAS, G.R. & D.J. MORRE. 1987. Subcellular distribution and biosynthesis of rat liver gangliosides. Biochim. Biophys. Acta **921:** 599–614.
4. ZORATTI, M. & I. SZABO. 1995. The mitochondrial permeability transition. Biochem. Biophys. Acta **1241:** 139–176.
5. GUNTER, T.E., K.K. GUNTER, S.S. SHEU & C.E. GAVIN. 1994. Mitochondrial calcium transport: physiological and pathological relevance. Am. J. Physiol. **267:** C313–C339.
6. DUCHEN, M.R., O. MCGUINESS, L.A. BROWN & M. CROMPTON. 1993. On the involvement of a cyclosporin A sensitive mitochondrial pore in myocardial reperfusion injury. Cardiovasc. Res. **27:** 1790–1794.
7. CASTEDO, M. *et al.* 1996. Sequential acquisition of mitochondrial and plasma membrane alterations during early lymphocyte apoptosis. J. Immunol. **157:** 512–521.
8. GREEN, D.R. & J.C. REED. 1998. Mitochondria and apoptosis. Science **281:** 1309–1312.
9. KRISTAL, B.S, B.K. PARK & B.P. YU. 1996. 4-Hydroxyhexenal is a potent inducer of the mitochondrial permeability transition. J. Biol. Chem. **271:** 6033–6038.

Interactions between Melatonin, Reactive Oxygen Species, and Nitric Oxide

D. K. LAHIRI[a] AND C. GHOSH

Department of Psychiatry, Institute of Psychiatric Research, Indiana University School of Medicine, 791 Union Drive, Indianapolis, Indiana 46202-4887, USA

INTRODUCTION

Nitric oxide (NO), which is a labile and highly reactive substance, is enzymatically generated by nitric oxide synthase (NOS).[1,2] NO can also be produced by a group of compounds known as NO donors such as sodium nitroprusside (SNP). NO plays an important role as a neuronal messenger involved in neurotransmitter release, long-term potentiation, and gene transcription. Neurons produce NO mainly by a calcium-dependent activation of constitutive neuronal NOS. Neurons also appear to express inducible NOS.[3] Glial cells synthesize NO in a calcium-independent manner via induction of NOS.[4] The endothelial form of NOS is also present in CNS and is associated with the brain vasculature.[5] As compared to the control, NOS activity was elevated significantly in brain microvessels from patients with Alzheimer's disease (AD).[6] Since reactive oxygen species (ROS) are involved in AD, our aim is to determine the role of free radicals and superoxides in NO production by SNP or NOS. Using a cell culture system and a NO donor, we have studied how melatonin treatment influences the release of SNP-mediated nitric oxide metabolites (NOx) and regulates the activity of NOS. Our results indicate that melatonin reduces the accumulation of SNP-mediated NOx. NOS activity is reduced by SNP, and melatonin restores the SNP-mediated inhibition of NOS activity in a murine neuroblastoma cell line, suggesting the role of melatonin as a free-radical scavenger.

EXPERIMENTAL PROCEDURES

Materials

Carboxyl-PTIO, L-NAME, melatonin, and superoxide dismutase-1 (SOD-1) were purchased from Sigma (St. Louis, MO). Media and sera were obtained from Life Technologies (Gaithersburg, MD). [^{3}H]-Arginine was procured from Amersham (Arlington Heights, IL).

Cell Culture and Treatment of Cells with Different Agents

N1E-115 cells were cultured in Dulbeccos modified Eagle medium (DMEM) with 10% FBS in the presence of 5% CO_2. Cells were plated at a density of 1×10^6

---

[a]Corresponding author. Phone: 317-274-2706; fax: 317-274-1365.
e-mail: DLAHIRI@IUPUI.EDU

cells per plate in the complete medium 24 hours prior to the experiment. On the day of the experiment, the medium was changed to a serum-free medium and the cells were incubated for 48 h with SNP (30 µM) alone and in the presence of either melatonin (25 µg/ml), C-PTIO (30 µM) or SOD-1 (100 units/ml). The cells were harvested at 48 h, and used for NOS assay.

Measurement of the Level of NOx

Levels of NOx released in the culture medium were measured at 6, 12, 24, 32, and 48 h with an equal amount of conditioned medium (100 µl) using the Griess reagent.[7] The plain culture medium incubated under the same condition was used as blank, and NOx accumulation in this medium was subtracted from that of the conditioned medium at each time point. Sodium nitrite solution in culture medium was used as a standard.

Measurement of NOS

NOS activity in cell lysate was measured by monitoring the formation of L-[^{3}H] citrulline from L-[^{3}H] arginine using a commercially available kit (Stratagene, La Jolla, CA). Briefly, the cells were harvested, resuspended in the homogenization buffer, lysed by sonication, and centrifuged at $10,000 \times g$ for 5 min at 4°C. Rat cerebellar extracts were similarly prepared from the rat cerebellum. Protein was measured in the supernatant, which was used to assay for NOS activity.

RESULTS AND DISCUSSION

Melatonin Decreases the Production of NOx in N1E-115 Cells

As expected, SNP was found to release NOx when incubated with the cell culture medium alone, and this value was subtracted in subsequent experiments. When the SNP-containing medium was added to the N1E-115 cells, there was an increased release of NOx in the conditioned medium in a time-dependent manner. For example, at 30 µM SNP, NOx release was 3.82 ± 0.4 µM at 12 h, 5.51 ± 0.9 µM at 24 h, and 5.93 ± 0.2 µM at 32 h (FIG. 1). The increase observed was significantly higher than the spontaneous NOx release from SNP in the medium. To evaluate the pathway involved in NOx production, effects of an antioxidant, a free radical scavenger, and a NOS inhibitor were studied. The coincubation of SNP and melatonin in N1E-115 cells resulted in a significant drop in NOx release at all time points studied (e.g., 32% and 29.4% drops at 12 and 24 h, respectively). The action of melatonin is not clearly understood but most likely is due to its scavenging role of hydroxyl ions, which react with peroxynitrite to form NOx. Melatonin has recently been shown to scavenge peroxynitrite anion and nitric oxide directly.[8] The specificity of NOx production from SNP was tested by treating the N1E-115 cells with C-PTIO, which inhibited significantly the release of NOx (e.g., 31% and 46% declines at 12 and 24 h, respectively) (FIG. 1). It is known that C-PTIO, which is a stable nitric oxide radical scavenger, reacts with free NO radical in a stoichiometric manner. Addition of SOD-1 to the cells produced a sharp decrease in NOx accumulation (e.g., 59% and 29% reductions

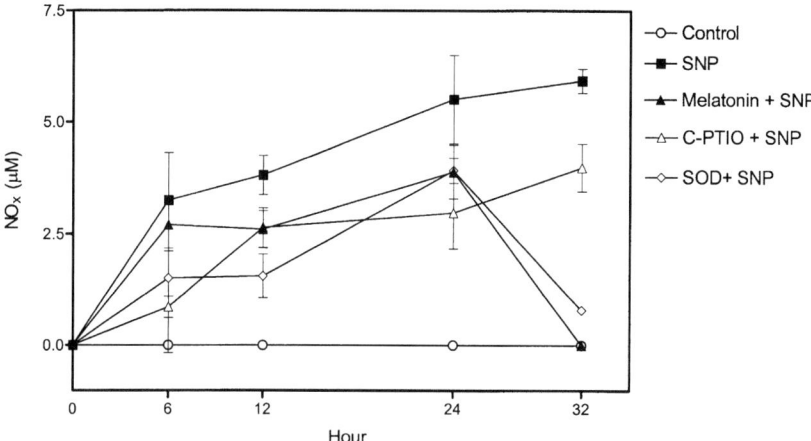

FIGURE 1. Time course of SNP-mediated NOx release in N1E-115 cells in the presence of melatonin and other agents. N1E-115 cells (1×10^6) were incubated in the presence of SNP (30 μM) alone and in combination with either melatonin (25 μg/ml), NO inhibitor carboxyl-PTIO (30 μM), or superoxide dismutase (100 units/ml). Levels of NOx were measured in the conditioned medium at 6, 12, 24, and 32 h. Values are mean ± SEM from triplicate experiments.

at 12 and 24 h, respectively). The reduced level of inhibition by SOD-1 at the latter time point was due to the inactivation of the enzyme over prolonged incubation. Nitric oxide reacts with superoxide radical to give rise to peroxynitrite.[2,9] SOD is a scavenger of superoxide and prevents peroxynitrite formation. The presence of SOD thus should decrease the NOx production as measured by the Griess assay. The involvement of NOS in SNP-mediated NO release was examined using L-NAME, a well-studied NOS inhibitor. The incubation of the N1E-115 cells with L-NAME did not change the amount of NOx production by SNP (data not shown). This observation suggests that the action of SNP does not involve NOS. A similar finding was reported previously.[10] When melatonin, C-PTIO or SOD-1 was incubated alone with the cells, there was very little NOx accumulation, which was subtracted from the total release when either of these agents was coincubated with SNP. The measurement of LDH demonstrated that compared to control, none of the agents described above showed a significant level of toxicity in cell culture under the conditions used here (data not shown).

Melatonin Inhibits NOS Activity in Rat Brain Extracts (in Vitro)

As expected, the rat cerebellar extracts displayed a significant level of NOS activity (FIG. 2A). When these extracts were incubated with SNP, there was a significant loss (72%) of NOS activity. This loss of the enzymatic activity was due to the feedback inhibition from NO that was simultaneously produced chemically by SNP. Interestingly, when the rat brain extracts were incubated with melatonin only, there was a 52% inhibition of NOS, suggesting that the conversion of L-arginine to L-citrulline by NOS involves melatonin via an unidentified pathway. It is possible that the

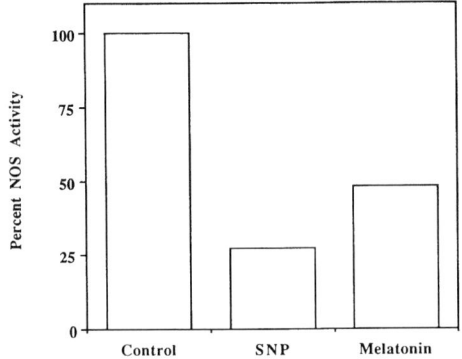

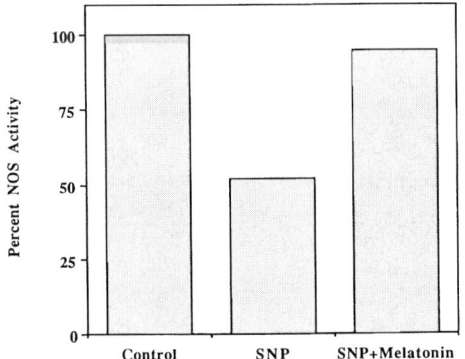

FIGURE 2. Effect of SNP and melatonin on the activity of NOS *in vitro* and *in vivo* (cell culture). **A.** Rat cerebellar extracts were used to assay for NOS activity in the presence of either SNP (3 mM) or melatonin (3 mM) alone as indicated. NOS activity was measured by monitoring the formation of L-[^{3}H]-citrulline from L-[^{3}H]-arginine. Both SNP and melatonin were found to inhibit the activity of NOS in cell-free extracts. **B.** N1E-115 cells were cultured in the presence of either SNP or SNP + melatonin as indicated. After 48-h treatment, the cells were harvested, sonicated in the homogenization buffer, and centrifuged at 11,000 × g. The supernatant was assayed for NOS activity, which was measured as described in the text. Levels of NOx were also measured with these treated cultures as described in the legend to FIGURE 1.

inhibition of NOS activity by melatonin was due to its binding with calmodulin, which is otherwise needed, for NOS activity.[11] Similar findings have recently been reported in which constitutive rat cerebellar NOS was shown to be inhibited by melatonin.[11] Thus the modes of inhibitory action of SNP and melatonin on NOS activity are different.

Melatonin Prevents SNP-Mediated Inhibition of NOS Activity (in Vivo)

The neuroblastoma cell line N1E-115 was chosen for this study because it expresses significant levels of endogenous NOS. When N1E-115 cells were treated with SNP in serum-free DMEM medium, no change in NOS activity was detected (data not shown). The basal level of NOS activity was very low after 48 h of serum deprivation. When the cells were incubated with SNP in DMEM medium containing 10% FBS, the basal level of NOS activity was significantly elevated. For this, N1E-115 cells were cultured for 48 h in the regular medium in the presence of either SNP or SNP plus melatonin. When cell lysates were assayed for NOS, we observed that, compared to control, SNP-treated cells showed a loss in NOS activity, which was expected from FIGURE 2A. Cotreating the cells with melatonin could abolish this inhibitory action of SNP, and thus the cotreatment of melatonin restored the NOS activity, suggesting that the feedback inhibition by SNP can be neutralized by melatonin. Our results demonstrate that nonenzymatic release of NOx could inhibit the activity of NOS resulting in decreased accumulation of NOx. This is in agreement with previous reports showing a feedback inhibition of NOS activity with excess NOx.[12] Melatonin has recently been shown to act as a neuroprotective agent and a promoter of neuronal differentiation.[13] There is gradual increase of ROS production during the aging process and also in neurodegenerative disorders. Our results showing the ability of melatonin to reduce ROS buildup, coupled with its role as a scavenger for hydroxyl radicals, warrant further research for developing melatonin as a preventing treatment for age-related disorders and for Alzheimer's disease.

SUMMARY

Accumulation of reactive oxygen species is critical for the neuropathology of Alzheimer's disease. Melatonin hormone, an antioxidant, could play a key role in aging and senescence. Nitric oxide, a biologically active unstable radical, is synthesized by nitric oxide synthase when converting L-arginine to L-citrulline. We have investigated whether the treatment of cultured cells with melatonin could possibly reduce the release of free radicals and other ROS. We assayed NO indirectly by measuring the level of its stable end products, nitrite/nitrate (NOx), using the Griess reagent. When the neuroblastoma cells such as N1E-115 were treated with a NO donor such as sodium nitroprusside (SNP), a significant level of NOx was detected in a time- and dose-dependent manner in the conditioned medium compared to the untreated cells or SNP-containing media. In neuroblastoma cells, the release of NOx as mediated by SNP was significantly inhibited by treatment with (i) carboxy-PTIO, a NO scavenger; (ii) SOD-1, superoxide dismutase; and (iii) melatonin. In these cells SNP-mediated NOx release was mediated by superoxide ions and/or free radicals that can be inhibited by melatonin. The ROS-scavenging function of melatonin along with its neuroprotective and neurodifferentiating role can be utilized for the prevention of neurodegenerative disorders such as AD.

ACKNOWLEDGMENTS

This work was supported by NIH grant AG14882-01 to D.K.L.

REFERENCES

1. BREDT, D.S. & S.H. SNYDER. 1992. Nitric oxide, a novel neuronal messenger. Neuron **8:** 3–11.
2. ZHANG, J. & S.H. SNYDER. 1995. Nitric oxide in the nervous system. Annu. Rev. Pharmacol. Toxicol. **35:** 213–233.
3. MINC-GOLOMB, D. *et al.* 1996. In vivo expression of inducible nitric oxide synthase in cerebellar neurons. J. Neurochem. **66:** 1504–1509.
4. GALEA, E., D.L. FEINSTEIN & D.J. REIS. 1992. Induction of calcium-independent nitric oxide synthase activity in primary rat glial cultures. Proc. Natl. Acad. Sci. USA **89:** 10945–10949.
5. MURPHY, S. & D. GUZYBICKI. 1996. Glial NO. Normal and pathological roles. Neuroscientist **2:** 90–99.
6. BENZING, W.C. & E.J. MUFSON. 1995. Increased number of NADPH-d–positive neurons within the substantia innominata in Alzheimer's disease. Brain Res. **670:** 351–355.
7. GREEN, L.C. *et al.* 1982. Analysis of nitrate, nitrite and ^{15}N nitrate in biological fluids. Anal. Biochem. **126:** 131–138.
8. REITER, R.J. *et al.* 1999. The oxidant/antioxidant network: role of melatonin. Biol. Signals Recept. **8:** 56–63.
9. BECKMAN, J.S. *et al.* 1990. Apparent hydroxyl radical production by peroxynitrite: implication for endothelial injury from nitric oxide and superoxide. Proc. Natl. Acad. Sci. USA **87:** 1620–1624.
10. YU, O. & D. CHUANG. 1996. Inhibition of excitatory amino acid–induced phosphoinositide hydrolysis as a possible mechanism of nitroprusside neurotoxicity. J. Neurochem. **66:** 346–354.
11. POZO, D. *et al.* 1997. Inhibition of cerebellar nitric oxide synthase and cyclic GMP production by melatonin via complex formation with calmodulun. J. Cell. Biochem. **65:** 430–432.
12. MONCADA, S. & A. HIGGS. 1993. The L-arginine–nitric oxide pathway. N. Engl. J. Med. **329:** 2002–2012.
13. SONG, W. & D.K. LAHIRI. 1997. Melatonin alters the metabolism of the beta-amyloid precursor protein in the neuroendocrine cell line PC12. J. Mol. Neurosci. **9:** 75–92.

Effect of Oxidative Stress on DNA Damage and β-Amyloid Precursor Proteins in Lymphoblastoid Cell Lines from a Nigerian Population

D. K. LAHIRI,[a,e] Y. XU,[b] J. KLAUNIG,[b] O. BAIYEWU,[c] A. OGUNNIYI,[c] K. HALL,[a] H. HENDRIE,[a] AND A. SAHOTA[d]

[a]*Departments of Psychiatry and* [b]*Pharmacology and Toxicology, Indiana University School of Medicine, Indianapolis, Indiana 46202, USA*

[c]*Department of Medicine, University of Ibadan, Ibadan, Nigeria*

[d]*Department of Genetics, Rutgers University, Piscataway, New Jersey 08854, USA*

INTRODUCTION

Neurodegenerative changes are a characteristic feature of the aging process. These changes may result from damage to intracellular macromolecules by reactive oxygen species (ROS) generated during oxidative phosphorylation and other enzymatic reactions.[1] The ability to repair macromolecular damage also decreases with age.[2] As one of the most metabolically active organs, the brain is considered to be most vulnerable to oxidative stress. Age is a well-established risk factor for late-onset sporadic Alzheimer's disease (AD), and the extent of neurodegenerative changes is significantly greater in AD patients than in controls. There are alterations in several markers of oxidative stress in brain tissue from AD patients, suggesting a link between increased oxidative stress and cell damage.[1] Since mitochondria are a major source of ROS, it has been hypothesized that increased cellular damage in AD may be the result of a genetic defect that occurred in a somatic cell at an early stage of development.[3] Such a defect may promote mitochondrial DNA damage or impair DNA repair.[4] Somatic mutations in mitochondrial (or nuclear) genes encoding enzymes of energy metabolism may also explain the pathological changes.[5] AD is primarily a neuronal disease, but the occurrence of a somatic mutation in early embryogenesis could explain the systemic features of this disorder.[3,6] It is on the basis of observations like these that ROS has been implicated in the pathogenesis of AD. Brain autopsy studies have shown that the density of plaques is lower in elderly Nigerians compared with Caucasian populations.[7] The age-adjusted prevalence of AD in an elderly population group in Ibadan (Nigeria) was significantly lower compared with an African-American population group in Indianapolis (Indiana).[8] The ε4 allele of apolipoprotein E (APOE) is a major risk factor for the development of late-

---

[e]Address correspondence to: Debomoy K. Lahiri, Ph.D., Department of Psychiatry, The Institute of Psychiatric Research, Indiana University Medical Center, 791 Union Drive, Indianapolis, IN 46202-4887. Phone: 317-274-2706; fax: 317-274-1365.

e-mail: dlahiri@iupui.edu

onset sporadic AD in Caucasian and Japanese populations.[9] Our studies have shown that APOE ε4 was a weak risk factor for AD in the Indianapolis group, but it was not a risk factor in the Ibadan group.[10] The conversion of the soluble amyloid-β (Aβ) peptides into the insoluble amyloid present in neuritic plaques is accelerated in the presence of APOE ε4,[11] and free radical reactions are involved in this process.[12] These observations suggest that the lower density of plaques in brains from Nigerian subjects may be related to reduced oxidative stress, but this remains to be established in AD patients.

Because of the problems associated with obtaining brain tissue, molecular analyses on AD have been carried out in lymphoblast or fibroblast cell lines.[13] The use of non-neuronal tissue is also justified in view of the alterations that have been demonstrated in fibroblast and other cell types from AD patients.[6] We have established transformed lymphoblast cell lines from several probable AD cases and non-demented controls from Indianapolis and Ibadan. The APOE genotypes of these individuals are known. In this preliminary study, we investigated the effects of endogenous and induced oxidative stress on total (nuclear and mitochondrial) DNA damage in eight cell lines (five probable AD and three controls) from the Ibadan group. The effect of oxidative stress on the level of a putative protein marker for AD (β-amyloid precursor protein, APP) was also examined.[14] The level of the heat shock protein (HSP-70) was used as a control.

EXPERIMENTAL PROCEDURES

Materials

Acrylamide, alkaline phosphatase, nuclease P1, sodium dodecyl sulfate (SDS), *t*-butyl peroxide and other chemicals were obtained from Sigma (St. Louis, MO). The growth medium, serum and other cell culture reagents were procured from Life Technologies (Gaithersburg, MD).

Cell Culture

Cells were grown in RPMI 1640 containing 7.5% fetal calf serum. For each cell line, 10 ml aliquots (about 7×10^6 cells) were used in triplicate: one each for total DNA (nuclear and mitochondrial) isolation, nuclear DNA isolation, and protein expression studies. This was performed before and after exposing the cells to 200 μM *t*-butyl peroxide for 4 hours. This chemical generates hydrogen peroxide, which is known to damage DNA through free radical formation.

Oxidative DNA Damage

Total DNA was isolated from untreated and treated samples by extracting with phenol-chloroform. For nuclear DNA, nuclei were first isolated by gentle homogenization. The DNA was digested with nuclease P1 and alkaline phosphatase and then analyzed by HPLC for 8-hydroxy-2′-deoxyguanosine (OH8dG), a marker for oxidative DNA damage. The level of oxidative DNA damage was expressed as μmol OH8dG per mol dG. Mitochondrial DNA damage was taken to be the difference between the total and nuclear DNA.

TABLE 1. Diagnosis, gender and APOE genotyping of different subjects

Subject/Cell line	Gender	Age	Diagnosis	APOE
1586	F	88	NDEM	3/3
1588	F	81	DEM	4/4
1589	F	115	DEM	3/3
1590	F	88	DEM	2/3
1593	F	73	CI	3/3
1597	F	85	DEM	3/4
2198	M	97	DEM	3/3
2205	M	70	NDEM	3/3

NOTES: Cell lines, derived from these subjects, were used in subseqent studies (TABLES 2 and 3). Clinical diagnosis was based on DSM-III and NINDS/ADRDA criteria (see Hendrie et al., 1995). NDEM: non-demented; DEM: demented; CI: cognitively impaired.

TABLE 2. Levels of 8-hydroxy-2′-deoxyguanosine (OH8dG) and 2′-deoxyguanosine (dG), and the ratio of OH8dG/dG, in untreated and treated cell lines

	Untreated			Treated		
Subject/Cell Line	OH8dG fmol	dG pmol	Ratio $\times 10^{-6}$	OH8dG fmol	dG pmol	Ratio $\times 10^{-6}$
1586	9.12	1007.07	9.06	9.03	439.29	20.56
1588	8.30	2032.81	4.08	16.95	1024.32	16.55
1589	15.93	2037.93	7.82	5.79	1024.75	5.65
1590	6.26	1794.72	3.49	7.18	1796.34	4.00
1593	13.94	1166.60	11.95	9.79	842.85	11.62
1597	9.73	767.37	12.68	9.03	1221.66	7.39
2198	13.96	1633.54	8.55	10.38	289.83	35.81
2205	5.09	454.59	11.20	20.27	1599.69	12.67

NOTES: DNA was digested with nuclease P1 and alkaline phosphatase and analyzed by HPLC. Levels of OH8dG were determined by electrochemical detection and dG by UV detection. The level of oxidative damage was expressed as the OH8dG/dG ratio.

Protein Levels

The untreated and treated samples were centrifuged and the conditioned medium saved. The cell pellets were re-suspended in buffer containing detergents and protease inhibitors. The cell suspension was sonicated, and the protein concentration determined. Thirty µg protein from the cell lysate was separated on a 10% polyacrylamide gel containing SDS. Immunodetection of specific bands was performed as previously described.[14] For the detection of APP and its derivatives, primary antibody mAb22C11 (Roche Molecular Biochemicals, Indianapolis, IN) was used. This antibody recognizes all mature forms of APP found in cell membranes, the carboxyl-truncated soluble forms secreted into the medium, and the APP-like protein (APLP). The 6E10 monoclonal antibody, which recognizes residues 1–28 of the APP-specific extra-membranous region of the Aβ sequence, was also used to confirm the APP band (data not shown). Primary antibody N27F3-4 (Sigma) was used for the detection of HSP-70/72, which served as an internal control.

TABLE 3. Qualitative changes in the OH8dG/dG ratio and the expression of APP and HSP following treatment of cell lines with *tert*-butyl peroxide

Subject/ Cell Line	Diagnosis	OH8dG/dG Ratio	APP	HSP
1586	NDEM	↑	–	↓
1588	DEM	↑	↓	↓
1589	DEM	–	–	–
1590	DEM	–	↓	–
1593	CI	–	–	–
1597	DEM	↓	↓	↓
2198	DEM	↑	↓	↓
2205	NDEM	–	↓	–

NOTES: The ratio of OH8dG/dG and the level of APP, HSP proteins are increased (↑), decreased (↓), or unchanged (–) in the treated cell lines relative to controls.

RESULTS

The clinical status and APOE genotypes of the patients; the levels of OH8dG, dG, and the OH8dG/dG ratio; and qualitative changes in APP and HSP-70/72 expression are shown in TABLES 1–3. The levels of APP and HSP-70/72, before and after *t*-butyl peroxide treatment, were measured semiquantitatively from Western blots. These changes are summarized in TABLE 3. In this preliminary study, we did not detect a significant difference in the OH8dG/dG ratio in total DNA in cell lines from patients or controls, with or without treatment with *t*-butyl peroxide. The ratio for the untreated group was in the range 3.8 to 13.2, and similar values were obtained in the treated group (range 3.9 to 37.1). There was also little or no change in APP and HSP-70/72 expression following oxidative stress. Our sample size was too small to make any correlation between these changes and APOE genotypes.

We also carried out oxidative stress studies in lymphoblastoid cell lines from two autopsy-confirmed cases of AD, six cases of probable AD, and three controls (all of Caucasian origin). We observed a 2- to 3-fold increase in OH8dG formation in total DNA, but not nuclear DNA, following *t*-butyl peroxide treatment in the patient group, but not in the control group (Sahota *et al.*, unpublished data). This suggested that there was increased mitochondrial damage in cell lines from AD patients compared with controls following oxidative stress.

DISCUSSION

ROS have been implicated in the pathogenesis of AD and other neurodegenerative diseases.[15] Although alterations in several markers of oxidative stress have been reported in AD patients, a cause and effect relationship between ROS and AD remains to be established. Impaired energy metabolism,[16] mutations in genes encoding enzymes of oxidative phosphorylation,[5] and increased mitochondrial DNA damage and/or defective repair strongly suggest a role for oxidative stress in the pathogenesis of AD.[4]

Our study of dementia in two populations of African origin suggested that the prevalence of AD was significantly lower in Nigerian blacks compared with African-Americans.[8] Non-demented elderly subjects from Nigeria also had a lower density of senile plaques compared with Caucasian subjects.[7] We have shown that APOE ε4, the gene product of which has been shown to promote free radical formation,[12] is not associated with AD in Nigerian blacks.[10] The lower density of plaques in brain tissue and the lower prevalence of AD may be related to reduced oxidative stress in this population.

In this preliminary study, we did not detect a difference in endogenous or induced DNA damage in cell lines from patients or controls from Ibadan, and this effect was independent of APOE genotype. We have observed an increase in mitochondrial DNA damage in cell lines from a small number of probable AD patients of Caucasian origin (Sahota *et al.*, unpublished data). Based on the cell culture studies, the absence of oxidative DNA damage in the Ibadan subjects may be related to the lower prevalence of AD in this population. Our findings also indicate that APP and its derivatives are expressed not only in neuronal cells but also in lymphoblastoid cell lines. Lymphoblast cell lines may prove to be useful tools for evaluating oxidative stress status, mitochondrial mutations, AD protein markers, and the risk for AD in epidemiological and clinical studies.

SUMMARY

The ε4 allele of apolipoprotein E (APOE) is strongly associated with late-onset Alzheimer's disease (AD) in Caucasian populations, but our studies suggest that APOE ε4 is not a risk factor for AD in Nigerian blacks and is a weak risk factor in African-Americans. The prevalence of AD is lower in Nigerians than in African-Americans. Increased oxidative damage to macromolecules in brain tissue by reactive oxygen species (ROS) has been reported in AD. Here we examined the effects of endogenous and induced oxidative stress on total (nuclear and mitochondrial) DNA damage in lymphoblastoid cell lines (5 probable AD and 3 controls) from Ibadan, Nigeria. Cells were exposed to 200 µM *t*-butyl peroxide (a generator of ROS) for 4 hours. Total DNA was isolated and digested with nuclease P1 and alkaline phosphatase. DNA fragments were separated by HPLC and the levels of 8-hydroxy-2′-deoxyguanosine (OH8dG, an indicator of DNA damage) and deoxyguanosine (dG) determined. We did not detect a significant difference in the OH8dG/dG ratio in untreated or treated cell lines in the two groups, and this was independent of APOE genotype. We also examined, by Western blotting, the level of β-amyloid precursor protein (APP) which is involved in AD. The level of the heat shock protein (HSP-70) was examined as a control. There was a slight decrease in levels of APP and HSP-70 following treatment. Studies in cell lines from Caucasian subjects have shown an increase in mitochondrial DNA damage following oxidative challenge. Our preliminary results suggest that African populations are less vulnerable to chemical-induced oxidative DNA damage.

ACKNOWLEDGMENTS

This work was supported by a pilot grant from the Indiana Alzheimer Disease Research Center and NIH grants AG 00956 and AG14882. We acknowledge the AD patients and control subjects for providing blood samples for cell lines. The cell lines were established under the aegis of the Alzheimer Disease National Cell Repository, which is located at the Indiana University Medical Center. We thank Cheryl Halter and Diana McLucas for assistance with cell lines and experimental work, respectively.

REFERENCES

1. GÖTZ, M.E. et al. 1994. Oxidative stress: free radical production in neural degeneration. Pharmac. Ther. **63**: 37–122.
2. BOERRIGTER, M.E.T.I. et al. 1992. DNA repair and Alzheimer's disease. J. Gerontol. (Biol. Sci). **47**: B177–B184.
3. HARMAN, D. 1995. Free radical theory of aging. Age **18**: 97–119.
4. YAKES, F.M. & B. VAN HOUTEN. 1997. Mitochondrial DNA damage is more extensive and persists longer than nuclear DNA damage in human cells following oxidative stress. Proc. Natl. Acad. Sci. USA **94**: 514–519.
5. DAVIS, R.E. et al. 1997. Mutations in mitochondrial cytochrome c oxidase genes segregate with late-onset Alzheimer disease. Proc. Natl. Acad. Sci. USA **94**: 4526–4531.
6. BAKER, A.C. et al. 1988. Systemic manifestations of Alzheimer's disease. Age **11**: 60–65.
7. OSUNTOKUN, B.O. et al. 1994. βA4-amyloid in the brains of non-demented Nigerian Africans. Lancet **343**: 56.
8. HENDRIE, H.C. et al. 1995. Prevalence of Alzheimer's disease and dementia in two communities: Nigerian Africans and African Americans. Am. J. Psychol. **152**: 1485–1492.
9. FARRER, L.A. et al. 1997. Effects of age, sex, and ethinicity on the association between apolipoprotein E genotype and Alzheimer disease. New Eng. J. Med. **278**: 1349–1356.
10. SAHOTA, A. et al. 1997. Apolipoprotein E-associated risk for Alzheimer disease in the African-American population is genotype-dependent. Ann. Neurol. **42**: 659–661.
11. SCHMECHEL, D.E. et al. 1993. Increased amyloid β-peptide deposition in a cerebral cortex as a consequence of apolipoprotein E genotype in late-onset Alzheimer disease. Proc. Natl. Acad. Sci. USA **90**: 9649–9653.
12. HARRIS, M.E. et al. 1995. Direct evidence of oxidative injury produced by the Alzheimer's β-amyloid peptide (1–40) in cultured hippocampal neurons. Exp. Neurol. **31**: 193–202.
13. PARSHAD, R. et al. 1996. Fluorescent light-induced chromatid breaks distinguish Alzheimer disease cells from normal cells in tissue culture. Proc. Natl. Acad. Sci. USA **93**: 5146–5150.
14. LAHIRI, D.K. & M.R. FARLOW. 1996. Differential effect of tacrine and physostigmine on the secretion of the beta-amyloid precursor protein in cell lines. J. Mol. Neurosci. **7**: 41–49.
15. HALLIWELL, B. 1992. Oxygen radicals as key mediators in neurological disease: fact or fiction? Ann. Neurol. **S32**: 10–15.
16. BOWLING, A.C. et al. 1993. Age-dependent impairment of impairment of mitochondrial function in primate brain. J. Neurochem. **60**: 1964–1967.

Evidence for Energy Failure following Irreversible Traumatic Brain Injury

STEFAN M. LEE,[a] MONICA D. WONG, AMIR SAMII, AND DAVID A. HOVDA

Division of Neurosurgery and Brain Injury Research Center, UCLA School of Medicine, Los Angeles, California 90095, USA

INTRODUCTION

Recent findings from our laboratory have indicated that a traumatic brain injury (TBI) induces an immediate increase in cerebral glucose utilization lasting up to several hours postinjury.[1] This acute period of intense metabolic activity is accompanied by a marked increase in lactate production, which is indicative of hyperglycolysis.[2] This study assessed the ATP content within injured tissue acutely following two models of experimental brain injury: (1) fluid percussion (FP), which induces a reversible cellular injury,[1] and (2) controlled cortical impact (CCI), which results in irreversible cellular damage.[3]

MATERIALS AND METHODS

Fifty-seven male Sprague Dawley rats (150–200 g) were injured under 1–2% enflurane (in 100% O_2) inhalation anesthesia. Following a FP (2.25 atm) or CCI (1.7 m/s, 2.0 mm penetration) injury, animals were randomly divided for measurement of ATP content or glucose metabolism (CMR_{glc}) using standard ^{14}C-2-deoxy-D-glucose autoradiographic methods.[4] Twenty-four additional animals were FP- or CCI-injured and processed for hemotoxylin and eosin (H&E) histology at distinct time points after injury (30 min, 2 h, and 24 h, $n = 4$ at each time point) to assess the viability of cells in parietal cortex (primary injury site).

For ATP measurements, animals were killed at 30 min or 2 h post-TBI by focused microwave irradiation (3.5 kW for 3 s) using a Thermex (New York, NY) model 4101 microwave apparatus as described by Delaney and Geiger.[5] The two cerebral hemispheres were removed from the skull, and approximately 3-mm cubic chunks were dissected from parietal cortices of both hemispheres. The tissue was weighed and stored at −70°C until extraction. For ATP extraction, the tissue was homogenized for 2 min in distilled water and centrifuged for 10 min. The ATP content in the supernatant was measured using an ATP Assay Mix (Kit #FL-AA, Sigma, St. Louis, MO). Aliquots of each sample were first diluted 1:200 with distilled water after which 0.1 ml of the diluted samples were aliquoted into 7-ml scintillation vials. Aliquots of ATP assay mix were diluted 1:50 with the ATP assay mix dilution buffer

---

[a]Address for correspondence: Stefan M. Lee, Ph.D., Division of Neurosurgery, 18-228 NPI, UCLA School of Medicine, Los Angeles, CA 90095. Phone: 310-794-1886; fax: 310-794-2147.
e-mail: lee@surgery.medsch.ucla.edu

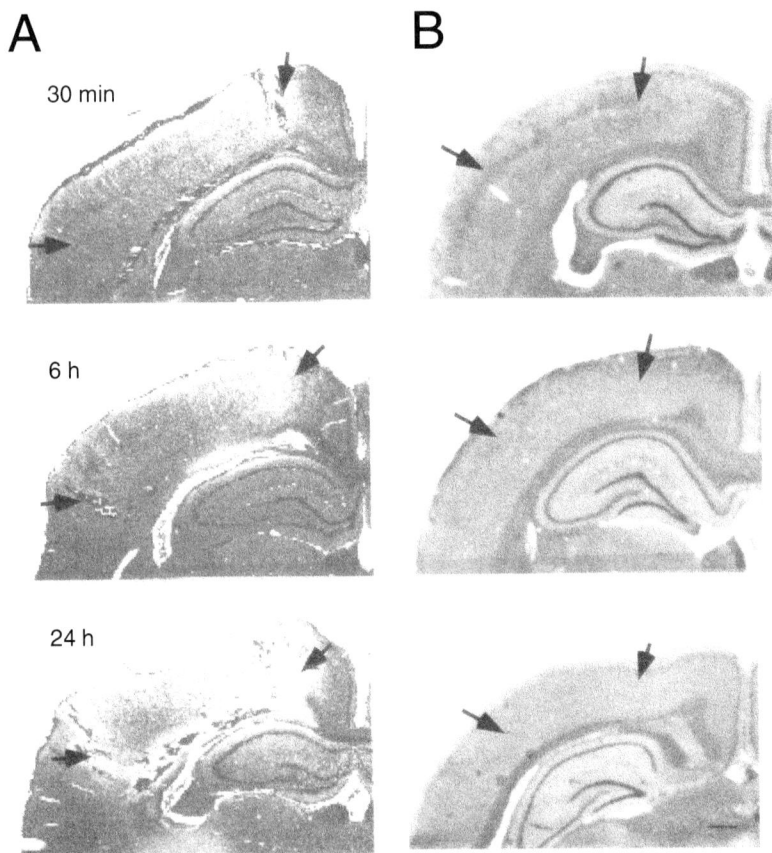

FIGURE 1. Representative photomicrographs of H&E-stained coronal sections through the primary injury site 30 min, 2 h, and 24 h following CCI (**A**) or FP (**B**) injuries. Tissue used for ATP assay and CMR_{glc} measurements was taken from the primary injury site (demarcated by *arrows*). Note that first signs of marked degeneration (*blanched areas*) were seen only approximately 24 h postinjury in the CCI-injured case and were absent following FP injuries. *Scale bar* = 1.0 mm.

and 0.1 ml was rapidly aliquoted into scintillation vials. The vials were swirled briskly and immediately placed in a scintillation counter set for a wide-open detection. The scintillation counts were initiated after 5 min and each vial was counted for 25 seconds. After the vials were counted, they were reversed in order and counted again to determine a geometric mean.[6]

The calculation of ATP values from scintillation CPM values are detailed by Reiger.[6] The two CPM values (first count and reversed count) were multiplied and the square root of this value, Cm, represented the mean CPM for each sample. The sample ATP values were determined from the ATP standard curves (10^{-7} to 10^{-10} M), which were generated separately for each case, and presented as µmol ATP per g wet tissue weight.

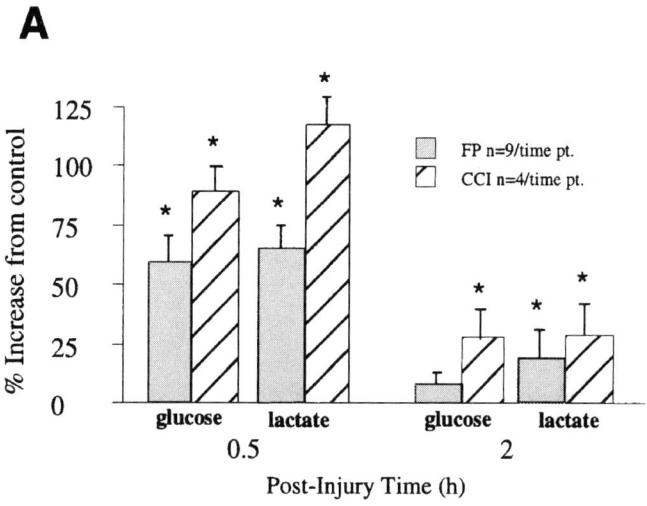

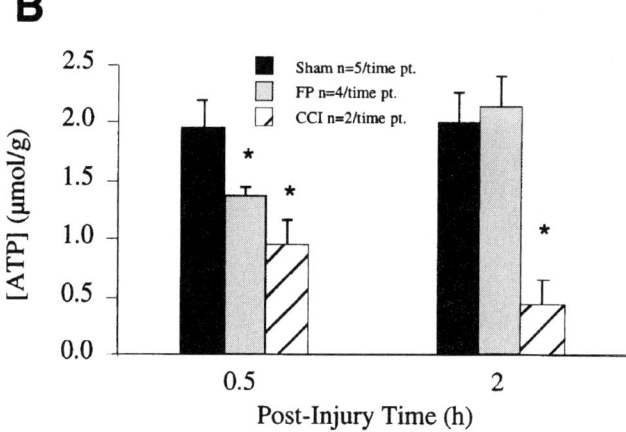

FIGURE 2. Changes in (**A**) glucose utilization and lactic acid levels and (**B**) ATP content following FP and CCI injuries measured 30 min and 2 h postinjury. As reported previously,[1] TBI results in an acute increase in glucose utilization, which was transient for reversible insults (FP injury), but more prolonged for irreversible brain damage (CCI injury). The high levels of lactic acid seen 30 min postinjury support the conclusion that increased glucose utilization primarily feeds anaerobic glycolysis. Note that ATP production was markedly reduced for both types of injuries within the primary injury site soon after trauma, but the reversibly injured tissue quickly recovered to control values by 2 hours. Lactic acid values are adapted from Ref. 2. * $p < 0.05$ compared to sham controls (two-tailed t-test).

RESULTS

Morphological analysis revealed first signs of degeneration following CCI injury—e.g., irregularly shaped pyramidal neurons, shrunken cells, darkly stained cell bodies—at 16–24 hours postinjury (FIG. 1A). In contrast, no detectable cell loss was seen within the primary injury site following a fluid percussion (FIG. 1B).

The mean (± SEM) ATP content for sham-injured cases was 1.96 ± 0.24 µmol/g. As shown in FIGURE 2, the reversible cellular injury (FP injury) led to 31.1% decrease in ATP content at 30 min postinjury, which recovered to sham-injured values by 2 hours. In contrast, the irreversible cellular injury (CCI injury) showed a marked decrease in ATP content (> 50%) which persisted for at least the first 2 h postinjury.

COMMENTS

A hallmark of both experimental and human traumatic brain injury is the immediate increase in glucose utilization, which may last for several minutes to several days.[1,7] The results from this study suggest that a crucial indicator of cellular recovery is the duration of this injury-induced increase in glucose metabolism and the extent of resulting cellular depletion of ATP.

ACKNOWLEDGMENTS

Research reported here was supported by NINDS Grants NS30308, NS27544, and NS37363; and the Lind Lawrence Foundation.

REFERENCES

1. YOSHINO, A. *et al.* 1991. Dynamic changes in local cerebral glucose utilization following cerebral concussion in rats: Evidence of a hyper- and hypometabolic state. Brain Res. **561:** 106–119.
2. KAWAMATA, T. *et al.* 1995. Lactate accumulation following concussive brain injury: The role of ionic fluxes induced by excitatory amino acids. Brain Res. **674:** 196–204.
3. SUTTON, R.L. *et al.* 1993. Unilateral cortical contusion injury in the rat: vascular disruption and temporal development of cortical necrosis. J. Neurotrauma **10:** 135–149.
4. SOKOLOFF, L. *et al.* 1977. The [^{14}C]deoxyglucose method for the measurement of local cerebral glucose utilization: Theory, procedure, and normal values in the conscious and anesthetized albino rat. J. Neurochem. **28:** 897–916.
5. DELANEY, S.M. & J.D. GEIGER. 1996. Brain regional levels of adenosine and adenosine nucleotides in rats killed by high-energy focused microwave irradiation. J. Neurosci. Methods **64:** 151–156.
6. REIGER, D. 1997. Batch analysis of the ATP content of bovine sperm, oocytes, and early embryos using a scintillation counter to measure the chemiluminescence produced by the luciferin-luciferase reaction. Anal. Biochem. **246:** 67–70.
7. BERGSNEIDER, M. *et al.* 1997. Cerebral hyperglycolysis following severe human traumatic brain injury: a positron emission tomography study. J. Neurosurg. **86:** 247-251.

Regulation of Mitochondrial Gene Expression in Differentiated PC12 Cells

LI-ING LIU,[a] S.I. RAPOPORT,[a] AND K. CHANDRASEKARAN[b,c]

[a]*Section on Brain Physiology and Metabolism, National Institute on Aging, NIH, Bethesda, Maryland 20892, USA*

[b]*Department of Anesthesiology, University of Maryland, Baltimore, Maryland 21201, USA*

INTRODUCTION

Clinical studies using positron emission tomography have shown that reduced cerebral glucose utilization is an early manifestation of Alzheimer's disease (AD).[1] Decreased brain-energy metabolism can precede and predict the pattern of neuropsychological abnormalities that appear later.[1] We observed corresponding decreases in expression of mitochondrial genes encoding for oxidative energy metabolism in postmortem brains from AD patients.[2] By Northern blot analysis, these brains showed a decrease of about 50% in levels of mRNA of mitochondrial DNA (mtDNA)-encoded subunits I–III of cytochrome oxidase (COX), compared with control brains. The amount of mtDNA-encoded 12S rRNA was not altered in AD brains compared with control brains. The amount of mtDNA also was unchanged, suggesting a mechanism of downregulation of COX expression rather than loss of mitochondria. In this project we investigated the mechanism of regulation of mtDNA gene expression using a morphological variant of rat pheochromocytoma (PC12S) cells.

METHODS AND MATERIALS

Cell Culture

We used a morphological variant of rat pheochromocytoma PC12 cells, called PC12S, that retains the ability to grow when attached to plastic tissue-culture dish.[3] Nerve growth factor (NGF) was added at a concentration of 50 ng/ml. After addition of NGF for 5 days, the morphology of PC 12S cells resembled that of sympathetic neurons.[3] All experiments were conducted on cells that had been maintained in the presence of NGF for 10 days.

Chemicals

All reagents and chemicals used were of the highest grade available from Sigma Chemical Co. (St. Louis, MO, USA). Ouabain was dissolved in water and monensin was dissolved in 95% ethanol. When ethanol was used as a solvent, appropriate con-

---

[c]Corresponding author. Phone: 410-706-3418; fax: 410-706-2550.
e-mail: kchandra@anesthlab.ab.umd.edu

trol experiments were conducted using the vehicle alone. Ethanol concentrations were always < 0.01%.

Experimental Procedure

PC12S cells grown (on 60 × 15 mm dishes) in presence of NGF for 10 days were treated either with ouabain (1 mM, final concentration), monensin (0. 1 μM, final concentration), or with vehicle (water in case of ouabain or alcohol in case of monensin). After various periods of time, cells were washed with Dulbecco's Phosphate Buffered Saline (DPBS), total RNA was isolated using the TRIzol reagent and subjected to Northern blot analysis with a mtDNA-encoded COX III cDNA probe and then with the control probe of β-actin cDNA.[2] The level of RNA hybridized was quantified using an image analysis program (NIH image 1.54). The ratio of COX III mRNA to β-actin mRNA was calculated.[2]

RESULTS

The level of β-actin mRNA and the ratio of mitochondrial DNA-encoded cytochrome oxidase subunit III mRNA (COX III mRNA) to β-actin mRNA showed no significant change in PC12S cells treated with vehicle during the 24-hour time period (data not shown). Addition of 1 mM ouabain (inhibitor of Na/K-ATPase) caused

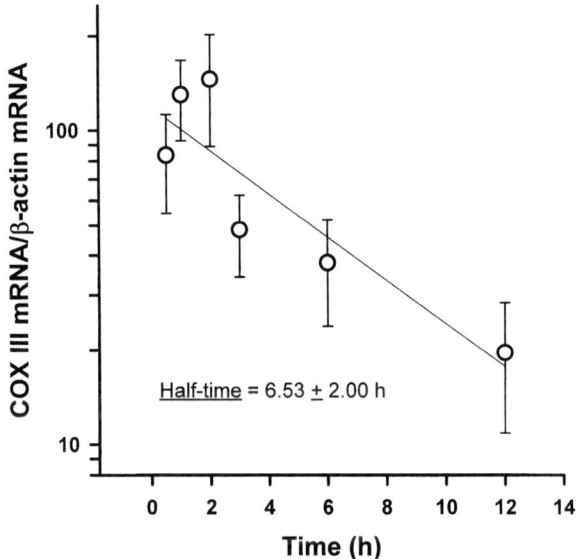

FIGURE 1. The time course of changes in COX III mRNA levels. The COX III mRNA and β-actin mRNA levels of PC12S cells exposed to 1 mM ouabain for the indicated periods were determined by Northern blot analysis and quantified by image analysis of autoradiograms. The relative change in COX III mRNA/β-actin mRNA ratio was related to COX III mRNA/β-actin mRNA ratio of zero time samples. Each point is the mean ± SEM of three to five separate experiments.

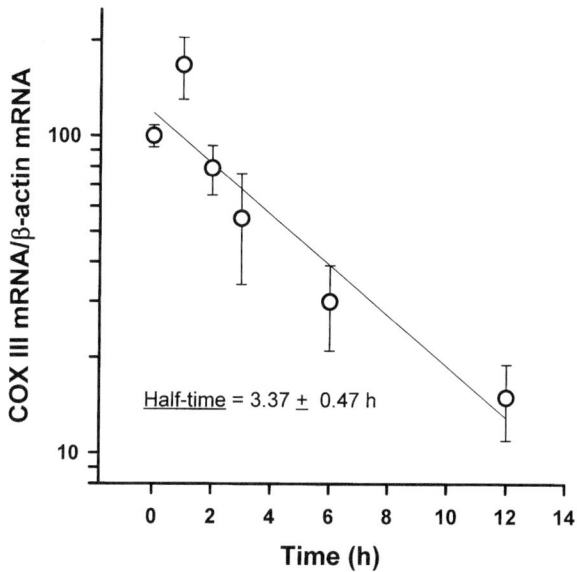

FIGURE 2. The time course of changes in COX III mRNA levels. The COX III mRNA and β-actin mRNA levels of PC12S cells exposed to 0.1 μM monensin for the indicated periods were determined by Northern blot analysis and quantified by image analysis of autoradiograms. The relative change in COX III mRNA/β-actin mRNA ratio was related to COX III mRNA/β-actin mRNA ratio of zero time samples. Each point is the mean ± SEM of three to five separate experiments.

a significant decrease in COX III mRNA levels but was preceded by a transient increase (from 30 min to 1 h). A 50% decrease in levels of COX III mRNA was observed in PC12S cells after 6 hours of ouabain treatment (FIG. 1). Under the same conditions the level of β-actin mRNA remained unchanged. In separate experiments, addition of monensin (0.1 μM) (sodium ionophore) also decreased significantly the levels of COX III mRNA, again preceded by an initial sharp increase. A 50% decrease in levels of COX III was observed after 3 hours of monensin treatment (FIG. 2). In both ouabain- and monensin-treated cells, levels of COX III mRNA reached control levels within 6 to 8 hours once the drugs were washed away (data not shown). No significant decrease in the amount of mtDNA was observed in Southern blot analysis of total DNA isolated from both ouabain- and monensin-treated cells.

DISCUSSION

Regulation of mitochondrial respiration by energy demand (level of ADP) is well documented. We measured levels of mtDNA-encoded cytochrome oxidase subunit III mRNA in differentiated PC12S cells under conditions of decreasing and increasing energy (ATP) demand. Since the Na/K-ATPase (sodium pump) is a major consumer of energy in excitable cells, we used ouabain (a specific inhibitor of Na/K-

ATPase) to decrease and monensin (a sodium ionophore) to increase energy consumption in the form of ATP. Both ouabain and monensin decreased significantly the level of mtDNA-encoded COX III mRNA. This decrease in COX III mRNA occurred in absence of changes in mtDNA and β-actin mRNA. It was also unrelated to the energetic status (ATP/ADP ratio) of the cell. These results suggest a new Na^+-dependent regulatory mechanism of mitochondrial gene expression that overrides the normal regulation by energy demand.[4]

REFERENCES

1. RAPOPORT, S.I. 1991. Positron emission tomography in Alzheimer's disease in relation to disease pathogenesis: a critical review. Cerebrovasc. Brain Metab. Rev. **3:** 297–335.
2. CHANDRASEKARAN, K. *et al.* 1994. Impairment in mitochondrial cytochrome oxidase gene expression in Alzheimer's disease. Mol. Brain Res. **24:** 336–340.
3. FUKUYAMA, R. *et al.* 1993. Nerve growth factor-induced neuronal differentiation is accompanied by differential induction and localization of the amyloid precursor protein (APP) in PC 12 cells and variant PC 12S cells. Mol. Brain Res. **17:** 17–22.
4. ENRIQUEZ, J.A. *et al.* 1996. The synthesis of mRNA in isolated mitochondria can be maintained for several hours and is inhibited by high levels of ATP. Eur. J. Biochem. **237:** 601–610.

Does Dopamine Contribute to Striatal Damage Caused by Impaired Mitochondrial Function?

WILLIAM F. MARAGOS,[a,b,e] REBEKAH J. JAKEL,[a] M. DATHAN CHESNUT,[a] JAMES W. GEDDES,[c] AND LINDA P. DWOSKIN[d]

[a]*Department of Neurology,* [b]*Department of Anatomy and Neurobiology,* [c]*Sanders Brown Center on Aging, and* [d]*School of Pharmacy, University of Kentucky Medical Center, Lexington, Kentucky 40536-0284, USA*

Huntington's disease (HD) is a heredofamilial disorder characterized clinically by involuntary and often incapacitating "choreiform" movements, dementia, and psychiatric disturbances.[1] Neuropathological changes include neuronal loss and inclusions, gliosis, and atrophy. Although several brain regions including the cerebral cortex, hypothalamus, and cerebellum show these changes, the basal ganglia are predominantly affected.

There is increasing evidence that diminished energy metabolism is a factor in the pathophysiology of HD. In rodents, administration of 3 nitropropionic acid (3NP) and intrastriatal injections of malonate (both inhibitors of the mitochondrial enzyme succinate dehydrogenase) replicate many of the neuropathological abnormalities seen in HD and have thus been used to model this disorder.[2,3] The mechanism of action of these compounds has been postulated to occur via the indirect activation of excitatory amino acid (EAA) receptors and induction of "cell death" pathways. Despite the insights gained from studies investigating 3NP and malonate toxicity, the preferential vulnerability of the striatum is unclear. We hypothesize that one factor contributing to the preferential striatal damage is the presence of the neurotransmitter dopamine (DA). DA is present in high concentrations in the striatum and under certain circumstances may be neurotoxic. In this study, we have investigated the effects of (1) dopamine deafferentation and (2) MAO inhibition on the striatal damage produced by 3NP and malonate. We also evaluated whether mitochondrial inhibition altered dopamine uptake into synaptosomes.

METHODS

Experiment 1: Effects of 6 Hydroxydopamine (6OHDA) Lesions

Malonate Lesions. Eight rats received 6OHDA (8 μg/4 μl), and 10 received sterile saline injections into the right medial forebrain bundle (mfb). Two weeks later, animals in which apomorphine induced >100 turns were injected into the right striatum with 1.0 μl of malonate (0.8 μmol/μl). Animals survived 5 days, were sacrificed, and lesion volumes determined.

---

[e]Address for correspondence: University of Kentucky Medical Center. Department of Neurology, Kentucky Clinic, Room L-445, Lexington, Kentucky 40536-0284. Phone: 606-323-2522; fax: 606-323-5943.

3NP Lesions. Four animals were injected with 6OHDA (12 µg/2 µl) into the right substantia nigra (SN) and four with sterile saline. One week later, rats were administered 20 mg/kg i.p. daily for 4–8 days. Animals were perfused with 4% paraformaldehyde, and every 6th striatal section was mounted and analyzed. Tyrosine hydroxylase immunohistochemistry was performed on sections taken through the substantia nigra to verify the efficacy of the 6OHDA lesion.

Experiment 2: Effects of MAO Inhibition

Malonate Lesions. Eight animals received i.p. injections of both deprenyl and clorgyline (5 mg/kg each in sterile saline), and eight control animals were treated with sterile saline for three consecutive days. On the fourth day, all animals were injected with 1.5 µl of malonate and allowed to survive 3 days, at which time the animals were euthanized, their brains rapidly removed, frozen, and tissue damage assessed.

3NP Lesions. Four animals were injected on three consecutive days with both deprenyl and clorgyline (5 mg/kg each in sterile saline) and 4 animals with sterile saline. On day four, animals began daily i.p. injections of 3NP (20 mg/kg in sterile saline, pH 7.4) while continuing to receive clorgyline and deprenyl. On day 8 of 3NP treatment, animals were sacrificed and lesion volumes determined.

Experiment 3: Effects of Mitochondrial Toxins on Dopamine Uptake

The method of Masserano *et al.*[4] was used. Briefly, striatal synaptosomes (20 µg protein/500 µl sample) were incubated for 10 min with 5×10^5 dpm/ml [3H]-dopamine at 37°C in the presence of varying concentrations of the succinate dehydrogenase (complex II) inhibitors 3NP and malonate as well as the complex IV inhibitor azide. Nonspecific binding was determined in the presence of 10 µM nomifensine. Experiments with each compound were replicated in triplicate.

RESULTS

Experiment 1

In the control group, the seven animals that survived both sham mfb and striatal malonate injections had striatal lesions that were histologically similar to those previously described.[2,3] In this group of animals the average lesion volume was 2.3 ± 0.43 mm^3. In the group of animals receiving 6OHDA injections into the mfb all but one animal had apomorphine-induced rotation. The average malonate lesion volume in these animals was 0.8 ± 0.28 mm^3 ($p < 0.01$; TABLE 1).

In the animals treated with 3NP, tyrosine hydroxylase staining of the substantia nigra confirmed the efficacy of the prior 6OHDA lesion of dopaminergic neurons (not shown). In these four animals, the lesion volume induced by 3NP on the non-6OHDA–lesioned side was 15.5 ± 1.5 mm^3 and on the injected side 6.9 ± 1.7 mm^3 ($p < 0.01$; TABLE 1).

TABLE 1. Effects of various treatments on malonate- and 3NP-induced striatal lesion volumes

Treatment	Lesion Volume (mm^3)		% Protection
	Control	Experimental	
6OHDA			
Malonate	2.3 ± 0.4	0.8 ± 0.3^a	65
3NP	15.5 ± 1.5	6.9 ± 1.7^a	55
MAO Blockade			
Malonate	26.4 + 2.2	15.3 + 1.5^b	42
3NP	15.5 + 4.0	3.8 + 1.6^b	75

$^a p < 0.01$; taken from Ref. 5. $^b p < 0.001$; taken from Ref. 6.

Experiment 2

In the vehicle-treated animals, the severity of malonate-induced tissue destruction was homogenous, and the average lesion volume was 26.4 ± 2.2 mm^3. In the group of animals treated with MAO inhibitors, the lesions appeared qualitatively similar to the control animals but were significantly smaller (15.3 ± 1.5 mm^3, $p < 0.001$; TABLE 1).

In the control group treated with 3NP, the average lesion volume was 15.5 ± 4.0 mm^3, while the average lesion volume of the clorgyline/deprenyl-treated group was 3.8 ± 1.6 mm^3 ($p < 0.001$; TABLE 1).

Experiment 3

The addition of mitochondrial toxins to synaptosomes inhibited [^{3}H]dopamine uptake in a dose-dependent manner (FIG. 1). The rank order of potency was similar to their toxicity profile *in vitro*, with azide being the most potent uptake inhibitor and malonate the least.

DISCUSSION

In the first experiment, we demonstrated that the removal of striatal dopamine was neuroprotective against lesions generated by either systemic or direct intrastriatal administration of the mitochondrial toxins 3NP and malonate, respectively. Our data support the observations that dopamine enhanced methyl-malonate toxicity in striatal neurons in culture[7] and support the findings of Reynolds *et al.*, who recently reported neuroprotection against 3NP toxicity following dopamine depletion.[8] As neuronal death caused by mitochondrial poisons is thought to involve the activation of EAA receptors,[2,3] our findings are consistent with other laboratories that have shown dopamine depletion to confer neuroprotection against striatal damage induced by the EAA agonists N-methyl-D-aspartate (NMDA)[9] and quinolinic acid[10] and the non-NMDA agonist kainic acid.[9]

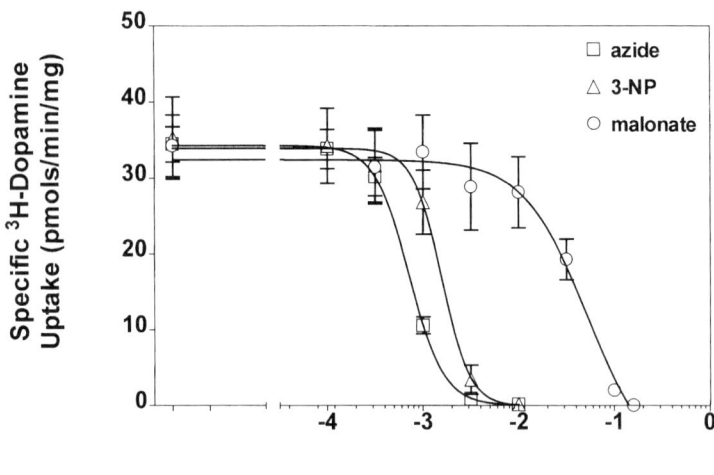

FIGURE 1. Dose-dependent inhibition of [^{3}H]dopamine uptake by mitochondrial toxins. The rank order of potency is azide > 3NP > malonate, and the respective IC$_{50}$s are 7.2×10^{-4}, 1.6×10^{-3}, and 5.1×10^{-2} M.

Although the mechanism of dopamine toxicity has not been completely determined, the production of oxygen-based free radicals have been implicated.[11,12] In the second experiment, inhibition of MAO using the combination of clorgyline and deprenyl also resulted in significant protection against malonate and 3NP-induced striatal lesions. MAO is an important enzyme that metabolizes dopamine to dihydroxyphenylacetic acid and hydrogen peroxide (H_2O_2).[13] Since H_2O_2 can be converted to the highly toxic hydoxyl radical, it is possible that the protection we observed following MAO blockade was due to decreased H_2O_2 (and ultimately hydroxyl radical) production. Treatment of animals with the dopamine precursor L-DOPA have been shown to result in the increased production of hydroxyl radicals.[14]

We have shown that mitochondrial toxins inhibit the uptake of [^{3}H]dopamine (FIG. 1). Although we are uncertain whether this phenomenon is specific to the dopamine transporter, blockade of dopamine uptake could result in enhanced dopamine metabolism by MAO (hence, the production of H_2O_2) and is consistent with the observation that systemic administration of 3NP increased dopamine turnover.[2] We concluded from the findings presented herein that dopamine and/or one of its metabolites contribute(s) to striatal damage caused by impaired energy metabolism and may thus be involved in the striatal damage seen in HD.

REFERENCES

1. MARTIN, J.B. & J.F. GUSELLA. 1986. Huntington's disease. Pathogenesis and management. N. Engl. J. Med. **315:** 1267–1276.
2. BEAL, M.F. *et al.* 1993. Neurochemical and histological characterization of striatal excitotoxic lesions produced by the mitochondrial toxin 3-nitropropionic acid. J. Neurosci. **13:** 4181–4192.

3. GREENE, J.G. et al. 1993. Inhibition of succinate dehydrogenase by malonic acid produces an excitotoxic lesion in rat striatum. J. Neurochem. **61:** 1151–1154.
4. MASSERANO, J.M. et al. 1994. Effects of chronic cocaine administration on [^{3}H]dopamine uptake in the nucleus accumbens, striatum and frontal cortex of rats. J. Pharmacol. Exp. Ther. **270:** 133–141.
5. MARAGOS, W.F. et al. 1998. 6-Hydroxydopamine injections into the nigrostriatal pathway attenuate striatal malonate and 3-nitropropionic acid lesions. Exp. Neurol. **154:** 637–644.
6. MARAGOS, W.F. et al. 1999. Clorgyline and deprenyl attenuate striatal malonate and 3-nitropropionic acid lesions. Brain Res. **834:** 168–172.
7. MCLAUGHLIN, B.A. et al. 1998. Toxicity of dopamine to striatal neurons in vitro and potentiation of cell death by a mitochondrial inhibitor. J. Neurochem. **70:** 2406–2415.
8. REYNOLDS, D.S. et al. 1998. Dopamine modulates the susceptibility of striatal neurons to 3-nitropropionic acid in the rat model of Huntington's disease. J. Neurosci. **18:** 10116–10127.
9. CHAPMAN, A.G. et al. 1989. Excitotoxicity of NMDA and kainic acid is modulated by nigrostriatal dopaminergic fibers. Neurosci. Lett. **107:** 256–260.
10. BUISSON, A. et al. 1991. Nigrostriatal pathway modulates striatum vulnerability to quinolinic acid. Neurosci. Lett. **131:** 257–259.
11. ROSENBERG, P.A. 1988. Catecholamine toxicity in cerebral cortex in dissociated cell culture. J. Neurosci. **8:** 2887–2894.
12. HASTINGS, T.G. et al. 1996. Role of oxidation in the neurotoxic effects of intrastriatal dopamine injections. Proc. Natl. Acad. Sci. USA **93:** 1956–1961.
13. SPINA, M.B. & G. COHEN. 1989. Dopamine turnover and glutathione oxidation: implications for Parkinson disease. Proc. Natl. Acad. Sci. USA **86:** 1398–1400.
14. SMITH, T.S. et al. 1994. L-dopa increases nigral production of hydroxyl radicals *in vivo*: potential L-dopa toxicity? NeuroReport **5:** 1009–1011.

Discordance between Traditional Pathologic and Energy Metabolic Changes in Very Early Alzheimer's Disease

Pathophysiological Implications

SATOSHI MINOSHIMA,[a,e] DONNA J. CROSS,[a] NORMAN L. FOSTER,[b] THOMAS R. HENRY,[c] AND DAVID E. KUHL[a,d]

Departments of [a]Internal Medicine, [b]Neurology, and [d]Radiology, The University of Michigan, Ann Arbor, Michigan 48109, USA

[c]Department of Neurology, Emory University, Atlanta, Georgia 30322, USA

INTRODUCTION

Recent pathologic and metabolic studies of very early Alzheimer's disease suggest a discordance in the distribution between traditional pathologic changes and glucose metabolic reduction. Traditional pathologic changes in very early stages of AD are shown to occur in the medial temporal lobe, specifically in the entorhinal cortex.[1] In contrast, very early glucose metabolic reductions occur in the lateral association cortices and posterior cingulate cortex.[2] This study addresses the possible pathophysiological mechanisms explaining such discordance between metabolic and pathologic changes using positron emission tomography (PET).

GLUCOSE METABOLIC REDUCTION IN VERY EARLY ALZHEIMER'S DISEASE

A longitudinal study of 23 patients with "isolated memory impairment" identified 10 patients who developed probable AD during a clinical follow-up. Glucose metabolic PET imaging with 2-[^{18}F]fluoro-2-deoxy-D-glucose (FDG) was performed for those patients before the clinical diagnosis of dementia was made. Those images were analyzed retrospectively. The mean glucose metabolic map in 10 elderly normal controls showed intense metabolic activity in the posterior cingulate cortex. This activity disappeared in very early AD patients. Statistical subtraction images showed the most severe glucose metabolic reduction in the posterior cingulate cortex followed by the parietotemporal and frontal association cortices, but relatively sparing the primary sensorimotor cortex, occipital cortex, and cerebellum. Glucose metabolism in the medial temporal cortex was also relatively preserved. These findings are contrary to pathological evidence of very early involvement of the medial temporal cortex in AD.

[e]Address for correspondence: Satoshi Minoshima, M.D., Ph.D., Department of Internal Medicine, The University of Michigan, B1G412 University Hospital, Ann Arbor, Michigan 48109-0028. Phone: 734-764-3485; fax: 734-936-8182
e-mail: satoshim@umich.edu

DOES THE LOSS OF ENTORHINAL EFFERENTS EXPLAIN SEVERE METABOLIC REDUCTION IN THE POSTERIOR CINGULATE AND ASSOCIATION CORTICES IN VERY EARLY AD?

Neurofibrillary tangles, a traditional pathologic change, occur in a cell body of the neuron. FDG uptake primarily occurs in the region of the synapse.[3] We hypothesized that the loss of entorhinal efferents caused by AD pathology may have caused glucose metabolic reduction in remote areas such as the posterior cingulate cortex. To test this hypothesis, we compared resting cerebral perfusion scans obtained with [^{15}O]water (under the assumption of flow-metabolic coupling) before and after left temporal lobectomy that removed the hippocampus and a substantial portion of entorhinal cortex in nine patients with refractory temporal lobe epilepsy (postoperative 68 ± 17 days). Temporal lobectomy resulted in reduced activities in remote areas including the posterior cingulate cortex and thalamus. When compared to the pattern of very early AD, reductions in the posterior cingulate cortex, a part of the thalamus, and a part of the lateral temporal cortex overlapped between two conditions. However, the removal of medial temporal cortex did not result in reduced activities in the parietotemporal and frontal association cortices, where a substantial metabolic reduction was seen in very early AD. In addition, the profile of the magnitude in percent reduction induced by the removal of the medial temporal cortex was different from that observed in very early AD. These findings indicate that severe metabolic reduction observed in the posterior cingulate cortex in very early AD may be explained in part by the loss of entorhinal efferents due to AD pathology, but more diffuse involvement of association cortices in very early AD cannot be attributed to the entorhinal pathology alone.

DO CHOLINERGIC DEFICITS EXPLAIN SEVERE METABOLIC REDUCTION IN THE CEREBRAL CORTEX IN AD?

Basal forebrain cholinergic neurons are known to degenerate severely in AD.[4] We tested a hypothesis that cholinergic deficit may have caused severe metabolic reduction in the cerebral cortex in AD. Acetylcholinesterase activity (AChE) in the brain was measured by PET imaging using N-[^{11}C]methyl-4-piperidinyl propionate (PMP)[5] and compared to glucose metabolic activity (CMRglc) measured by FDG PET imaging as well as cerebral atrophy measured by 3D Spoiled Gradient Echo Imaging (3D-SPGR) MR imaging in nine AD patients. There was an inconsistency between the patterns of AChE and CMRglc reduction. Reduction in AChE involved the entire cerebral cortex including the primary sensorimotor cortex. CMRglc reduction and cerebral atrophy primarily involved association cortices sparing the primary sensorimotor cortex. Multivariate correlational analysis (principal components analysis) revealed four latent common factors (eigenvalues > 1) among changes in AChE, CMRglc, and atrophy. The first component was dominated by changes in AChE that had a relatively uniform involvement of the entire cerebral cortex. However, these changes were not intercorrelated with changes in CMRglc or atrophy. The second to fourth components represented intercorrelated changes in CMRglc within the frontal association cortex, parietotemporal cortex, and anterior tempral cortex, respectively. These CMRglc changes were also intercorrelated with atrophy, but not with

AChE changes. Intercorrelation between changes in CMRglc and atrophy indicates the presence of common pathologic processes that contribute to both measurements.

SUMMARY

These results suggest that neither the loss of entorhinal efferents nor cholinergic deficit explains all the metabolic features seen in very early AD. Given recent immunohistological evidence of massive glutamatergic synaptic alteration in early AD cortex[6] and insights into neuronal and glial mechanisms of glucose metabolism,[7] very early metabolic changes in AD probably reflect a significant impairment of glycolytic activities in the cortico-cortical glutamatergic systems in a preclinical stage of the disease. However, the exact mechanisms of such impairment in these neurons are yet to be determined.

REFERENCES

1. BRAAK, H. & E. BRAAK. 1991. Neuropathological staging of Alzheimer-related changes. Acta Neuropathol. (Berl.) **82:** 239–259.
2. MINOSHIMA, S. et al. 1997. Metabolic reduction in the posterior cingulate cortex in very early Alzheimer's disease. Ann. Neurol. **42:** 85–94.
3. KADEKARO, M., A.M. CRANE & L. SOKOLOFF. 1985. Differential effects of electrical stimulation of sciatic nerve on metabolic activity in spinal cord and dorsal root ganglion in the rat. Proc. Natl. Acad. Sci. USA **82:** 6010–6013.
4. WHITEHOUSE, P.J. et al. 1981. Alzheimer disease: evidence for selective loss of cholinergic neurons in the nucleus basalis. Ann. Neurol. **10:** 122–126.
5. KUHL, D.E. et al. 1999. In vivo imaging of cerebra acetylcholinesterase activity in aging and Alzheimer's disease. Neurology **52:** 691–699.
6. MASLIAH, E. et al. 1991. Cortical and subcortical patterns of synaptophysinlike immunoreactivity in Alzheimer's disease. Am. J. Pathol. **138:** 235–246
7. MAGISTRETTI, P.J. & L. PELLERIN. 1996. Cellular mechanisms of brain energy metabolism. Relevance to functional brain imaging and to neurodegenerative disorders. Ann. N.Y. Acad. Sci. **777:** 380–387

Low Frequencies of Mitochondrial DNA Mutations Cause Cardiac Disease in the Mouse

J. L. MOTT,[a] D. ZHANG,[a] P. L. FARRAR,[b] S. W. CHANG,[a] AND
H. P. ZASSENHAUS[a,c]

Departments of [a]Molecular Microbiology and Immunology, and [b]Comparative Medicine, St. Louis University School of Medicine, St. Louis, Missouri 63104, USA

INTRODUCTION

One possible cause of age-related disease is the accumulation of sporadic mutations within the mitochondrial genome, which disrupts mitochondrial function. Support for this hypothesis includes the observations that mutations of mitochondrial DNA (mtDNA) are found in aged[1] and diseased tissues.[2] Because this support is only correlative, we have developed a mouse model to assess whether sporadic mtDNA mutations cause cardiac disease. MtDNA is replicated by the nuclear-encoded DNA polymerase γ, which contains a functional 3′–5′ exonuclease necessary for proofreading. To generate random mutations, we created transgenic mice that express a proofreading-deficient mtDNA polymerase in the heart. Mitochondrial point and deletion mutations are detectable by PCR as early as 1 week after birth. A few weeks after mutations can be detected, mice develop clinical signs of congestive heart failure. Thus, our model demonstrates that random mtDNA mutations can cause disease.

METHODS

Transgenic Mice

The generation of the transgenic mice will be published in detail elsewhere. Briefly, the codon for aspartic acid residue 181, which is critical for 3′–5′ exonuclease activity, was mutated to code for alanine, and this mutant gene was placed behind the cardiac-specific promotor for α-myosin heavy chain. This construct was used to generate transgenic mice on an FVB/N background, and expression was confirmed by RT-PCR.

Mutation Detection

Point mutations were detected by PCR using a primer pair in which one primer was a perfect complement to the mitochondrial genome, while the other contained

---

[c]Address for correspondence: Peter Zassenhaus, Department of Molecular Microbiology and Immunology, St. Louis University School of Medicine, 1402 South Grand Boulevard, St. Louis, Missouri 63104-1081. Phone: 314-577-8444; fax: 314-773-3403.
zassenp@slu.edu

base mismatches (misprimed PCR). Conditions used were 30 cycles of 1 minute each at 94°, 54°, and 72°C. The misprimed product was then sequenced to establish its identity.

Deletion mutations were detected in cardiac mtDNA of 11-week-old mice by long-PCR (30 cycles, 10-second melt at 92°, 5-minute extension at 68°) using primers that were separated by about 16 kbp. The products formed by this "short-cycle, long-PCR" were diluted 1000-fold and reamplified using primers that were nested 187 bases inside of the original primers.

Pathology

Hearts were retrograde-perfused sequentially with 100 mM cadmium chloride to arrest the heart in diastole, phosphate-buffered saline to remove residual Cd^{++}, and 10% buffered formalin to fix the tissue. Excised hearts were photographed, immersion fixed for at least 24 hours, and sectioned perpendicular to their long axis on a vibratome. Sections were mounted on glass slides and captured to computer with a Spot camera. Alternatively, immersion-fixed hearts were embedded in plastic and thin-sectioned for histological staining using toluidine blue.

RESULTS AND DISCUSSION

Expression of the transgene was cardiac-specific and detectable only after birth (data not shown). Mutations were detected by misprimed PCR, at increasing levels starting about 1–2 weeks after birth (FIG. 1A). Some mutations within the target sequence of the imperfect primer improved base pairing and allowed amplification of a product; thus, the appearance of a product suggested mutations. Because this assay was PCR based and the presumptive mutations were within the primer region, sequencing determined only where the primers annealed (FIG. 1A), but not the mutation that allowed annealing. Deletions are another common mutation of mtDNA that accumulate with age. In yeast, their frequency increases when mtDNA is replicated by a proofreading-deficient polymerase.[3] By 11 weeks of age the level of deletion mutations was also elevated in transgenic mice (FIG. 1B, lane 3), but not in controls (lane 1). The possibility that the products resulted from mispriming was excluded by reamplifying the deletion products with a pair of nested primers (lanes 2, 4). If the products were mispriming events, then nested PCR should yield no product. Instead, the nested profile was the same as the original, except that each product was smaller by approximately 200 base pairs (FIG. 1B, lane 4).

Transgenic mice as young as four weeks of age showed clinical signs of congestive heart failure, including tachypnea, lethargy, and severe systemic edema. Excised hearts from transgenic mice were much larger than littermate control hearts (FIG. 2A, B), and showed marked ventricular dilation (FIG. 2A) and atrial enlargement (FIG. 2B, arrows), with little or no hypertrophy. Histological analysis showed minimal inflammation and sporadic degenerating myocytes in transgenic hearts (FIG. 2C). For some years it has been known that specific inherited mutations within the mitochondrial genome can be pathogenic; the diseases are collectively known as mitochondrial encephalomyopathies.[4] Now, for the first time, we show that random mutations accumulating over time also cause disease. It is of interest that the level

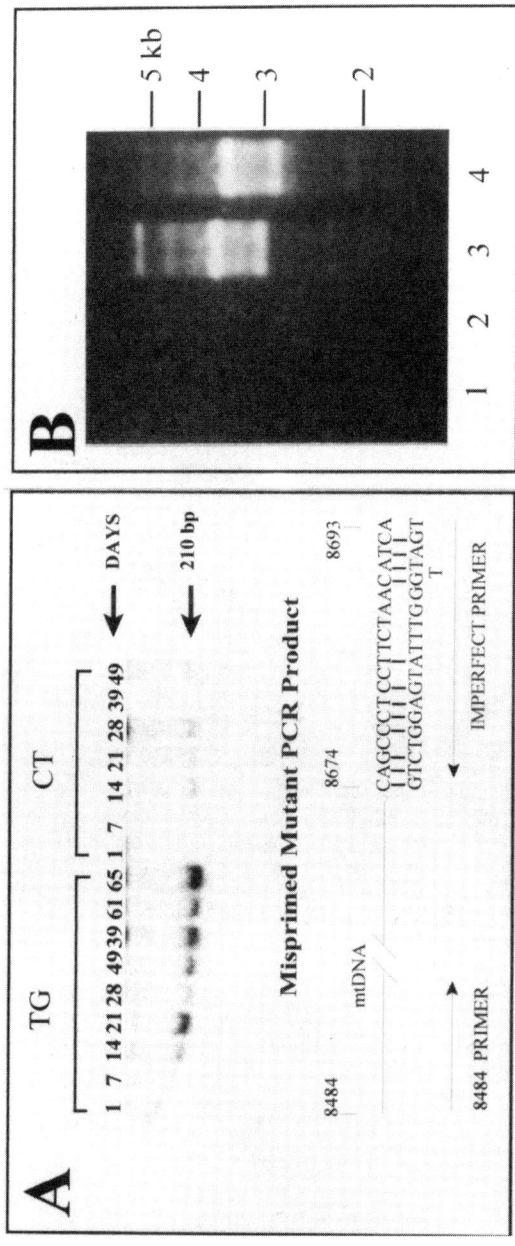

FIGURE 1. Mutations accumulate in the hearts of transgenic mice. (**A**) Misprimed PCR showed that mutations began to accumulate between 7 and 14 days after birth, and were increased at all times thereafter. One primer was a perfect complement to the mtDNA at bases 8484–8504. The 3' region of the other primer had 80% complementarity to nucleotides 8683–8674 and perfect complementarity to region 13098–13078 (5' tgattgggtttatgaggtcg) of the mitochondrial genome. (**B**) Whole-cell DNA was isolated from 11-week-old control (*lane 1*) and transgenic (*lane 3*) mouse hearts, and deletion fragments were detected as outlined in the methods section. *Lanes 1* and *3* used 30–33 bp primers directed to nucleotides 1953–1924 and 2473–2505, and whole-cell DNA as the template. *Reactions 2* and *4* used the diluted products of *reactions 1* and *3* as template and used 23–25-bp primers that annealed to positions 1794–1772 and 2501–2525, which were nested 187 bases within the primers from *lanes 1* and *3*.

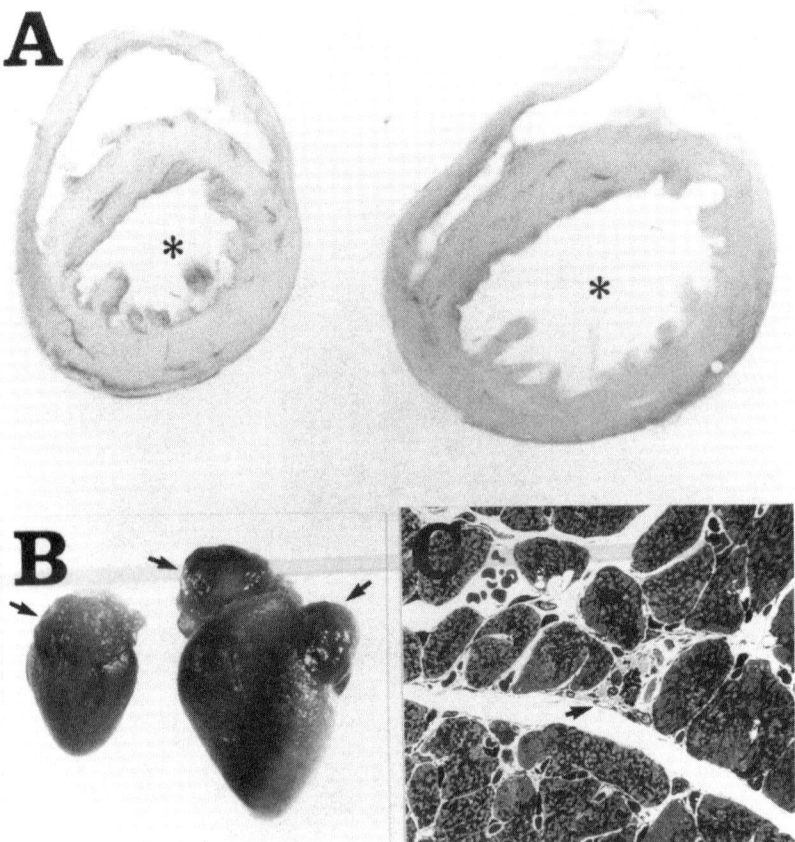

FIGURE 2. Transgenic mice display cardiac pathology. (**A**) Side-by-side comparison of cross-sectioned hearts from 4-week-old littermates showed bilateral ventricular dilation. The normal heart on the *left* was from a nontransgenic mouse, and the enlarged heart on the *right* was from a transgenic mouse that expressed the proofreading-deficient mtDNA polymerase; note the markedly increased left ventricular chamber size (*). (**B**) Enlarged hearts also showed marked atrial dilation *(arrows)*. (**C**) Light microscopy of thin-sectioned tissue stained with toluidine blue revealed a dying myocyte *(arrow)* surrounded by healthy myocytes in the ventricle of a transgenic mouse.

of mutations in our model, measured by automated sequencing of mtDNA (0.01% at 4 weeks, data not shown), is similar to the levels found in aged human tissues.[1]

ACKNOWLEDGMENTS

This work was supported by grants to H.P.Z. from the American Heart Association (9650390N), the Alzheimer Association (RG2-96-094), and the NIH Institute of Aging (RO1AG15710-01). We thank Grace Denniger and Zehua Feng for technical assistance and Dr. Jan Ryerse for the thin section analysis. J.L. Mott and D. Zhang contributed equally to this work.

REFERENCES

1. LIU, V.W.S., C. ZHANG & P. NAGLEY. 1998. Mutations in mitochondrial DNA accumulate differentially in three different human tissues during ageing. Nucleic Acids Res. **26:** 1268–1275.
2. OZAWA, T. 1995. Mitochondrial DNA mutations associated with aging and degenerative diseases. Exp. Gerontol. **30:** 269–290.
3. FOURY, F. & S. VANDERSTRAETEN. 1992. Yeast mitochondrial DNA mutators with deficient proofreading exonucleolytic activity. EMBO J. **11:** 2717–2726.
4. GRAEBER, M.B. & U. MÜLLER. 1998. Recent developments in the molecular genetics of mitochondrial disorders. J. Neurol. Sci. **153:** 251–263.

Effects of the Solvent 1,2,4-Trimethylcyclohexane on Respiratory Burst in Human Neutrophil Granulocytes

A Chemiluminescence and Electron Paramagnetic Resonance Spectrometry Study

O. MYHRE,[a,d,e] T. A. VESTAD,[b] E. SAGSTUEN,[b] H. AARNES,[c] AND F. FONNUM[a,d]

[a]*Division for Protection and Materiel, Norwegian Defence Research Establishment, N-2027 Kjeller, Norway*

Departments of [b]*Physics and* [c]*Biology, University of Oslo, Oslo, Norway*

[d]*VISTA (The Norwegian Academy of Science and Letters/Statoil), Oslo, Norway*

INTRODUCTION

During the respiratory burst in neutrophil granulocytes oxygen consumption and hexose monophosphate shunt activity increase, with the subsequent generation of oxygen reduction products such as $O_2^{\bullet-}$, H_2O_2, and HOCl.[1] Dearomatized White Spirit causes elevation of reactive oxygen species, for example, in rat brain, kidney and liver.[2] In this report, we have investigated the effects of the solvent 1,2,4-trimethylcyclohexane (TMCH), which is present in Dearomatized White Spirit, on respiratory burst in human neutrophil granulocytes. Particular attention was paid to the importance of Ca^{2+} and phospholipase A_2 (PLA_2) in TMCH-induced respiratory burst.

MATERIALS AND METHODS

Chemicals. 2,7-dichlorofluorescein diacetate (DCF-DA), 7,7-dimethyl-(5Z,8Z)-eicosadienoic acid (DEDA), N-formyl-met-leu-phe (fMLP), methanol (MeOH), and phorbol 12-myristate 13-acetate (PMA) were purchased from Sigma Chemical Co. (USA). 1,2,4-Trimethylcyclohexane (TMCH) was obtained from Aldrich Chemical Company (USA). 5–Diethoxyphosphoryl–5–methyl–1–pyrroline–N–oxide (DEPMPO) was from Oxis International, Inc. (USA).

Separation of Human Granulocytes. Human neutrophil granulocytes were separated from human EDTA blood by the standard density-gradient centrifugation method.[3] The cells were exposed to TMCH previously dissolved in methanol (final methanol concentration 0.5 % v/v).

---

[e]Address for correspondence: O. Myhre, Norwegian Defence Research Establishment, Division for Protection and Materiel, P.O. Box 25, N-2027 Kjeller, Norway. Phone: + 47 63807827; fax: + 47 63807811.
e-mail: Oddvar.Myhre@ffi.no

Fluorescent Spectrometry. The reaction mixture (250 µl) contained (final concentration): DCF-DA (2 µM), TMCH (40 µM), the PLA_2 inhibitor DEDA[4] (4 µM, when included), and 2×10^5 cells in HEPES-buffered (20 mM) Hank's balanced salt solution (HBSS, pH 7.4) with 5 mM glucose. When Ca^{2+}-free medium was used, 2 mM EGTA was included in the reaction. The cells were incubated with DCF-DA at 37°C for 15 min. Following centrifugation, the extracellular medium with DCF-DA was exhanged with fresh buffer, and intracellular chemiluminescence was measured after 30 min. The control mixtures contained methanol instead of TMCH, DEDA (when studying the involvement of PLA_2), and Ca^{2+}-free media (when studying the importance of extracellular Ca^{2+}). Addition of methanol to the cell suspensions had a small, insignificant decreasing effect on the respiratory burst.

Electron Paramagnetic Resonance (EPR) Spectrometry. The reaction mixture (200 µl) contained (final concentration): DEPMPO (20 mM), TMCH (40 µM), DEDA (4 µM, when included), and 8×10^5 cells in HEPES-buffered (20 mM) HBSS with 5 mM glucose. The control mixtures contained methanol instead of TMCH, and contained DEDA when studying the involvement of PLA_2.

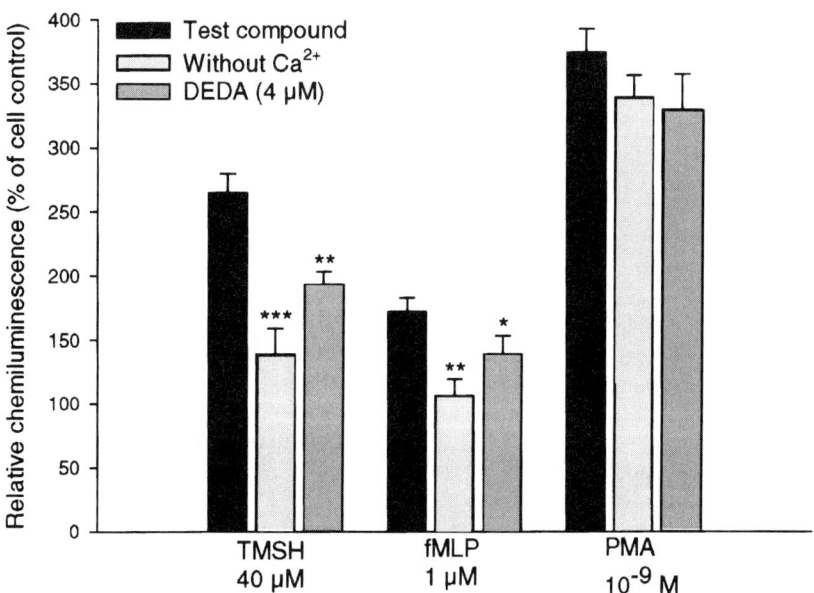

FIGURE 1. The effect of Ca^{2+}-free medium (HEPES-buffered HBSS, pH 7.4) and the PLA_2 inhibitor DEDA on granulocyte respiratory burst in response to the test compounds TMCH, fMLP, and PMA. Values are mean ± SD of three independent experiments. Student's *t* test, paired two samples for means, was performed for the data presented. *** $p < 0.001$, ** $p < 0.005$, * $p < 0.05$ that values differ from those of the test compounds.

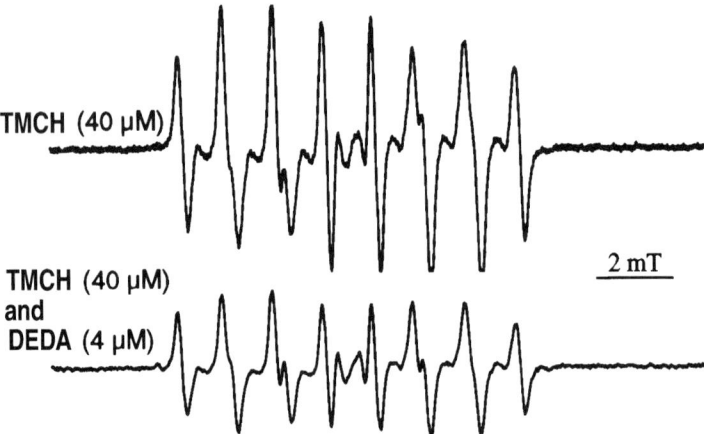

FIGURE 2. EPR spectra obtained 70 min after addition of TMCH to the granulocytes. The *lower spectrum* shows the effect of the PLA$_2$ inhibitor DEDA. The control experiments gave very small responses (not shown). The EPR spectrum of a spin adduct exhibits a hyperfine splitting pattern, from which the free radical can be identified; and the presenting spectra shows mainly the superoxide adduct of DEPMPO (DEPMPO-OOH). The samples were measured using a flat cell in vertical orientation. EPR spectrometer settings: microwave power, 12.5 mW; modulation frequency, 100 kHz; modulation amplitude, 2.0 G, sweep width, 17 mT; time constant, 327 ms; conversion time, 163 ms; sweep time, 168 s; receiver gain, 5e5; size, 1024.

RESULTS

Human granulocytes showed increased respiratory burst when exposed to TMCH, measured as elevated chemiluminescence and by EPR spectrometry. fMLP and PMA were included as positive controls (FIG. 1). In the absence of extracellular Ca^{2+}, the chemiluminescence in response to TMCH and fMLP was reduced by 72% and 81%, respectively (FIG. 1). Pretreatment of the granulocytes with DEDA reduced the TMCH,- and fMLP-induced respiratory burst by 41%, and 39%, respectively (FIGS. 1 and 2). Ca^{2+}-free medium and DEDA had small, insignificant decreasing effects on the PMA-stimulated cells.

DISCUSSION

Low concentrations of TMCH stimulate respiratory burst in human granulocytes. Removal of extracellular Ca^{2+} decreased chemiluminescence in response to TMCH and fMLP. This indicates that TMCH activates respiratory burst in a Ca^{2+}-dependent pathway (FIG. 1). DEDA reduced the effect of TMCH significantly (FIGS. 1 and 2). PLA$_2$ can be activated by Ca^{2+} [5] and protein kinase C (PKC),[6] probably via mitogen-activated protein kinase (MAPK).[7] PLA_2 are shown to be involved in the mechanism of activation of neutrophils by polychlorinated biphenyls.[8] In conclusion, our results indicate that Ca^{2+} activation of PLA_2 is involved in the TMCH-induced activation of respiratory burst. The consequence is a rapid activation of the general immune system and unessecary production of free radicals that can be harmful to the tissue.

REFERENCES

1. EDWARDS, S.W. 1996. The O_2^- generating NADPH oxidase of phagocytes: structure and methods of detection. Methods Enzymol. **9:** 563–577.
2. LAM, H.R. *et al.* 1994. Three weeks exposure of rats to dearomatized white spirit modifies indices of oxidative stress in brain, kidney, and liver. Biochem. Pharmacol. **47:** 651–657.
3. BØYUM, A. *et al.* 1991. Separation of leucocytes: improved cell purity by fine adjustment of gradient medium density and osmolality. Scand. J. Immunol. **34:** 697–712.
4. COHEN, N. *et al.* 1984. Analogs of arachidonic acid methylated at C-7 and C-10 as inhibitors of leukotriene biosynthesis. Prostaglandins **27**(4): 553–562.
5. CLARK, J.D. *et al.* 1991. A novel arachidonic acid–selective cytosolic PLA_2 contains a Ca(2+)-dependent translocation domain with homology to PKC and GAP. Cell **65**(6): 1043–1051.
6. GLASER, K.B. *et al.* 1993. Phospholipase A_2 enzymes: regulation and inhibition. (Review.) Trends. Pharmacol. Sci. **14**(3): 92–98.
7. LIN, L.-L. *et al.* 1993. $cPLA_2$ is phosphorylated and activated by MAP kinase. Cell **72**(2): 269–278.
8. TITHOF, P.K. *et al.* 1996. Phospholipase A_2 is involved in the mechanism of activation of neutrophils by polychlorinated biphenyls. Environ. Health Perspect. **104:** 52–58.

Neuronal RNA Oxidation in Alzheimer's Disease and Down's Syndrome

AKIHIKO NUNOMURA,[a,b] GEORGE PERRY,[a,c] KEISUKE HIRAI,[a,d] GJUMRAKCH ALIEV,[a] ATSUSHI TAKEDA,[a] SHIGERU CHIBA,[b] AND MARK A. SMITH[a]

[a]*Institute of Pathology, Case Western Reserve University, Cleveland, Ohio 44106, USA*

[b]*Department of Psychiatry and Neurology, Asahikawa Medical College, Asahikawa 078-8510, Japan*

[d]*Pharmaceutical Research Laboratories I, Pharmaceutical Research Division, Takeda Chemical Industries Ltd., Osaka 532-8686, Japan*

We have found a marked accumulation of an oxidized nucleoside, 8-hydroxyguanosine (8OHG), derived from cytoplasmic RNA within the cerebral neurons of patients with Alzheimer's disease (AD) and Down's syndrome (DS). In DS, the increase in neuronal 8OHG temporally precedes the predictable increase of amyloid-β (Aβ) deposition by 10 years. Increased Aβ deposition is associated with a decrease in relative 8OHG in both AD and DS. These findings support Aβ as a compensatory response that reduces oxidative stress.

MATERIALS AND METHODS

Cerebral cortex was obtained at autopsy from 22 cases of AD (ages 57–93 years, average 78.2) and 17 controls (ages 62–86 years, average 73.6); as well as 22 cases of DS (ages 0.3–65 years, average 22.5) and 10 controls (ages 0.3–64 years, average 23.6) with similar postmortem intervals (2–27 h). Six-µm-thick sections were prepared from tissue fixed in methacarn (AD and its control) or in 10% phosphate-buffered formalin (DS and its control) and embedded in paraffin. To detect oxidized nucleosides (RNA-derived 8OHG and DNA-derived 8-hydroxydeoxyguanosine), Aβ deposition, and neurofibrillary tangles (NFT) immunocytochemically we used monoclonal antibodies, 1F7, 4G8, and AT8, respectively. Immunostaining was performed by the peroxidase-antiperoxidase method with diaminobenzidine as cosubstrate. Relative scale measurements of 8OHG[1] and the area of Aβ deposition[2] were performed using a Quantimet 570C Image Processing and Analysis system (Leica).

RESULTS AND DISCUSSION

Immunoreactivity for oxidized nucleosides with monoclonal antibody 1F7 was prominent in the cytoplasm and, to a lesser extent, in the nucleus in neurons within

---

[c]Address for correspondence: Dr. G. Perry, Institute of Pathology, Case Western Reserve University, 2085 Adelbert Road, Cleveland, Ohio 44106. Phone: 216-368-2488; fax: 216-368-8964. e-mail: gxp7@po.cwru.edu

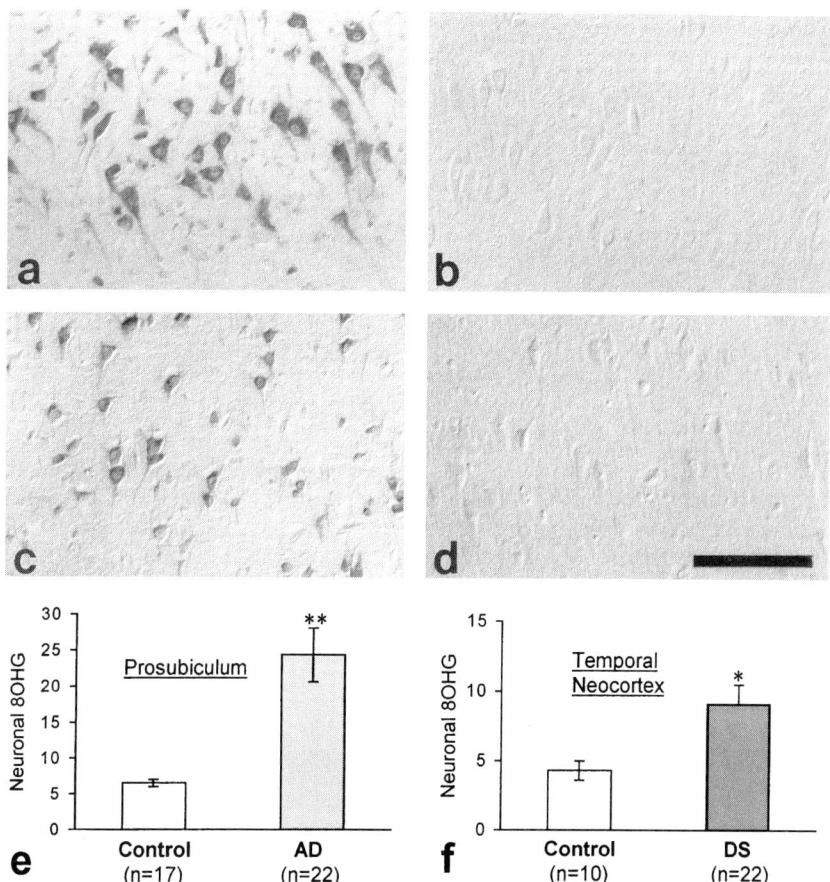

FIGURE 1. A marked accumulation of oxidized RNA nucleoside, 8OHG, in the cytoplasm of cerebral neurons of AD and DS. Immunostaining with 1F7 antibody in the stratum pyramidale of the hippocampal CA1 field of a 76-year-old case of AD (**a**) and an 80-year-old control case (**b**); and in the layer III of the temporal neocortex of a 25-year-old case of DS (**c**) and a 31-year-old control case (**d**). *Scale bar* = 100 μm. Relative intensity measurements of neuronal 8OHG show that there is a significant increase in AD (**$^{**}p < 0.0001$** by Mann Whitney U test) compared with age-matched controls in the prosubiculum (**e**) as well as a significant increase in DS (*$p < 0.02$) in the temporal neocortex (**f**). Values shown are means ± SEM.

the temporal neocortex as well as hippocampal formation from AD and DS cases (FIG. 1 a,c). As previously described,[1] such immunoreaction was diminished greatly by RNase but not by DNase treatments. In marked contrast to the high levels of oxidized RNA in cerebral neurons in AD and DS, age-matched controls showed low levels of immunoreactivity with 1F7 (FIG. 1 b,d). Relative density measurements indicate a significant increase in neuronal RNA oxidation in AD and DS (FIG. 1 e,f). Of note, individuals with DS showed significant elevations in neuronal 8OHG in

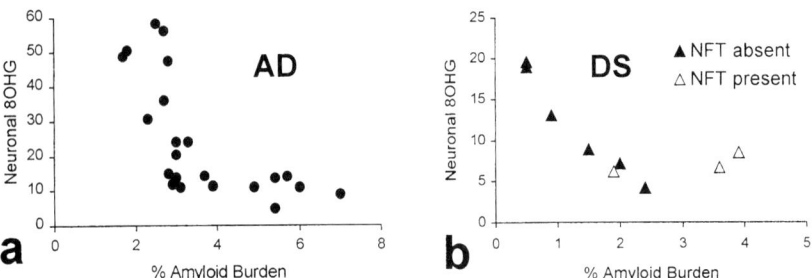

FIGURE 2. The extent of Aβ deposition is associated with a decrease in relative 8OHG in both 22 cases of AD (**a**) and nine cases of DS (**b**). In six DS cases (▲) (ages 17–32 years) bearing Aβ deposition and free of NFT, there is an inverse relationship ($r = -0.97, p < 0.001$) between the levels of neuronal 8OHG and Aβ deposition (**b**).

their teens and 20s, which decrease after age 30, coincident with the accumulation of Aβ deposits. In fact, in both AD and DS, the extent of Aβ deposition was associated with a decrease in neuronal 8OHG levels (FIG. 2a,b).

That RNA oxidation is a prominent feature of cerebral neurons in AD and DS is compatible with previous studies showing significant oxidative stress in AD[1,3,4] and DS.[5–7] The inverse relationship between neuronal RNA oxidation and Aβ deposition in both AD and DS, together with the temporal primacy of oxidative stress to Aβ deposition in DS, strongly support Aβ as a compensatory response that protects neurons from reactive oxygen species—i.e., Aβ appears to act as an antioxidant.

REFERENCES

1. NUNOMURA, A. *et al.* 1999. RNA oxidation is a prominent feature of vulnerable neurons in Alzheimer's disease. J. Neurosci. **19:** 1959–1964.
2. HYMAN, B.T., K. MARZLOFF & P.V. ARRIAGADA. 1993. The lack of accumulation of senile plaques or amyloid burden in Alzheimer's disease suggests a dynamic balance between amyloid deposition and resolution. J. Neuropathol. Exp. Neurol. **52:** 594–600.
3. SMITH, M.A. *et al.* 1996. Oxidative damage in Alzheimer's. Nature **382:** 120–121.
4. SAYRE, L.M. *et al.* 1997. 4-Hydroxynonenal-derived advance lipid peroxidation end products are increased in Alzheimer's disease. J. Neurochem. **68:** 2092–2097.
5. BUSCIGLIO, J. & B.A. YANKNER. 1995. Apoptosis and increased generation of reactive oxygen species in Down's syndrome neurons *in vitro*. Nature **378:** 776–779.
6. JOVANOVIC, S.V., D. CLEMENTS & K. MACLEOD. 1998. Biomarkers of oxidative stress are significantly elevated in Down syndrome. Free Radic. Biol. Med. **25:** 1044–1048.
7. ODETTI, P. *et al.* 1998. Early glycoxidation damage in brains from Down's syndrome. Biochem. Biophys. Res. Commun. **243:** 849–851.

Ca^{2+}-Dependent Permeability Transition and Complex I Activity in Lymphoblast Mitochondria from Normal Individuals and Patients with Huntington's or Alzheimer's Disease

ALEXANDER PANOV,[a] TRACY OBERTONE, JULIE BENNETT-DESMELIK, AND J. TIMOTHY GREENAMYRE

Department of Neurology, Emory University, Atlanta, Georgia 30322, USA

INTRODUCTION

Alzheimer's disease (AD) and Huntington's disease (HD) are neurodegenerative diseases with distinct clinical and morphological manifestations and genetic etiologies. Presently it is hypothesized that a common feature of different neurodegenerative diseases is impairment of mitochondrial energy metabolism in brain cells. Mitochondria play a key role in glutamate excitotoxicity and other forms of cell death via apoptotic or necrotic pathways.[1,2] The proposed mechanisms by which mitochondria induce cell death are the Ca^{2+}-dependent disruption of mitochondrial electrical membrane potential $(\Delta\Psi)$[1,3] and opening of the pore with nonspecific permeability, known as the mitochondrial permeability transition pore (mPTP).[4]

Studies on mitochondrial neuromuscular diseases[5] have shown the usefulness of evaluating mitochondrial function in mitochondria isolated from muscle biopsies or from patient-derived lymphoblasts grown in culture. The latter approach may be particularly useful for understanding pathogenesis of neurodegenerative diseases (NDD). The purpose of this work was to study the functional parameters of mitochondria from human lymphoblasts, particularly the Ca^{2+}-dependent mPTP in patients with HD or AD, and elderly individuals without NDD.

METHODS

The human lymphoblastoid cells (HLB) were obtained from the Leslie Thompson's laboratory at the University of California, Irvine. The cells were grown in rotary bottles at 37°C in RPMI 1640 medium supplemented with 10% FBS. The HLB mitochondria (HLBM) were isolated by the method described in ref. 6 except that cells were disrupted without the use of digitonin. Mitochondrial functions were stud-

---

[a]Address for correspondence: Alexander Panov, Department of Neurology, Emory University, 1639 Pierce Drive, WMB Suite 6000, Atlanta, GA 30322. Phone: 404-727-9799; fax: 404-727-3157.
e-mail: apanov@emory.edu

TABLE 1. Functional parameters of lymphoblast mitochondria and submitochondrial particles from aged control individuals and patients with Alzheimer's or Huntington's disease

Functional Parameter	Control	Alzheimer's Disease	Huntington's Disease
Mitochondria			
Membrane potential ($\Delta\Psi$) in mV	-162.1 ± 5.6	-164.5 ± 2.7	-168.4 ± 3.4
Calcium capacity (nmol Ca^{2+}/mg)	102.1 ± 7.9	114.8 ± 4.4	63.5 ± 5.5
Calcium capacity with 0.5 µM CsA	196.2 ± 11.0	234.1 ± 30.0	191.7 ± 12.1
Submitochondrial particles			
Complex I: V_{MAX} (nm decylubiquinone (DB) reduced/min/mg protein)	33.2 ± 7.5	27.5 ± 1.1	41.3 ± 6.8
K_M for NADH (µM)	4.4 ± 1.3	1.5 ± 0.12	6.3 ± 0.8
Complex I + complex III (nm cyt. C^{3+} reduced/min/mg SMP)	93.6 ± 18	92.6 ± 20.7	85.0 ± 8.8
Complex II + complex III (nm cyt. C^{3+} reduced/min/mg SMP)	19.4 ± 6	11.6 ± 2.3	15.3 ± 4
Complex III (nm cyt. C^{3+} reduced/min/mg SMP)	72.8 ± 8	51.1 ± 7	48.5 ± 5.5
Complex IV (nm cyt. C^{2+} oxidized/min/mg SMP)	98.5 ± 32	103.3 ± 36	135 ± 28

ied using incubation medium containing: sucrose 250 mM, glycyl-glycin 3 mM, pH 7.2, KH_2PO_4 1 mM, succinate 10 mM, mitochondria 0.5 mg (final volume 1.0 ml). The mitochondrial membrane potential ($\Delta\Psi$) was estimated with a tetraphenyl phosphonium-sensitive electrode. Measurements of activities of complexes I, I + III, II + III, III, IV in submitochondrial particles (SMP) were performed as described elsewhere.[6]

RESULTS AND DISCUSSION

Lymphoblast mitochondria (LBM) from control elderly individuals, and patients with AD or HD showed the same values of $\Delta\Psi$ in the range of -160 to -170 mV with succinate as a substrate (see TABLE 1). Gradual loading of the mitochondria with Ca^{2+} showed that with the control and AD LBM depolarization occurs abruptly (see FIG. 1) at the total calcium load, designated as *calcium capacity* (CC). Control and AD LBM had CCs of 102.1 ± 7.9 and 114.8 ± 4.4 nmol Ca^{2+}/mg mitochondrial protein correspondingly, while the CC of HD mitochondria was significantly lower (63.5 nmol Ca^{2+}/mg; TABLE 1). The mitochondrial depolarization presumably resulted from the opening mPTP, which was supported by the fact that cyclosporin A (CsA) at 0.5 µM roughly doubled the calcium load necessary for depolarization of mitochondria. The dynamics of depolarization of HD mitochondria during Ca^{2+} loading was variable. With LBM from a patient with the early onset of HD (with a CAG repeat of 65) the depolarization occurred even at the first addition of calcium

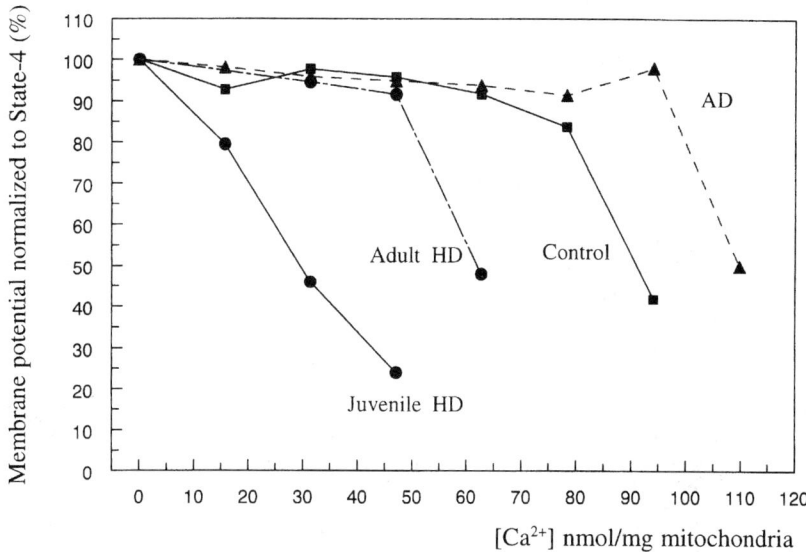

FIGURE 1. Dynamics of $\Delta\Psi$ changes during Ca^{2+} loads of lymphoblast mitochondria from control individuals and patients with Huntington's or Alzheimer's disease.

(see FIG. 1), indicating an inability to restore membrane potential. Further, in some experiments with LBM from HD patients, we did not observe an increase in CC in the presence of CsA. Preliminary results suggest the variability with HD patients depends on the length of the polyglutamine stretch in the mutated huntingtin protein.

Estimation of activities of the respiratory complexes in submitochondrial particles (SMP) did not reveal significant differences in the activities of complexes I + III, II + III, complex III, or complex IV (TABLE 1). The respiratory complex activities for the elderly control individuals, which are age-matched for the AD cases, are 2- to 5-fold lower than those published in the literature for the people below age 60.[5] This likely reflects normal age-related loss of mitochondrial functions.[5]

CONCLUSIONS

When compared to the elderly control individuals and patients with Alzheimer's disease, lymphoblast mitochondria from patients with Huntington's disease show significantly lower calcium capacity in induction of the mPTP. In some HD mitochondria, a graded early depolarization was observed during Ca^{2+} loads, rather than the normal abrupt threshold one. These data indicate that HD lymphoblast mitochondria are more susceptible to the Ca^{2+}-induced mPTP than are mitochondria from control individuals or patients with AD. This provides more support for the systemic nature of the impairment of the mitochondrial functions in patients with HD. In comparison with age-matched controls, LBM from patients with Alzheimer's disease do not show any specific changes in respiratory chain enzyme activities or Ca^{2+}-induced mPTP.

ACKNOWLEDGMENT

Research reported here was supported by the Huntington's Disease Society of America.

REFERENCES

1. GREENE, J.G. & J.T. GREENAMYRE. 1996. Bioenergetics and glutamate excitotoxicity. Prog. Neurobiol. **48:** 613–634.
2. BEAL, M.F. 1998. Mitochondrial dysfunction in neurodegenerative diseases. Biochim. Biophys. Acta **1366:** 211–223.
3. NICHOLLS, D.G. & S.L. BUDD. 1998. Mitochondria and glutamate excitotoxicity. Biochim. Biophys. Acta **1366:** 97–112.
4. ZAMZAMI, N. et al. 1996. Mitochondrial control of nuclear apoptosis. J. Exp. Med. **183:** 1533–1544.
5. WALLACE, D.C. 1997. Mitochondrial DNA mutations and bioenergetic defects in aging and degenerative diseases. In The Molecular and Genetic Basis of Neurological Disease, 2d edit. R.N. Rosenberg et al., Eds. :237–269. Butterworth-Heinemann. Boston, Oxford, Johannesburg, Melbourne.
6. TROUNCE, I.A. et al. 1996. Assessment of mitochondrial oxidative phosphorylation in patient muscle biopsies, lymphoblasts, and transmitochondrial cell lines. Methods Enzymol. **264:** 484–509.

Mitochondrial Porin, a Novel Target to Prevent Ischemia-Induced Neurodegeneration?

J. L. PEREZ VELAZQUEZ,[a] M. V. FRANTSEVA, D. HUZAR, C. GUEZURIAN, AND P. L. CARLEN

Playfair Neuroscience Unit, Toronto Western Hospital, 399 Bathurst Street, Toronto, Ontario M5T 2S8, Canada

The mitochondrial permeability transition (MPT) is considered to represent one of the final events that results in irreversible damage and subsequent cell death.[1] The MPT has been described in a variety of cell systems and occurs as a consequence of oxidative insults and excitotoxicity.[2] Recent reports indicate that the permeability transition could also occur in neurons, since it has been inferred from experiments using isolated brain mitochondria in the presence of elevated calcium and in dissociated neuronal or glial cultures exposed to glutamate or N-methyl-D-aspartate.[3-6]

Using an *in vitro* ischemia model, we obtained evidence for prolonged mitochondrial depolarization in rat organotypic hippocampal brain slices during reperfusion. Techniques for culturing embryonic brain slices were prepared as described previously.[7] The experiments were carried out after 7–14 days *in vitro*. Hypoxia-hypoglycemia was initiated by superfusing slices for 8 minutes with glucose-free artificial cerebrospinal fluid (ACSF) aerated with 95% N_2/5% CO_2, sucrose (10 mM) was added to the solution to maintain osmolarity.[7] The organotypic hippocampal cultures were loaded with the mitochondrial dye rhodamine123 (R123, 15 µM), which reflects the mitochondrial potential and is used as an indicator of the MPT. R123 fluorescence increased briefly in CA1 neurons for 1–2 minutes before becoming diffuse and disappearing abruptly during reperfusion in 5 of 6 slices, or during the ischemic insult (1 of 6). Similar results were obtained when slices were loaded with the ratiometric mitochondrial dye JC-1 ($n = 4$). The loss of mitochondrial potential in CA1 neurons during hypoxia-hypoglycemia and reperfusion was prevented by cyclosporin-A (CsA, $n = 4$). CsA is also a potent blocker of the MPT[1] and prevents mitochondrial depolarization induced by N-methyl-D-aspartate.[3]

We then used a novel strategy to identify the molecular mechanisms underlying mitochondrial dysfunction, based on CsA-affinity chromatography of mitochondrial proteins from injured and control brain tissue. Mitochondria were isolated from brain slices and mitochondrial proteins were purified on a CsA affinity column.[8] CsA is the most potent inhibitor of mitochondrial dysfunction, in particular the MPT,[9] and therefore we hypothesized that it may interact with proteins involved in the permeability transition after mitochondria were subjected to manipulations that promote the MPT. Mitochondrial porin was specifically eluted with 0.42 mM CsA from the affinity column using proteins from ischemic brain mitochondria ($n = 5$),

---

[a]Address for correspondence: Jose Luis Perez Velazquez, Playfair Neuroscience Unit, McL 12-413, Toronto Western Hospital, 399 Bathurst Street, Toronto, Ontario, M5T 2S8, Canada. Phone: 416- 603-5040; fax: 416-603-5745.
e-mail: jlpv@playfair.utoronto.ca

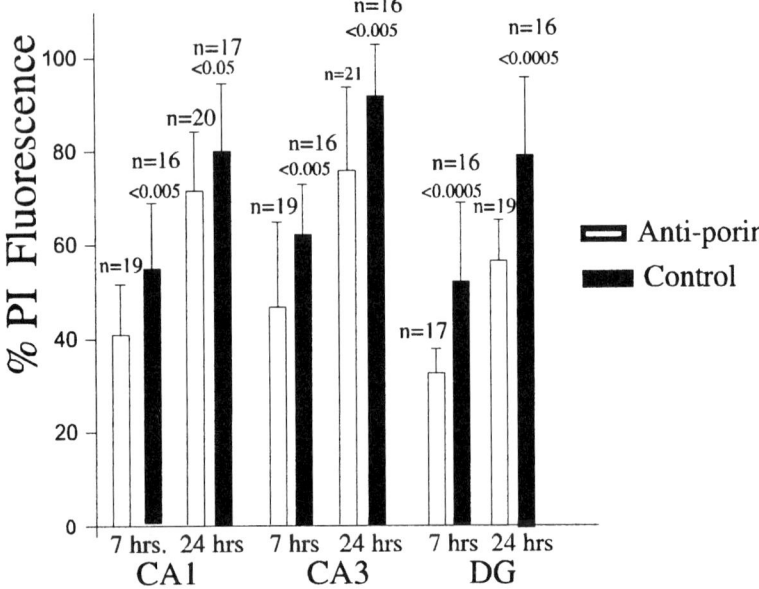

FIGURE 1. Anti-porin antibodies decrease ischemia-induced cell death. Organotypic hippocampal slices were subjected to a 1 h ischemic insult as described in Frantseva *et al.* and stained with propidium iodide (PI) 7 and 24 h later to reveal cell death. At 7 and 24 h after the insult, cellular death in the three main areas of the hippocampus (CA3, CA1, and DG) was lower in slices treated with the anti-porin antibody (20–30 μg/ml, obtained from Calbiochem; see also ref. 10). The antibody was present 2 h before the insult and throughout the experiment until the last images were taken 24 h later. The slices were incubated with the antibody and subjected to osmotic changes to improve the loading of the antibody into cells via osmotic lysis of pinocytic vesicles (Influx Pinocytic cell-loading reagent, Molecular Probes). Cell death was quantified as percentage of the maximal fluorescence for each slice (100% cell death), which was achieved by submerging the slices at 4°C for 24 h and staining with PI (see Ref. 11 for technical details). Initial fluorescence background (before the ischemic episode) for each slice was subtracted from the values at 7 h, 24 h, and final images. Images were acquired by a digital Sensys-CCD camera (Photometrics, Tucson, AZ) and were stored and analyzed using the Axon Imaging Workbench (Axon Instruments). A rhodamine filter (510–560/590 nm) was used to visualize PI staining. All images were taken through a 5× objective. *Bars* represent means ± SD, and p values were obtained using the Student's t-test.

or from mitochondria exposed to oxidative stress in the presence of t-butyl-hydroperoxide that were used as a positive control ($n = 5$). Porin was eluted only 3 of 10 times from control mitochondria. Anti-porin antibodies (Calbiochem, see ref. 10) introduced into individual neurons via patch electrodes (10 μg/ml) prevented mitochondrial depolarization and electrophysiological deterioration of hippocampal neurons during hypoxia-reperfusion ($n = 9$), as measured by simultaneous fluorescence imaging (R123) and whole-cell recordings. To control for specificity, similar experiments were performed with heat-inactivated antibody, resulting in deterioration of the neuronal membrane characteristics (input resistance and membrane potential) during reperfusion. R123 fluorescence also decreased significantly during reperfusion, as occurred in the experiments without the antibody, the mean fluorescence intensity during reperfusion was $35 \pm 28\%$ of control ($n = 5$, $p < 0.02$ compared with antibody group).

By staining the slices with propidium iodide, we further assessed whether the anti-porin antibodies prevented cellular death after ischemia in organotypic hippocampal cultures.[11] The results of three experiments (see FIG. 1 for experimental details) confirmed a significant decrease in cellular death after ischemia-reperfusion by treatment with the anti-porin antibody.

These observations provide biochemical and functional evidence that porin is directly involved in mitochondrial dysfunction[12,13] and neuronal impairment during ischemia-reperfusion, and indicate that porin could be a novel therapeutic target to prevent cellular degeneration.

REFERENCES

1. ZORATTI, M. & I. SZABO. 1995. The mitochondrial permeability transition. Biochem. Biophys. Acta **1241:** 139–176.
2. CROMPTON, M. *et al.* 1987. Evidence for the presence of a reversible calcium dependent pore activated by oxidative stress in heart mitochondria. Biochem. J. **245:** 915–918.
3. NIEMINEN, A.L. *et al.* 1996. Cyclosporin-A delays mitochondrial depolarisation induced by N-methyl-D-aspartate in cortical neurons: evidence of the mitochondrial permeability transition. Neuroscience **75** (4): 993–997.
4. SCHINDER, A.F. *et al.* 1996. Mitochondrial dysfunction is a primary event in glutamate neurotoxicity. J. Neurosci. **16** (19): 6125–6133.
5. WHITE R.J. & I.J REYNOLDS. 1996. Mitochondrial depolarization in glutamate-stimulated neurons: an early signal specific to excitotoxin exposure. J. Neurosci. **16** (18): 5688–5697.
6. KRISTAL, B.S. & J.M. DUBINSKY. 1997. Mitochondrial permeability transition in the central nervous system: induction by calcium cycling-dependent and independent pathways. J. Neurochem. **69** (2): 524–538.
7. PEREZ VELAZQUEZ, J.L. *et al.* 1997. In vitro ischemia promotes glutamate-mediated free radical generation and intracellular calcium accumulation in hippocampal pyramidal neurones. J. Neurosci. **17** (23): 9085–9094.
8. NICOLLI A., *et al.* 1996. Interactions of cyclophilin with the mitochondrial inner membrane and regulation of the permeability transition pore, a cyclosporin A-sensitive channel. J. Biochem. **271**(4): 2185–2192.
9. BERNARDI, P. *et al.* 1994. Recent progress on regulation of the mitochondrial permeability transition pore; a cyclosporin-sensitive pore in the inner mitochondrial membrane. J. Bioenerg. Biomembr. **26:** 509–517.
10. BABEL, D. *et al.* 1991. Production and characterization of eight monoclonal mouse antibodies against the human VDAC "porin 31HL" and their application for histotopological studies in human skeletal muscle. Biol. Chem. Hoppe-Seyler **372:** 1027–1034.
11. FRANTSEVA, M.V. *et al.* 1999. A submersion method to induce hypoxic damage in organotypic hippocampal cultures. J. Neurosci. Methods **89:** 25–31.
12. BEUTNER, G. *et al.* 1998. Complexes between porin, hexokinase, mitochondrial creatine kinase and adenylate translocator display properties of the mitochondrial permeability transition pore. Biochim. Biophys. Acta **1368** (1): 7–18.
13. CROMPTON, M. *et al.* 1998. Cyclophilin-D binds strongly to complexes of the voltage-dependent anion channel and the adenine nucleotide translocase to form the permeability transition pore. Eur. J. Biochem. **258** (2): 729–735.

# Bcl-2 and p53: Role in Dopamine-Induced Apoptosis and Differentiation

S. PORAT AND R. SIMANTOV[a]

Department of Molecular Genetics, Weizmann Institute, Rehovot 76100, Israel

INTRODUCTION

Dopaminergic neurons are confined to well-defined nuclei in the brain, and they play a central role in motor activity, cognition, hormone secretion, and several normal and pathological behaviors. It has been found that dopamine has a developmental role, in addition to its activity in neuronal communication. Depletion of dopamine, e.g., inhibits the growth and the innervation of neurons in the striatum.[1] In the midbrain, dopaminergic neurons regulate the expression of several neuropeptides, including substance P, enkephalin and neuropeptide Y. The developmental effects of dopamine are similar to the role of other neurotransmitters in the developing brain, such as acetylcholine, serotonin and GABA. It is now well established that dopamine is also neurotoxic, and it induces cell cycle arrest, intranucleosomal DNA fragmentation and apoptosis in cultured neurons.[2,3] It appears, therefore, that dopamine has a developmental role in the embryonic brain, but under certain experimental conditions, and apparently also *in vivo,* it can damage cells and cause cell death. In the current study we further analyze the conditions under which dopamine induces cell differentiation or apoptosis, and determined the involvement of p53 and Bcl-2 genes in these processes.

MATERIALS AND METHODS

Human neuronal NMB cells, culture conditions, dopamine treatment, cell viability, DNA fragmentation assay, dopamine uptake and other methods were described before.[3] Transfection with pcDNA-Bcl-2 (kindly provided by J. Tschopp) was performed with the calcium phosphate method. p53 immunoblotting was performed with anti p53 antibodies, kindly provided by M. Oren (Weizmann Institute).

RESULTS

When cultured with low concentrations of dopamine, 10–50 µM, NMB cells developed a larger cell body, a spindle shape, and extended processes. Concurrent with these morphological changes, the cell's growth rate was decreased and the uptake of ^{3}H-dopamine was increased. These morphological and neurochemical

---

[a]Corresponding author. Phone: 972-8934-2110; fax: 972-8934-4108.
e-mail: lgrabi@weizmann.weizmann.ac.il

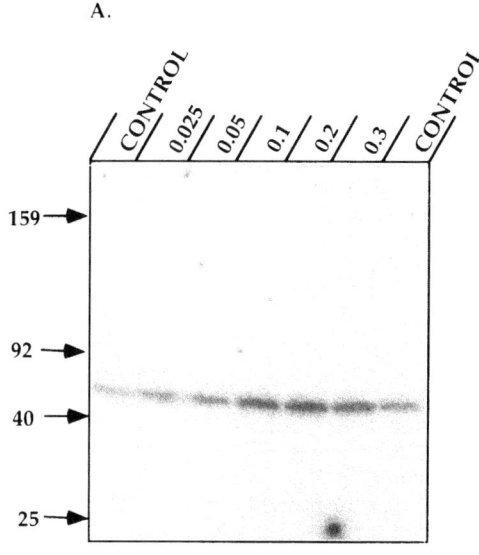

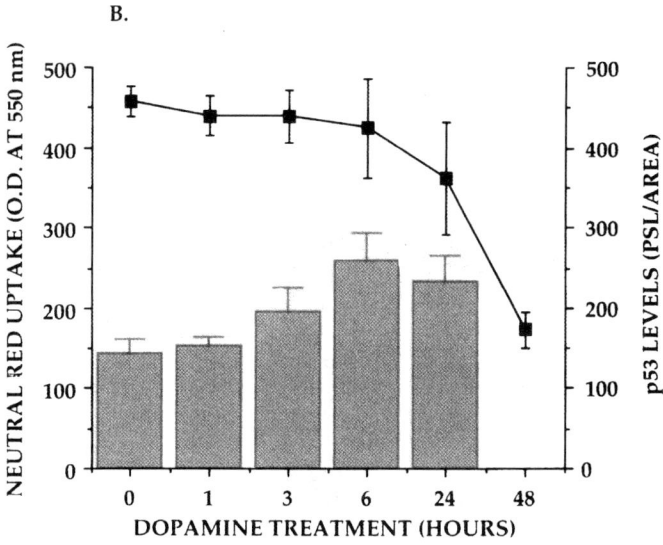

FIGURE 1. p53 levels in dopamine treated NMB cells: dose- and time-dependent effect. **A:** p53 immunoblotting in cells treated with 0.025–0.3 mM dopamine. **B:** Time course of p53 immunoblotting in cells treated with 0.3 mM dopamine.

changes suggest that at low concentrations dopamine induces differentiation of NMB cells. At higher concentrations, 0.05–0.3 mM, dopamine induces apoptosis, as reported previously.[3]

In an attempt to verify the relationship between differentiation and apoptosis, two other compounds that induce differentiation, dibutyryl-cAMP and DMSO were analyzed. Both compounds induced morphological changes, inhibited cell proliferation, and increased both ^{3}H-dopamine uptake and DAT immunoreactivity (to be published). DMSO also increased the number of dead cells, assessed with neutral red, while dibutyryl-cAMP had no such effect. Dibutyryl-cAMP and DMSO did not induce apoptosis, as verified morphologically or by the DNA fragmentation assay.

The dose-dependent effect of dopamine on differentiation and apoptosis raises the possibility that different concentrations of dopamine metabolites or free radicals differentially activate these two pathways, by damaging the DNA. Being a key element in cell response to DNA damage, p53 could play as a major factor in response to dopamine. Treatment with different concentrations of dopamine showed a dose-dependent increase in p53, about 3-fold with 0.2 mM (FIG. 1A). FIGURE 1B shows a time-dependent increase in p53 immunoreactivity upon treatment with 0.3 mM dopamine, with a maximum effect after 6 hours. This kinetics suggests that the increase in p53 preceded cell death. Whether p53 plays a role in differentiation of NMB cells was assessed by treatment with 10, 25 or 50 mM dopamine; p53 level in these cells was 95%, 113%, and 126% of controls, respectively, clearly lower than the effect of 0.1–0.3 mM dopamine. Dibutyryl-cAMP and DMSO, which both induced differentiation but not apoptosis, did not alter p53 levels.

As Bcl-2 contains positive p53 response elements, and it is located downstream to p53 in the apoptosis pathway, we tested whether Bcl-2 influence dopamine-induced apoptosis. FIGURE 2 hows that NMB cells transfected with pcDNA-Bcl-2 are less sensitive to the neurotoxic effect of dopamine, tested at three different concentrations.

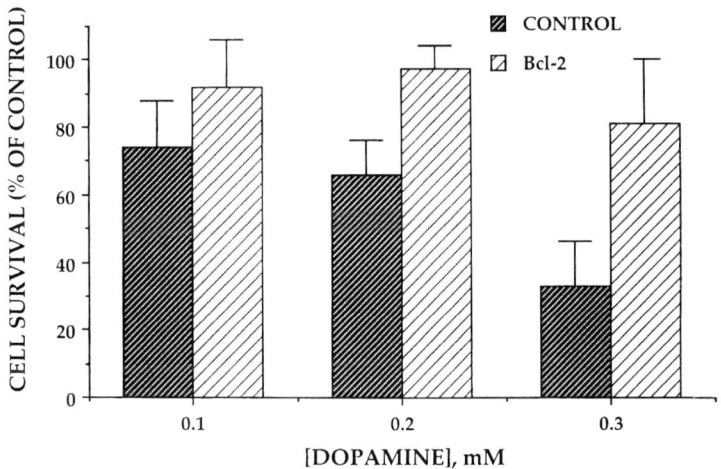

FIGURE 2. Decreased dopamine toxicity in pcDNA-Bcl-2 transfected NMB cells. Cell viability was determined in control or transfected cells, with three different concentrations of dopamine.

SUMMARY

The fate of a neuron in the developing brain to multiply, differentiate, or die in an apoptotic manner depends on the expression of genes that are involved in regulating the cell cycle. Recent studies determined the involvement of several genes, including cyclin A and B2, in dopamine-induced apoptosis in cultured chick sympathetic neurons.[4] Another gene that plays a role in apoptosis and differentiation of neurons, oligodendrocytes and PC12 cells is p53. It is also known that DNA damage increases p53 levels, triggering repair or apoptosis in response to moderate or severe damage, respectively. NMB cells express active and inducible forms of p53, thus being particularly suitable to analyze the role of this gene in dopamine-induced apoptosis and differentiation. The main observation of this work is that low concentrations of dopamine induce differentiation while high concentrations induce apoptosis, and that concentrations of dopamine that induce apoptosis increased p53 levels. The peak increase in p53 was within 3–6 h, before cell death. Thus, treatment with a high dopamine concentration may result in oxidation products and/or free radicals that heavily damage DNA, thus increasing p53 levels and initiating a cascade of events leading to apoptosis. Lower concentrations of dopamine apparently have a milder damaging effect on the DNA and induce growth arrest and differentiation.

In various systems Bcl-2 inhibits cell death, being apoptotic or necrotic. Bcl-2, and other members of the family, such as Bax, are located downstream to p53 in the apoptotic pathway,[5] and they contain negative or positive p53 response elements. Bcl-2 also protects cells by acting as antioxidant. Neuronal differentiation may be accompanied with an increase in Bcl-2, though it was suggested that the role of Bcl-2 in differentiation is less critical than in apoptosis.[6] Herein, Bcl-2 was found to inhibit dopamine neurotoxicity. Whether the expression of Bcl-2 is regulated by different dopamine concentrations, or by dibutyryl-cAMP and DMSO, remains to be determined.

ACKNOWLEDGMENTS

The study was supported by Israel Antidrug Authority, and Israel Ministry of Health.

REFERENCES

1. GRAYBIEL, A.M. 1990. Neurotransmitters and neuromodulators in the basal ganglia. Trends Neurosci. **13:** 244–254.
2. ZIV, I. et al. 1994. Dopamine induces apoptosis-like cell death in cultured chick sympathetic neurons—a possible novel pathogenetic mechanism in Parkinson's disease. Neurosci. Lett. **170:** 136–140.
3. SIMANTOV, R. et al. 1996. Dopamine-induced apoptosis in human neuronal cells: inhibition by nucleic acids antisense to the dopamine transporter. Neuroscience **74:** 39–50.
4. SHIRVAN, A. et al. 1997. Two waves of cyclin B and proliferating cell nuclear antigen expression during dopamine-triggered neuronal apoptosis. J. Neurochem. **69:** 539–549.
5. THOMPSON, C.B. 1995. Apoptosis in the pathogenesis and treatment of disease. Science **267:** 1456–1462.
6. GREENLUND, L.J. et al. 1995. Role of Bcl-2 in the survival and function of developing and mature sympathetic neurons. Neuron **15:** 649–661.

Mitochondria Mediate Nitric Oxide–Induced Cell Death

ANNA BAL-PRICE,[a] VILMA BORUTAITE, AND GUY C. BROWN

Department of Biochemistry, University of Cambridge, Tennis Court Road, Cambridge, CB2 1QW, United Kingdom

INTRODUCTION

In the central nervous system, nitric oxide (NO) at physiological concentrations acts as an intracellular messenger, but at higher concentrations it can initiate a neurotoxic cascade leading to cell death.[1] NO has been implicated in a wide range of pathological processes in the brain and other tissues, including ischemia, stroke, and neurodegenerative diseases.[2,3] It has been proposed that a number of the physiological and pathological effects of NO may be mediated by suppression of mitochondrial functions, particularly by inhibition of cytochrome oxidase.

We have shown that astrocytes and macrophages activated with cytokines and endotoxin to express the inducible NO synthase produce up to 1 μM NO and inhibit their own cellular respiration and that of coincubated cells via the NO inhibition of cytochrome oxidase.[4,5] We have shown that NO causes glutamate release from synaptosomes via inhibition of synaptosomal respiration,[6] and kills cerebellar granule neurons by excitotoxic mechanisms.

In the present studies we have used macrophages and PC12 cells as a model system to investigate the relation between mitochondrial inhibition and cell death. We tested whether NO and other inhibitors of mitochondrial respiratory chain can induce apoptotic or necrotic cell death, and whether this is related to opening of the mitochondrial permeability transition pore and subsequent activation of caspases.

MATERIAL AND METHODS

Culture of rat pheochromacytoma cells (PC12) was maintained at 37°C (5% CO_2) in DMEM medium supplemented with heat-inactivated horse serum (10% v/v), fetal bovine serum (5% v/v), and 100 U/ml penicilin, and 100 μg/ml streptomycin. Murine J774 macrophages were maintained in DMEM supplemented with 10% fetal bovine serum, 100 U/ml pencillin, and 100 μg/ml steptomycin.

---

[a]Corresponding author. Phone: +44 1223 333342; fax: +44 1223 333345.
e-mail: akp26@mole.bio.cam.ac.uk

TABLE 1. The induction of necrosis or apoptosis by NO donors or inhibitors of mitochondrial respiration in PC12 cells

Treatment	Necrotic Cells (%)	Apoptotic Cells (%)
Control	13.00 ± 1.40	0.10 ± 0.10
SNAP	83.80 ± 4.30	0.56 ± 0.10
SNAP + glucose	19.05 ± 1.80	19.45 ± 1.40
SNAP + glucose + z-VAD	14.80 ± 2.30	0.09 ± 0.1
NOC-18	97.80 ± 0.80	0.10 ± 0.01
NOC-18 + glucose	17.30 ± 1.80	25.50 ± 2.4
NOC-18+ glucose + z-VAD	8.30 ± 1.10	0.50 ± 0.10
Myxothiazol	92.50 ± 3.40	0.00
Myxothiazol + glucose	11.20 ± 3.60	10.38 ± 2.20
Azide	97.23 ± 1.50	0.00
Azide + glucose	15.20 ± 3.10	8.23 ± 2.80

NOTE: In all studies, medium was changed to glucose-free 1 day before experiments, and to fresh medium (with or without 20 mM glucose) containing NO donors (1 mM SNAP or 1 mM NOC-18) or mitochondrial inhibitors (2 µM myxothiazol or 2 mM azide) for 24 h. After this time the cell death was analyzed by Trypan blue or propidium iodide staining (necrotic cells) or by using chromatin dye Hoechst 33342 (apoptotic cells).

RESULTS AND DISCUSSION

The inhibition of mitochondrial respiratory chain by 2 mM azide or 2 µM myxothiazol resulted in necrosis of PC12 cells, as characterized by apparently intact chromatin (Hoechst 33342 staining) and loss of plasma membrane integrity (Trypan blue and propidium iodine staining). The number of necrotic cells was greatly reduced when cells were cultured in a medium with 20 mM glucose (TABLE 1). A similar effect was observed when cells were treated for 24 h with the NO donors SNAP (S-nitroso-N-acetylpenicillamine) or NOC-18 (DETA NONOate). In the presence of glucose, SNAP and NOC-18 induced apoptosis, which was completely prevented by 100 µM z-VAD, a broad-spectrum caspase inhibitor. In a glucose-free medium, incubation of cells with NO donors resulted in necrosis (TABLE 1).

Apoptotic cell death is mediated by specific proteases—caspases. We investigated whether inhibition of mitochondrial respiration by azide, myxothiazol, and NO donors causes activation of caspases, particularly caspase 3 and caspase 8-9, in macrophages. After 4-h incubation of macrophages with NO donors there was significant increase of caspase 3– and caspase 8-9–like activity, and it increased more during 8-h incubation (TABLE 2). Azide and myxothiazol were less efficient in causing activation of caspases, as there was no change in activity after 4-h incubation of macrophages with the inhibitors. After 8-h incubation with azide and myxothiazol only caspase 8-9–like activity increased to the same level as in the case of NO donors. Caspase 3–like proteases were activated only partially. Cyclosporin A completely prevented caspase 8-9 activation in the presence of NO donors and mitochondrial inhibitors (TABLE 2). Activation of caspase 3–like proteases in the presence of NO donors also was prevented by cyclosporin A, but it exerted only partial protection in the case of azide (TABLE 2).

TABLE 2. The effects of NO donors and inhibitors of mitochondrial respiration on the activity of caspases in macrophages

Treatment	4-h Incubation		8-h Incubation	
	Caspase-3	Caspase 8-9	Caspase-3	Caspase 8-9
Control	0.03 ± 0.01	0.12 ± 0.05	0.05 ± 0.01	0.12 ± 0.06
SNAP	0.16 ± 0.05	0.20 ± 0.02	0.35 ± 0.03	0.39 ± 0.09
NOC-18	0.22 ± 0.03	0.26 ± 0.08	0.31 ± 0.01	0.42 ± 0.1
Myxothiazol	0.05 ± 0.02	0.13 ± 0.06	0.13 ± 0.06	0.29 ± 0.08
NaN$_3$	0.06 ± 0.02	0.08 ± 0.02	0.19 ± 0.09	0.32 ± 0.11
SNAP + CsA			0.05 ± 0.02	0.05 ± 0.03
Myxothiazol + CsA			0.07 ± 0.01	0.10 ± 0.01
NaN$_3$ + CsA			0.16 ± 0.04	0.12 ± 0.02

NOTE: Macrophages (2–3 × 10^7 cells) were incubated with 1 mM SNAP, 1 mM NOC-18, 2 mM NaN$_3$ or 2 μM myxothiazol, and 3 μM cyclosporin A (where indicated). Cells were lysed in the buffer containing 100 mM HEPES (pH 7.4), 10% sucrose, 0.1% CHAPS, 1 mM EDTA, 10 mM DTT, 1 μM PMSF, 10 μg/ml pepstatin, and 10 μg/ml leupeptin, and then were centrifuged at 10000 × g × 10 min. Supernatants (100 μg total protein) were incubated with 100 μM DEVD-amc or IETD-amc for 60 min. Substrate cleavage was determined fluorimetrically (excitation at 380 nm, emission at 460 nm). Activity of caspases is expressed as nmol of substrate/min/mg protein. Mean ± SE of 3–6 experiments are presented.

Our data indicate that prolonged exposure of cells to NO might cause cell death by opening of a mitochondrial permeability transition pore and subsequent activation of caspases. The mode of cell death—necrotic or apoptotic—seems to be determined by cellular level of ATP.

REFERENCES

1. HUNOT, S. *et al.* 1996. Nitric oxide synthase and neuronal vulnerability in Parkinson's disease. Neuroscience. **72:** 355–363.
2. REIF, D.W. 1993. Delayed production of nitric oxide contributes to NMDA-mediated neuronal damage. Neuropharmacol. Neurotoxicol. **4:** 566–568.
3. WEIKERT, S. 1997. Rapid Ca^{2+}-dependent NO-production from central nervous system cells in culture measured by NO-nitrite/zone chemoluminescence. Brain Res. **748:** 1–11.
4. BROWN, G.C. *et al.* 1995. Nitric oxide produced by activated astrocytes rapidly and reversibly inhibits cellular respiration. Neuroscience Lett. **193:** 201–204.
5. BROWN, G.C. *et al.* 1998. Transcellular regulation of cell respiration by nitric oxide generated by activated macrophages. FEBS Lett. **439:** 321–324.
6. MCNAUGHT, K.S.P. & G.C. BROWN. 1998. Nitric oxide causes glutamate release from brain synaptosomes. J. Neurochem. **70:** 1541–1546.

Induction of Apoptosis and Necrosis in Human Neuroblastoma Cells by Cholesterol Oxides

MARIE LUISE RAO,[a,d] DIETER LÜTJOHANN,[b] MICHAEL LUDWIG,[c] AND HEIKE KÖLSCH[a,c]

Departments of [a]Psychiatry, [b]Clinical Pharmacology, and [c]Clinical Biochemistry, University of Bonn, Bonn, Germany

Oxidative stress leads to extensive loss of neuronal cells and plays an important role in neurodegenerative disorders. Nerve cells synthesize cholesterol, and there is evidence for the occurrence of low-density lipoprotein (LDL) in the cerebrospinal fluid (CSF) that could be subjected to oxidative modification.[1] Cholesterol oxides and cholesterol have been associated with the occurrence of atherosclerosis. Neuronal cell death such as that associated with Alzheimer's disease, is also accompanied by an increased release of cholesterol, which is subject to oxidation, resulting in the formation of 24S-hydroxy-cholesterol.[2] It has been shown previously that the neurotoxic potential of cholesterol oxides differs with respect to the position of the oxygen in the cholesterol molecule, thus leading predominantly to an apoptotic process (24-hydroxy-cholesterol[3]) or to necrotic cell damage (25-hydroxy-cholesterol[4]). We investigated the different neurotoxic processes elicited by cholesterol oxides on the human neuroblastoma cell line SH-SY5Y.

We tested the influence of cholesterol, 7-keto-cholesterol, 7α-hydroperoxy-cholesterol, and 24-hydroxy-cholesterol on SH-SY5Y human neuroblastoma cells. We recorded cell morphology, Trypan blue exclusion, LDH release, DNA fragmentation, increase in intracellular calcium, calpain activation, and the generation of reactive oxygen species (ROS).

Our results show that cholesterol does not exert any effect on the cells within 48 h of treatment. Exposure to 7-keto-cholesterol, 7α-hydroperoxy-cholesterol, and 24-hydroxy-cholesterol causes 50% of neuronal cells to die within 30, 7, and 18 h, repectively, and causes an elevation in LDH release (FIG. 1). Thus, cholesterol oxides possess neurotoxic properties, but cell death occurs via different pathways. DNA fragmentation occurs after 24 h (TABLE 1) and is detectable only after incubation with 24-hydroxy-cholesterol, suggesting that the early action of this cholesterol oxide leads to activation of an apoptotic pathway. Incubation with 7α-hydroperoxy-cholesterol and 24-hydroxy-cholesterol is accompanied by intracellular calcium increase and subsequent activation of calpain. Treatment with 7-keto-cholesterol effects neither intracellular calcium nor calpain activity. Exposure to 7α-hydroperoxy-cholesterol or 24-hydroxy-cholesterol generates ROS, but the other cholesterol oxides do not show any effect.

These results suggest that native cholesterol does not influence the viability of human neuroblastoma cells. Cholesterol oxides, which are generated *in vivo* and are

---

[d]Address for correspondence: Department of Psychiatry, University of Bonn, Sigmund-Freud-Strasse 25, Bonn D-53105, Germany. Phone: +49 228 287 6125; fax: +49 228 287 6097.
  e-mail: m.l.rao@uni-bonn.de

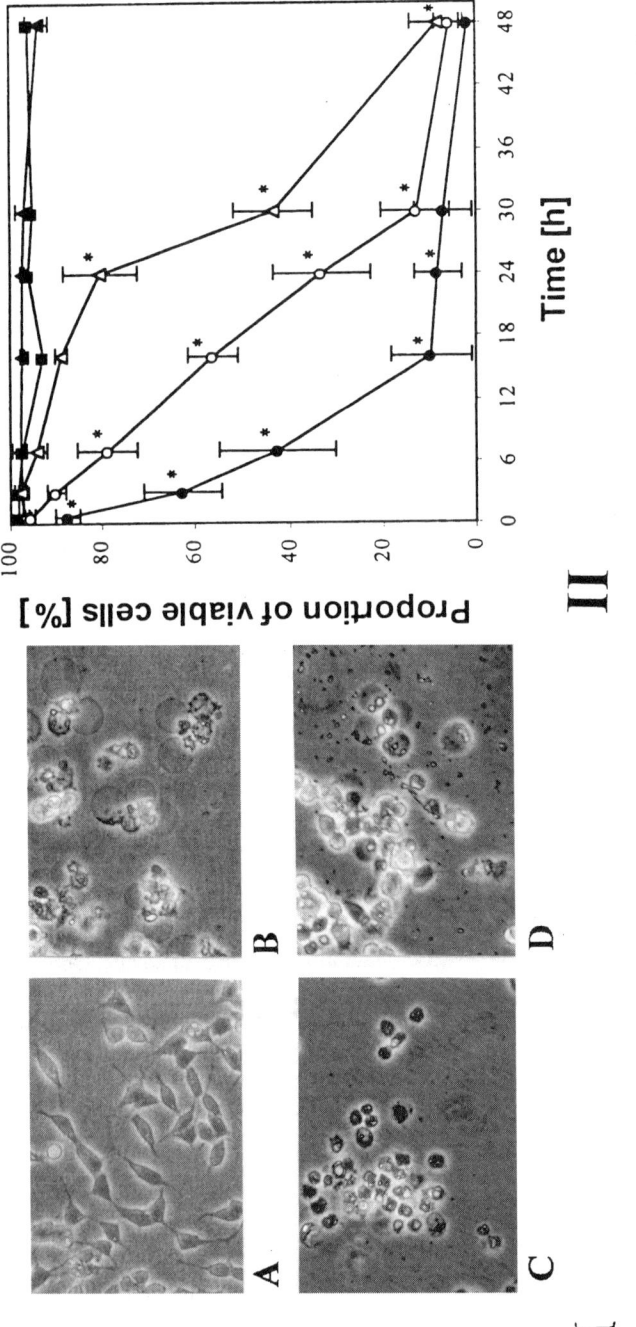

FIGURE 1. Neurotoxicity of cholesterol oxides. (**I**) Neuroblastoma cells SH-SY5Y were incubated for 24 h without additive (**A**), and with each 50 μM of (**B**) 7-keto-cholesterol, (**C**) 7α-hydroperoxy-cholesterol, and (**D**) 24-hydroxy-cholesterol. (**II**) Test of viability with Trypan blue staining. (■) without additive, (▲) with cholesterol, (●) 7α-hydroperoxy-cholesterol, (△) 7 keto-cholesterol, and (○) 24-hydroxy-cholesterol; *$p < 0.01$.

TABLE 1. Effect of treatment of human neuroblastoma cells SH-SY5Y with 50 μM cholesterol oxides

Effect	7-Keto-Cholesterol	7α-Hydroperoxy-Cholesterol	24-Hydroxy-Cholesterol
50% reduction in cell viability	30 h	7 h	18 h
Time of significant LDH release	24 h	16 h	16 h
DNA fragmentation	−	−	+
Increase in intracellular calcium	−	+	(+)
Calpain activation	−	+	+
ROS generation	−	+	(+)
Presumed neurotoxic Process	necrotic	necrotic	apoptotic

present in food, become toxic to these cells; the effects depend on the position of the oxygen molecule in the cholesterol moiety, in that 7- or 25-oxidized cholesterols ultimately provoke necrotic processes,[4] while 24-hydroxy-cholesterol, as inferred from our study, promotes an apoptotic process. These cholesterol oxides appear to be candidates in the etiology of neurodegenerative disorders. In the human 24-hydroxy-cholesterol is mainly of cerebral origin;[2] it has been speculated that it may be used as a marker for pathological changes in the brain.[6] In this context the importance of our *in vitro* results on the neurotoxicity of 24-hydroxy-cholesterol have been substantiated by the observation of elevated 24-hydroxy-cholesterol concentrations *in vivo*—namely, in serum of Alzheimer patients.[5]

REFERENCES

1. PAPASSOTIROPOULOS, A., M. LUDWIG, W. NAIB-MAJANI & G.S. RAO. 1996. Induction of apoptosis and secondary necrosis in rat dorsal root ganglion cell cultures by oxidized low density lipoprotein. Neurosc Lett. **209:** 33–36.
2. LÜTJOHANN, D, O. BREUER, G. AHLBORG, I. NENNESMO, A. SIEDEN, U. DICZFALUSY & I. BJÖRKHEM. 1996. Cholesterol homeostasis in human brain: evidence for an age-dependent flux of 24S-hydroxycholesterol from the brain into the circulation. Proc. Natl. Acad. Sci. USA **93:** 9799–9804.
3. KÖLSCH, H., D. LÜTJOHANN, A. TULKE, I. BJÖRKHEM & M.L. RAO. 1999. The neurotoxic effect of 24-hydroxycholesterol on SY-SY5Y human neuroblastoma cells. Brain Res. **818:** 171–175.
4. CHANG, J.Y., K.D. PHELAN & J.A. CHAVIS. 1998. Neurotoxicity of 25-OH-cholesterol on sympathetic neurons. Brain Res. Bull. **45:** 615–622.
5. PAPASSOTIROPOULOS, A., D. LÜTJOHANN, S. LOCATELLI, M. BAGLI, M.L. RAO, I. BJÖRKHEM, K. VON BERGMANN & R. HEUN. 1998. Oxidized products of brain cholesterol metabolism are potential markers of Alzheimer's disease. Soc. Neurosci. 24: 1270.
6. LÜTJOHANN, D., A. PAPASSOTIROPOULOS, I. BJÖRKHEM, S. LOCATELLI, M. BAGLI, R.D. OEHRING, U. SCHLEGEL, F. JESSEN, M.L. RAO, K. VON BERGMANN & R. HEUN. 1999. Plasma 24S-hydroxy-cholesterol (cebrosterol) is increased in Alzheimer and vascular demented patients. J. Lipid Res. In press.

Measurement of Oxidative DNA Damage in the Human p53 and PGK1 Gene at Nucleotide Resolution[a]

HENRY RODRIGUEZ[b,c] AND STEVEN A. AKMAN[d]

[b]Biotechnology Division, National Institute of Standards and Technology, Gaithersburg, Maryland 20899, USA

[d]Department of Cancer Biology, Comprehensive Cancer Center of Wake Forest University, Winston-Salem, North Carolina 27157, USA

INTRODUCTION

DNA damage induced by ROS[e] is an important intermediate in the pathogenesis of human conditions such as cancer and aging.[1] The mutational spectra of H_2O_2[2] and the transition metal ions Fe, Cu,[3,4] and Cr have been studied in model systems, but the relationship of induced DNA damage to these spectra remains unknown.

We mapped ROS-induced DNA base modifications sensitive to Nth and Fpg proteins.[6] Using a modified ligation-mediated polymerase chain reaction (LMPCR) technique, we investigated the *in vivo* and *in vitro* frequencies of DNA base modifications caused by ROS in the human *p53* and *PGK1* genes.[6–8]

MATERIALS AND METHODS

In Vivo H_2O_2 Treatment of Human Skin Fibroblasts. Human male foreskin fibroblasts were treated, harvested, and the DNA isolated as previously described.[9]

In Vitro Metal Ion-Ascorbate-H_2O_2 Treatment. Dialyzed DNA was treated with $CuCl_2$, $FeCl_3$, or $K_2Cr_2O_7$ as previously described.[8]

Ligation-Mediated Polymerase Chain Reaction. Digestion of treated DNA with Nth and Fpg proteins and the LMPCR technique (FIG. 1) has been described in detail elsewhere.[6] Key steps are:

---

[a]Certain commercial equipment, instruments, materials, or companies are identified in this paper to specify adequately the experimental procedure. Such identification does not imply recommendation or endorsement by the National Institute of Standards and Technology, nor does it imply that the materials or equipment identified are the best available for the purpose.

[c]Address for correspondence: Dr. Henry Rodriguez, Biotechnology Division, National Institute of Standards and Technology, 100 Bureau Drive, Stop 8311, Gaithersburg, Maryland 20899-8311. Phone: 301-975-2578; fax: 301-975-8505.

    e-mail: henry.rodriguez@nist.gov

    Web: http://www.nist.gov

[e]ABBREVIATIONS: *PGK1*, PhosphoGlycerate Kinase; LMPCR, ligation-mediated polymerase chain reaction; ROS, reactive oxygen species; H_2O_2, hydrogen peroxide; Fpg, FormamidoPyrimidine DNA glycosylase; Nth, Endonuclease III.

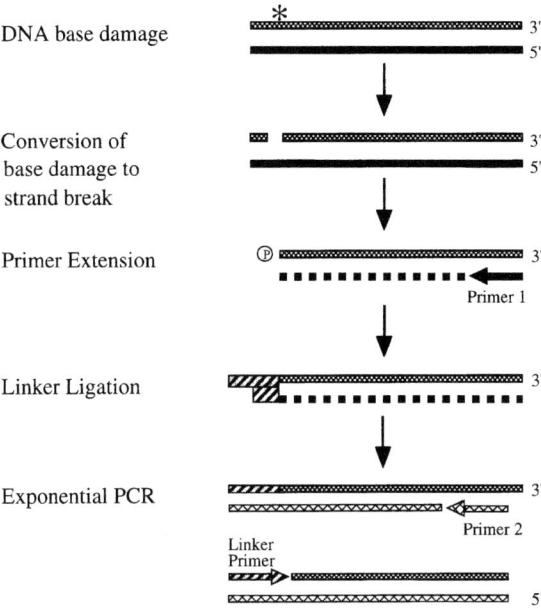

FIGURE 1. Schematic representation of the steps in DNA base damage mapping by LMPCR.

(1) conversion of a modified base into a strand break, chemically or enzymatically, followed by primer extension of an upstream primer 1 to generate blunt ends;
(2) ligation of a universal asymmetric double-strand linker;
(3) PCR amplification using a second upstream primer 2 with a downstream linker primer; and
(4) separation of DNA fragments on a sequencing gel, transfer to a nylon membrane, and hybridization with a radiolabeled probe.

CONCLUSION

The distribution of oxidative damage induced in exons 5 and 9 of human *p53* and the promoter region of human *PGK1* gene was assessed. A representative autoradio-

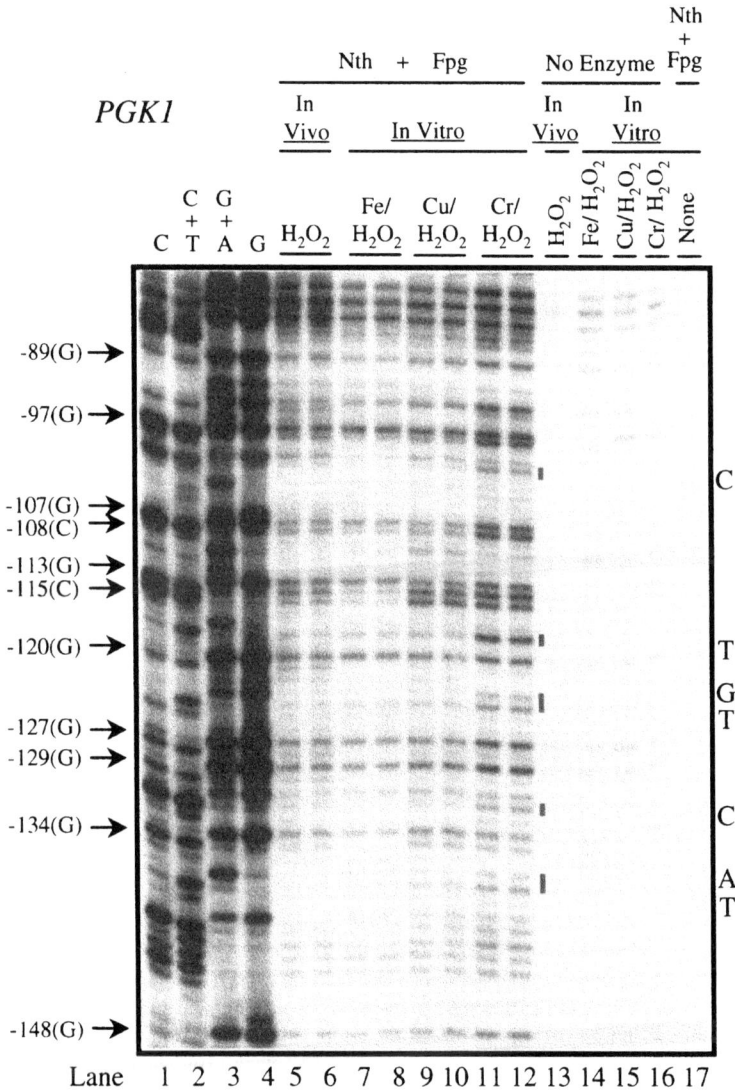

FIGURE 2. LMPCR analysis of damage induced in the promoter region of human *PGK1* using primer set A (transcribed strand). *Lanes 1–4*, DNA treated with standard Maxam-Gilbert cleavage reactions. *Lanes 5–6, 13*, DNA recovered from intact human foreskin fibroblasts exposed to 50 mM H_2O_2. *Lanes 7–8, 14*, dialyzed genomic DNA treated with 100 µM Fe(III)/100 µM ascorbate/5 mM H_2O_2 in the presence of 0.3 M sucrose. *Lanes 9–10, 15*, DNA treated with 50 µM Cu(II)/100 µM ascorbate/5 mM H_2O_2 in the presence of 1 mM potassium phosphate buffer. *Lanes 11–12, 16*, DNA treated with 100 µM Cr(VI)/100 µM ascorbate/5 mM H_2O_2. *Lane 17*, DNA incubated in potassium phosphate buffer and digested with Nth and Fpg proteins. The DNA in lanes 5–12 was digested with Nth and Fpg proteins

gram indicating the damage distributions induced in the *PGK1* gene is shown in FIGURE 2.

The nucleotide-resolution maps of DNA base damage induced *in vitro* in the presence of Cu(II), Fe(III), or Cr(VI) transition metal ions were similar to the *in vivo* base damage induced by H_2O_2. The *in vitro* similarity suggest a model in which the local binding site occupancy rate and the local geometry of the metal ion-DNA-peroxo coordination complex determine the damage event. The principal determinant of a damaging event occurring at any position is DNA sequence context. Guanine was the most heavily modified base. The triplet d(pCGC) was the principal hotspot sequence with guanine stretches also hit.[8]

REFERENCES

1. AMES, B.N. 1987. Oxidative DNA damage, cancer, and aging. Ann. Intern. Med. **107:** 526–545.
2. MORAES, E.C., S.M. KEYSE, M. PIDOUX & R.M. TYRRELL. 1989. The spectrum of mutations generated by passage of a hydrogen peroxide damaged shuttle vector plasmid through a mammalian host. Nucleic Acids Res. **17:** 8301–8312.
3. AKMAN, S.A., G.P. FORREST, J.H. DOROSHOW & M. DIZDAROGLU. 1991. Mutation of potassium permanganate and hydrogen peroxide-treated plasmid pz189 replicating in cv-1 monkey kidney cells. Mutat. Res. **261:** 123–130.
4. LOEB, L.A., E.A. JAMES, A.M. WALTERSDORPH & S.J. KLEBANOFF. 1988. Mutagenesis by the autoxidation of iron with isolated DNA. Proc. Natl. Acad. Sci. USA **85:** 3918–3922.
5. TKESHELASNVILI, L.K., T. MCBRIDE, K. SPENCE & L.A. LOEB. 1991. Mutation spectrum of copper-induced DNA damage. J. Biol. Chem. **266:** 6401–6406.
6. RODRIGUEZ, H., R. DROUIN, G.P. HOLMQUIST, T.R. O'CONNOR, S. BOITEUX, J. LAVAL, J.H. DOROSHOW & S.A. AKMAN. 1995. Mapping of copper/hydrogen peroxide-induced DNA damage at nucleotide resolution in human genomic DNA by ligation-mediated polymerase chain reaction. J. Biol. Chem. **270:** 17633–17640.
7. RODRIGUEZ, H., R. DROUIN, G.P. HOLMQUIST & S.A. AKMAN. 1997. A hot spot for hydrogen peroxide-induced damage in the human hypoxia-inducible factor 1 binding site of the PGK 1 gene. Arch. Biochem. Biophys. **338:** 207–212.
8. RODRIGUEZ, H., G.P. HOLMQUIST, R. D'AGOSTINO, Jr., J. KELLER & S.A. AKMAN. 1997. Metal ion-dependent hydrogen peroxide-induced DNA damage is more sequence specific than metal specific. Cancer Res. **57:** 2394–2403.
9. DROUIN, R., H. RODRIGUEZ, S.W. GAO, Z. GEBREYES, T.R. O'CONNOR, G.P. HOLMQUIST & S.A. AKMAN. 1996. Cupric ion ascorbate hydrogen peroxide-induced DNA damage: DNA-bound copper ion primarily induces base modifications. Free Radical Biol. Med. **21:** 261–273.

---

after treatment; the DNA in lanes 13–16 was incubated in digestion buffer alone after treatment. Positions of high damage frequency bases are marked with *arrows* to the left of lane 1. The sequence of positions heavily damaged in the presence of chromium, but not copper or iron, is denoted by *rectangles* to the right of lane 12.

The Glucose Paradox in Cerebral Ischemia
New Insights

AVITAL SCHURR,[a,d] RALPHIEL S. PAYNE,[a] MICHAEL T. TSENG,[b]
JAMES J. MILLER,[c] AND BENJAMIN M. RIGOR[a]

[a]*Brain Attack Research Laboratory, Department of Anesthesiology,*
[b]*Department of Anatomical Sciences & Neurobiology, and*
[c]*Department of Pathology & Laboratory Medicine,*
University of Louisville, School of Medicine, Louisville, Kentucky 40292, USA

INTRODUCTION

Several hypotheses have been proposed for the mechanistic basis of brain damage following cerebral ischemia. Disruption of cerebral energy metabolism is the hallmark of this disorder in which there is an insufficient supply of oxygen and glucose. Since the most efficient energy producing metabolic pathway is the oxidative conversion of glucose to CO_2 and H_2O, any disruption in the supply of these substrates would thus significantly cripple ATP production. A large increase in glycolytic glucose consumption (Pasteur effect) and an equally large increase in lactate production are typical responses of the brain to ischemia.

The urgent need for large supplies of glucose during the ischemic period, on one hand, and the reported aggravation of ischemic neuronal damage by preischemic glucose loading (hyperglycemia),[1–4] on the other, have presented a paradox that has baffled investigators for over two decades. This phenomenon has been summoned constantly to support one of the most enduring mechanistic hypotheses of ischemic brain damage, namely, the lactic acidosis hypothesis. The concept of "more glucose, more lactic acid, greater ischemic damage" became the dogma, despite lack of direct cause-effect relationships between lactic acid levels and ischemic brain damage and data that refute a damaging role for acidosis in cerebral ischemia.[5–7] The present study revisited the glucose paradox issue of cerebral ischemia.

METHODS

Cardiac Arrest–Induced Transient Global Cerebral Ischemia (TGI) in the Rat

The rat cardiac arrest model of TGI was employed using Sprague-Dawley male rats as described elsewhere.[8] Cardiac arrest was produced by 7 min of chest compression in rats fasted for 24 hours.

---

[d]Address for correspondence: Dr. Avital Schurr, Brain Attack Research Laboratory, Department of Anesthesiology, University of Louisville School of Medicine, Louisville, KY 40292. Phone: 502-852-6544; fax: 502-852-6056.
e-mail: A0schu01@ulkyvm.louisville.edu

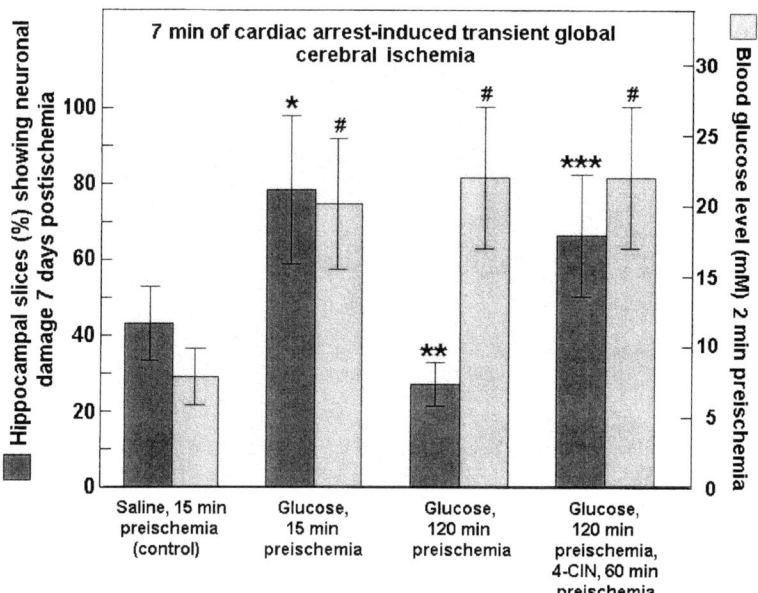

FIGURE 1. The effect of glucose (2 g/kg, i.p.) and its time of administration prior to cardiac arrest–induced transient global cerebral ischemia (TGI) on the level of blood glucose and on the degree of delayed neuronal damage as assessed by electrophysiologic means 7 days postischemia. Also shown is the effect of administration of the lactate transporter inhibitor, α-cyano-4-hydroxycinnamate (4-CIN), 60 min prior to TGI. *Significantly different from either control or 120 min preischemic glucose loading ($p < 0.0001$); **significantly different from control ($p < 0.05$) using X^2-test. #Significantly different from control ($p < 0.0006$) using t-test.

Preischemic Glucose Loading and Postischemic Delayed Neuronal Damage

Blood glucose levels 2 min prior to TGI and brain lactate and glucose levels, and protein concentration in brain tissue samples at the end of 7-min TGI were determined[9–11] in rats (3 in each group in duplicates) loaded with glucose (2 g/kg, i.p.) either 15 min or 120 min preischemia and compared to their levels in the brains of control, saline-injected rats. Delayed hippocampal neuronal damage in area CA1 was assessed by electrophysiologic means and a blinded histologic confirmation[12,13] on selected specimens in three groups of rats:

(1) saline (i.p.) 15 min prior to TGI (6 rats),
(2) glucose (2 g/kg, i.p.) 15 min prior to TGI (5 rats), and
(3) glucose (2 g/kg, i.p.) 120 min before TGI (8 rats).

Inhibition of Lactate Utilization Postischemia by α-Cyano-4-Hydroxycinnamate (4-CIN) in Vivo

To test the effect of lactate transport inhibition on the degree of ischemic damage, rats were administered 4-CIN (90 mg/kg, i.p.) 60 min prior to a 5-min cardiac arrest-

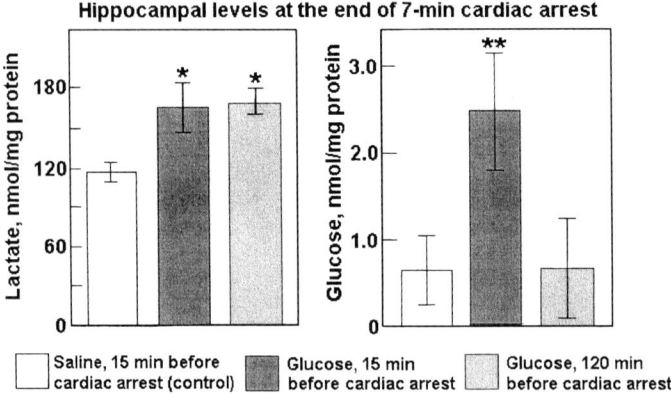

FIGURE 2. The levels of lactate and glucose at the end of 7-min cardiac arrest-induced transient cerebral ischemia (TGI) in the hippocampus of rats injected (i.p.) with either saline or glucose (2 g/kg) 15 min prior to TGI or with glucose (2 g/kg) 120 min prior to TGI. Rats were removed from the compression apparatus after 5 min of compression, decapitated, their brains removed quickly, one hippocampus was dissected out and cut into two halves, each placed in 0.5 ml of ice cold 8% perchloric acid. This procedure, from decapitation to placement of the tissue in perchloric acid, lasted 2 min. Thus, when combined with 5 min of TGI, each brain was exposed to 7 min of ischemia. After homogenization, the homogenate was neutralized with 0.25 ml ice cold $KHCO_3$ (2 M) and centrifuged. Lactate and glucose determinations were done on the supernatant and the pellet was used for protein determination. *Significantly different from control ($p < 0.002$); **significantly different from both control and glucose, 120 min before cardiac arrest ($p < 0.004$) using t-test.

induced TGI. This shorter than standard (7 min) TGI period was chosen since preliminary experiments indicated that mortality rate of 4-CIN-treated rats exposed to 7 min of cardiac arrest is very high. In a separate set of preliminary experiments (not shown), 4-CIN (90 mg/kg) was shown to cross the blood-brain barrier and to inhibit lactate transport when administered (i.p.) 60 min prior to testing its effect.

RESULTS AND DISCUSSION

Loading rats fasted for 24 h with glucose (2 g/kg, i.p.) 15 min prior to a 7-min TGI significantly increased the degree of delayed neuronal damage measured 7 days postischemia (FIG. 1). It also made these rats hyperglycemic, doubling the blood glucose level in comparison with nine control, saline-injected rats. When the same dose of glucose was injected to rats 120 min preischemia, the amount of delayed neuronal damage measured 7 days postischemia was significantly smaller than that measured either in rats loaded with glucose 15 min preischemia or in control, saline-injected rats (FIG. 1). The rats loaded with glucose 120 min preischemia also had blood glucose levels significantly higher than those measured in control, saline-injected rats.

As shown in FIGURE 2, the brain levels of lactate in the two glucose-loaded groups (3 rats per group) at the end of 7-min cardiac arrest–induced TGI were significantly higher than those measured in three control, saline-injected rats. Brain glucose levels in all three groups were very low; however, rats injected with glucose 15 min preischemia showed significantly higher levels of brain glucose than the other two groups. Thus, preischemic hyperglycemia had two opposing effects on the outcome of global cerebral ischemia. Glucose loading shortly (15 min) preischemia, as has been reported previously,[1–4] increased significantly the degree of delayed neuronal damage in comparison to the damage measured under euglycemic (control) conditions. Glucose loading 120 min preischemia significantly decreased the degree of delayed neuronal damage as compared to both hyperglycemic conditions induced 15 min preischemia and control, euglycemic conditions.

These results refute the hypothesis that the aggravation of ischemic delayed neuronal damage by hyperglycemia is due to excess lactic acidosis. While both hyperglycemic groups of rats had similar levels of brain lactate after 7-min TGI, the outcome in the two groups 7 days post-TGI was significantly different.

A short (5 min) episode of TGI produced only a small degree ($8.4 \pm 8.4\%$) of delayed hippocampal neuronal damage in control, untreated rats as measured 7 days postischemia. In rats pretreated with 4-CIN (90 mg/kg, i.p.), 5 min of TGI produced a significantly greater degree ($90 \pm 16\%$) of delayed neuronal damage. Rats injected with 4-CIN without exposure to TGI exhibited noticeable lethargy that lasted for approximately 2 h without any apparent residual effect or any sign of neuronal damage 7 days postinjection as measured both electrophysiologically and histologically. When ten rats, loaded with glucose 120 min preischemia, were also administered 4-CIN (90 mg/kg, i.p.) 60 min preischemia, five rats did not recover from cardiac arrest–induced TGI. In the five rats that did survive, the neuroprotective effect of glucose was almost completely abolished (FIG. 1). These data stipulate that brain oxidative lactate utilization postischemia reduces the degree of delayed CA1 neuronal damage in the hippocampus. Blockade of this utilization through inhibition of lactate transport into neurons,[9–11,14] significantly increases the degree of such damage. 4-CIN's ability to nullify the beneficial effect of glucose, when loaded 120 min preischemia, also supports the notion that the hyperglycemia-induced increase in brain lactate production is responsible for the observed reduction in the ischemic damage.

SUMMARY

The present *in vivo* findings that lactate, accumulated during an ischemic episode, is an essential aerobic energy substrate during the initial postischemic period are in full agreeement with our *in vitro* findings.[8–10] Moreover, the beneficial effects of hyperglycemia are also in agreement with our and others' *in vitro* results that have demonstrated a neuroprotective effect of glucose against hypoxic damage.[15] The aggravation of ischemic delayed neuronal damage by glucose loading 15 min prior to the ischemic insult is likely the result of glucose induction of a short-acting (30 to 60 min) systemic factor (hormonal?) that, when combined with an ischemic insult, potentiates the ischemic damage.

ACKNOWLEDGMENT

This study was supported in part by a grant from the Jewish Hospital Foundation, Louisville, Kentucky.

REFERENCES

1. MYERS, R.E. & S. YAMAGUCHI. 1977. Nervous system effects of cardiac arrest in monkeys. Preservation of vision. Arch. Neurol. **34:** 65–74.
2. GINSBERG, M.D., F.A. WELSH, & W.W. BUDD. 1980. Deleterious effect of glucose pretreatment on recovery from diffuse cerebral ischemia in the cat. I. Local cerebral blood flow and glucose utilization. Stroke **11:** 347–354.
3. WELSH, F.A., M.D. GINSBURG, W. RIEDER & W.W. BUDD. 1980. Deleterious effect of glucose pretreatment on recovery from ischemia in the cat. II. regional metabolite levels. Stroke **11:** 355–363.
4. REHNCRONA, S. I. ROSÉN & B.K. SIESJÖ. 1981. Brain lactic acidosis and ischemic cell damage. 1. Biochemistry and neurophysiology. J. Cereb. Blood. Flow Metab. **1:** 297–311.
5. SIESJÖ, B.K., H.-C. NORDSTROM & S. REHNCRONA. 1977. Metabolic aspects of cerebral hypoxia-ischemia. *In* Tissue Hypoxia and Ischemia. M. Reivich, R. Coburn, S. Lahiri & B. Chance, Eds. : 261–269. Plenum Press. New York, NY.
6. LJUNGGREN, B., K. NORBERG & B.K. SIESJÖ. 1974. Influence of tissue acidosis upon restitution of brain energy metabolism following total ischemia. Brain Res. **77:** 173–186.
7. GISSELSSON, L., M.-L. SMITH & B.K. SIESJÖ. 1999. Hyperglycemia and focal brain ischemia. J. Cereb. Blood Flow Metab. **19:** 288–297.
8. REID, K.H., C. YOUNG, A. SCHURR, M. TSENG, R.S. PAYNE, P. KEELEN, J.J. MILLER & V. IYER. 1996. Audiogenic seizures following global ischemia induced by chest compression in Long-Evans rats. Epilepsy Res. **23:** 195–209.
9. SCHURR, A., R.S. PAYNE, J.J. MILLER & B.M. RIGOR. 1997. Brain lactate, not glucose, fuels the recovery of synaptic function from hypoxia upon reoxygenation: an in vitro study. Brain Res. **744:** 105–111.
10. SCHURR, A., R.S. PAYNE, J.J. MILLER & B.M. RIGOR. 1997. Brain lactate is an obligatory aerobic energy substrate for functional recovery after hypoxia: further in vitro validation. J. Neurochem. **69:** 423–426.
11. LOWRY, O.H., N.J. ROSENBROUGH, A.L. FARR & R.J. RANDALL. 1951. Protein measurement with the folin phenol reagent. J. Biol. Chem. **193:** 265–275.
12. SCHURR, A., R.S. PAYNE, K.H. REID, V. IYER, M.T. TSENG, M.M. LI, S.-A. CHAN, C. YOUNG, J.J. MILLER, & B.M. RIGOR. 1995. Cardiac arrest-induced global cerebral ischemia studied in vitro. Life Sci. **57:** 2425–2430.
13. LI, M.M., R.S. PAYNE, K.H. REID, M.T. TSENG, B.M. RIGOR & A. SCHURR. 1999. Correlates of delayed neuronal damage and neuroprotection in a rat model of cardiac arrest-induced cerebral ischemia. Brain Res. In press.
14. SCHURR, A. & B.M. RIGOR. 1998. Brain anaerobic lactate production: a suicide note or a survival kit? Dev. Neurosci. **20:** 348–357.
15. SCHURR, A., C.A. WEST, K.H. REID, M.T. TSENG, S.J. REISS & B.M. RIGOR. 1987. Increased glucose improves recovery of neuronal function after cerebral hypoxia in vitro. Brain Res. **421:** 135–139.

Interactions of Chloromethyltetramethylrosamine (Mitotracker Orange™) with Isolated Mitochondria and Intact Cells

LUCA SCORRANO,[a] VALERIA PETRONILLI,[a] RAFFAELE COLONNA,[a] FABIO DI LISA,[b] AND PAOLO BERNARDI[a,c]

[a]*Consiglio Nazionale delle Ricerche Unit for the Study of Biomembranes, and the Department of Biomedical Sciences and*
[b]*Department of Biological Chemistry, University of Padova, I-35121 Padova, Italy*

Activation of the apoptotic program by cytochrome c and apoptosis-inducing factor (AIF) requires release of these intermembrane proteins into the cytosol, and the mechanism(s) by which this occurs is the subject of intense investigation. One of the most studied targets is the mitochondrial permeability transition (PT). The sudden increase of permeability of the inner membrane to solutes has been extensively studied.[1] A specific issue is whether or not a mitochondrial PT is a requisite for the release of apoptogenic proteins. This is an issue that has generated conflicting results and has become a major controversy in the literature. A specific, PT-independent release mechanism has been proposed based on the finding that cytochrome c release occurs without measurable decrease of the mitochondrial membrane potential ($\Delta\Psi_m$).[2–4] This point remains very controversial. Others have reported that cytochrome c release matches membrane depolarization,[5] and that it is mediated by a PT.[6,7] This has been indicated as the major determinant of AIF release.[8,9] Many of these studies were performed with the "fixable" probe chloromethyltetramethyl rosamine (CMTMRos) to monitor changes of $\Delta\Psi_m$ *in situ*. We have carried out a characterization of the interactions of this probe with mitochondria.

In the experiments of FIGURE 1, rat liver mitochondria were incubated in an isotonic KCl medium and energized with glutamate plus malate. Mitochondrial permeability to solutes was measured from the changes of absorbance at 620 nm.[10] Following the addition of a small Ca^{2+} load unable to induce changes of membrane permeability *per se*, EGTA was added to prevent Ca^{2+} redistribution among mitochondria with different stability, and increasing concentrations of CMTMRos were added. Increasing fractions of mitochondria became swollen in a concentration-dependent fashion (traces a to d). The swelling induced by the highest concentration of CMTMRos (5 µM) was prevented by cyclosporin A (CsA), indicating that the PT was the underlying cause. The dependence of the fraction of mitochondria having PT

---

[c]Address for correspondence: Prof. Paolo Bernardi, Dipartimento di Scienze Biomediche Sperimentali, Viale Giuseppe Colombo 3, I-35121 Padova, Italy. Phone: + 39-049-827-6063; fax: +39-049-827-6361.
 e-mail: bernardi@civ.bio.unipd.it

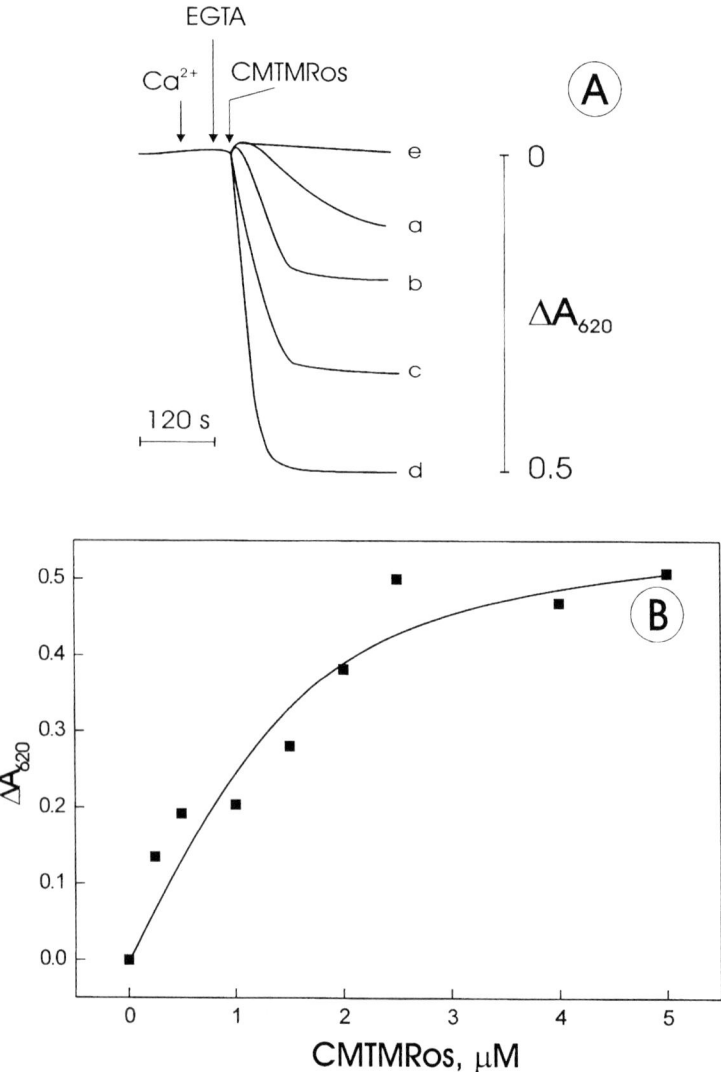

FIGURE 1. CMTMRos induces opening of the permeability transition pore in isolated mitochondria. The incubation medium contained 150 mM KCl, 10 mM Tris-MOPS, 1 mM Pi-Tris, 10 µM EGTA-Tris, 0.25 µg/ml oligomycin and 5 mM glutamate-Tris plus 2.5 mM malate-Tris. Final volume 2 ml, pH 7.4, 25°C. The experiments were started by the addition of 1 mg of rat liver mitochondria (*not shown*) prepared by standard centrifugation techniques. Where indicated 5 µM Ca^{2+}, 0.5 mM EGTATris and 0.25 µM (*trace a*), 1 µM (*trace b*), 2 µM (*trace c*), or 5 µM (*traces d,e*) CMTMRos were added. In the experiments of *trace e* 2 µM CsA was present. Data in **B** were obtained from **A**. Only selected traces have been reported in **A** for the sake of clarity.

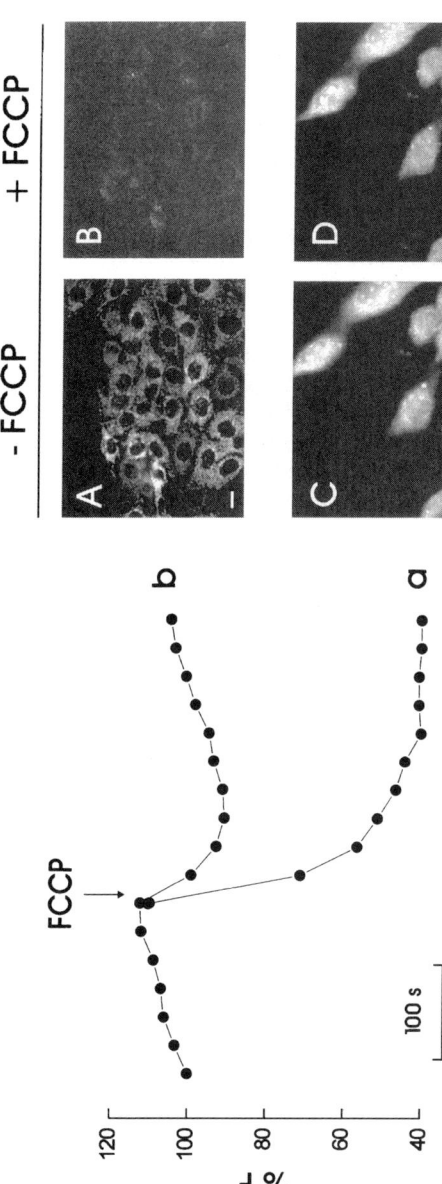

FIGURE 2. TMRM and CMTMRos response to FCCP in intact cells. MH1C1 rat hepatoma cells were seeded on uncoated 22 mm diameter round glass coverslips and grown for two days in Ham's F-10 nutrient mixture supplemented with 20% fetal calf serum in a tissue culture CO_2 incubator. Cells were loaded for 45 min at 37°C with 50 nM TMRM (**A**, **B** and *trace a*) or 50 nM CMTMRos (**C**, **D** and *trace b*) dissolved in 0.5 ml of Hank's balanced salt solution (HBSS). Cell fluorescence images were acquired with an Olympus IMT-2 inverted microscope, equipped for epifluorescent illumination with a Xenon light source (75 W), a 12 bit digital cooled CCD camera (Micromax, Princeton Instruments), and excitation/detection wavelength filter settings at 568/585 longpass. Images were collected with exposure times of 80 msec using a 40×, 1.3 NA oil immersion objective (Nikon). Data were acquired and analyzed using Metamorph software (Universal Imaging). Clusters of several mitochondria (10 to 30) were identified as regions of interest (RoI), whereas background was taken from areas not containing cells. Sequential digital images were acquired every 30 s for 510 sec, and the average fluorescence intensity of all the RoIs, of the nuclear regions, and of the background was recorded and stored for subsequent analysis. *Left panel:* changes of mean fluorescence intensity over RoIs minus background, where fluorescence intensity is normalized to the initial RoIs mean fluorescence for comparative purposes. Where indicated (*arrow*), 2 μM FCCP was added. The normalized fluorescence changes prior to FCCP addition refer to CMTMRos fluorescence while the corresponding, stable readings of TMRM fluorescence have been omitted for clarity. *Right panel:* **A**, **C** and **B**, **D** correspond to the first and last frame of the acquisition sequence, respectively. *Bar*, 17μm.

as a function of the concentration of CMTMRos is depicted in panel B. Increasing concentrations of CMTMRos caused profound respiratory inhibition, owing to depletion of pyridine nucleotides following the PT (results not shown). Since a residual respiratory inhibition was also observed in the presence of CsA (results not shown), we conclude that CMTMRos is both a respiratory inhibitor and a PT inducer in isolated mitochondria. In keeping with these observations, CMTMRos was very effective at depolarizing isolated mitochondria (results not shown).

In the experiments of FIGURE 2 we compared the fluorescence changes induced by the uncoupler carbonylcyanide-p-trifluoromethoxyphenyl hydrazone (FCCP) in cultures of MH1C1 cells grown on coverslips and loaded with the same concentration (50 nM) of tetramethyl rhodamine (TMRM) or CMTMRos. In either case the fluorescence displayed a punctate pattern typical of a mitochondrial distribution, which was stable with time. The addition of FCCP was followed by the expected fluorescence decrease with TMRM, while no relevant changes were recorded with CMTMRos. We found a small FCCP-induced fluorescence decrease at 25 nM CMTMRos but not at any higher concentration tested (up to 150 nM, results not shown).

The main conclusion of these experiments is that CMTMRos is a powerful inducer of the mitochondrial PT, which causes mitochondrial depolarization. Since most of the probe binds covalenty to –SH groups *via* its chloromethyl moiety, depolarization may not be detected by the probe itself. The major implication of this study is that CMTMRos cannot be used as a $\Delta\Psi_m$ probe in living cells, a consideration that has profound implications for the current debate on the role of mitochondria in cell death, and in particular on the mechanism of cytochrome c release in apoptosis. A PT caused by CMTMRos could have contributed to cause cytochrome c release, and the ensuing depolarization may simply not have been reported by the fluorescence changes of the probe (FIG.2).[3,4,11] Our results invalidate conclusions about the time sequence of events between mitochondrial depolarization and cytochrome c release that are based on the use of CMTMRos.

REFERENCES

1. BERNARDI, P. 1999. Mitochondrial transport of cations: channels, exchangers and permeability transition. Physiol. Rev. **79:** 1127–1155.
2. YANG, J., X. LIU, K. BHALLA, C.N. KIM, A.M. IBRADO *et al.* 1977. Prevention of apoptosis by Bcl-2: release of cytochrome c from mitochondria blocked. Science **275:** 1129–1132.
3. BOSSY-WETZEL, E., D.D. NEWMEYER & D.R. GREEN. 1998. Mitochondrial cytochrome c release in apoptosis occurs upstream of DEVD-specific caspase activation and independently of mitochondrial transmembrane depolarization. EMBO J. **17:** 37–49.
4. LI, H., H. ZHU, C.J. XU & J. YUAN. 1998. Cleavage of BID by caspase 8 mediates the mitochondrial damage in the Fas pathway of apoptosis. Cell **94:** 491–501.
5. HEISKANEN, K.M., M.B. BHAT, H.-W. WANG, J. MA & A.-L. NIEMINEN. 1999. Mitochondrial depolarization accompanies cytochrome c release during apoptosis in PC6 cells. J. Biol. Chem. **274:** 5654–5658.
6. PASTORINO, J.G., S.T. CHEN, M. TAFANI, J.W. SNYDER & J.L. FARBER. 1998. The overexpression of Bax produces cell death upon induction of the mitochondrial permeability transition. J. Biol. Chem. **273:** 7770–7775.
7. BRADHAM, C.A., T. QIAN, K. STREETZ, C. TRAUTWEIN, D.A. BRENNER & J.J. LEMASTERS. 1998. The mitochondrial permeability transition is required for tumor necrosis

factor alpha-mediated apoptosis and cytochrome *c* release. Mol. Cell. Biol. **18:** 6353–6364.
8. SUSIN, S.A., N. ZAMZAMI, M. CASTEDO, T. HIRSCH, P. MARCHETTI *et al.* 1996. Bcl-2 inhibits the mitochondrial release of an apoptogenic protease. J. Exp. Med. **184:** 1331–1341.
9. SUSIN, S.A., H.K. LORENZO, N. ZAMZAMI, I. MARZO, B.E. SNOW *et al.* 1999. Molecular characterization of mitochondrial apoptosis-inducing factor. Nature **397:** 441–446.
10. PETRONILLI, V., C. COLA, S. MASSARI, R. COLONNA & P. BERNARDI. 1993. Physiological effectors modify voltage sensing by the cyclosporin A-sensitive permeability transition pore of mitochondria. J. Biol. Chem. **268:** 21939–21945.
11. FINUCANE, D.M., E. BOSSY-WETZEL, N.J. WATERHOUSE, T.G. COTTER & D.R. GREEN. 1999. Bax-induced caspase activation and apoptosis via cytochrome *c* release from mitochondria is inhibitable by BCL-xL. J. Biol. Chem. **274:** 2225–2233.

Attenuation of Neuronal Death by NMDA and Oxygen-Glucose Deprivation in Cortical Neurons Maintained in High Glucose

SO YOUNG SEO,[a,b] EUN YOUNG KIM,[a] HARRIET KIM,[b] ILO JOU,[a] AND BYOUNG JOO GWAG[a,c]

[a]*Department of Pharmacology, Ajou University School of Medicine, Suwon, Kyungkido, Korea*

[b]*Department of Food and Nutrition, College of Human Ecology, Seoul National University, Seoul, Korea*

Maintaining appropriate levels of glucose is essential for brain metabolism, and imbalance in glucose metabolism may result in catastrophic dysfunction of the nervous system.[1,2] Hypoglycemia can produce neuronal injury through mechanisms involving activation of glutamate receptors sensitive to N-methyl-D-aspartate.[3] Hyperglycemia also appears to be harmful to brain under certain pathological conditions. Systemic administration of high glucose aggravates degenerative events following hypoxic-ischemia or epilepsy.[4,5] Several studies raise the possibility that increasing glucose entry to neurons may be beneficial against excitotoxicity and hypoxic-ischemia.[6,7]

We examined whether exposure to high glucose concentrations would influence the severity of excitotoxic neuronal death *in vitro*. Mixed cortical cell cultures, containing both neurons and glia, were prepared from brains of fetal ICR mice at 14–15 d gestation as previously described.[8] Cortical cell cultures (DIV 12) maintained in 25 mM glucose showed neuronal cell body swelling within 2 h and late death within 24 h following 10-min exposure to 100–1000 µM NMDA. The similar pattern of NMDA neurotoxicity was observed in cortical cell cultures placed in 100 mM glucose for 24 h prior to exposure to NMDA (TABLE 1). Cortical neurons maintained continuously in 100 mM glucose for 12–14 d were highly resistant to NMDA toxicity. Cortical cultures maintained in 25 mM or 100 mM glucose for 12–14 d showed the similar levels of vulnerability following 24-h exposure to 10 µM AMPA or 40 µM kainate. This suggests that prolonged exposure to high glucose protects neurons from injuries induced by NMDA, but neither AMPA nor kainate.

Recognizing the property of the NMDA receptor complex as a major route of Ca^{2+} entry, the preferential neuroprotective effect of high glucose against NMDA likely involves detoxification of Ca^{2+} accumulated in neurons. To determine this possibility, we used fura-2 fluorescence microphotometry to analyze $[Ca^{2+}]_i$ in cortical cell cultures. An increase in $[Ca^{2+}]_i$ was detected immediately, peaked within 2 min, and was sustained over 10 min in neurons following exposure of cortical cul-

[c]Address for correspondence: Department of Pharmacology, Ajou University School of Medicine, San 5, Wonchondong, Paldalgu, Suwon, Kyungkido 442-749, Korea. Phone: 82-331-219-5063; fax: 82-331-219-5069.

e-mail: bjgwag@madang.ajou.ac.kr

TABLE 1. Effects of high glucose on excitotoxicity in cortical cell cultures

Types of Injuries	Chronic Exposure[d]		Acute Exposure	
	25 mM Glucose	100 mM Glucose	25 mM Glucose	100 mM Glucose[e]
NMDA[a] 30 (μM)	10.01 ± 8.79	8.61 ± 3.60	11.01 ± 6.48	9.61 ± 5.97
100	58.45 ± 9.59	19.74 ± 3.44*	62.34 ± 10.42	60.85 ± 6.64
300	84.12 ± 4.97	28.51 ± 8.98*	79.54 ± 7.76	85.64 ± 12.34
1000	95.85 ± 3.82	50.68 ± 5.45*	99.11 ± 15.64	97.46 ± 8.76
AMPA[b] 10	73.95 ± 4.58	69.59 ± 5.45		
Kainate 40 (μM)	55.84 ± 8.60	54.46 ± 15.40		
OGD[c] 60 (min)	68.57 ± 11.58	19.24 ± 7.45*		

NOTE: Neuronal death was assessed by measuring LDH efflux into the bathing medium, mean ± SEM (n = 12–20 culture wells per each condition), scaled to the mean LDH value released after 24-h exposure to 500 μM NMDA (=100). *, significant difference from relevant control group (cultures grown in 25 mM glucose), at $p < 0.05$ using analysis of variance and Student-Neuman-Keuls's test.
[a]Cortical cell cultures at 12–14 days *in vitro* (DIV) were exposed to different doses of NMDA for 10 min.
[b]Cultures (DIV 12–14) were continuously exposed AMPA or kainate.
[c]Cultures (DIV 14–16) were deprived of oxygen and glucose (OGD).
[d]Cultures were maintained continuously in 25 mM or 100 mM glucose.
[e]Cultures were placed in 100 mM glucose for 24 h prior to exposure to NMDA.

tures grown in 25 mM glucose to 100 μM NMDA (FIG. 1). Interestingly, NMDA-induced accumulation of $[Ca^{2+}]_i$ was significantly reduced in neurons cultured and maintained in 100 mM glucose. The exact mechanisms underlying the protective effect of high glucose need to be delineated. The mitochondrial membrane potential, which was enhanced in cortical neurons maintained at high glucose levels (data not shown), may attribute to reducing $[Ca^{2+}]_i$ accumulation and death following activation of NMDA receptors.

We tested if cortical neurons maintained at high glucose levels would show differential vulnerability to deprivation of oxygen and glucose. Cortical cell cultures maintained in 25 mM glucose showed a 60–80% neuronal loss 24 h following deprivation of oxygen and glucose for 60 min, whereas cultures maintained in 100 mM glucose had 20–30% neuronal loss following the same injury.

The present findings that prolonged exposure to high glucose levels protects mature cortical neurons from injuries induced by NMDA or deprivation of oxygen-glucose do not conflict with toxic effects of high glucose on hypoxic-ischemic brain damage. In the latter experiments, glucose was administered within 1 h before onset of hypoxic-ischemic injury. It is possible that maneuvers enhancing mitochondrial function (e.g. high glucose levels) may have a therapeutic potential for treating stroke or neurodegenerative diseases coupled to NMDA neurotoxicity.

ACKNOWLEDGMENT

Work reported in this paper was supported by KOSEF through BDRC at Ajou University (B.J.G.).

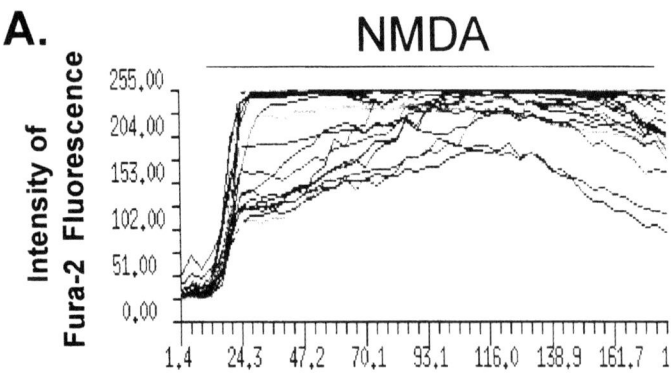
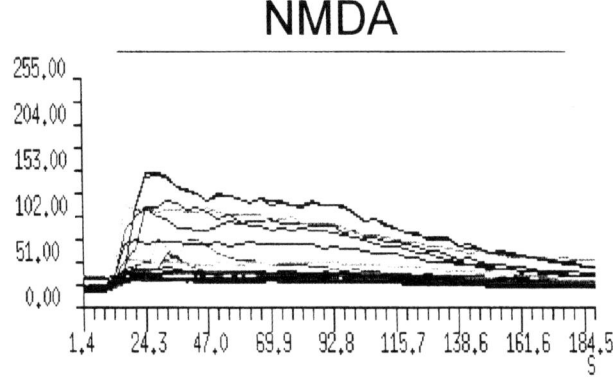
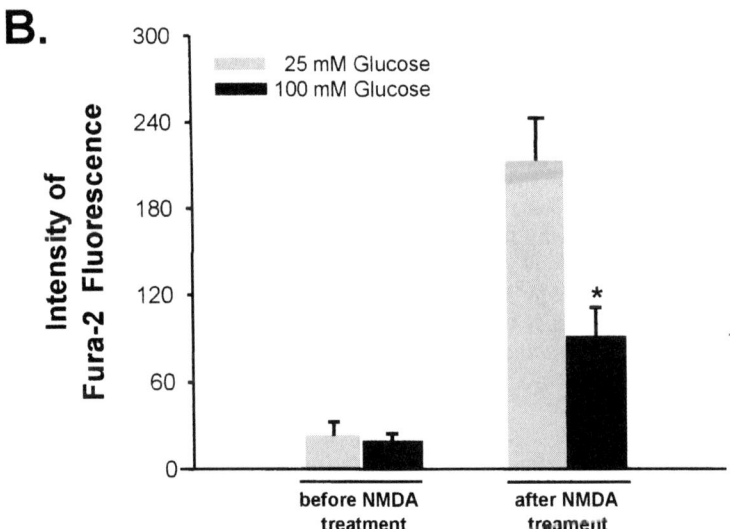

REFERENCES

1. SIESJO, B.K. 1988. Hypoglycemia, brain metabolism, and brain damage. Diabetes Metab. Rev. **4:** 113–144.
2. ERECINSKA, M. & I.A. SILVER. 1989. ATP and brain function. J. Cereb. Blood. Flow Metab. **9:** 2–19.
3. MONYER, H., M.P. GOLDBERG & D.W. CHOI. 1989. Glucose deprivation neuronal injury in cortical culture. Brain Res. **483:** 347–354.
4. LI, P.A., L. GISSELSSON, J. KEUKER, J. VOGEL, M.L. SMITH, W. KUSCHINSKY & B.K. SIESJO. 1998. Hyperglycemia-exaggerated ischemic brain damage following 30 min of middle cerebral artery occlusion is not due to capillary obstruction. Brain Res. **804:** 36–44.
5. LIN, B., M.D. GINSBERG & R. BUSTO. 1998. Hyperglycemic exacerbation of neuronal damage following forebrain ischemia: microglial, astrocytic and endothelial alterations. Acta Neuropathol. **96:** 610–620.
6. VANNUCCI, R.C., R.M. BRUCKLACHER & S.J. VANNUCCI. 1996. The effect of hyperglycemia on cerebral metabolism during hypoxic-ischemia in the immature rat. J. Cereb. Blood Flow Metab. **16:** 1026–1033.
7. LAWRENCE, M.S., G.H. SUN, D.M. KUNIS, T.C. SAYDAM, R. DASH, D.Y. HO, R.M. SAPOLSKY & G.K. STEINBERG. 1996. Overexpression of the glucose transporter gene with a herpes simplex viral vector protects striatal neurons against stroke. J. Cereb. Blood. Flow Metab. **16:** 181–185.
8. NOH, J.S. & B.J. GWAG. 1997. Attenuation of oxidative neuronal necrosis by a dopamine D1 agonist in mouse cortical cell cultures. Exp. Neurol. **146:** 604–608.

FIGURE 1. High glucose reduces NMDA-induced accumulation of $[Ca^{2+}]_i$. **A:** Traces of neuronal $[Ca^{2+}]_i$ following exposure to 100 μM NMDA in cortical cell cultures (DIV 12–14) maintained in 25 mM (*top panel*) and 100 mM (*bottom panel*) glucose. Fura-2 was added 20 min prior to exposure to NMDA. **B:** Fluorescence signal of fura-2 was analyzed before and 1 min after exposure to 100 μM NMDA, mean ± SEM (n = 25–30 neurons randomly chosen for each condition). *, significant difference between values from cultures in 25 mM and 100 mM glucose at $p < 0.05$ using Independent-Samples T Test.

Astrocyte Nitric Oxide Causes Neuronal Mitochondrial Damage, but Antioxidant Release Limits Neuronal Cell Death

R. STONE,[a] V. C. STEWART,[a] R. D. HURST,[b] J. B. CLARK,[a] AND
S. J .R. HEALES[a,c,d]

[a]*Department of Neurochemistry, Institute of Neurology, Queen Square, London, England*

[b]*Department of Biological and Biomedical Sciences, University of West England, Bristol, England*

[c]*Department of Clinical Biochemistry, National Hospital, Queen Square, London, England*

INTRODUCTION

Excessive generation of nitric oxide (NO), by astrocytes, has been proposed to be an important cause of the neuronal mitochondrial damage that may occur in conditions such as Parkinson's disease, Alzheimer's disease, and multiple sclerosis.[1] Previous studies have revealed that neurons, when cultured alone, are particularly susceptible to exposure to NO, resulting in marked damage to the mitochondrial respiratory chain and cell death.[2] Using a system whereby NO-generating astrocytes were cocultured with neurons, we recently demonstrated NO-mediated damage to the neuronal mitochondrial respiratory chain.[3] Under these coculture conditions there was no evidence of neuronal cell death. In view of this observation, we have considered the possibility that neuronal damage is minimized due to the concomitant release, by astrocytes, of antioxidants such as reduced glutathione (GSH). In order to begin to address this hypothesis, we monitored the ability of cultured astrocytes to release GSH.

METHODS

Primary cultures of astrocytes were prepared from neonatal Wistar rats as described by Tabernero *et al.*[4] After 13 days of culture, the cells were seeded at a density of 2.5×10^5 cells ml^{-1}. On day 14 the L-valine–containing minimal essential medium was replaced with minimal medium.[5] At set time points (0–8 hours), the experiment was terminated by removal of the astrocyte-conditioned media. This media was then frozen immediately, by immersion in liquid nitrogen, and stored at −70°C until analysis. For GSH analysis, the media was thawed and mixed (1:1) with 15 mM

---

[d]Address for correspondence: Dr. S. J. R. Heales, Department of Neurochemistry, Institute of Neurology, Queen Square, London WC1N 3BG, England. Phone: +44-171-837-3611, ext. 4208; fax: +44-171-833-1016.
 e-mail: sheales@ion.ucl.ac.uk

ortho-phosphoric acid. Following mixing and centrifugation (15,000 × g for 6 minutes), 20 µl of supernatant was injected onto an HPLC. Separation and electrochemical detection were based on the method described by Riederer *et al.*[6] In order to ascertain whether oxidation of GSH to GSSG had occurred, some of the media was pretreated with glutathione reductase (1.0 units) and NADPH (80 mM) for 10 minutes prior to the addition of *ortho*-phosphoric acid. GSH stability in both conditioned (previously astrocyte-exposed) and unconditioned minimal medium was also determined by adding GSH (5 µM) to both media, sampling at set time points and treating as above. In all cases results are expressed as mean ± SEM.

RESULTS

Astrocytes were found to release reduced GSH. The GSH concentration in the extracellular medium increased in a linear manner over the eight-hour time period, accumulating to 1.13 ± 0.12 µM (FIG. 1). GSSG was not formed in the extracellular media at any time point, as incubation with glutathione reductase and NADPH did not alter the amount of GSH detected. The stability of 5 µM GSH in both conditioned and unconditioned minimal medium was monitored over a 5-hour time period. GSH was found to be more stable in astrocyte-conditioned minimal medium than in unconditioned minimal medium (FIG. 2). More specifically, after a 5-hour incubation in astrocyte-conditioned minimal medium, 5 µM GSH was found to be preserved and was 43% of its initial concentration. However, 5 µM GSH added to unconditioned minimal media was very unstable, decaying to 0.4 ± 0.18 µM instantaneously, and dropping further after 2 hours to 0.14 ± 0.12 µM. This loss of GSH was due to oxidation to GSSG, as incubation with glutathione reductase and NADPH restored the GSH concentration to within 75% of the initial GSH concentration.

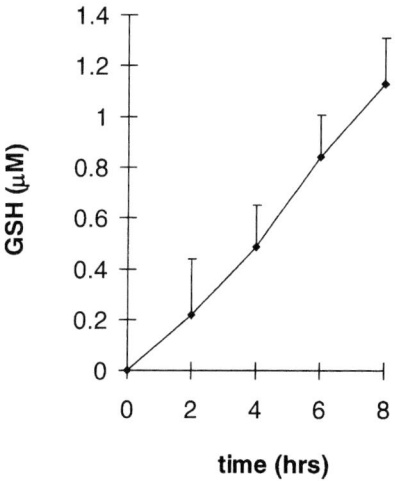

FIGURE 1. Release of GSH by astrocytes.

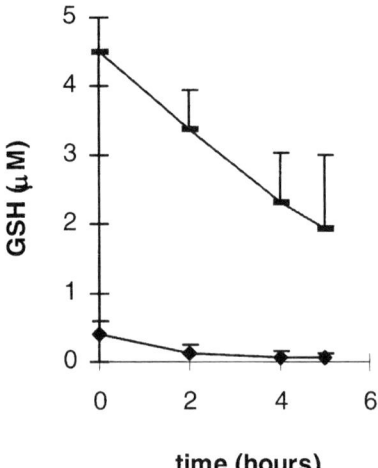

FIGURE 2. Reduced glutathione (5 µM) stability in conditioned versus unconditioned minimal medium. *Diamonds,* unconditioned minimal medium; *rectangles,* conditioned minimal medium.

CONCLUSIONS

Astrocytes release GSH into the extracellular medium. This observation supports the previous findings of Dringen et al.,[5] who monitored total glutathione (GSH + GSSG) release from astrocytes. The data presented here suggest that GSH is released from astrocytes and is *not* oxidized to GSSG over an incubation period of at least eight hours. In contrast, GSH added to minimal medium (unconditioned and cell free) is rapidly oxidized to GSSG, but in conditioned medium GSH oxidation is impaired. These findings imply that astrocytes, in addition to releasing GSH, release a factor that prevents extracellular GSH from oxidation. The release and preservation of GSH may be important in limiting neuronal damage from astrocyte-derived NO.

ACKNOWLEDGMENTS

Rebecca Stone is the recipient of a prize studentship from the Brain Research Trust (UK). Simon Heales is the recipient of a travel grant from the Wellcome Trust (UK).

REFERENCES

1. HEALES, S.J.R., J.P. BOLAÑOS, V.C. STEWART *et al.* 1999. Nitric oxide, mitochondria and neurological disease. Biochim. Biophys. Acta **1410:** 215–228.
2. BOLAÑOS, J.P., S.J.R. HEALES, S. PEUCHEN *et al.* 1996. The role of glutathione in protecting neuronal mitochondrial activity from nitric oxide mediated damage. Relevance for neurodegenerative diseases. Free Radical Biol. Med. **21:** 995–1001.

3. STEWART, V.C., J.M. LAND, J.B. CLARK, J.B. *et al.* 1998. Pretreatment of astrocytes with interferon-α/β prevents neuronal mitochondrial respiratory chain damage. J. Neurochem. **70:** 432–434.
4. TABERNERO, A., J.P. BOLAÑOS & J.M. MEDINA. 1993. Lipogenesis from lactate in rat neurons and astrocytes in primary culture. Biochem. J. **294:** 635–638.
5. DRINGEN, R., O. KRANICK & B. HAMPRECHT. 1997. The gamma-glutamyl transpeptidase inhibitor acivicin preserves glutathione released by astroglial cells in culture. Neurochem. Res. **22:** 727–733.
6. RIEDERER, P., E. SOFIC, W.D. RAUSCH *et al.* 1989. Transition metals, ferritin, glutathione and ascorbic acid in parkinsonian brains. J. Neurochem. **52:** 515–520.

Mechanisms of Selective Neuronal Cell Death due to Thiamine Deficiency

KATHRYN G. TODD[a] AND ROGER F. BUTTERWORTH

Neuroscience Research Unit, Centre Hospitalier de l'Université de Montréal (Campus Saint-Luc), Montréal, Québec, Canada

INTRODUCTION

Vitamin B_1 (thiamine) deficiency results in one of two well-defined neurological disorders—namely, a mixed sensory-motor neuropathy or Wernicke's encephalopathy (also referred to as the Wernicke-Korsakoff Syndrome, WKS). WKS is characterized clinically by ophthalmoplegia, ataxia, and memory loss, and is encountered in disorders of gross malnutrition including chronic alcoholism, gastrointestinal carcinoma, and AIDS.[1,2] Thiamine deficiency in alcoholics results from inadequate dietary intake, impaired gastrointestinal absorption, and depletion of liver stores of the vitamin.[3-5] Furthermore, it has been demonstrated that thiamine deficiency contributes to the pathogenesis of alcoholic peripheral neuropathy.[6] In spite of widespread fortification of various foods with thiamine, WKS and alcoholic peripheral neuropathy remain important health care problems. Following thiamine administration to patients with WKS, a prompt improvement in ocular signs is observed. However, improvements in motor coordination and memory performance are often minimal. It has been suggested that these last two symptoms are the consequence of histopathologically defined lesions in diencephalic and brainstem structures in patients with WKS.[1,7]

ROLE OF THIAMINE IN CENTRAL NERVOUS SYSTEM FUNCTION

In brain, as in peripheral tissues, thiamine occurs mainly in the form of its diphosphate ester (TDP). TDP is cofactor for three major enzymes involved in cellular glucose and energy metabolism—namely, the pyruvate dehydrogenase complex (PDHC), α-ketoglutarate dehydrogenase (α-KGDH), and transketolase (TK).

Over the last several years, studies from our unit have focused on the effects of thiamine deficiency *in vivo* on the rates of depletion of thiamine esters, and on activities of TDP- dependent enzymes in discrete areas of brain. Use was made of the rat treated with the central thiamine antagonist pyrithiamine, a well-validated experimental animal model that is widely used and that recapitulates the clinical and neuropathological features of WKS.[8,9] Onset of neurological symptoms—ataxia, loss of righting (vestibulocerebellar) reflexes—in these animals occurs when TDP concen-

---

[a]Corresponding author. Present address: Kathryn G. Todd, Ph.D., Neurochemical Research Unit, Department of Psychiatry, 1E7.44 W. Mackenzie Health Sciences Center, University of Alberta, Edmonton, Alberta T6G 2R7, Canada. Phone: 780-492-760;4 fax: 780-492-6841.
e-mail: kgtodd@ualberta.ca

TABLE 1. Vulnerable and nonvulnerable brain regions in experimental thiamine deficiency compared to WKS in humans

Brain Region	Human WKS	Experimental thiamine Deficiency (Pyrithiamine Model)
Vulnerable regions		
Thalamic nuclei		
Medial dorsal n.	+	+
Medial geniculate n.	+	+
Inferior colliculus	+	+
Inferior olivary n.	+	+
Mammillary bodies	+	±
Lateral vestibular n.	+	+
Nonvulnerable regions		
Caudate n.	−	−
Frontoparietal cortex	−	−
Hippocampus	−	−

trations fall below 15% of control values.[10] Reductions in TDP concentrations are accompanied by selective early reductions in activity of α-KGDH.[4,11]

Furthermore, these selective reductions in α-KGDH due to thiamine activity (a) become apparent prior to the onset of neurological symptoms and (b) are noted in vulnerable diencephalic structures such as thalamic nuclei prior to their occurrence in nonvulnerable structures such as the cerebral cortex.[10] α-KGDH is a rate-limiting tricarboxylic acid cycle enzyme. Therefore, not surprisingly, reductions in α-KGDH activity in brain due to thiamine deficiency have metabolic consequences, including a decreased flux of substrates through the cycle and a consequent accumulation of alanine and lactate.[12,13] On the bases of these findings, we proposed that reduction in activity of α-KGDH constitutes the principal metabolic defect responsible for the neurological consequences of thiamine deficiency;[14] this hypothesis was subsequently confirmed by others.[15] Reductions in activities of α-KGDH were subsequently reported by us in autopsied brain tissue from alcoholic patients with neuropathologically confirmed WKS.[16]

NEURONAL CELL DEATH IN THIAMINE DEFICIENCY

Neuropathologic evaluation of the brains of patients with WKS reveals a highly selective and reproducible pattern of neuronal cell loss involving primarily diencephalic and brainstem structures. Bilateral symmetrical lesions are consistently observed in thalamic nuclei, inferior colliculi, inferior olivary nuclei, mammillary bodies, and lateral vestibular nuclei.[1,7] Other major structures, including cerebral cortex, caudate nuclei, and hippocampi manifest little (if any) significant neuronal loss in WKS (TABLE 1). The pyrithiamine-treated rat model of WKS also results in neuronal loss in diencephalic and brainstem structures.[8,11]

Brain structures that manifest selective vulnerability to thiamine deficiency are those with high thiamine turnover rates[17] and those that demonstrate, under normal physiological conditions, high rates of oxidative metabolism as revealed by studies of local cerebral glucose utilization using variously labeled ^{14}C-glucoses.[18] In view of these observations and of our findings of selectively decreased activities of α-KGDH in vulnerable brain regions in thiamine deficiency, deficits in mitochondrial oxidation and their consequences are likely causes of neuronal cell death due to thiamine deficiency. In support of this, studies in cultured neuroblastoma cells reveal that the major cause of cellular dysfunction in thiamine deficiency is mitochondrial uncoupling.[19] Probable mechanisms implicated in neuronal cell death and that are a direct consequence of mitochondrial dysfunction include impairments in cerebral energy metabolism,[20] focal lactic acidosis,[12] glutamate receptor–mediated excitotoxicity,[21] and oxidative damage,[22,23] all of which may be implicated in the pathogenesis of neuronal cell death due to thiamine deficiency (FIG. 1).

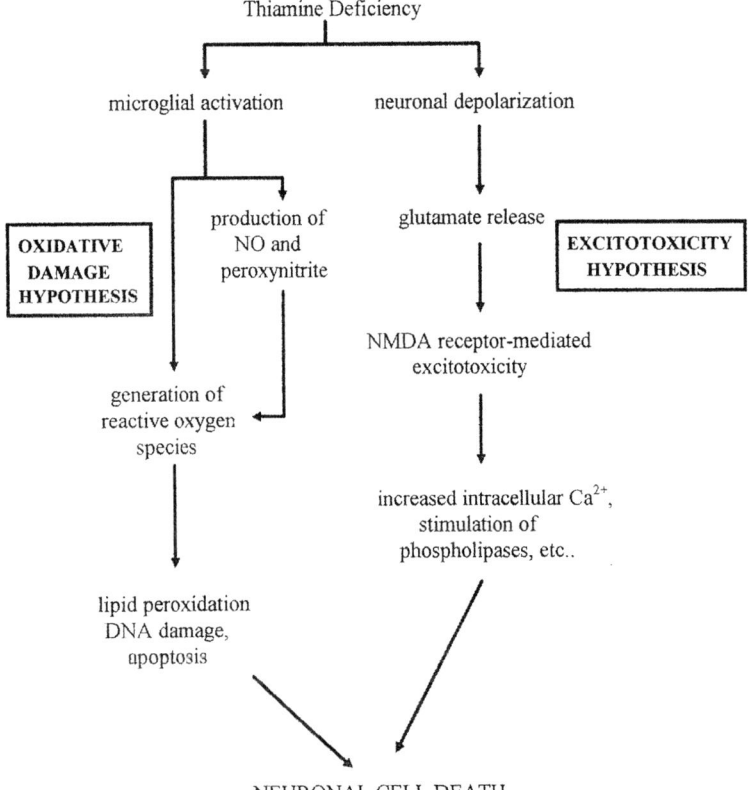

FIGURE 1. Schematic representation of the oxidative and excitotoxicity hypotheses of neuronal cell death in thiamine deficiency.

NMDA RECEPTOR–MEDIATED EXCITOTOXICITY

Armstrong-James *et al.*[24] first drew attention to the marked similarity in the neuropathological characteristics of the thalamic and olivary lesions in thiamine deficiency to those observed following intrathalamic administration of glutamate. This report prompted us to initiate a series of studies to evaluate the role of glutamate excitotoxicity as a possible cause of neuronal cell death in thiamine deficiency. In order to confirm that excitotoxic mechanisms are implicated in neuronal cell death, it is necessary to demonstrate:

(a) increased extracellular concentrations of glutamate in thalamus and inferior olive (regions that ultimately manifest the neuropathological characteristics of excitotoxic brain damage);

(b) a significant reduction in neuronal cell death in these regions following the administration of appropriate glutamate receptor antagonists.

Substantial progress has been made in these investigations, and initial reports have been published.[25] Using the technique of *in vivo* cerebral microdialysis, we demonstrated increased extracellular concentrations of glutamate in thalamic nuclei but not in spared brain structures such as frontoparietal cortex. MK801, a selective antagonist of the N-methyl-D-aspartate (NMDA) subclass of glutamate receptors, was shown to be neuroprotective in thiamine-deficient animals. These effects were noted at advanced stages of thiamine deficiency and were, to some extent, confounded by the appearance of convulsions, as was the case in an earlier study by Langlais and Mair.[26] We concluded that NMDA receptor–mediated excitotoxicity is a late-stage event in this disorder.

EARLY MICROGLIAL ACTIVATION IN BRAIN IN THIAMINE DEFICIENCY

We recently demonstrated that microglial activation (increased immunolabeling with antibodies against ED1, a selective macrophage/microglial marker) is the initial cellular response to thiamine deficiency. Microglial activation occurs by day 8 of the thiamine deficiency treatment paradigm (FIG. 2), some two days ahead of blood-brain barrier breakdown and neuronal cell loss.[27] Similar findings of early microglial activation have been reported in vulnerable hippocampal regions in experimental cerebral ischemia.[28]

This early microglial activation in vulnerable brain structures in thiamine deficiency has the potential to be an important trigger in the cascade of cell death mechanisms since microglia produce neurotoxic substances, including reactive oxygen species (ROS) and nitric oxide (NO).[29–31] NO, in turn, may promote oxidative damage via the production of peroxynitrite, a toxic oxidant anion.[32]

OXIDATIVE DAMAGE DUE TO THIAMINE DEFICIENCY

Recently, free radicals have been reported to be involved in the brain damage associated with WKS. Increased reactive oxygen species were reported in the thalamus

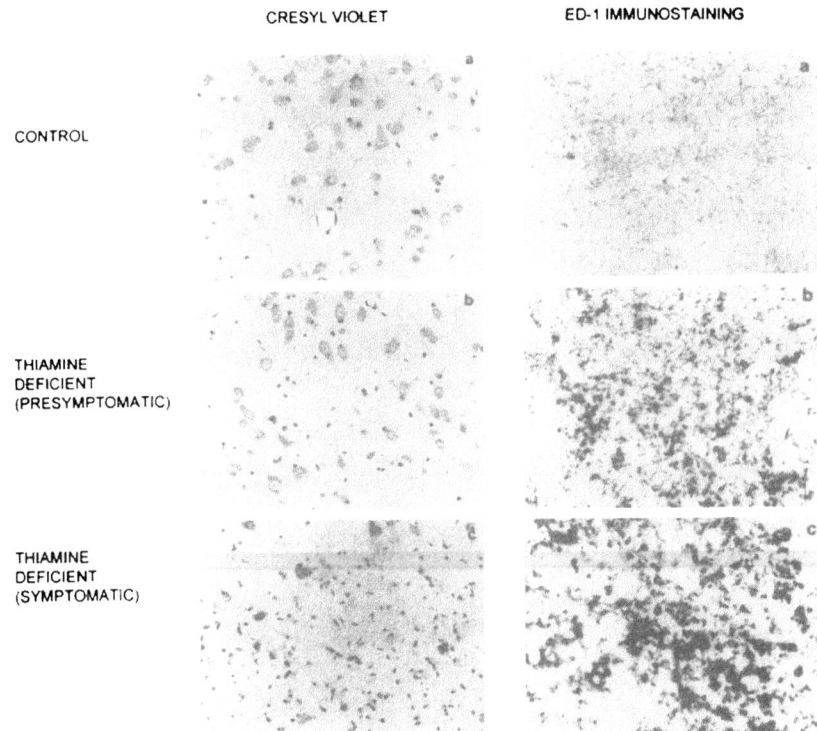

FIGURE 2. Increased ED1 immunolabeling indicative of microglial activation occurs selectively in brain structures that ultimately manifest neuronal cell loss and is evident by day 8 of thiamine deficiency, prior to breakdown of the blood-brain barrier and neuronal cell death. (Modified from Todd & Butterworth.[27])

of pyrithiamine-treated rats.[33] The source of increased free radical production in thiamine deficiency remains unknown, but possible sources include increased extracellular glutamate levels and NMDA receptor activation, disturbances in the blood-brain barrier, and proliferation of microglia and astrocytes. Recent results from our laboratories showed early (prior to neuronal cell loss) increased immunohistochemical evidence of microglial activation in thiamine-deficient rats.[34] We also reported increased immunolabeling of superoxide dismutase during the same time period as the microglial activation,[35] and reduced thiamine deficiency–induced neuronal cell loss after administration of the radical scavenging drug l-deprenyl.[23] These studies have now been extended, and our results showed increased expression of mRNA for endothelial nitric oxide synthase, the enzyme responsible for generation of the NO free radical.[36] Taken together, these data support the hypothesis that free radical species including the superoxide anion and NO play a role in thiamine-induced brain damage.

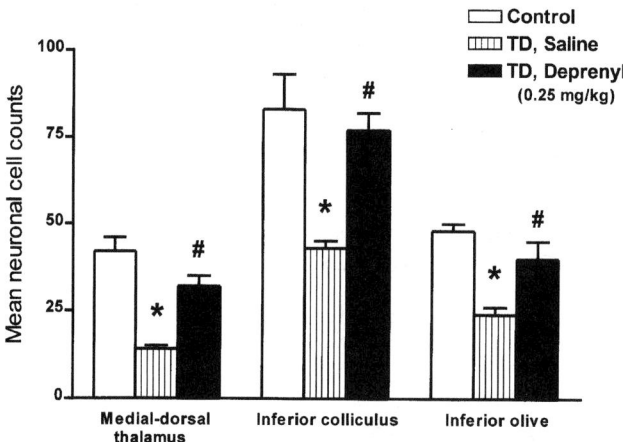

FIGURE 3. Graphic representation of the neuroprotection offered by the MAO inhibitor and free radical scavenger l-deprenyl (0.25 mg/kg/day). Significant cell survival was observed in thiamine-deficient animals treated with l-deprenyl as compared to the vehicle-treated thiamine-deficient animals *, significantly different from control (non–thiamine-deficient animals), $p < 0.05$; #, significantly different from saline-treated thiamine-deficient animals, $p < 0.05$.

SUMMARY

Multiple mechanisms contribute to the selective brain lesions observed in WKS and experimental thiamine deficiency. Recent evidence of early microglial activation and increased free radical production suggest that oxidative stress processes play an important early role in the brain damage associated with thiamine deficiency.

REFERENCES

1. VICTOR, M. et al. 1989. The Wernicke-Korsakoff Syndrome and Related Neurologic Disorders Due to Alcoholism and Malnutrition, 2nd edit. Davies. Philadelphia, PA.
2. BUTTERWORTh, R.F. et al. 1991. Thiamine deficiency and Wernicke's encephalopathy in AIDS. Metab. Brain Dis. **6:** 207–212.
3. THOMPSON, S.D. et al. 1970 GABA-transaminase and glutamic acid decarboxylase changes in the brain of rats treated with pyrithiamine. Neurochem. Res. **10:** 1653–1660.
4. BUTTERWORTH, R.F. 1993. Drug and Alcohol Rev. **12:** 315
5. KRIL, J.J. et al. 1997.The cerebral cortex is damaged in chronic alcoholics. Neuroscience **79:** 983–998.
6. D'AMOUR, M.L. et al. 1994. Pathogenesis of alcoholic peripheral neuropathy: direct effect of ethanol or nutritional deficit? Metab. Brain Dis. **9:** 23–27.
7. HARPER, C.G. et al. 1997. Nutritional and metabolic disorders. In Greenfield's Neuropathology, 6th edit.: 601–655. Arnold. London.
8. TRONCOSO, J.C. et al. 1981. Model of Wernicke's encephalopathy. Arch. Neurol. **38:** 350–354.

9. HEROUX, M. *et al.* 1992 Neuromethods, Vol. 22: Animal Models of Neurological Disease II : 95–131. Humana Press. Clifton, NJ.
10. HEROUX, M. *et al.* 1995. Regional alterations of thiamine phosphate esters and of thiamine diphosphate–dependent enzymes in relation to function in experimental Wernicke's encephalopathy. Neurochem. Res. **20:** 87–93.
11. LEONG, D. *et al.* 1996. Neuronal cell death in Wernicke's encephalopathy: pathophysiologic mechanisms and implications for PET imaging. Metab. Brain Dis. **11:** 71–79.
12. HAKIM, A.M. 1984. The induction and reversibility of cerebral acidosis in thiamine deficiency. Ann. Neurol. **16:** 673–679.
13. BUTTERWORTH, R.F. *et al.* 1989. Effect of pyrithiamine treatment and subsequent thiamine rehabilitation on regional cerebral amino acids and thiamine-dependent enzymes. J. Neurochem. **52:** 1079–1084.
14. BUTTERWORTH, R.F. *et al.* 1986. Activities of thiamine dependent enzymes in two experimental models of thiamine-deficiency encephalopathy II: α-KGDH. Neurochem. Res. **11:** 567–577.
15. ELNAGEH, K.M. *et al.* 1988. Effect of a deficiency of thiamine on brain pyruvate dehydrogenase: enzyme assay by three different methods. J. Neurochem. **51:** 1482–1489.
16. BUTTERWORTH, R.F. *et al.* 1993. Thiamine dependent enzyme changes in the brains of alcoholics: relationship to the Wernicke-Korsakoff Syndrome. Alcohol Clin. Exp. Res. **17:** 1084–1088.
17. RINDI, G. *et al.* 1980. Thiamine content and turnover rates of some rat nervous regions using labelled thiamine as tracer. Brain Res. **181:** 369–380.
18. HAWKINS, R.A. *et al.* 1985. Oxidative metabolism. Am. J. Physiol. **248:** C170.
19. BETTENDORFF, L. *et al.* 1995. Thiamine deficiency–induced partial necrosis and mitochondrial uncoupling in neuroblastoma cells are rapidly reversed by addition of thiamine. J. Neurochem. **65:** 2178–2184.
20. AIKAWA, H. *et al.* 1984. Low energy levels in thiamine-deficient encephalopathy. J. Neuropath. Exp. Neurol. **43:** 276–287.
21. TODD, K.G. *et al.* 1997. Immunohistochemical evidence that superoxide dismutase is upregulated in experimental thiamine deficiency [abstr.]. Soc. Neurosci. Abstr. **23:** S69
22. TODD, K.G. *et al.* 1997. Microglial activation: the initial cellular response in experimental thiamine deficiency [abstr.]. J. Neurochem. **70:** S64.
23. TODD, K.G. *et al.* 1998. Increased neuronal cell survival after l-deprenyl treatment in experimental thiamine deficiency. J. Neurosci. Res. **52:** 240–246.
24. ARMSTRONG-JAMES, M. *et al.*, 1988. Low energy levels in thiamine-deficient encephalopathy. J. Neuropathol. Exp. Neurol. **43:** 276–287.
25. TODD, K.G. *et al.* 1998. Evaluation of the role of NMDA-mediated excitotoxicity in the selective neuronal loss in experimental Wernicke encephalopathy. Exp. Neurol. **149:** 130–138.
26. LANGLAIS, P.J. *et al.* 1990. Protective effects of the glutamate antagonist MK801 on pyrithiamine-induced lesions and amino acid changes in rat brain. J. Neurosci. **10:** 1664–1674.
27. TODD, K.G. *et al.* 1999. Early microglial response in experimental thiamine deficiency: an immunohistochemical analysis. Glia **25:** 190–198.
28. STREIT, W.J. *et al.* 1988. Functional plasticity of microglia: a review. Glia **1:** 301–307.
29. CALINGASAN, N. *et al.* 1999. Induction of nitric oxide sythase and microglial responses precede selective cell death induced by chronic impairment of oxidative metabolism. Am. J. Neuropathol. In press.
30. CHAO, C.C. *et al.* 1992. Activated microglia mediate neuronal cell injury via a nitric oxide mechanism. J. Immunol. **149:** 2736.

31. DAWSON, V.L. *et al.* 1994. Expression of inducible nitric oxide synthase causes delayed neurotoxicity in primary mixed neuronal-glial cortical cultures. Neuropharmacol. **33:** 1425.
32. MERRILL, J. *et al.* 1996. Nitric Oxide, The Role of Glia in Neurotoxicity: 263–281. CRC Press. Boca Raton, FL.
33. LANGLAIS, P.J. *et al.* 1997. Increased cerebral free radical production during thiamine deficiency. Metab. Brain Dis. **12:** 137–143.
34. TODD, K.G. *et al.* 1999. Selective neuronal cell death mechanisms in thiamine deficiency I: evidence for early microglial activation [abstr.]. Collected Abstracts of Oxidative/Energy Metabolism in Neurodegenerative Disorders: P44. New York Academy of Sciences. New York, NY.
35. TODD, K.G. *et al.* 1999. Selective neuronal cell death mechanisms in thiamine deficiency III: evidence for early increased superoxide dismutase (CuZn-SOD) immunoreactivity and late stage glutamate excitotoxicity [abstr.]. Collected Abstracts of Oxidative/Energy Metabolism in Neurodegenerative Disorders: P46. New York Academy of Sciences. New York, NY.
36. DESJARDINS, P. *et al.* 1999. Selective neuronal cell death mechanisms in thiamine deficiency II: expression of nitric oxide synthase isoforms in the thalamus of the thiamine deficient rat [abstr.]. Collected Abstracts of Oxidative/Energy Metabolism in Neurodegenerative Disorders: P45. New York Academy of Sciences. New York, NY.

Inhibition of α-Ketoglutarate Dehydrogenase due to H_2O_2-Induced Oxidative Stress in Nerve Terminals

LASZLO TRETTER AND VERA ADAM-VIZI[a]

Department of Medical Biochemistry, Semmelweis University of Medicine, P.O. Box 262, Budapest H-1444, Hungary

INTRODUCTION

Oxidative stress, which is an important factor in the pathogenesis of various neurodegenerative diseases and ischemia, can be induced by H_2O_2, a membrane-permeable reactive oxygen intermediate. Recently we reported that principal functions, such as maintenance of Na^+ and Ca^{2+} gradients[1,2] or mitochondrial membrane potential[3,4] are impaired in the early stage of H_2O_2-induced oxidative stress in nerve terminals. ATP level and the ATP/ADP ratio are also significantly decreased shortly after exposure to the oxidant.[2] A detailed study on the mitochondrial membrane potential ($\Delta\Psi m$) has shown that $\Delta\Psi m$ is maintained in the presence of the oxidant. For this the contribution of H^+ efflux via the F_0F_1-ATPase is necessary. When oligomycin, an inhibitor of the mitochondrial ATPase, is also present, a significant decrease in $\Delta\Psi m$ is induced by H_2O_2.[3,4] This shows that the oxidant has a target in the mitochondria by which it limits the capacity of the respiratory chain to maintain $\Delta\Psi m$. Given the relative resistance of the respiratory chain components to H_2O_2,[5] a possibility emerges that the oxidant might decrease the amount of reducing equivalents available for the respiratory chain, and this could be responsible for depolarization of mitochondria. In the nerve terminals most of the energy is produced in the process of aerobic oxidation of glucose.[6] Therefore, dehydrogenases in the citric acid cycle play a key role in determining the NADH/NAD ratio. The present study shows that the activity of α-ketoglutarate dehydrogenase, a crucial enzyme in the tricarboxylic acid cycle (TCA cycle), and the steady state NAD(P)H level are decreased under H_2O_2-induced oxidative stress. This could be an important mechanism by which oxidative stress impairs the function of mitochondria.

MATERIALS AND METHODS

Preparation of Synaptosomes. Isolated nerve terminals (synaptosomes) were prepared from brain cortex of guinea pigs as detailed previously.[7] Incubations were done in a medium containing 140 mM NaCl, 3 mM KCl, 2 mM $MgCl_2$, 2 mM $CaCl_2$, 10 mM PIPES (pH 7.38), and 10 mM glucose at 37°C.

---

[a]Corresponding author. Phone: 361-266 2773; fax: 361-267 0031.
e-mail: av@puskin.sote.hu

TABLE 1. Effect of H_2O_2 on citric acid cycle dehydrogenases

	Isocitrate dehydrogenase activity (nmol NADH/mg protein)	α-Ketoglutarate dehydrogenase activity (nmol NADH/mg protein)
Control	59.8 ± 3.6	6.8 ± 0.27
0.1 mM H_2O_2	58.2 ± 4.2	4.08 ± 0.25[a]
0.5 mM H_2O_2	54.6 ± 4.1	3.5 ± 0.28[a]

NOTE: Synaptosomes (0.5 mg/ml protein) were incubated in the presence or absence of H_2O_2. The activity of dehydrogenases was detected by measuring the formation of NADH. Data are mean ± SEM values for four independent determinations.
[a]Significantly different from the control ($p < 0.05$).

Determination of Activity of Isocitrate Dehydrogenase. NAD-specific isocitrate dehydrogenase was assayed spectrophotometrically according to the method described by Plaut.[8]

Determination of Activity of α-Ketoglutarate Dehydrogenase. α-ketoglutarate dehydrogenase was assayed as described by Sanadi.[9]

Steady State NAD(P)H Quantification. The intrasynaptosomal NAD(P)H level was measured fluorimetrically in the dual emission mode in a PTI Deltascan fluorescence spectrophotometer using 344-nm excitation wavelength with emission at 550- and 460-nm wavelengths. Changes in NAD(P)H concentration were quantified using a calibration curve of externally added NADH (1–3 nmol).

Materials. Standard laboratory chemicals were obtained from Sigma.

RESULTS

Steady state NAD(P)H level in synaptosomes was investigated after H_2O_2-evoked oxidative stress by monitoring the fluorescence at 358/460-nm excitation/emission wavelengths, respectively. FIGURE 1/A shows that H_2O_2 applied in different concentrations (10–500 μM) decreased the NAD(P)H level. On administration of rotenone, which blocks the oxidation of NADH by inhibiting the NADH dehydrogenase (complex I) in mitochondria, an increase in NAD(P)H fluorescence was observed (FIG. 1/B). The new eqiuilibrium attained after addition of rotenone reflects the maximum available amount of NAD(P)H in nerve terminals. FIGURE 1 shows that, in addition to decreasing the steady state NAD(P)H level, H_2O_2 also diminished the fluorescence signal induced by rotenone (2 μM). The decrease in the rotenone-induced fluorescence signal was proportional to the concentration of the oxidant. In order to examine whether decrease in the NAD(P)H level could be related to inhibition of key dehydrogenases in the TCA-cycle, the activities of isocitrate dehydrogenase and α-ketoglutarate dehydrogenase were measured after exposure of nerve terminals to H_2O_2. Data in TABLE 1 show that the activity of NAD^+-dependent isocitrate dehydrogenase (ICDH) was unchanged even after addition of 0.5 mM H_2O_2. The activity of α-ketoglutarate dehydrogenase was significantly inhibited in the presence of 0.1 or 0.5 mM H_2O_2.

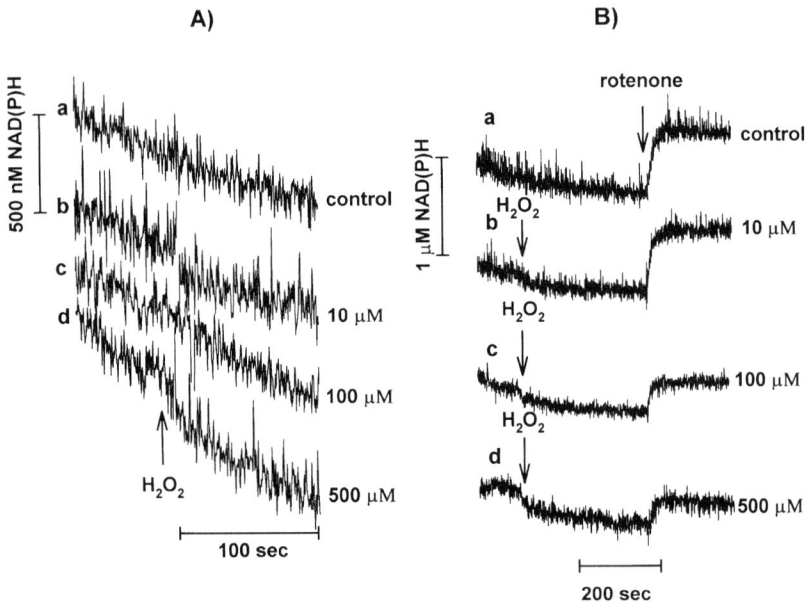

FIGURE 1. H_2O_2-induced change in the steady state NAD(P)H level in synaptosomes. **(A)** H_2O_2 was added at different concentrations (10, 100, and 500 µM) to synaptosomes (0.5 mg protein/ml), and NAD(P)H fluorescence was followed as described in MATERIALS AND METHODS (b, c, d). **(B)** Rotenone (2 µM) was added 300 seconds after challenging synaptosomes with various concentrations of H_2O_2 (10, 100, and 500 µM.) In controls recordings were made in the absence of H_2O_2. *Traces* are representative of four independent determinations.

DISCUSSION

NAD(P)H/NAD(P) ratio is one of the most important indicators of intracellular redox conditions. H_2O_2 in isolated nerve endings energized by glucose, induced a decrease in the steady state NAD(P)H level (FIG. 1A). While monitoring the fluorescence at 344 nm no distinction could be made between changes in NADH or NADPH concentrations. Taking into account that the NADH/NAD(P)H ratio in the brain is 10:1,[10] fluorescence changes most likely indicate changes in the NADH level. Rotenone, by eliminating the oxidation of NADH by complex I, shifted the NADH/NAD^+ equilibrium to a maximally reduced state (FIG. 1B, control curve), and previous application of H_2O_2 concentration dependently decreased the rotenone-induced signal (Fig 1B, traces b, c, d). This might be an indication of decreased NADH production in the TCA-cycle under H_2O_2-induced oxidative stress. We found that α-ketoglutarate dehydrogenase, a key enzyme in the TCA cycle, is sensitive to oxidative stress; a significant inhibition of this enzyme was induced by the oxidant at 0.1 mM concentration. In contrast, the activity of NAD^+-dependent isocitrate dehydrogenase was unaltered by H_2O_2. It has been reported that aconitase could also be inhibited by reactive oxygen species.[11] The question arises whether inhibition of α-ketoglutarate dehydrogenase could be responsible for the decreased NADH level observed during H_2O_2-induced oxidative stress. This enzyme appears to play a crucial role in providing NADH for complex I and determining the rate of respiration.[12,13] It appears likely that H_2O_2, by inhibiting α-ketoglutarate dehydrogenase, limits the amount of NADH available for complex I, giving rise to an impaired respiratory capacity. This could be the underlying mechanism for the oxidant-induced mitochondrial depolarization.[3,4] The importance of α-ketoglutarate dehydrogenase is reinforced by the finding[14] that the TCA cycle could be fueled by glutamate via transamination to α-ketoglutarate. This could be a critical process, especially when normal metabolic substrates are not available in sufficient amounts to fuel the TCA cycle.

In summary, it is suggested that under H_2O_2-induced oxidative stress the activity of α-ketoglutarate dehydrogenase in the TCA cycle is inhibited, resulting in decreased production of NADH. This could be critical in limiting the capacity of the respiratory chain and the maintenance of mitochondrial membrane potential.

ACKNOWLEDGMENT

Thanks are expressed to Ms. K. Takacs and K. Zolde for excellent technical assitstance. This work was supported by grants (OTKA, ETT, MKM, and MTA) to V. A.-V.

REFERENCES

1. TRETTER, L. & V. ADAM-VIZI. 1996. Early events in free radical–mediated damage of isolated nerve terminals: Effect of peroxides on membrane potential and intracellular Na^+ and Ca^{2+} concentrations. J. Neurochem. **66:** 2057–2066.
2. TRETTER L., C. CHINOPOULOS & V. ADAM-VIZI. 1997. Enhanced depolarization-evoked calcium signal and reduced ATP/ADP ratio are unrelated events induced by oxidative stress in synaptosomes. J. Neurochem. **69:** 2529–2537.
3. CHINOPOULOS, C., L. TRETTER & V. ADAM-VIZI. 1999. Depolarization of in situ mitochondria due to hydrogen peroxide–induced oxidative stress in nerve terminals: inhibition of α-ketoglutarate dehydrogenase. J. Neurochem. **73:** 220–228.
4. CHINOPOULOS, C. & V. ADAM-VIZI. 1999. Depolarization of in situ mitochondria by hydrogen peroxide in nerve terminals. Ann. N.Y. Acad. Sci. This volume.
5. ZHANG, Y., O. MARCILLAT, C. GIULIVI, L. ERNSTER & K.J.A. DAVIES. 1990. The oxidative inactivation of mitochondrial electron transport chain components and ATPase. J. Biol. Chem. **265:** 16330–16336.
6. ERECINSKA, M., D. NELSON, J. DEAS & I.A. SILVER. 1996. Limitation of glycolysis by hexokinase in rat brain synaptosomes during intense ion pumping. Brain Res. **726:** 153–159.
7. ADAM-VIZI, V. & E. LIGETI. 1986. Calcium uptake of synaptosomes as a function of membrane potential under different depolarizing conditions. J. Physiol. (Lond.) **372:** 363–377.
8. PLAUT, G.W.E. 1969. Isocitrate dehydrogenase from bovine heart. *In* Methods in Enzymology, Vol. XIII. J.M. Lowenstein, Ed.: 34–42. Academic Press. NewYork.
9. Sanadi, D.R. 1969. α-Ketoglutarate dehydrogenase from pig heart. *In* Methods in Enzymology, Vol. XIII. J.M. Lowenstein, Ed.: 52–55. Academic Press. New York.
10. SIES, H. 1982. Nicotinamide nucleotide compartmentation. *In* Metabolic Compartmentation. H. Sies, Ed.: 205–231. Academic Press. London.

11. PATEL M., B.J. DAY, J.D. CRAP, I. FRIDOVICH & J.O. MCNAMARA. 1996. Requirement for superoxide in excitotoxic cell death. Neuron. **16:** 345–355.
12. COONEY, G.J., H. TAEGTMEYER and E.A. NEWSHOLME. 1981. Tricarboxylic acid cycle flux and enzyme activities in the isolated working rat heart. Biochem. J. **200:** 701–703.
13. MORENO-SANCHEZ, R., B.A. HOGUE & R.A. HANSFORD. 1990. Influence of NAD-linked dehydrogenase activity on flux through oxidative phosphorylation. Biochem. J. **268:** 421–428.
14. YUDKOFF, M., D. NELSON, Y. DAIKHIN & M. ERECINSKA. 1994. Tricarboxylic acid cycle in rat brain synaptosomes. J. Biol. Chem. **269:** 27414–27420.

ATPases of Synaptic Plasma Membranes and Vesicles from Rat Cerebral Cortex during Aging and Hypoxia

R. F. VILLA,[a] A. D'ANGELO, AND A. GORINI

Laboratory of Neurochemistry and Molecular Medicine, Division of Pharmacology and Pharmacological Biotechnologies, Department of Physiological-Pharmacological Cellular-Molecular Sciences, University of Pavia, Piazza Botta, 11-1, 27100 Pavia, Italy

The enzymatic systems of ATPases located in synaptosomes, synaptic plasma membranes, and synaptic vesicles play a key role in the regulation of presynaptic nerve ending homeostasis and postsynaptic activities. Na^+, K^+-ATPase is involved in neurotransmitter turnover and in postsynaptic activities, while low- and high-affinity Ca^{++}-ATPase and Ca^{++}, Mg^{++}-ATPase of synaptic plasma membranes and intrasynaptic organelles pump Ca^{++} ions into synaptic structures or out of nerve endings, to maintain synaptic Ca^{++} homeostasis; Mg^{++}-ATPase of synaptic plasma membranes is an ectoenzyme involved in the hydrolysis of ATP to adenosine, while the enzyme located on synaptic vesicles modulates the turnover of neurotransmitters.

The adaptation to aging, to normoxia-hypoxia cycles for 4 weeks (12 h normoxia and 12 h of "mild" and "severe" hypoxia, N_2O_2 90:10 and 91.5:8.5, respectively), and the relationships between each individual ATPase linked to energy-utilizing systems of functional areas have been studied during aging on synaptosomes, synaptic plasma membranes, and synaptic vesicles of frontal cerebral cortex of 3-, 12-, and 24-month-old rats; the maximum rate (V_{max}) of Na^+, K^+-ATPase, Mg^{++}-ATPase, Ca^{++}, Mg^{++}-ATPase, and low- and high-affinity Ca^{++}-ATPase were evaluated.

Purified synaptosomes (SY), synaptic plasma membranes (SPM), and synaptic vesicles (SV) were obtained from rat frontal cerebral cortex according to Cotman and Matthews[1] as modified by Gurd *et al.*[2] and as described in details by Gorini *et al.*[3] Briefly, the rats, aged 3, 12, 24, months were anesthetized and sacrificed by ether; the frontal cerebral cortex was isolated, and the homogenate was diluted to 7-10% (w/v) for differential centrifugation.[1–3] The crude nuclear fraction was removed at 3.0×10^3 $g \times$ min (2×10^7 $\omega^2 t$), the combined supernatants were centrifuged at 132×10^3 $g \times$ min (94×10^7 $\omega^2 t$); the crude mitochondrial pellet containing synaptosomes was applied to a two-step discontinuous Ficoll-sucrose gradient 7.5 and 13% (w/v). After centrifugation in a Sorvall Ultracentrifuge OTD65B, rotor AH-650, at 283×10^4 $g \times$ min (213×10^8 $\omega^2 t$) the synaptosomal fraction was obtained at the interface of the 7.5–13% Ficoll-sucrose layer, and an aliquot was used for enzyme assays. The band of synaptosomes was collected and centrifuged at 268×10^4 $g \times$ min (189×10^8 $\omega^2 t$). The synaptosomal pellet was osmotically lysed in 5 volumes of 6 mM Tris-HCl, pH 8.1, for 1.5 hours. After osmotic shock, the lysate was

---

[a]Corresponding author. Fax: 39-0382-506391.
e-mail: rfvilla@ipv36.unipv.it

centrifuged at 94.3×10^4 g × min (666×10^7 ω^2t), the pellet in 0.32 M sucrose was applied on a discontinuous sucrose gradient of 25, 32.5, 35, and 38% sucrose (w/w), pH 7.4, and centrifuged at 566×10^4 g × min (400×10^8 ω^2t) in a rotor AH-650; the bands at the interface of 10–25% sucrose (synaptic vesicles) and of 25–32.5% (synaptic plasma membranes) were collected by aspiration, diluted, and centrifuged at 26×10^4 g × min (189×10^8 ω^2t). The pellets of synaptic vesicles and synaptic plasma membranes were resuspended in a small volume of 0.32 M sucrose (pH 7.4) for the assay of activity of enzymes.

The purity of the different subcellular fractions was previously determined,[3] and ATPases' activities were determined by measuring the Pi released from the hydrolysis of ATP in the presence of different ions and enzymes; ATPases' activities were expressed as μmoles of Pi relased $\times h^{-1} \times$ (mg of protein)$^{-1}$ of subcellular fraction tested.[4–8] The protein concentration was determined according to Lowry et al.[9]; the ANOVA test for multiple comparisons evaluated the interactions between the animals during aging and hypoxia for each cellular subfraction and for each individual enzyme activity, and post hoc tests were used to control statistical evaluations by Tukey's and Dunnett's tests.

The catalytic properties of these enzyme systems of frontal cerebral cortex in normal 3-, 12-, and 24-month-old animals during the steady state markedly differ according to their subneuronal localization—i.e., on synaptosomes, synaptic plasma membranes, or synaptic vesicles. The same enzyme may be located on different subneuronal structures and thus the metabolic role of each individual ATPase is determined by its subcellular *in vivo* localization.

Results about aging indicate that the enzyme catalytic activity of ATPase tends to decrease (FIG. 1), suggesting the hypothesis that the modifications of ATPase undergone by adaptation to aging exert important physiopathological modifications of responsiveness of the nerve endings and that these modifications of enzyme activities

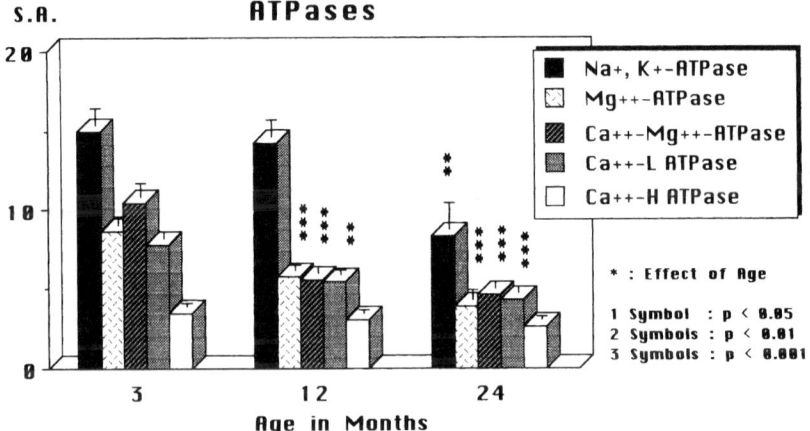

FIGURE 1. Effect of age on specific activities of indicated ATPases in synaptic plasma membranes from frontal cerebral cortex.

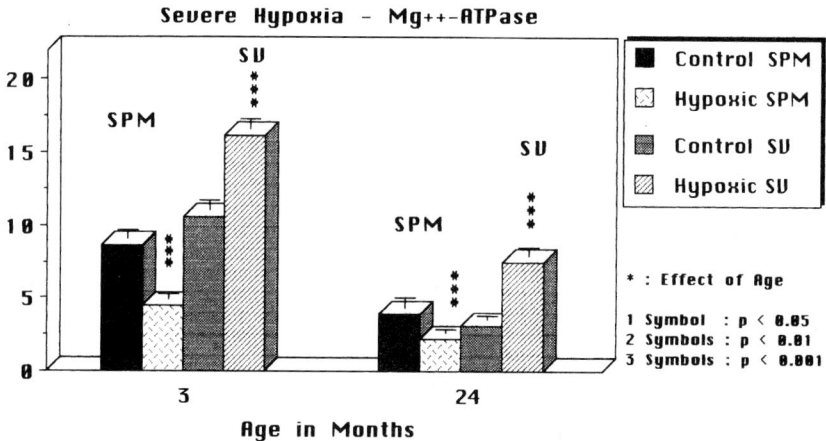

FIGURE 2. Effect of severe hypoxia on specific activity of Mg^{++}-ATPase during aging in synaptic plasma membranes (SPM) and synaptic vesicles (SV) from frontal cerebral cortex.

reflect the bioenergetic state of the cerebral tissue with respect to the energy demand at each single age. Mild and severe hypoxia cycles decreased catalytic activities of all ATPases similarly, except of Mg^{++}-ATPase: after mild hypoxia Mg^{++}-ATPase activity on synaptic plasma membranes increased in 3-, 12-, and 24-month-old rats; severe hypoxia-normoxia cycles on Mg^{++}-ATPase showed (FIG. 2) opposite changes on synaptic plasma membranes (decrease of activity) or in synaptic vesicles (increase of activity). Na^+, K^+-ATPase, Mg^{++}-ATPase, Ca^{++}, Mg^{++}-ATPase, and low- and high-affinity Ca^{++}-ATPase decreased during aging and hypoxia on synaptic plasma membranes. Cerebral concentration and content of synaptic plasma membrane proteins was increased by aging, and the results may be influenced by this observation in that many defective noncatalytic proteins may be formed during aging, and hypoxia affected specific proteins, as previously demonstrated.[10]

Our findings confirm the distinct catalytic properties of these ATPases, which appear to be not a subtle, unclear problem of microheterogeneity, but a possible relevant factor in aging processes in both the physiological and pathological brain. These data indicate also the adaptations of ATPase systems and their plasticity to aging and hypoxia, although the enzyme's subsynaptosomal localization may play an important role. This microheterogeneity should be considered in the evaluation of pharmacodynamic effects of drugs at subneuronal levels.

REFERENCES

1. COTMAN, C.S. & D.A. MATTHEWS. 1971. Synaptic plasma membranes from rat brain synaptosomes. Biochim. Biophys. Acta **249:** 416–423.
2. GURD, J.W. *et al.* 1974. Isolation and partial characterization of rat brain synaptic plasma membranes. J. Neurochem. **22:** 281–290.
3. GORINI, A., B. GHIGINI & R.F. VILLA. 1996. Acetylcholinesterase activity of synaptic plasma membranes during ageing: effect of L-acetylcarnitine. Dementia **7:** 147–154.

4. LIN, S.C. & E.L. WAY. 1982. Calcium-activated ATPases in presynaptic nerve endings. J. Neurochem. **39:** 1641–1651.
5. LIN, S.C. & E.L. WAY. 1982. A high affinity Ca^{2+}-ATPase in enriched nerve-ending plasma membranes. Brain Res. 235: 387–392.
6. MICHAELIS, E.K. et al. 1983. High affinity Ca^{2+}-stimulated Mg^{2+}-dependent ATPase in rat brain synaptosomes, synaptic membranes and microsomes. J. Biol. Chem. **258:** 6101–6108.
7. PALAYOOR, S.T., T.N. SEYFRIED & D.T. BERNARD. 1986. Calcium ATPase activities in synaptic plasma membranes of seizure-prone mice. J. Neurochem. **46:** 1370–1375.
8. SHALLOM, J.M. & S.S. KATYARE. 1985. Altered synaptosomal ATPase activity in rat brain following prolonged in vivo treatment with nicotine. Biochem. Pharmacol. **34:** 3445–3449.
9. LOWRY, O.H. et al. 1951. Protein measurement with the Folin phenol reagent. J. Biol. Chem. **193:** 265–275.
10. VILLA, R.F. et al. 1991. Effect of hypoxia on protein composition of synaptic plasma membranes from cerebral cortex during ageing. Neurochem. Res. **16:** 827–832.

Mitochondrial Compartmentation at the Cellular Level: Astrocytes and Neurons

HELLE S. WAAGEPETERSEN,[a] URSULA SONNEWALD,[b] HONG QU,[b] AND ARNE SCHOUSBOE[a,c]

[a]*PharmaBiotec Research Center, Department of Pharmacology, Royal Danish School of Pharmacy, DK-2100 Copenhagen, Denmark*

[b]*Department of Pharmacology and Toxicology, Norwegian University of Science and Technology (NTNU), N-7489 Trondheim, Norway*

INTRODUCTION

Vulnerability to excitotoxic stimuli is well documented, and certain subsets of cells are more affected than others (see contributions from authors in this volume). It is accepted that heterogeneity on the cellular level is an important factor in the analysis of neurodegeneration. Compartmentation within a cell might also play an important role. The concept of metabolic compartmentation in the brain emerged from early studies of the product precursor relationship of glutamine and glutamate using ^{14}C-labeled substrates for brain metabolism.[1] The existence of glutamate pools in neurons and glia having different turnover rates was firmly established. The use of ^{13}C-labeled substrates and MRS for metabolic analysis of monotypic neurons and astrocytes in culture has expanded the concept of compartmentation. In the individual cell types heterogeneity of mitochondrial TCA cycle metabolism exists, reflected in different pools of glutamate with different turnover rates.[2] The present study was performed to investigate whether thiopental interferes with glutamate metabolism in astrocytes. In neurons, exposure to a depolarizing concentration of K$^+$ was used to assess the stimulating effect of depolarization on neuronal TCA cycle metabolism monitored by ^{13}C labeling in glutamate.

RESULTS AND DISCUSSION

Cultured astrocytes were incubated with [U-^{13}C]glutamate in the absence or presence of thiopental (1 mM). The amounts of [U-^{13}C]glutamine, [U-^{13}C]aspartate and [U-^{13}C]lactate, synthesized from [U-^{13}C]glutamate were decreased in the presence of thiopental (FIG. 1A). Surprisingly, the amounts of [1,2,3-^{13}C]glutamate and [3,4-^{13}C]aspartate originating from TCA cycle intermediates leaving the cycle after the second turn were unchanged and [1,2,3-^{13}C]glutamine was only slightly decreased

---

[c]Address for correspondence: Prof. A. Schousboe, D.Sc., PharmaBiotec Research Center, Department of Pharmacology, Royal Danish School of Pharmacy, 2 Universitetsparken, DK-2100, Copenhagen, Denmark. Phone: +45 3530 63; fax: +45 3530 60.
e-mail: as@mail.dfh.dk

ABBREVIATIONS: GC/MS, gas chromatography/mass spectrometry; MRS, magnetic resonance spectroscopy; TCA, tricarboxylic acid.

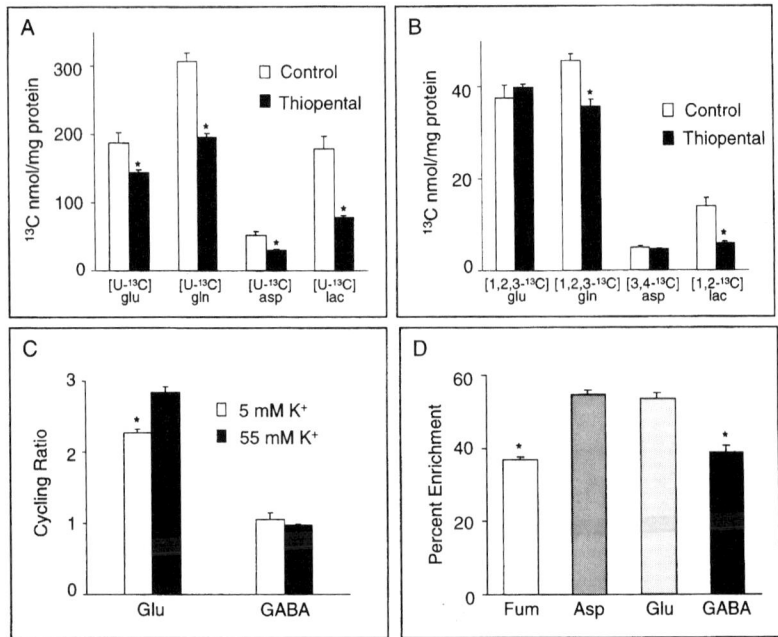

FIGURE 1. Cerebral cortical astrocytes and neurons were cultured as described earlier.[5,6] The astrocytes were incubated for 2 h in a medium containing 0.5 mM [U-^{13}C]glutamate and 3 mM glucose in the absence or presence of 1 mM sodium thiopental. The incubation media and ethanol extracts of the cells were analyzed by MRS as previously described.[7] *Upper panels* show amounts of ^{13}C (nmol/mg protein) of uniformly labeled metabolites synthesized from [U-^{13}C]glutamate (**A**) and metabolites obtained after the second turn of the TCA cycle (**B**) in the absence or presence of thiopental. Glutamate and aspartate were quantified from MR spectra of the cell extracts and glutamine and lactate from the media. Results are averages ± SEM, $n = 6$ for the control group and $n = 4$ for the thiopental group. Statistically significant differences (ANOVA) between the control group and the thiopental group are indicated with an *asterisk* ($p < 0.05$).

The cultured neurons were incubated in 0.5 mM [U-^{13}C]glucose and subsequently repetitively exposed to 5 or 55 mM K$^+$ for 1 h and cell extracts were analyzed by MRS and GC/MS as previously described.[8,9] *Lower panels*: **C** shows cycling ratios of glutamate and GABA, which were obtained from analysis of MR spectra by dividing total amounts of label in C-3 with the amounts of isotopomers giving rise to the 4,5 doublet of glutamate C-4 and 1,2 of GABA C-2, respectively. **D** shows the percent enrichment of ^{13}C in fumarate, aspartate, glutamate and GABA measured by GC/MS. Results are averages ± SEM, $n = 3$. Statistically significant differences (ANOVA) between resting and depolarizing conditions (**C**) are indicated with an *asterisk* ($p < 0.025$). Additionally, statistically significant differences (ANOVA) between percent enrichment of fumarate and aspartate and glutamate and GABA, respectively, (**D**) are indicated with *asterisks* ($p < 0.001$). Fum, fumarate; Asp, aspartate; Glu, glutamate. Asp, aspartate; Glu, glutamate; Gln, glutamine; Lac, lactate.

Original data used for **A** and **B** are from Qu *et al.* (1999) and those used for **C** and **D** are from Waagepetersen *et al.* (1999; 2000).

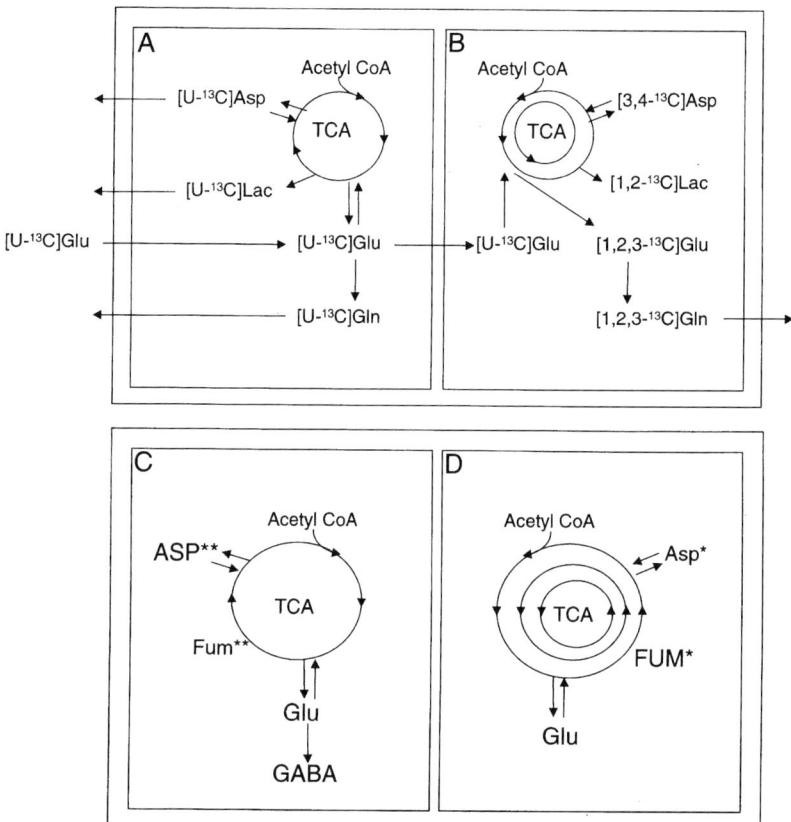

FIGURE 2. A model illustrating mitochondrial heterogeneity in astrocytes (**A, B**) and neurons (**C, D**). In astrocytes a mitochondrion (**A**) in a milieu of a high glutamate concentration has a low TCA cycle turnover, and another mitochondrion (**B**) with a higher turnover has a lower glutamate concentration. In neurons mitochondrial heterogeneity is illustrated by one mitochondrion (**C**) with a low turnover, a highly labeled large aspartate pool, and a close association with synthesis of transmitter GABA; and a mitochondrion (**D**) with a slightly labeled large fumarate pool and a higher TCA cycle turnover, which is affected by depolarization. TCA, tricarboxylic acid; Fum, fumarate; Asp, aspartate; Glu, glutamate; Gln, glutamine; Lac, lactate.

(FIG. 1B). Such variations indicate cellular compartmentation, possibly reflecting heterogeneity of the intracellular glutamate concentration, which may affect TCA cycle turnover rates differently in mitochondria located in different parts of the cell (FIG. 2A,B). It is conceivable that mitochondria close to the cell wall (FIG. 2A) have a large capability to sequester glutamate thereby removing it from possible release *via* reversal of uptake and representing a buffering capacity. Mitochondria in regions distant from the cell wall (FIG. 2B) and surrounding neurons responsible for glutamate release, might experience a constant, low glutamate concentration. Since transamination is dependent on substrate concentrations it is apparent that a high glutamate concentration will lead to rapid exit from the TCA cycle, and a low glutamate concentration will lead to repeated cycling of intermediates.

Cultured cerebral cortical neurons (GABAergic) were incubated with [U-^{13}C]glucose and subsequently repetitively depolarized by 55 mM K$^+$. A cycling ratio of glutamate which is indicative of TCA cycle activity was increased when the neurons were exposed to depolarizing conditions. This was not observed with regard to neurotransmitter GABA (FIG. 1C). This strongly suggests compartmentation of TCA cycle metabolism in cerebral cortical neurons being reflected by different glutamate pools one of which is used for GABA synthesis (FIG. 2C) and another which is affected by depolarization (FIG. 2D). The percent enrichment in fumarate was much lower than that in aspartate, which is metabolically linked to fumarate *via* oxaloacetate (FIG. 1D). This can only be explained by the existence of multiple TCA cycles with different turnover rates (FIG. 2C,D). Those with low turnover are quantitatively the most important for labeling of aspartate (FIG. 2C), whereas those with high turnover have the largest fumarate pool (FIG. 2D). Neuronal TCA cycle activity has previously been shown to be associated with multiple compartments exhibiting differences in cycling ratios.[2-4]

Mitochondrial heterogeneity can be caused by various conditions such as different enzyme compositions or substrate concentrations. It could in the present study be demonstrated that different glutamate concentrations lead to heterogeneous mitochondrial metabolism in astrocytes, and mitochondria might have different functions in distinct regions of a cell. In neurons compartmentation of amino acid synthesis as well as energy production is likely to exist.

ACKNOWLEDGMENTS

The expert technical assistance by Ms. Lone Petersen is cordially acknowledged. This research was supported by the Danish State Biotechnology Programs (1991-95 and 1995-99); the Danish Medical Research Council (9700761); NORMOX NorFa Grant; the Research Council of Norway (SIP 782008.01); the Lundbeck, NOVO Nordisk, Blix and SINTEF Foundations; and the Foundation for Special Medical Applications (RiT).

REFERENCES

1. BERL, S. & D.D. CLARKE. 1983. The metabolic compartmentation concept. *In* Glutamine, Glutamate and GABA in the Central Nervous System. L. Hertz, E. Kvamme, E.G. McGeer & A. Schousboe, Eds.: 205–217. Alan R. Liss, Inc. New York.
2. SONNEWALD, U., L. HERTZ & A. SCHOUSBOE. 1998. Mitochondrial heterogeneity in the brain at the cellular level. J. Cereb. Blood Flow Metab. **18**: 231–237.
3. WAAGEPETERSEN, H.S., I.J. BAKKEN, O.M. LARSSON, U. SONNEWALD & A. SCHOUSBOE. 1998. Comparison of lactate and glucose metabolism in cultured neocortical neurons and astrocytes using ^{13}C NMR spectroscopy. Dev. Neurosci. **20**: 310–320.
4. WAAGEPETERSEN, H.S., I.J. BAKKEN, O.M. LARSSON, U. SONNEWALD & A. SCHOUSBOE. 1998. Metabolism of lactate in cultured GABAergic neurons studied by ^{13}C-NMR spectroscopy. J. Cereb. Blood Flow Metab. **18**: 109–117.

5. HERTZ, E., A.C.H. YU, L. HERTZ, B.H.J. JUURLINK & A. SCHOUSBOE. 1989. Preparation of primary cultures of mouse cortical neurons. *In* Dissection and Tissue Culture Manual for the Central Nervous System. A. Shahar, J. De Vellis, A. Vernadakis & B.A. Haber, Eds.: 183–186. Alan R Liss. New York.
6. HERTZ, L., B.H.J. JUURLINK, E. HERTZ, H. FOSMARK & A. SCHOUSBOE. 1989. Preparation of primary cultures of mouse (rat) astrocytes. *In* Dissection and Tissue Culture Manual for The Central Nervous System. A. Shahar, J. De Vellis, A. Vernadakis, B.A. Haber, Eds.: 105–108. Alan R Liss. New York.
7. QU, H., E. FÆRØ, P. JØRGENSEN, O. DALE, S.E. GISVOLD, G. UNSGÅRD & U. SONNEWALD. 1999. Decreased glutamate transport and metabolism in cultured astrocytes in the presence of thiopental; studied by ^{13}C MR spectroscopy. Biochem. Pharmacol. **58:** 1075–1080.
8. WAAGEPETERSEN, H.S., U. SONNEWALD, O.M. LARSSON & A. SCHOUSBOE. 1999. Synthesis of vesicular GABA from glutamine involves TCA cycle metabolism in neocortical neurons. J. Neurosci. Res. **57:** 342–349.
9. WAAGEPETERSEN, H.S., U. SONNEWALD, O.M. LARSSON & A. SCHOUSBOE. 2000. Compartmentation of TCA cycle metabolism in cultured neocortical neurons revealed by ^{13}C MRS. Neurochem. Int. In press.

Detection of Respiratory Chain Defects in Cultivated Skin Fibroblasts and Skeletal Muscle of Patients with Parkinson's Disease

FALK ROBERT WIEDEMANN,[a,c] KIRSTIN WINKLER,[a] HARTMUT LINS,[a] CLAUS-W. WALLESCH,[a] AND WOLFRAM S. KUNZ[b]

[a]*Department of Neurology, University of Magdeburg, Leipziger Strasse 44, 39120 Magdeburg, Germany*

[b]*Department of Epileptology, University of Bonn, 53105 Bonn, Germany*

Parkinson's disease (PD) is associated with a degeneration of dopaminergic neurons. The cause of this neuronal death is still unknown. Since 1989,[1,2] the hypothesis of mitochondrial respiratory chain dysfunction as a possible basis for the neuronal death has been intensively debated. It is expected that the proposed mitochondrial defect as a possible pathogenic factor for PD should include mitochondria in other cell populations than neurons. Reported data obtained from different extracerebral tissues are controversial.[3,4] Therefore, the direct proof of a suspected generalized mitochondrial dysfunction in PD is still missing. However, in muscle homogenates of patients with mitochondrial myopathies the mitochondrial defect is not always detectable with conventional enzyme assays for the determination of respiratory chain enzyme complexes I + III (NADH:cytochrome *c* reductase), II + III (succinate:cytochrome *c* reductase), and IV (cytochrome *c* oxidase).[5] Therefore, we assessed the applicability of different *in situ* methods for the detection of putative mitochondrial respiratory chain defects in extracerebral tissues of PD patients. The functional behavior of mitochondria in digitonin-permeabilized skin fibroblasts and in saponin-permeabilized skeletal muscle fibers of PD patients was studied applying respiration experiments with different substrates using metabolic control analysis. Furthermore, we determined the redox state of the mitochondrial NAD system measuring the laser-excited autofluorescence of NAD(P)H and fluorescent flavoproteins in permeabilized skeletal muscle fibers.[6]

Fifteen patients with PD (4 female and 11 male, age range 37 to 78 years, mean 58 years) had a skeletal muscle and skin biopsy. The patients had had symptomatic PD for 2 to 13 years. The median UPDRS was 20 (range 4 to 56). All patients had the akinetic-rigid type of the disease. Skeletal muscle samples from diagnostic biopsies of 32 patients with discrete myopathic EMG abnormalities but no biopsy evidence for a manifest myopathy (17 male, 15 female, age range 29 to 72 years) were used as control, and all patients gave written informed consent. The study was approved by the Ethics Committee of Magdeburg University. Bundles of fibers from fresh M. vastus lateralis were used for preparation of saponin-permeabilized fibers.

[c]Address for correspondence: Dr.med. Falk Robert Wiedemann, Universität Magdeburg, Klinik für Neurologie, Leipziger Strasse 44, 39120 Magdeburg, Germany. Phone: +49-391-671-5232; fax: +49-391-671-5216.

e-mail: falk.wiedemann@medizin.uni-magdeburg.de

TABLE 1. Enzyme pattern in skeletal muscle homogenate and maximal respiration of saponin-permeabilized skeletal muscle fibers

	Control Group ($n = 32$)	PD Patients ($n = 15$)
Citrate synthase (CS)	11.6 ± 2.3	17.43 ± 4.8 $p < 0.01$
Cytochrome c oxidase/CS	0.20 ± 0.09	0.18 ± 0.07 NS
NADH: cytochrome c reductase/CS	0.24 ± 0.10	0.17 ± 0.03 $p < 0.05$
Succinate: cytochrome c reductase/CS	0.14 ± 0.05	0.14 ± 0.03 NS
Glutamate + malate respiration	8.22 ± 2.02	9.53 ± 2.78 NS
Pyruvate + malate respiration	8.41 ± 1.90	9.88 ± 2.51 NS
Octanoylcarnitine + malate respiration	5.83 ± 1.84	6.11 ± 1.70 NS
Succinate respiration	11.25 ± 3.10	12.74 ± 2.88 NS
Flux control coefficient of complex I	0.24 ± 0.04	0.34 ± 0.09 $p < 0.01$
Flux control coefficient of complex IV	0.20 ± 0.05	0.28 ± 0.07 $p < 0.01$

NOTE: CS activity in U/g wet weight; respiration in nmol O_2/min/mg dry weight. NS = not significant; n = number of patients.

The saponin treatment was performed by incubation of the fiber bundles in a solution containing 50 µg/ml saponin as described in Reference 5. The skin specimens were taken from the incision during muscle biopsy. Fibroblasts were cultured as described.[7]

As shown in TABLE 1, the elevated activity of citrate synthase points to a markedly increased content of mitochondria in the skeletal muscle of PD patients. Despite the high citrate synthase activity, the oxygen consumption of the permeabilized fibers was only slightly increased. After normalization to citrate synthase activity, the activity of the NADH:cytochrome c reductase in muscle homogenate from PD patients was significant lower ($p < 0.05$) than in control muscle. In contrast, the normalized activities of the respiratory chain complexes II + III and IV did not show a significant decrease. The application of inhibitor titrations and calculation of flux control coefficients from titration curves was previously described for the detection of subtle oxidative phosphorylation defects on saponin-skinned muscle fibers from patients with mitochondrial diseases[5] and amyotrophic lateral sclerosis.[8] To determine the flux control coefficient of mitochondrial respiratory chain complexes I and IV, we used the specific inhibitors amytal and azide, respectively. Working with this highly sensitive method, we found a marked complex I deficiency in the skeletal muscle of 7 PD patients, a combined complex I + IV deficiency in 6 patients, and an

TABLE 2. Investigation of respiratory chain function in digitonin-permeabilized skin fibroblasts

	Maximal Respiration of Permeabilized Skin Fibroblasts		Flux Control Coefficient	
	Glutamate + Malate	Succinate + Rotenone	Complex I	Complex IV
Control group $n = 14$	5.71 ± 2.16	11.2 ± 2.09	0.21 ± 0.05	0.24 ± 0.04
PD patients $n = 14$	5.89 ± 3.00 NS	8.67 ± 3.54 $p < 0.05$	0.28 ± 0.11 $p < 0.05$	0.34 ± 0.07 $p < 0.01$

NOTE: The results are presented as mean ± SD. Significant changes were assessed by Student's t-test. NS = not significant.

isolated complex IV defect in 1 patient. We analyzed the mitochondrial function in saponin-permeabilized muscle fibers using measurements of autofluorescence of NAD(P)H and fluorescent flavoproteins.[6] This method allows the determination of redox state changes of mitochondrial pyridine nucleotides on substrate, ADP, or inhibitor additions. In comparison to the control group, the redox states of the NAD(P)H-system for the substrates glutamate/malate, octanoylcarnitine/malate, and pyruvate/malate were significantly increased ($p < 0.05$) in the muscle of the PD patients (data not shown).

Summarizing, the obtained results support the presence of a respiratory chain defect in skeletal muscle of PD patients. The increased content of mitochondria in PD muscle tissue might be due to adaptational changes. This adaptation presumably causes the problems of the detection of the complex I deficiency in muscle tissue.

To confirm the existence of the mitochondrial defect in another extracerebral tissue we investigated the maximal respiration rate of digitonin-permeabilized skin fibroblasts obtained from the 14 PD patients. As shown in TABLE 2, no difference was detectable in the maximal oxygen uptake rates with the complex I specific substrates glutamate + malate. The maximal respiration with succinate + rotenone was clearly diminished in the fibroblasts of the PD patients. This seems to be inconsistent with the data obtained on skeletal muscle, but it should be considered that the *in vivo* control of respiration by complex IV is very high in skin fibroblasts but low in skeletal muscle, as recently reported by Villani and coworkers.[9] We additionally applied metabolic control analysis for the investigation of mitochondrial function in digitonin-permeabilized skin fibroblasts. The calculation of the flux control coefficients showed a statistically significant increase of both the complex I and the complex IV flux control coefficients in the fibroblasts of PD patients. This is in agreement with the detected mitochondrial defect in PD skeletal muscle. Identical deficiencies of respiratory chain enzymes were detected in both skeletal muscle and cultivated skin fibroblasts from 7 of the 14 PD patients. Heterogeneous distributed defects were detected in skeletal muscle and skin fibroblasts of the 7 other patients. This heterogeneous pattern points to the existence of stochastically distributed respiratory chain defects in extracerebral tissues of patients with PD. In addition, the results obtained with cultivated skin fibroblasts suggest that the deficiencies of mitochondrial respiratory chain enzymes in extracerebral tissues of patients with Parkinson's disease are independent from therapeutic influences.

ACKNOWLEDGMENT

The generous support by a grant from the Deutsche Parkinson-Vereinigung e.V. (Neuss) is gratefully acknowledged.

REFERENCES

1. PARKER, W.D., JR. et al. 1989. Abnormalities of the electron transport chain in idiopathic Parkinson's disease. Ann. Neurol. **26:** 719–723.
2. MIZUNO, Y. et al. 1989. Deficiencies in the complex I subunits of respiratory chain in Parkinson's disease. Biochem. Biophys. Res. Commun. **163:** 1450–1455.
3. SINGER, T. P. et al. 1995. Deficiency of NADH and succinate dehydrogenases in degenerative diseases and myopathies. Biochim. Biophys. Acta **1271:** 211–219.
4. HAAS, R. H. et al.1995. Low platelet mitochondrial complex I and II/III activity in early untreated Parkinson's disease. Ann. Neurol. **37:** 714–722.
5. KUZNETSOV, A.V. et al. 1997. Application of inhibitor titrations for the detection of oxidative phosphorylation defects in saponin-skinned muscle fibers of patients with mitochondrial diseases. Biochim. Biophys. Acta **1360:** 142–150.
6. KUNZ, W.S. et al. 1997. Detection of mitochondrial defects by laser fluorimetry. Mol. Cell. Biochem. **174:** 97–100.
7. KUNZ, D. et al. 1995. Oxygraphic evaluation of mitochondrial function in digitonin-permeabilized mononuclear cells and cultured skin fibroblasts of patients with chronic progressive external ophthalmoplegia. Biochem. Mol. Med. **54:** 105–111.
8. WIEDEMANN, F.R. et al. 1998. Impairment of mitochondrial funktion in skeletal muscle of patients with amyotrophic lateral sclerosis. J. Neurol. Sci. **156:** 65–72.
9. VILLANI, G. et al. 1998. Low reserve of cytochrome c oxidase capacity in vivo in the respiratory chain of a variety of human cell types. J. Biol. Chem. **273:** 31829–3136.

Prevention of Neurodegeneration by a Neuroprotective Radical Scavenger

HIROSHI YASUDA[a]

Research Institute of Life Science, Ishibashi, Tochigi 329-0512, Japan

INTRODUCTION

Active oxygens are believed to contribute to neurodegenerative disorders in the Alzheimer's, Parkinson's, Huntington's, and cerebrovascular diseases.[1] A neuroprotective radical scavenger, nizofenone,[2] is used for reducing ischemic brain infarction due to delayed vasospasm after subarachnoid hemorrhage, improving acute neurologic outcome of the patients.[3,4] This article reports on the antioxidative properties of the drug and the possibility of treatment of neurodegenerative diseases by neuroprotective radical scavengers.

RESULTS

Incubation of brain or liver mitochondria with ascorbic acid and Fe induced a marked formation of malondialdehyde, a lipid peroxidation product, and decrease in the turbidity (FIG. 1), indicating that the mitochondria were disintegrated by active oxygens. Nizofenone at 10 µM or more inhibited the malondialdehyde formation, and complete inhibition was observed at 100 µM. The antioxidative activity of the drug was almost comparable to that of vitamin E, a physiological antioxidant. The anti-oxidative property of nizofenone was further examined using a neuronal cell line N18-RE-105, which is sensitive to glutamate-induced oxidative neuro-toxicity.[5] Incubation of the neurons with 5 mM glutamate for 24 hours resulted in a marked decrease in the survival rate to about 10%. The addition of nizofenone produced an increase in the surviving cells, in a concentration-dependent manner. The glutamate-induced oxidative neurotoxicity was significantly reduced by 3 µM nizofenone, and its neuroprotective activity was also almost comparable to that of vitamin E.

In rat brains subjected to a temporary global ischemia for 15 min by four vessel occlusion, the hippocampus lactate levels were increased to about fourfold during ischemia. Interestingly, renewed blood supply induced a more marked increase in the lactate levels to 2 times higher than that during the ischemic period (FIG. 2). Nizofenone (10 mg/kg, i.p.) completely inhibited the second lactate increase after reperfusion, with a slight attenuation of the primary increase phase during ischemia. The neuroprotective effect of nizofenone was confirmed in the pathological study. Seven days after the transient ischemic event, there was a marked decrease in the

---

[a]Address for correspondence: Hiroshi Yasuda, Ph.D., Research Institute of Life Science, Snow Brand Milk Products Co. Ltd., 519 Shimo-Ishibashi, Ishibashi, Tochigi-ken 329-0512 Japan. Phone: 0285-52-1335; fax: 0285-53-1314.
 e-mail: hiro-yasuda@snowbrand.co.jp

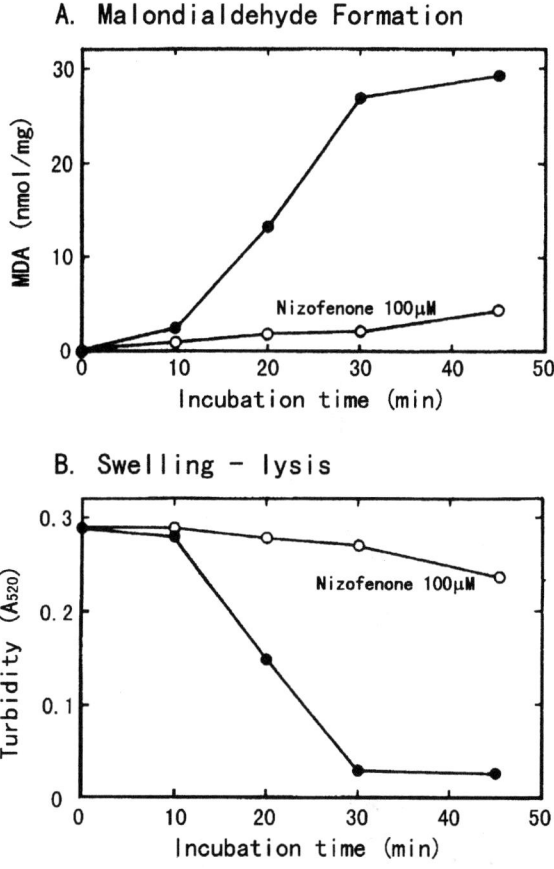

FIGURE 1. Active oxygen-induced disintegration of rat liver mitochondria: (**A**) malondialdehyde formation and (**B**) swelling-lysis of mitochondria. Rat liver mitochondria (0.40 mg protein/ml) were incubated in 25 mM Tris-HCl buffer (pH 7.4), 175 mM KCl, 100 μM ascorbic acid and 20 μM $FeSO_4$.

number of surviving pyramidal neurons to 38% of that found in the sham-operated control group (in the hippocampus CA1 region). In the nizofenone-pretreated group, the survival rate of pyramidal neurons was 95%, indicating that the delayed neuronal cell death was almost completely prevented.

DISCUSSION

The present experiments with isolated brain mitochondria suggest that the organelles are extremely susceptible to active oxygens. Brain ischemia produces a total reduction of mitochondrial electron transport components such as flavin-, heme- and

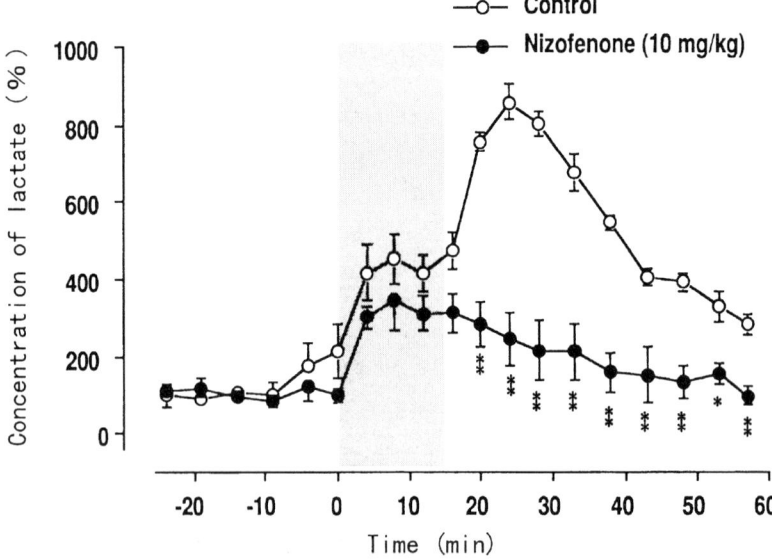

FIGURE 2. Ischemia/reperfusion-induced lactate accumulation in rat hippocampus. Rats were subjected to temporary global ischemia for 15 min induced by four-vessel occlusion. Nizofenone (10 mg/kg i.p.) was administered 30 min before occlusion.

nonheme iron-proteins and ubiquinone, a decreased ability for oxidative phosphorylation, and swelling of mitochondria. These changes are accompanied by a liberation of fatty acids from membranous phospholipids and their accumulation, which in turn cause structural and functional impairment.[6] The impaired mitochondria in a reduced state are liable to react with molecular oxygen at sites other than cytochrome oxidase, leading to the formation of active oxygens such as superoxide or hydrogen peroxide, which act as generators of radical reactions.[7] Mitochondrial superoxide generation is stimulated by Ca or the loss of cytochrome c, one principal member of the electron transport chain and a potent apoptosis inducer, as recently reported.[8] Mitochondria are a primary source of active oxygen generation and also a primary target itself in ischemia/reperfusion injury.

An interesting finding is that hippocampus lactate concentrations increased to higher levels after reperfusion more than during ischemia. This finding means that aerobic energy metabolism in mitochondria (ATP formation due to oxidative phosphorylation) does not recover after reperfusion and so anaerobic glycolysis is further stimulated in order to cover brain energy deficiency, even though cerebral blood flow is restored and sufficient oxygen is supplied. The postischemic lactate increase would result from the radical-induced impairment of mitochondria.[2,9] In addition, the reperfusion-induced lactate increase was completely suppressed by neuroprotective radical scavenger nizofenone, supporting that the disorder of mitochondria function is peroxidatively induced after renewed blood supply.

Nizofenone (10 mg/kg) prevented the postischemic increase in the hippocampus lactate levels and also prevented the ischemia/reperfusion-induced neuronal injury in the CA1 pyramidal cells. These findings suggest that the prevention of postischemic reperfusion injury brings about effective neuroprotection.

ACKNOWLEDGMENT

The author sincerely acknowledges his colleagues for their collaboration.

REFERENCE

1. MURPHY, A.N. et al. 1999. Mitochondria in neurodegeneration: bioenergetic function in cell life and death. J. Cereb. Blood Flow Metab. **19:** 231–245.
2. YASUDA, H. & A. NAKAJIMA. 1993. Brain protection against ischemic injury by nizofenone. Cerebrovasc. Brain Metab. Rev. **5:** 264–276.
3. SAITO, I. et al. 1983. A double-blind clinical evaluation of the effect of nizofenone on delayed ischemic neurological deficits following aneurysmal rupture. Neurol. Res. **5:** 29–47.
4. OHTA, T. et al. 1986. Nizofenone administration in the acute stage following subarachnoid hemorrhage. Results of a multi-center controlled double-blind clinical study. J. Neurosurg. **64:** 420–426.
5. MURPHY, T.H. et al. 1989. Glutamate toxicity in a neuronal cell line involves inhibition of cystine transport leading to oxidative stress. Neuron **2:** 1547–1558.
6. YASUDA, H. et al. 1985. Biphasic liberation of arachidonic and stearic acids during cerebral ischemia. J. Neurochem. **45:** 168–172.
7. YASUDA, H. et al. 1981. Cerebral protective effect and radical scavenging action. J. Neurochem. **37:** 934–938.
8. WALLACE, D.C. 1999. Mitochondrial diseases in man and mouse. Science **283:** 1482–1488.
9. MATSUMOTO, Y. et al. 1994. Nizofenone, a neuroprotective drug, suppresses glutamate release and lactate accumulation. Eur. J. Pharmacol. **262:** 157–161.

Conference Summary

Mitochondria, Neurodegenerative Diseases, and Selective Neuronal Vulnerability

JOHN P. BLASS[a]

Burke Medical Research Institute, Cornell University,
785 Mamaroneck Avenue, White Plains, New York 10605, USA

This conference has covered a variety of aspects of oxidative and energy metabolism in relation to diseases of the nervous system. In general, the articles are in-depth discussions of specific and often specialized topics. The authors presume that the reader is familiar with the specific pathways being discussed. To facilitate the use of this volume by readers who do not work primarily on oxidative/energy metabolism, the pathways are diagrammed in FIGURES 1–3: glycolysis (FIG. 1), the Krebs tricarboxylic acid cycle (FIG. 2), and electron transport (FIG. 3). Mitochondrial metabolism of Ca^{2+} is described in Chapter 1 (Nichols *et al.*) and the metabolism of reactive oxygen species in Chapter 2 (Fridovich).

MITOCHONDRIA

The mitochondrial metabolism of free radicals and of Ca^{2+} are critical not only for the cell as a whole but also for mitochondria themselves. Fridovich (Duke) reviewed the metabolism of reactive oxygen species (ROS). Overexcitation of neurons by the neurotransmitter glutamate or by analogues such as NMDA typically kills neurons which have NMDA receptors. Nichols (Dundee) presented evidence that the death of these cells is mediated in part by excessive ROS production by calcium-overloaded but still polarized mitochondria.

A much-studied but still poorly understood property of mitochondria is the *permeability transition* (PT). Treatment of mitochondria with supraphysiological concentrations of Ca^{2+} leads to the opening of a pore through which solutes of up to 1,500 daltons can move freely. The mitochondria then swell and become nonfunctional. Cytochrome *c* is released, and it acts as a signal for cell death by apoptosis. These reactions are influenced by the products of the genes of the Bcl series. The PT was discussed by Reynolds (Pittsburgh) and Murphy (MitoKor).

Mitochondrial constituents are encoded on both mitochondrial DNA (mtDNA) and nuclear DNA (nDNA). mtDNA encodes components of the electron transport chain and tRNAs. mtDNA is much smaller than nDNA, and the complete sequence of mtDNA is known for humans as well as for other species. Shoffner (Scottish Rite Children's Medical Center) reviewed the biology of mtDNA and the complex relationships between abnormalities (mutations) in mtDNA and the diseases associated

---

[a]Phone: 914-597-2359; fax: 914-597-2757.
e-mail: jpblass@mail.med.cornell.edu

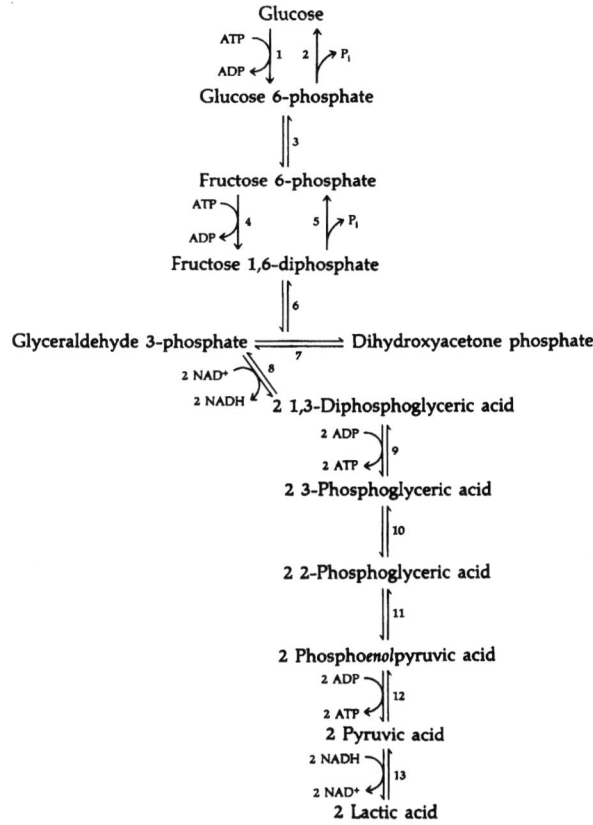

FIGURE 1. Glycolysis. (Reprinted from White et al.[4] by permission.)

with them. He also discussed abnormalities of electron transport associated with mutations in nDNA. About 95% or more of mitochondrial constituents are encoded on nuclear DNA (nDNA). Blass (Burke, Cornell) used the α-ketoglutarate dehydrogenase complex (KGDHC) as an example of a mitochondrial substructure encoded on nDNA genes. Abnormalities of KGDHC have been associated with a family of diseases, some rare and some common. In general, for KGDHC deficiencies as for many other disorders, the more severe the biochemical impairment the earlier the onset and the more severe the disease.

NEURODEGENERATIVE DISEASES

Mitochondrial disorders frequently present clinically as disorders of the nervous system. That observation is not surprising. The brain has a second-to-second dependence on oxidative/energy metabolism to maintain its function, and even mild impairments of oxidative/energy metabolism lead to clinically significant

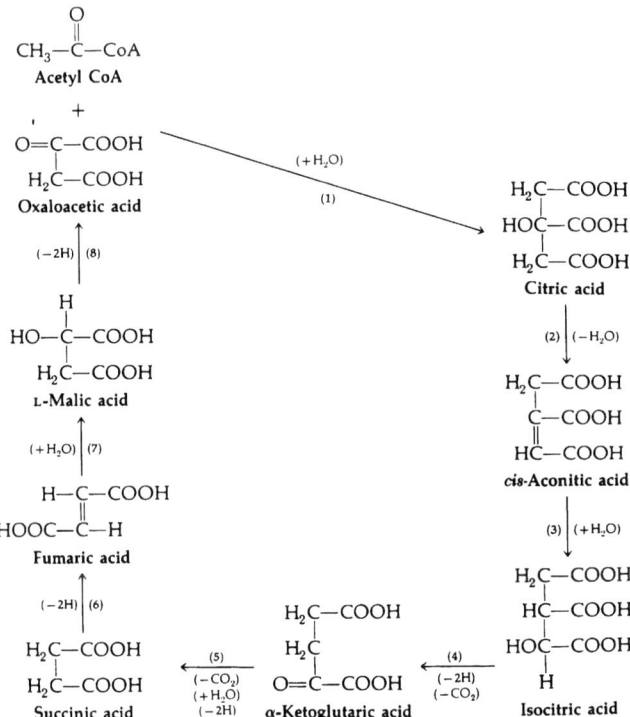

FIGURE 2. The Krebs tricarboxylic acid cycle. (Reprinted from White et al.[4] by permission.)

impairments.[1] The McEwen group (Rockefeller) reported that even a very mild impairment in energy metabolism can lead to significant structural changes in the brain. Their model was the effect of environmental stress on hippocampal neurons in rats who were diabetic due to treatment with streptozotocin.

Experimental model diseases were discussed by Gibson (Burke, Cornell) and Epstein (UCSF). Gibson reviewed models for studying disorders of KGDHC. These included animals with thiamine deficiency and various types of cell cultures including cells cultured from humans with Alzheimer's disease. Epstein (UCSF) demonstrated that the "clinical" phenotype produced by knocking out the gene for magnesium-dependent supraoxide dismutase (*SOD2*) in transgenic mice depended heavily on the genetic background (strain) of the mouse. In other words, the specifics of the "clinical syndrome" depended as much on the nature of the mouse as on the nature of the mutation.

The results of Epstein and coworkers and of others have fundamental implications for clinical genetics. Earlier research on inborn errors of metabolism often incorporated an implicit hypothesis of "one gene, one disease." This implicit assumption may have been related to the "one gene, one enzyme" hypothesis of Bea-

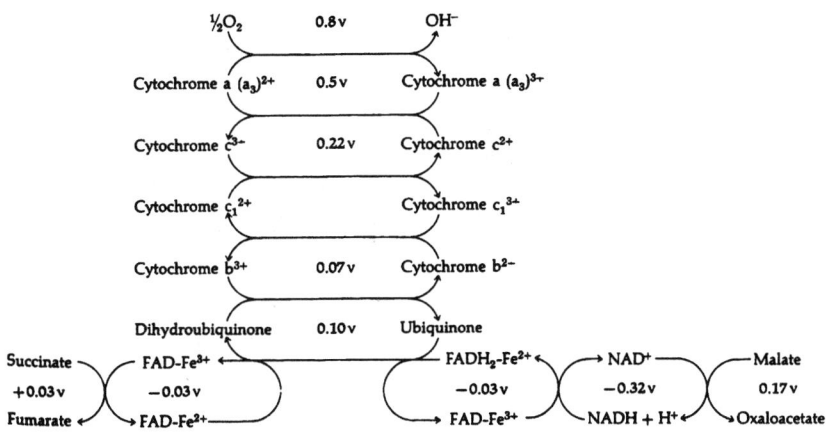

FIGURE 3. Electron transport. (Reprinted from White et al.[4] by permission.)

dle and Tatum. More recent discoveries such as differential splicing indicate that the "one gene, one enzyme" hypothesis was too simple. Analogously, data gathered over the last quarter century indicate that the "one gene, one disease" assumption is also too simple. Among many examples in clinical genetics are the variety of presentations of GM_2 gangliosidoses,[2] the variety of mtDNA related disorders discussed by Shoffner, and the KGDHC deficiency disorders discussed by Blass (see above). Association of a gene with a human disease is necessarily a descriptive observation. One can not achieve the clarity of inducing a mutation and observing its effects. In practice, acceptance of an association between a gene and a disease depends in part on the degree of statistical association and in part on the "plausibility" of the association. Controlled experimental manipulations in animals have the potential to be more compelling than studies in humans can be. In the experiments described by Epstein, the *same* genetic abnormality on the *same* promotor leads to different "clinical syndromes," depending upon the strain (genetic background) of the mouse. The data indicate that it is no longer suitable to assume in clinical investigation that a mutation in humans must be associated with only one clinical syndrome in order for the mutation to be considered significant.

Among human neurodegenerative diseases discussed at this conference, Alzheimer's disease (AD) was prominent. This is in accord with its high prevalence and significance in our aging population. The existence of defective oxidative/energy metabolism is well documented in AD. Kalaria (Newcastle upon Tyne) discussed the role of circulatory compromise in AD, and Rapoport (NIA) discussed the *in vivo* evidence for impaired oxidative/energy metabolism in AD patients and some of the pathological and enzymatic correlates. Mattson (Kentucky) described cellular and molecular mechanisms impairing brain metabolism in AD, including the role of oxidative stress (ROS). Gibson and Sheu and Blass discussed the role of KGDHC deficiency in AD. Smith and Perry and their coworkers (Case Western Reserve) put forward evidence that the accumulation of amyloid in AD is a defense mechanism against oxidative damage. Their proposal implies that oxidative damage may be at least as "seminal" an event in AD pathogenesis as is the accumulation of amyloid.

Huntington's disease (HD) was discussed from two viewpoints. HD is one of a growing number of diseases in which the genetic defect is expansion of the length of a series of CAG repeats in a gene, with the resulting insertion of expanded glutamine (Q) repeats in the encoded protein. The CAG_n/Q_n disorders are believed to be associated with a "toxic gain of function" in the protein containing the Q_n expansion. Burke (Duke) discussed the properties of Q_n proteins and the formation of intranuclear and intracytoplasmic aggregates in HD. Beal (Cornell) discussed the extensive evidence for defective brain oxygen/energy metabolism in HD. The mechanisms linking the genetic abnormality to the metabolic defect are not as yet known.

A number of other disorders were also discussed, including Parkinson's disease and cerebrovascular disease.

(SELECTIVE) CELL DEATH

Defects in oxidative/energy metabolism can lead not only to the dysfunction of neurons[1] but also to their death. Jenkins (Harvard) discussed the relationship of *in vivo* decreases in brain metabolism to aging and neurodegeneration, including a critical discussion of the methodologies used to carry out the *in vivo* studies in humans. Snider (Washington University, St. Louis) described the the two best recognized forms of cell death, necrosis and apoptosis.

An aspect of brain oxidative/energy metabolism which was repeatedly touched on in this conference is the nonuniform distribution of enzymes of oxidative/energy metabolism among brain cells and among different types of neurons. This finding bears directly on the question of how abnormalities in oxidative/energy metabolism can lead to specific types of brain damage and therefore specific neurological entities rather than just to overall "brain failure." The term *selective vulnerability* has been used to describe the phenomenon of a subpopulation of cells in the nervous system being relatively selectively affected by a generalized insult. It is now well established that the distribution of mitochondrial enzymes varies not only between glia and neurons but also among populations of neurons (see Sheu & Blass chapter for references). It is also well established that mitochondrial damage can lead to both necrotic and apoptotic cell death, including neuronal cell death (see Snider *et al.* chapter). Standard animal models of selective vulnerability are due to experimental impairments of energy metabolism. Examples include the loss of CA1 but not CA2 and CA3 neurons in rat hippocampus in the 4-vessel occlusion model,[3] and the limited and stereotyped location of brain lesions in thiamine deficiency (see Gibson chapter). These observations suggest the following hypothesis:

> Selective vulnerability in brain disorders may often be due to selective distribution of mitochondrial components in the vulnerable compared to the non-vulnerable brain cells.

These considerations suggest that continuation of studies like those described in this conference volume will be important both for understanding the brain and for helping people with diseases of the brain.

REFERENCES

1. GIBSON, G.E., W.A. PULSINELLI & J.P. BLASS. 1981. Brain dysfunction in mild to moderate hypoxia. Am. J .Med. **70:** 1247–1254.
2. MACQUEEN, G.M., P.I. ROSEBUSH & M.F. MAZUREK. 1998. Neuropsychiatric aspects of the adult variant of Tay-Sachs disease. J. Neuropsych. Clin. Neurosci. **10:** 10-19.
3. GAO, T.M., W.A. PULSINELLI & Z.C. XU. 1999. Changes in membrane properties of CA1 pyramidal neurons after transient forebrain ischemia in vivo. Neuroscience **90**(3)**:** 771–780.
4. WHITE, A., P. HANDLER & E.L. SMITH. 1973. Principles of Biochemistry. McGraw Hill. New York.

Index of Contributors

Aarnes, H., 358–361
Adam-Vizi, V., 269–272, 412–416
Akman, S.A., 382–385
Alberici, A.C., 294–297
Aliev, G., 362–364

Baic, C., 298–300
Baiyewu, O., 331–336
Bal-Price, A., 376–378
Beal, M.F., 203–213, 214–242
Bennett-Desmelik, J., 365–368
Benussi, L., 294–297
Benzycry, J., 309–313
Bernardi, P., 391–395
Binetti, G., 294–297
Blass, J.P., xi–xiii, 61–78, 434–439
Blum-Degen, D. 290–293
Boireau, A., 254–257
Bordier, F., 254–257
Borutaite, V., 376–378
Brown, A.M., 321–324
Brown, G.C., 376–378
Brownell, A.L., 214–242
Brustovetsky, N., 258–260
Budd, S.L., 1–12, 261–264
Burke, J.R., 192–202
Butterfield, D.A., 265–268
Butterworth, R.F., 404–411

Calingasan, N.Y., 79–94
Camandola, S., 154–175
Carlen, P.L., 286–289, 369–371
Carlson, E.J., 95–112
Castilho, R.F., 1–12
Chandrasekaran, K., 341–344
Chang, S.W., 353–357
Chen, Y.I., 214–242
Chesnut, M.D., 345–349
Chiba, S., 362–364

Chinopoulos, C., 269–272
Choi, D.W., 243–253
Clark, J.B., 400–403
Cohen, G., 273–278
Coimbra, M., 254–257
Colonna, R., 391–395
Cross, D.J., 350–352
Cruz, N.F., 279–281
Culmsee, C., 154–175

D'Angelo, A., 417–420
Davies, M., 304–308
Davis, R.E., 176–191
Di Lisa, F., 391–395
Dienel, G.A., 279–281
Duan, W., 154–175
Dubedat, P., 254–257
Dubinsky, J.M., 258–260
Dwoskin, L.P., 345–349

Epstein, C.J., 95–112

Farrar, P.L., 353–357
Ferrante, R.J., 314–320
Fonnum, F., 358–361
Foster, N.L., 350–352
Fox, D.A., 282–285
Frantseva, M.V., 286–289, 369–371
Fridovich, I., 13–18
Frölich, L., 290–293

Gasparini, L., 294–297
Geddes, J.W., 345–349
Ghidoni, R., 294–297
Ghosh, C., 325–330
Ghosh, S.S., 176–191
Gibson, G.E., 79–94
Gillespie, A.M., 95–112

Gorini, A., 417–420
Gottron, F.J., 243–253
Greenamyre, J.T., 365–368
Growdon, J.H., 294–297
Grünewald, T., 203–213
Guezurian, C., 369–371
Gwag, B.J., 396–399

Hall, K., 331–336
He, L., 282–285
Heales, S.J.R., 400–403
Hendrie, H., 331–336
Henry, T.R., 350–352
Higgins, D.S., 298–300
Hirai, K., 362–364
Holtzman, D., 309–313
Hovda, D.A., 337–340
Hoyer, S., 290–293, 301–303
Hoyt, K.R., 298–300
Huang, T.-T., 95–112
Hurst, R.D., 400–403
Huzar, D., 369–371

Imperato, A., 254–257

Jakel, R.J., 345–349
Jenkins, B.G., 214–242
Jou, I., 396–399

Kalaria, R.N., 113–125
Kapoor, R., 304–308
Kekelidze, T., 309–313
Kennedy, D.N., 214–242
Kesler, N., 273–278
Khait, I., 309–313
Kim, E.Y., 396–399
Kim, H., 396–399
Klaunig, J., 331–336
Klein, A.M., 314–320
Kölsch, H., 379–381
Koroshetz, W.J., 214–242
Kowall, N.W., 314–320
Kozy, H., 95–112
Kraft, E., 214–242

Kristal, B.S., 321–324
Kuestermann, E., 214–242
Kuhl, D.E., 350–352
Kunz, W.S., 426–429

Lahiri, D.K., 325–330 , 331–336
Lannert, H., 301–303
Lee, S.M., 337–340
Link, C.D., 265–268
Lins, H., 426–429
Lipton, S.A., 261–264
Liu, K., 279–281
Liu, L.-I., 341–344
Ludwig, M., 379–381
Lütjohan, D., 379–381

Magariños, A.M., 126–137
Makris, N.M., 214–242
Maragos, W.F., 345–349
Mattson, M.P., 154–175
McEwen, B.S., 126–137
Meunier, M., 254–257
Miller, J.J., 386–390
Miller, S.W., 176–191
Minoshima, S., 350–352
Moratto, D., 294–297
Mott, J.L., 353–357
Moussaoui, S., 254–257
Mulkern, R., 309–313
Murphy, A.N., 19–32
Myhre, O., 358–361

Nagai, Y., 192–202
Nguyen, T.V., 214–242
Nicholls, D.G., 1–12
Nitsch, R.M., 294–297
Nunomura, A., 362–364

Obertone, T., 365–368
Ogunniyi, A., 331–336
Onodera, O., 192–202

Panov, A., 365–368

INDEX OF CONTRIBUTORS

Park, L.C.H., 79–94
Parker, W.D., Jr., 176–191
Payne, R.S., 386–390
Pedersen, W.A., 154–175
Perez Velazquez, J.L., 286–289, 369–371
Perry, G., 362–364
Petronilli, V., 391–395
Poblenz, A.T., 282–285
Popp, D., 279–281
Porat, S., 372–375

Qu, H., 421–425

Raineri, I., 95–112
Rao, M.L., 379–381
Rapoport, S.I., 138–153, 341–344
Reagan, L.P., 126–137
Reynolds, I.J., 33–41
Riederer, P., 290–293
Rigor, B.M., 386–390
Rodriguez, H., 382–385
Rosas, H.D., 214–242
Rosen, B.R., 214–242

Sagstuen, E., 358–361
Sahota, A., 331–336
Samii, A., 337–340
Schousboe, A., 421–425
Schurr, A., 386–390
Scorrano, L., 391–395
Seo, S.Y., 396–399
Sheeman, B., 176–191
Sheu, K.-F. R., 61–78
Shoffner, J.M., 42–60
Simantov, R., 372–375
Smith, K.J., 304–308
Smith, M.A., 362–364

Snider, B.J., 243–253
Sonnewald, U., 421–425
Sorbi, S., 79–94
Stewart, V.C., 400–403
Stone, R., 400–403
Strittmatter, W.J., 192–202
Sulka, M., 298–300
Swerdlow, R.H., 176–191

Takeda, A., 362–364
Todd, K.G., 404–411
Togliatti, A., 309–313
Trabucchi, M., 294–297
Tretter, L., 412–416
Tseng, M.T., 386–390

Vensel, J., 298–300
Vestad, T.A., 358–361
Villa, R.F., 417–420

Waagepetersen, H.S., 421–425
Wallesch, C.-W., 426–429
Ward, M.W., 1–12
Wiedemann, F.R., 426–429
Winkler, K., 426–429
Wong, M.D., 337–340

Xu, Y., 331–336

Yasuda, H., 430–433
Yatin, S.M., 265–268

Zassenhaus, H.P., 353–357
Zhang, D., 353–357
Zhang, H., 79–94